D0971698

BASIC STATISTICS FOR BUSINESS AND ECONOMICS

Second Edition

The Wiley/Hamilton Series in
MANAGEMENT AND ADMINISTRATION

ELWOOD S. BUFFA, *Advisory Editor*
University of California, Los Angeles

PRINCIPLES OF MANAGEMENT: A MODERN APPROACH
FOURTH EDITION
Henry H. Albers

OPERATIONS MANAGEMENT: PROBLEMS AND MODELS
THIRD EDITION
Elwood S. Buffa

MODERN PRODUCTION MANAGEMENT: MANAGING THE OPERATIONS FUNCTION
FIFTH EDITION
Elwood S. Buffa

CASES IN OPERATIONS MANAGEMENT: A SYSTEMS APPROACH
James L. McKenney and Richard S. Rosenbloom

ORGANIZATIONS: STRUCTURE AND BEHAVIOR, VOLUME I
SECOND EDITION
Joseph A. Litterer

ORGANIZATIONS: SYSTEMS, CONTROL AND ADAPTATION, VOLUME II
Joseph A. Litterer

MANAGEMENT AND ORGANIZATIONAL BEHAVIOR: A MULTIDIMENSIONAL APPROACH
Billy J. Hodge and Herbert J. Johnson

MATHEMATICAL PROGRAMMING
SECOND EDITION
Claude McMillan

DECISION MAKING THROUGH OPERATIONS RESEARCH
SECOND EDITION
Robert J. Thierauf and Robert C. Klekamp

QUALITY CONTROL FOR MANAGERS & ENGINEERS
Elwood G. Kilpatrick

PRODUCTION SYSTEMS: PLANNING, ANALYSIS AND CONTROL
SECOND EDITION
James L. Riggs

COMPUTER SIMULATION OF HUMAN BEHAVIOR
John M. Dutton and William H. Starbuck

QUANTITATIVE BUSINESS ANALYSIS
David E. Smith

INTRODUCTION TO GAMING: MANAGEMENT DECISION SIMULATIONS
John G. H. Carlson and Michael J. Misshauk

PRINCIPLES OF MANAGEMENT AND ORGANIZATIONAL BEHAVIOR
Burt K. Scanlan

COMMUNICATION IN MODERN ORGANIZATIONS
George T. and Patricia B. Vardaman

THE ANALYSIS OF ORGANIZATIONS
SECOND EDITION
Joseph A. Litterer

COMPLEX MANAGERIAL DECISIONS INVOLVING MULTIPLE OBJECTIVES
Allan Easton

MANAGEMENT SYSTEMS
SECOND EDITION
Peter P. Schoderbek

ADMINISTRATIVE POLICY: TEXT AND CASES IN THE POLICY SCIENCES
Richard M. Hodgetts and Max S. Wortman, Jr.

THE ECONOMICS OF INTERNATIONAL BUSINESS
R. Hal Mason, Robert R. Miller and Dale R. Weigel

BASIC PRODUCTION MANAGEMENT
SECOND EDITION
Elwood S. Buffa

MANAGEMENT SCIENCE/OPERATIONS RESEARCH:
Elwood S. Buffa and James S. Dyer

OPERATIONS MANAGEMENT: THE MANAGEMENT OF PRODUCTIVE SYSTEMS
Elwood S. Buffa

PERSONNEL ADMINISTRATION AND HUMAN RESOURCES MANAGEMENT
Andrew F. Sikula

MANAGEMENT PRINCIPLES AND PRACTICES
Robert J. Thierauf, Robert Klekamp and Daniel Geeding

BASIC STATISTICS FOR BUSINESS AND ECONOMICS

Second Edition

Paul G. Hoel and Raymond J. Jessen
UNIVERSITY OF CALIFORNIA, LOS ANGELES

A Wiley/Hamilton Publication
JOHN WILEY & SONS
New York • Chichester • Brisbane • Toronto

This book was designed by William
Tenney, copyedited by Susan Gerstein
and set in 10 point Plantin by
Applied Typographic Systems. The
cover was designed by Kathy Trainor
and printing and binding was done by
Halliday Lithographers. Chuck Pendergast
and Jean Varven supervised production.

Library of Congress Cataloging in Publication Data:

Hoel, Paul Gerhard, 1905-
Basic statistics for business and economics.

(The Wiley/Hamilton series in management and administration)
Includes index.
1. Statistics. I. Jessen, Raymond James, 1910- joint author. II. Title.
HA29.H66 1977 519.5 76–54504
ISBN 0–471–40268-0

Printed in the United States of America

10 9 8 7 6 5 4 3

Preface

This book is designed for the beginning course in statistical methods for students in economics and in the management-administration fields of business and government. The emphasis is on explaining in a simple manner the fundamental ideas of statistical theory that are particularly appropriate for solving certain classes of problems in those fields.

Since the objective of this book is to explain basic ideas simply, illustrations of the theory have been kept simple, and usually involve ideas or experiences familiar to beginning students. Although a large share of the exercises involve actual business problems, many of them have been stripped of descriptive matter and simplified so that the student can quickly recognize the fundamental statistical problem present. Complex business problems are best treated after the student has acquired a satisfactory background of courses and experience in his chosen field.

In this second edition we cover essentially the same topics as before, but we have distributed them over three more chapters to improve the exposition and to permit an instructor greater flexibility in his choice of topics. Since the first edition did not contain enough material for most two-semester courses, we have expanded the coverage considerably in several chapters. This additional material, however, is organized in such a manner that it can be easily omitted without sacrificing continuity for a one-semester course.

The second edition contains approximately 50 percent more illustrations, examples, and problems than the first edition. This should make it easier for an instructor to assign problems and design tests.

The first eight or ten chapters are basic to statistical understanding in any field, whereas the remaining chapters involve topics some of which are particularly suitable for certain subject matter fields and others of which promote additional research techniques. For a one-semester course we recommend that these basic chapters be studied, but that the following sections be omitted: Chapter 3-sections 9, 10, 11; Chapter 5-sections 5, 6, 7; Chapter 6-section 6; Chapter 7-sections 4, 5, 6; Chapter 8-sections 6, 7, 8; Chapter 9-section 4; and Chapter 10-section 7. Additional topics should be selected from Chapters 12–17. A two semester course should include the preceding excluded sections and should include as many additional chapters as time and interest permit.

The final section of most chapters contains a number of review illustrations designed to test a student's understanding of that chapter. A student should attempt to solve the problems of those sections without first looking at the solutions, and then he should check his answers and methods against the solutions presented there.

The exercises at the end of each chapter have been labeled with the section number to which they belong to assist the instructor in selecting exercises for homework. Answers to the odd-numbered exercises are given in the appendix. A Solutions Manual that contains solutions to both even and odd numbered exercises is available from the publisher.

Paul G. Hoel
Raymond J. Jessen

Contents

About the Authors

Paul G. Hoel received his Ph.D. in mathematics from the University of Minnesota. He taught three years at Rose Polytechnic Institute, two years at Oregon State University, and thirty-two years at the University of California, Los Angeles.

He was a Fellow of the American-Scandinavian Foundation in 1936, and a Fulbright Research Fellow to Norway in 1953 and 1954. He was awarded an Outstanding Achievement Award by the University of Minnesota in 1972.

Dr. Hoel is also the author of INTRODUCTION TO MATHEMATICAL STATISTICS, 4th ed., ELEMENTARY STATISTICS, 4th ed., FINITE MATHEMATICS AND CALCULUS WITH APPLICATIONS TO BUSINESS, and is a co-author with Dr. Port and Dr. Stone on a three volume sequence: INTRODUCTION TO PROBABILITY THEORY, INTRODUCTION TO STATISTICAL THEORY, and INTRODUCTION TO STOCHASTIC PROCESSES.

Raymond J. Jessen is professor of business statistics in the School of Management, University of California, Los Angeles. He received his B.S. degree from University of California, Berkeley in Agricultural Economics and his Ph.D. from Iowa State University, Ames, in statistics. He was instructor, research associate and professor in economics and statistics at Iowa State University, and concurrently Agricultural Statistician for the U.S. Bureau of Agricultural Economics during the period 1938–57. For the years 1947–50 he was acting Director of the Statistical Laboratory and the first Head of the newly organized Department of Statistics at Iowa State University. He has been Project Director, General Analysis Corporation, Los Angeles, 1957–60, and its successor, CEIR, Inc., Beverly Hills, 1960–62.

Chapter 1 The Nature of Statistical Methods

1 INTRODUCTION

In these days of mass communication and information storage, which have been made possible by technical advances in business machines, the ability to understand and use information intelligently has become increasingly important in all fields of business enterprise. Not only is it important to know how to use available data properly but it is also essential to know how to collect the proper information for making decisions if such information is not available. Problems concerned with the collection, analysis, and interpretation of data lie in the domain of the field of statistics; therefore, it is essential for anyone making business decisions on the basis of data to possess a clear understanding of that field.

Several decades ago the main contribution of statistical methods to business problems was the collection of vital data and an efficient description of such data. The federal government, for example, has been doing this for years and employs a large number of statisticians whose principal duty is to design efficient ways of collecting and summarizing various kinds of information.

In more recent times, however, the study of statistics has concentrated rather heavily on the analysis of data and its use in decision making. This role is particularly important when a decision must be made on the basis of a limited amount of information.

The application of statistical methods during recent years has brought about radical changes in all phases of a business enterprise, from production and cost control to sales management. Various governmental agencies, from city to federal, have been forced to become increasingly sophisticated in their statistical treatment of problems, particularly with respect to long-range forecasting of needs and services.

Since business and public administrators are regularly required to make policy decisions by means of available information, or information that can be obtained on a limited budget, it is essential for individuals pursuing a career in those fields to have a rather substantial knowledge of statistical techniques. The increased use of statistical methods by administrators is part of the trend to base administrative decisions on as scientific a foundation as possible. Personal judgment, based on limited experience, is being replaced by knowledge acquired through surveys and experiments that have been designed to measure with precision the variables that are required for making an objective decision.

Several decades ago, statistical quality control methods were introduced in the production control division of various manufacturing firms and immediately resulted in large savings. This led to the introduction of statistical methods into many other phases of business activities for the purpose of increasing their efficiency, and today there are few phases that have not been affected by this trend.

Economists who are interested principally in constructing theories and mathematical models for economic activities must eventually test the validity of their theories by means of statistical methods. Problems of maximizing economic variables are mathematical in nature but become quite statistical when they are applied to actual economic situations. Thus, applied economics is heavily dependent on statistical techniques to suggest new theories and to justify tentative ones.

The trend toward a more scientific approach to business problems is partly an outgrowth of the large increase in interdisciplinary studies in universities during recent years. New fields such as management science, operations research, and systems control have evolved from these cooperative ventures. All such disciplines are heavily dependent on statistical theory and techniques for their success.

In spite of the wide range of fields of application and the diversity of the problems treated, it is possible to analyze the fundamental nature of statistical methods. In its simplest form, as it applies to unsophisticated problems, statistics is concerned with data that have been obtained from taking samples from some source and in using those data to draw certain conclusions about that source. For example, in attempting to determine the quality of a shipment of drugs, a few boxes of those drugs would be tested for quality of the entire shipment, or in attempting to determine whether the production of a new product would be a profitable venture, a sample of potential customers would be interviewed and the results of that study used to predict the potential sales of the product.

The set of measurements or counts (called observations) that is taken from some source for the purpose of obtaining information about the source is called a *sample*, whereas the source of those observations is called a *population*. In view of the preceding discussion, the fundamental role of elementary statistical methods may be described as drawing conclusions about populations by means of samples.

2 BRANCHES OF STATISTICS

Statistical methods are applied regularly to such diverse fields as agriculture, business, education, engineering, government, and medicine; therefore, there must be a central body of theory and method that is applicable to all those fields. However, beyond this basic material, there are numerous special techniques that have evolved to solve particular classes of problems that are of interest only in certain fields of application. This is particularly true of statistical methods for business applications.

To show the diversity of statistical subject matter and, at the same time, to display the more or less common elements that bind them together into a discipline, it is convenient to analyze statistics into four parts and give a brief description of the nature of each.

(*a*) Data Gathering. This is concerned with the determination of what to measure, the time and the extent of the data collection process, which includes the design of the survey or experiment, and the assessment and control of the errors that arise in such operations.

(*b*) Description. Here the problem is to extract from the data a few simple properties (such as various averages) that will adequately describe the underlying structure of the process being studied. This includes the problem of fitting appropriate mathematical models to the data.

(*c*) Statistical Inference. This is the problem of drawing conclusions about the population that was sampled to produce the data that have been gathered and described. As we stated earlier, statistical inference is the fundamental objective of most statistical investigations. In some business problems, however, the population may not always be clearly defined even though the inference is quite simple. Thus, in studying the effect of various size advertising budgets on profits, the decision might be to choose a budget of $50,000. The underlying population here is the population of potential customers.

(*d*) Interpretation and Decision. Although an experiment or survey is usually designed to obtain a statistical inference concerning a population, it is the province of the originator of the investigation to decide how that inference is to be used. This is a matter of judgment and does not constitute a part of the statistical problem. Thus, after a business executive has been presented with the results of a statistical investigation concerning the expected sales resulting from opening a new branch of the firm, he must decide whether to open the branch. There are investigations, however, in which the statistician essentially determines which is the best action to take with respect to several possible actions that are available and, in that sense, decision making is also a branch of statistics. The role of decision making has recently become increasingly important and popular in the solution of business problems. Such methods require a knowledge of elementary statistics as a basis, but many of them require fairly sophisticated mathematical and statistical techniques as well, and therefore they are best treated in a second course.

Many statistical problems involve only one or two of these branches, but most of them involve at least three. Although this book is concerned with all four, particular emphasis will be given to statistical inference.

3 ILLUSTRATIONS

This section describes a few problems of the type that statistical methods were designed to solve. It does not begin to cover the broad class of problems capable of being solved by statistical methods but rather illustrates a few of the simpler ones that can be solved by using the methods developed in this book. One problem is of academic interest, whereas the others are typical real-life problems.

(*a*) A television program sponsor wishes to know how popular his program is, compared to others at the same hour. In particular, he wishes to know what percentage of the television audience is viewing his program rather than some other. To satisfy him, an organization engaged in determining program popularity agrees to take a poll of the television audience at that hour to evaluate program choices. By using statistical methods, such an organization can decide how large a poll will be necessary in order to estimate, within any desired degree of accuracy, the percentage of the audience viewing this program.

(*b*) A medical research team has developed a new serum it hopes will help to prevent a common children's disease. It wishes to test the serum. In order to assist the researchers in carrying out such a test, a school system in a large city has agreed to inoculate half of the children in certain grades with the serum. Records of all children in those grades are kept during the following year. On the basis of the percentages of those children who contract the disease during that year, both for the inoculated group and for the remaining half, it is possible by statistical methods to determine whether the serum is really beneficial.

(*c*) An industrial firm is concerned about the large number of accidents occurring in its plant. In the process of trying to discover the various causes of such accidents, an investigator considers factors related to the time of day. He collects information on the number of accidents occurring during the various working hours of the day, and by using statistical methods he is able to show that the accident rate increases just before lunch and also just before quitting time. Further statistical studies then reveal some of the major contributing factors involved in those accidents.

One might be tempted to say that statistical methods are not needed in a problem such as this, and that all one needs to do is to calculate percentages and look at them to decide what is happening. If one has a large amount of properly selected data, such decisions will often be correct; however, the high cost of collecting data usually forces one to work with only small amounts, and it is precisely in such situations that statistical methods are needed to yield valid conclusions.

(*d*) A merchandizing firm wishes to determine the size of an advertising budget that will maximize the profit for one of its products. It decides to pick out three widely separated market areas and carry out promotional programs of varying intensity and costs that it believes are appropriate for that product. In doing so, it will need statistical methods on optimization to plan the study and estimate the best size budget to use.

(*e*) An instructor of an elementary statistics course is having difficulty convincing some of his students that the chances of winning from a slot machine are just as good immediately after someone has won some money as after a run of losses. For the purpose of convincing them, he, together with a few students of sterling character, performs the following experiment on a slot machine located in a private golf club. The machine is played for one hour, or until the combined resources of instructor and students are exhausted, whichever occurs first. A record is kept of the number of wins and losses that occur immediately after a win, together with the amounts won, and also of the number of wins and losses, and amounts won, immediately after a run of, say, five losses. With data of this type available, the instructor should be able to apply statistical methods to convince the skeptics of his wisdom in this matter. Since a run of bad luck might make it difficult to demonstrate this wisdom, unless the machine were played a long while, the instructor would be well advised to come amply supplied with cash. An experiment of this type should also convince the students that slot machines are designed to extract money from naïve individuals.

An analysis of the preceding illustrations shows that they properly belong to the field of statistics because all are concerned with drawing conclusions about some population of objects or individuals and they propose to do so on the basis of a sample.

It may also be observed that most of those problems fall into two general categories. They are concerned either with estimating some property of the population or with testing some hypothesis about the population. The first illustration, for example, is concerned with estimating the percentage of the television audience that is watching a particular program at a certain hour. The second illustration may be formulated as one of testing the hypothesis that the percentage of children contracting a disease is the same for inoculated children as for children receiving no inoculation. The fourth illustration, however, is not of this simple type but depends heavily on estimation techniques.

Most of the statistical methods explained in this book are those for treating problems of these two general types: estimating properties or testing hypotheses about populations. Many of the simpler problems of decision making can be formulated in this manner also; however, those that cannot will be treated separately in a later chapter following the material on estimation and hypothesis testing.

4 PROBABILITY

In the problem of estimating the percentage of a certain kind of television audience the solution will consist of a percentage based on the sample and a statement of the accuracy of the estimate, usually in the form, "The probability is .95 that the estimate will will be in error by less than 3 percent." Similarly, in problems involving the testing of some hypothesis the decision to accept or reject the hypothesis will be based on certain probabilities.

It is necessary to use probability in such conclusions because a conclusion based on a sample involves incomplete information about the population, and therefore it cannot be made with certainty. The magnitude of the probability attached to a conclusion represents the degree of confidence one should have in the truth of the conclusion. The basic ideas and rules of probability are studied in a later chapter; meanwhile it should be treated from an intuitive point of view. Thus the statement that the probability is .95 that an estimate will be in error by less than 3 percent should be interpreted as meaning that about 95 percent of such statements made by a statistician are valid and about 5 percent are not. In the process of studying statistical methods one will soon discover that probability is the basic tool of those methods.

Probability is an exceedingly interesting subject, even for those who have little liking for mathematics or quantitative methods. Many people enjoy some of the events associated with probability, if not the study itself; otherwise, how can one account for the large number of people who love to gamble at horse races, lotteries, cards, etc.? It may well be that it is their lack of probability sense that encourages them to gamble as they do. In any case, probability is used consciously or unconsciously by everyone in making all sorts of decisions based on uncertainty, and any student who wishes to be well educated, or to behave rationally, should have some knowledge of probability.

5 THE ROLE OF STATISTICS IN PLANNING INVESTIGATIONS

Many business-analysis problems arise from the fact that the necessary data for making decisions are collected by other agencies. Thus, cost data for a certain operation may come from the accounting department, sales data on the performance of a given product against competitors may come from a commercial marketing research firm, and price data from a government agency. An analyst may have to spend a considerable amount of time looking into the circumstances under which the data were obtained and take them into account in reporting his findings. It is often true that if he had been able to obtain the data himself or at least have control over the manner in which they were obtained, his task would have been much easier and his conclusions more accurate and relevant. Some knowledge of the difficulties involved in the analysis of data can be of great help in suggesting how data should be obtained. Good statistics like good health result largely from preventive measures rather than from a good treatment after the illness has set in.

6 THE ROLE OF STATISTICS IN MAKING DECISIONS

As was mentioned earlier, during the last few decades the field of probability decision making has become increasingly useful as a method for studying certain problems of business and economics. This field may be looked upon as a branch of statistics, and yet statistics may also be treated as a branch of it. It is in business applications that statistical decision making is finding some of its most interesting uses. For example, business management may have to decide whether to market a new product now on the basis of its judgment as to how well the product will sell or to wait a year after carrying out a rather expensive marketing survey. What is the best decision: be optimistic and market now, be pessimistic and reject the new product, or be neither and spend $50,000 on a marketing survey to assist in making a final decision later? Problems of this type are essentially statistical in nature and can be solved by using the proper statistical techniques.

7 THE HISTORY OF STATISTICS

Professional interest in the study of statistics is not very new and yet much of the theory of statistics that is being applied today is less than fifty years old. The American Statistical Association, which was founded in 1839, is actually the second oldest professional organization in the United States, being exceeded in seniority only by the American Philosophical Society. The ever-increasing development of highly quantitative and scientific methods in the social, behavioral, and administrative sciences has brought forth a strong interest in statistical methods to solve many of their problems. As a consequence, statistics is undergoing continuous adaptation and development; however, the basic elements that are common to all such fields have not changed and are needed for an understanding of the more sophisticated techniques that are being developed. It is the purpose of this book to explain those fundamental ideas and thereby provide a background for going on to the more advanced methods.

8 REFERENCES

For readers who wish to consult other books and articles to supplement the material in this book, we suggest the following:

(*a*) Books that may arouse an interest in statistics.

Hoff, D., *How to Lie with Statistics,* Norton (paperback), New York (1954). Many examples written in a frothy, Madison-Avenue style.

Monroney, M. J., *Facts from Figures,* Penguin Books (paperback), Baltimore (1956). Basic statistics in a light but accurate style.

Wallis, W. A., and Roberts, H. V., *The Nature of Statistics,* The Free Press (paperback), New York (1965). Many excellent examples illustrating the nature and philosophy of statistics.

(*b*) Books that give applications of statistics to real problems.

Sieloff, T. J., *Statistics in Action: Readings in Business and Economic Statistics,* Lansford Press (paperback), San Jose (1963). Reprints of well-selected articles in the literature of business and economic statistics.

Tanur, J. M., *Statistics: A Guide to the Unknown,* Holden-Day (paperback), San Francisco (1972). Reprints of articles in the literature from various fields.

(*c*) Books that present case studies using statistics.

Enrick, N. L., *Cases in Management Statistics,* Holt, Rinehart and Winston (paperback), New York (1962). Actual and simulated realistic cases are presented for solution.

O'Hara, J. B., and Clelland, R. C., *Effective Uses of Statistics in Accounting and Business,* Holt, Rinehart and Winston (paperback), New York (1964). Realistic cases based on the experience of the authors. Questions but no solutions.

(*d*) Books of statistical exercises and their solutions.

Perlman, R., *Problems in Statistics for Economics and Business Students,* Holt, Rinehart and Winston (paperback), New York (1963). Collection of exercises grouped by topic with illustrative problems at the beginning of each chapter. Solutions are given in the appendix.

Spiegel, M. R., *Schaum's Outline of Theory and Problems of Statistics,* McGraw-Hill (paperback), New York (1961). Some theory but mostly a collection of exercises, many of which are worked out as illustrations.

Chapter 2 The Description of Data

1 INTRODUCTION

This chapter studies the problem of how to describe data in a satisfactory manner for making statistical inferences. In Chapter 1 this phase of the analysis was listed as the second of the four parts of a statistical problem.

Although the data that are to be used for making an inference are usually a sample of some sort, it is always advisable to consider the circumstances under which the data were obtained before proceeding further with the investigation. For example, suppose a firm that produces a food item wishes to check on the water content of its product and has measured the water content in each of 150 packages of this item. Consider the following possible sources of the data.

(*a*) The 150 packages were the entire output of machine A on the afternoon shift of the preceding day.
(*b*) The 150 packages were selected from the 10,000 packages currently in the stock room.
(*c*) The 150 packages were selected by taking every 50th package off the production line during the past week.

In case (*a*) the data constitute a complete census of product quality for that particular machine during that period of time and should not be treated as a sample of product quality for the firm. In case (*b*) the data do represent a sample from the population consisting of the 10,000 stockroom packages; however, this sample may not represent the water content of packages now being turned out on the production line. Furthermore, unless more information is given as to how the stockroom sample was taken, there is no assurance that it is a satisfactory sample for measuring stockroom product quality. The data from (*c*) do represent a sample of current production and should be satisfactory for determining present product quality, if that is the objective. In case (*a*) the problem of describing the data is simple because a listing in order of magnitude of the water content will suffice to answer all pertinent questions such as: What percentage of this batch exceeds the allowable water content? In case (*b*), however, the objective will be to use the data to say something useful about the 10,000 packages in stock, whereas in case (*c*) the inference will be with respect to current production. It is possible to state the exact percentage of below-quality packages in the population under consideration in case (*a*); however, in (*b*) and (c) only estimates of those percentages can be given.

If an analysis such as the preceding one of available data shows that the data are satisfactory for making a valid statistical investigation of the type desired, then the next stage of the investigation is that of describing the data in an efficient manner. It should be clear, however, from the preceding discussion that it is far preferable to choose one's own sample in any statistical study rather than rely on someone else's data; therefore, consider the problem of how to choose a satisfactory sample before proceeding to the problem of data description.

Experience shows, given a population, that it is very difficult for an individual to select a sample that represents the population in miniature. For example, if a sample of 20 college students were to be selected by an individual to represent the distribution of hair color of such students, it is likely that too high a percentage of unusual hair colors would be included in the sample. Expert samplers of grain in the field, however, are frequently biased in the other direction because they tend to select samples that they feel are typical with the result that too few extreme values are obtained.

The problem of how to select a sample from a population so that valid conclusions about the population can be drawn from the sample can become quite complicated and often varies with the type of population. There is, however, a method of sampling, called *random sampling,* that does satisfy this requirement, although it is at times difficult or expensive to apply. A rather extensive discussion of this type of sampling will be given in Chapter 6; therefore, only a brief introduction to it is given here.

In its simplest form, when a single individual is to be selected from a population of individuals, the sampling is said to be *random* if each member of the population has the same chance of being chosen. Techniques used in games of chance are often employed to obtain such a sample. For example, at a large social affair at which a grand door prize is to be given away, it is customary to place all the numbered ticket stubs in a large container and then have a blindfolded individual select one ticket from the container, after the tickets have been thoroughly mixed. If, say, three individuals are to be selected from a population, the sampling will be random if every possible group of three individuals from the population has the same chance of being chosen. The preceding device that was employed to select one individual could also be used to select three individuals. Experience with such devices has shown that each individual, or group of individuals, is selected approximately the same number of times as every other individual, or group, when the experiment is repeated a large number of times, and therefore that they do conform to the requirement of showing no favoritism. In selecting a sample of students from a student body, choosing each tenth card from the files would undoubtedly be satisfactory as a practical method of random selection, provided the information desired about an individual is unrelated to his alphabetical position in the card file.

As was pointed out before, one reason for taking random samples is that samples selected in this manner tend to represent the population from which they are taken. This means, for example, that if the experiment of selecting 100 students from a student body were repeated a number of times and the percentage of students who worked part time was calculated for the entire sample, that percentage should tend to get increasingly close to the true percentage for the student body as the sampling continued. Other types of sampling in which personal judgment enters usually do not possess this desirable property.

As an illustration of this property of random samples generally representing the population being sampled, a random sampling experiment was carried out for an artificially constructed population. Suppose a large population of individuals can be classified into three groups. For example, it might be on the basis of age, those less than 30 years of age, those between 30 and 50 inclusive, and those over 50. Suppose further that the proportions of individuals in those three groups are $\frac{3}{6}$, $\frac{2}{6}$, and $\frac{1}{6}$, respectively. Then random sampling from such a population can be simulated by sampling from a population of 6 playing cards consisting of, say, 3 aces, 2 twos, and 1 three, provided the sampling is done properly. After each drawing of a card, the number is recorded and the drawn card is returned to the set. The cards are then thoroughly mixed and another drawing is made. This repeated mixing and returning of the drawn card to the set insures that the population proportions will remain the same. If the original population were very large the removal of some individuals from it by sampling would have no appreciable effect on the population proportions either. The preceding experiment was carried out 700 times in sets of 100 each. The accumulated results were reduced to percentages, calculated to the nearest decimal only, after each additional sample of 100 had been obtained. These random sample percentages, which are shown in

Table 1

Sample size	100	200	300	400	500	600	700
Aces	44	47	45.7	47.8	49.4	49.6	50.0
Twos	35	34	35.0	33.2	32.8	32.8	32.9
Threes	21	19	19.3	19.0	17.8	17.5	17.1

Table 1, should be compared with the population percentages of 50, $33\frac{1}{3}$, and $16\frac{2}{3}$, respectively, to see how well the samples represent the population being sampled.

The sample percentages appear to conform well to the population percentages as the sample size increases.

Hereafter, whenever a sample is to be taken it will be assumed that it will be obtained by a random sampling method, even though the word random is not used explicitly.

Now turn to the problem of studying properties of samples taken from populations. In this connection, although the word "population" would seem to refer to a group of human individuals, in statistics it refers to a group of individuals, human or not, or objects of any kind. In studying samples and populations it is often assumed that interest is centered on a single particular property of members of the population. Thus, one might be interested in the property of weight in a population of students at a university, or in the annual income in a population of factory workers, or in the percentage of iron in a population of meteorites. Samples are taken of individuals in a population, but then the property of interest is measured or counted for those individuals. If the property is one that is measured it is often denoted by the letter x, largely because a graph of the sample is usually placed on the x-axis of a coordinate system.

As an illustration of how one proceeds in the study of samples, consider the problem of what a physical education department at a university would do if it were interested in determining whether its male dormitory students were typical university students with respect to physical characteristics. In such a study it would undoubtedly wish to compare, as one source of information, the weight distribution of the dormitory students with that of nondormitory students. Now, weighing every male student on campus would certainly yield the desired information on weight distribution; however, this would become quite an undertaking in a large school at which such information is not required at registration time. The desired information, to sufficient accuracy, could be obtained much more easily by studying the weight distributions of samples of dormitory and nondormitory students.

Suppose then that a random sample of, say, 120 students has been obtained from the dormitory population. Since the only concern here is what to do with samples, the nondormitory sample can be ignored in this discussion–it would be treated in the same manner as the dormitory sample. Suppose, furthermore, that the weights of these 120 students have been recorded to the nearest pound and that they range from the lightest at 110 pounds to the heaviest at 218 pounds.

It is difficult to look at 120 measurements and obtain any reasonably accurate idea of how those measurements are distributed. For the purpose of obtaining a better idea of the weight distribution of the 120 students, it is therefore convenient to condense the data somewhat by classifying the measurements into groups. It will then be possible to graph the modified distribution and learn more about the original set of 120 measurements. This condensation will also be useful for simplifying the computations of various averages that need to be evaluated, particularly if fast computing facilities are not available. These averages will supply additional information about the distribution. Thus the purpose of classifying data is to assist in the extraction of certain kinds of useful information from the data.

The weight measurements considered here comprise an example of observations made on what is called a *continuous variable.* This name is applied to variables, such as length, weight, temperature, and time, that can be thought of as capable of assuming any value in some interval of values. Thus the weight of a student in the 140–150 pound range can be deemed capable of assuming any value in this range. Variables such as the number of automobile accidents during a day, the number of beetles dying when they are sprayed with an insecticide, or the number of children in a family are examples of what is called a *discrete variable.* For the purposes of this book, discrete variables can be considered as variables whose possible values are integers; hence they involve counting rather than measuring.

Since any measuring device is of limited accuracy, measurements in real life are actually discrete in nature rather than continuous; however, this should not deter one from thinking of such variables as continuous. Although the dormitory weights have been recorded to the nearest pound, they should be regarded as the values of a continuous variable, the values having been rounded off to the nearest integer. When a weight is recorded as, say, 152 pounds, it is assumed that the actual weight is somewhere between 151.5 and 152.5 pounds.

2 CLASSIFICATION OF DATA

The problem of classifying the data of a sample usually arises only for continuous variables because discrete variables by their very nature are naturally classified; therefore, consider the problem for the 120 dormitory weight measurements. What needs to be done is to place each weight in its proper weight class, for instance, between 130 and 140 pounds. Experience and theory indicate that for most types of data it is desirable to use from 10 to 20 classes, with the smaller number of classes for smaller quantities of data. With fewer than about

10 classes, too much sample detail is lost, whereas with more than about 20 classes computations become unnecessarily tedious. In order to determine boundaries between the various class intervals, it is necessary merely to know the smallest and largest measurements of the set. For the weight data, these are 110 and 218 pounds, respectively. Since the range of values, which is 108 pounds here, is to be divided into 10 to 20 equal intervals, the length of the class interval is first determined for those two extreme cases. If 10 intervals were chosen, the class-interval length would be $108/10 = 10.8$ pounds, whereas if 20 intervals were chosen, it would be $108/20 = 5.4$ pounds. Any convenient number between 5.4 and 10.8 may therefore be chosen. A class-interval length of 10 pounds will evidently be very convenient. Other class-interval lengths such as 6, 7, 8, or 9 would have been satisfactory also, although preference should be given to one of the larger intervals because 120 is not considered to be a large number of measurements. Since the first class interval should contain the smallest measurement of the set, it must begin at least as low as 110. Furthermore, in order to avoid having measurements fall on the boundary of two adjacent class intervals, it is customary to choose class boundaries to one-half unit beyond the accuracy of the measurements. Thus in this problem, with weights recorded to the nearest pound, it is satisfactory to choose the first class interval as 109.5–119.5, since 109.5 is one-half unit below the smallest measurement of 110, and it was agreed to use 10 pounds as the length of the class interval. This interval is certain to contain the smallest measurement, in view of the fact that a recorded weight of 110 pounds represents an actual weight between 109.5 and 110.5 pounds. The remaining class boundaries are determined by merely adding the class-interval length 10 repeatedly until the largest measurement, namely 218, is enclosed in the final interval. If 109.5–119.5 is chosen as the first class interval, there will be 11 class intervals, and the last class interval will turn out to be 209.5–219.5. When the class boundaries have been determined, it is a simple matter to list each measurement of the set in its proper class interval by merely recording a short vertical bar to represent it, as shown in Table 2(a). In doing this no attempt should be made to arrange the measurements in order of size; they are taken in their original order. When this listing has been completed and the number of bars for each class interval has been recorded, the data are said to have been classified in a *frequency table.*

It is assumed in such a classification that all measurements in a given class interval have been assigned the value at the midpoint of the interval. This midpoint value is called the *class mark* for that interval. Thus, for the first interval, 109.5–119.5, the class mark is 114.5, and any weight within this interval is assigned the value 114.5. The midpoint value of an interval is obtained by adding one-half the length of the interval to the number representing the left boundary of the interval. For example, the preceding first interval class mark of 114.5 was obtained by adding 5 to 109.5 because the length of the class interval is 10 and the left boundary of that interval is 109.5. Replacing all measurements in a given

Table 2

(a)		(b)	
Class boundaries	Frequencies	Class marks: x	Frequencies: f
109.5–119.5	/	114.5	1
119.5–129.5	////	124.5	4
129.5–139.5	~~////~~ ~~////~~ ~~////~~ //	134.5	17
139.5–149.5	~~////~~ ~~////~~ ~~////~~ ~~////~~ ~~////~~ ///	144.5	28
149.5–159.5	~~////~~ ~~////~~ ~~////~~ ~~////~~ ~~////~~	154.5	25
159.5–169.5	~~////~~ ~~////~~ ~~////~~ ///	164.5	18
169.5–179.5	~~////~~ ~~////~~ ///	174.5	13
179.5–189.5	~~////~~ /	184.5	6
189.5–199.5	~~////~~	194.5	5
199.5–209.5	//	204.5	2
209.5–219.5	/	214.5	1

interval by the midpoint value of that interval replaces a set of measurements by a new more convenient set whose values are only approximately equal to the original values. The approximations are, however, usually very good.

Table 2 illustrates the tabulation (*a*) and resulting frequency table (*b*) for a set of 120 weights of the type under consideration. The class marks are usually listed in such a table because they are the new values assigned to the measurements. The letter x is used to denote a class mark, and the letter f to denote the corresponding frequency. Subscripts on x and f designate the class interval. Thus $x_1, x_2, x_3, \ldots, x_{11}$ denote the class marks for the 11 class intervals in Table 2, and $f_1, f_2, f_3, \ldots, f_{11}$ denote the corresponding frequencies. For example, $x_2 = 124.5$, and $f_2 = 4$. The letter n is used to denote the total number of measurements. Since the sum of the frequencies for the various intervals must equal the total number of measurements, it follows that

$$n = f_1 + f_2 + \cdots + f_k$$

in which k denotes the number of class intervals in the frequency table.

As another illustration of how class boundaries and class marks are chosen, suppose a list of the hourly wages in 200 industrial plants yielded values in dollars of 2.90 to 3.74. These values possess a range of .84; hence if 10 classes were chosen, the length of the class interval would be $.84/10 = .084$. If 20 classes were chosen, the length would be $.84/20 = .042$. Since any convenient value between .042 and .084 may be selected, it follows that .05 is a natural choice for the length of the class interval. Wages are given to the nearest cent; therefore in order to include the lowest wage, $2.90, in the first class interval it will suffice to choose 2.895 as the left boundary of this interval. The right boundary is obtained by adding the class interval length to this value; hence it is given by the value $2.895 + .05 = 2.945$. The class mark is obtained by adding one-half the class-interval length to this same first boundary value; hence it is given by the value 2.895 + .025 = 2.92. Successive boundaries and class marks can now be obtained by adding .05 to the preceding ones. If this is done repeatedly until the maximum wage, 3.74, is included in an interval, it will be found that there are 17 intervals and that the boundaries of the last interval are 3.695 and 3.745.

Magazines and newspapers often indicate class intervals in a slightly different manner from that suggested here. They do not record actual class interval boundaries but rather noncontiguous boundaries. Thus they would indicate the first three class intervals in the preceding problem associated with Table 2 by 110–119, 120–129, and 130–139. When intervals are so indicated, the boundaries, as defined earlier, are ordinarily halfway between the upper and lower recorded boundaries of adjacent intervals. Thus, one would choose the halfway point between 119 and 120, namely 119.5, as the boundary between the first and second class intervals. Another method used by the media employs common boundaries but agrees that an interval includes measurements up to but not including the upper boundary. With this method, the first three class intervals would be indicated by 110–120, 120–130, and 130–140. A measurement that falls on a boundary is placed in the higher of the two intervals. These alternative methods are undoubtedly used because the reading public finds them easier to follow. If one knows the accuracy of measurement of the variable involved, there will be little difficulty in determining the correct boundaries and class marks for those two methods of classification. It is important to use the correct class marks; otherwise a systematic error will be introduced in many of the computations to follow.

3 GRAPHICAL REPRESENTATION

Frequency distributions are easier to visualize if they are represented graphically. For a discrete type of variable this representation is usually made by means of a *line chart.* For example, the sample distributions obtained in the earlier experiment of Table 1 would employ such a chart. One can graph either the actual frequencies obtained, or the relative frequencies, which are the frequencies divided by the total number of measurements, or the percentages, whichever is desired. Figure 1 shows such a line chart for the final percentages obtained in that experiment. In such charts one chooses any convenient units on the horizontal and vertical axis.

For continuous variables, a more useful type of graph is a graph called a *histogram.* The histogram for the frequency distribution of Table 2 is shown in Fig. 2. The class boundaries of Table 2 are marked off on the x-axis starting and finishing at any convenient points. The frequency corresponding to any class interval is represented by the height of the rectangle whose base is the interval in question. The vertical axis is therefore the frequency, or f, axis. Histograms

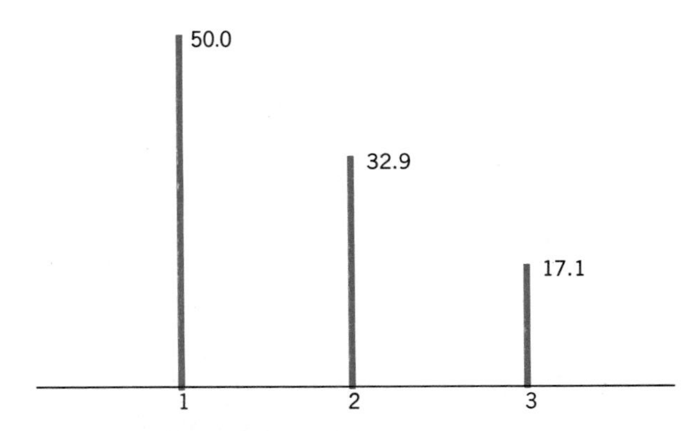

Figure 1 Line chart for percentages in a sampling experiment.

are particularly useful graphs for later work when frequency distributions of populations are introduced.

The histogram of Fig. 2 is typical of many frequency distributions obtained from data found in nature and industry. They usually range from a rough *bell-shaped* distribution, such as that in Fig. 3, to something resembling the right half of a bell-shaped distribution, such as that in Fig. 5. A distribution of the latter type is said to be *skewed.* Skewness refers to lack of symmetry with respect to a vertical axis. If a histogram has a long right tail and a short left tail, it is said to be skewed to the right; it is skewed to the left if the situation is reversed. The

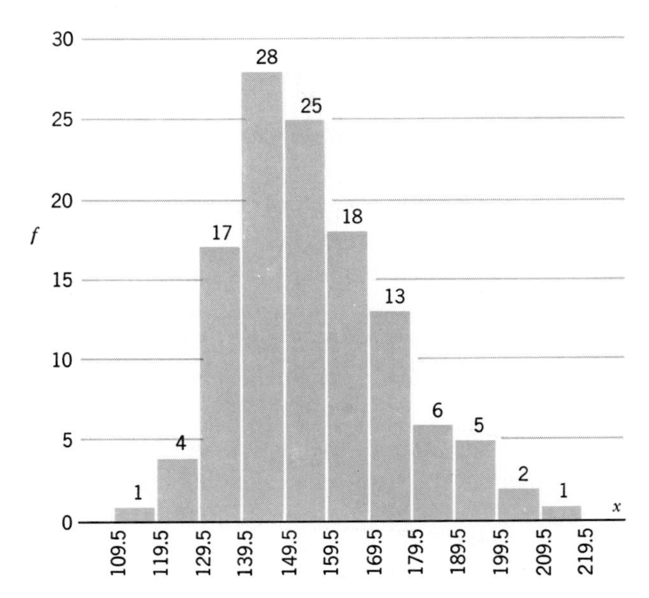

Figure 2 Distribution of the weights of 120 students.

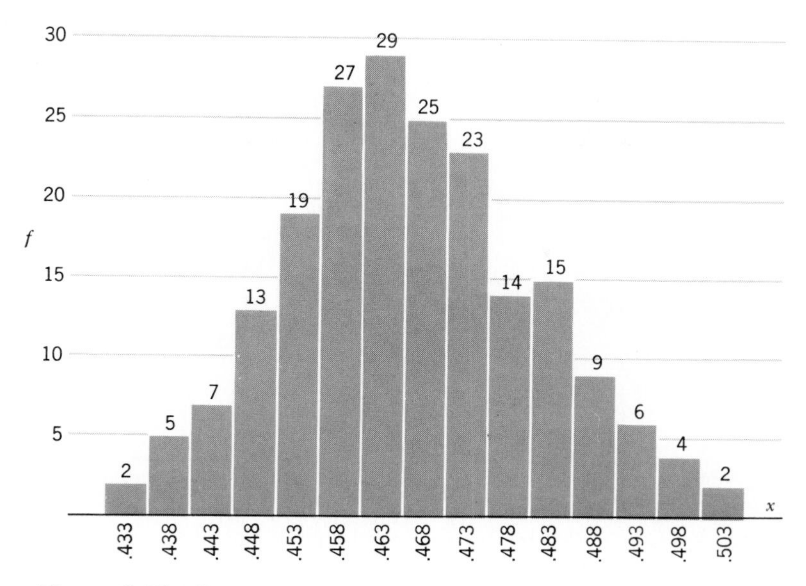

Figure 3 Distribution of the diameters in inches of 200 steel rods.

greater the unbalance, the greater the degree of skewness. It will be found, for example, that the following variables have frequency distributions that possess shapes of the type being discussed with approximately increasing degrees of skewness: stature, certain linear industrial measurements, the number of defective pieces in a box of manufactured parts, mortality age for certain children's diseases,

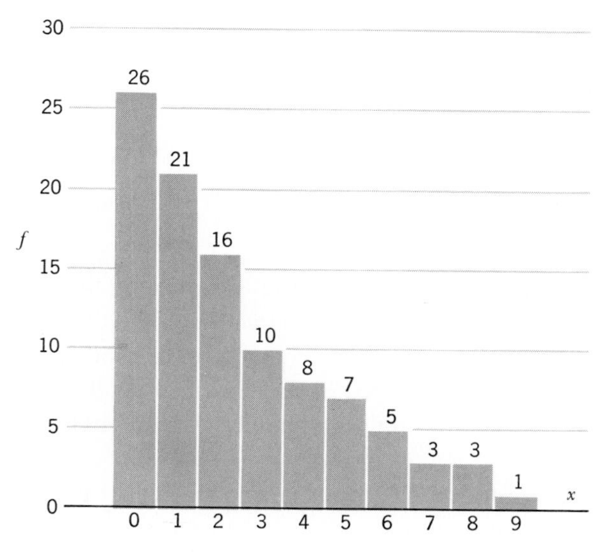

Figure 4 Distribution of the number of defective parts found in 100 boxes of such parts taken from a shipment.

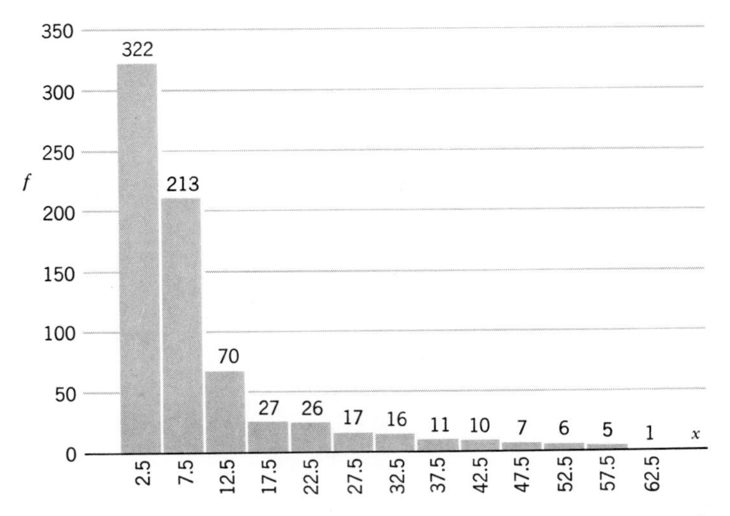

Figure 5 Distribution of 731 deaths from scarlet fever classified according to age.

and wealth. Figures 3, 2, 4, and 5 give the histograms for four typical distributions with increasing degrees of skewness. It should be noted that in Figs. 3, 4, and 5 the class marks rather than the class boundaries are specified on the x-axis. This is advantageous if it yields numbers with fewer digits to list along the x-axis (as in Fig. 3) or if those are the actually observed values (as in Fig. 4). A line chart such as that in Fig. 1 would have been more appropriate than the histogram of Fig. 4 for those data to show that the variable is naturally discrete and no condensation was required; however, the histogram is better at illustrating the degree of skewness in the distribution.

4 ARITHMETICAL DESCRIPTION

The principal reason for classifying data and drawing the histogram of the resulting frequency table is to determine the nature of the distribution. Some of the theory that is developed in later chapters requires that the distribution resemble the type of distribution displayed in Fig. 3; consequently, it is necessary to know whether one has this type of distribution before attempting to apply such theories to it.

Although a histogram yields a considerable amount of general information concerning the distribution of a set of sample measurements, more precise and useful information for studying a distribution can be obtained from an arithmetical description of the distribution. For example, if the histogram of weights for a sample of nondormitory students were available for comparison with the histogram of the dormitory sample, it might be difficult to state, except in very general terms, how the two distributions differ. Rather than compare the two weight distributions in their entirety, it might suffice to compare the average weights and the variation in weights of the two groups. Such descriptive quantities are called arithmetic because they yield numbers, as contrasted to a histogram, which is geometric in nature.

The nature of a statistical problem largely determines whether a few simple arithmetical properties of the distribution will suffice to describe it satisfactorily. Most of the problems that are encountered in this book are of the type that requires only a few properties of the distribution for its solution. For more advanced problems such a condensation of the information supplied by a distribution may not suffice. The situation here is similar to that arising when one discusses problems related to women's clothes and female beauty contests. A dress salesman, for example, might be satisfied to be told that a girl is 5 feet 7 inches tall and has measurements of 36–24–36 inches; however, a beauty contest judge would hardly be satisfied with such an arithmetical description. He most certainly would want to see the entire distribution.

In giving an arithmetical description of a frequency distribution, it is particularly useful in statistical inference to determine where the distribution is located, and to determine the extent to which the distribution spreads. Geometrically, this implies determining where the histogram is centered on the x-axis and the degree to which its area spreads out from there. There are numerous ways of measuring each of these two properties, several of which are described in section 8. The most useful measure for locating a distribution is the *mean* and the most useful measure for determining the degree of spread is the *standard deviation.* The next three sections are concerned with explaining these two measures.

5 THE MEAN

Suppose that a set of n sample values of a variable has been obtained from some population. These values are denoted by $X_1, X_2, \ldots, X_n$. This implies that X_1 is the first sample value obtained, and X_n the last one. Thus, for the sample of 120 dormitory weights, X_1 would be the weight of the first student selected and X_{120} the weight of the last student selected. The familiar average of this set of numbers, which is denoted by $\overline{X}$, is given by the formula

$$\overline{X} = \frac{X_1 + X_2 + \cdots + X_n}{n} \tag{1}$$

Since there are other averages also used in statistics, it is necessary to give $\overline{X}$ a special name. It is called the arithmetic mean, or, for brevity, the *mean.*

Now consider the problem of calculating the mean when the data have been classified into a frequency table, such as the weight measurements in Table 2. For such data it is assumed that each of the original weights has been replaced by the weight at the midpoint of the class interval in which it lies. As a result, for classified data the only values of X that arise in calculating the mean of a sample are the class mark values of the intervals. Suppose there are k class intervals in the frequency table. Let $x_1, x_2, \ldots, x_k$ denote the midpoint values of those k intervals and let $f_1, f_2, \ldots, f_k$ denote the corresponding frequencies in the frequency table.

To calculate the mean of the classified data by means of formula (1), it will suffice to add each classified value as many times as it occurs. For example, the value $x_4 = 144.5$ would be added 28 times in calculating the mean for the classified data of Table 2. In general, since x_1 occurs f_1 times it follows that x_1 must be added f_1 times, which is equivalent to multiplying x_1 by f_1. The same reasoning applies to the other class marks. The sum of all the measurements in a classified table is therefore the sum of all the products like $x_1 f_1$. The mean for such a table, which is denoted by $\bar{x}$, therefore assumes the form

$$\bar{x} = \frac{x_1 f_1 + x_2 f_2 + \cdots + x_k f_k}{n} \tag{2}$$

The preceding formulas can be written in much neater form if the summation symbol Σ (Greek sigma) is used. Since this symbol appears in other formulas also, it may be well to become acquainted with it now. It is merely a symbol that tells one to sum all the values of the quantity written after the symbol. Thus, formula (1) is written

$$\bar{X} = \frac{\sum_{i=1}^{n} X_i}{n} \tag{3}$$

The Σ symbol has n above it and $i = 1$ below it to indicate that all the numbers X_i should be added, starting with $i = 1$ and finishing with $i = n$. The symbol X_i denotes the ith sample value, where i is some integer from 1 to n. In words, this symbol merely says, "Sum all the values of X's, beginning with the first ($i = 1$) sample value and ending with the last ($i = n$) sample value." In this condensed notation, formula (2) can be written in the form

$$\bar{x} = \frac{\sum_{i=1}^{k} x_i f_i}{n} \tag{4}$$

Replacing each measurement of a sample by its class mark will usually yield a value of the mean that differs slightly from the mean of the original measurements; therefore, the value of the mean given by formula (4) is only an approximation to the correct value given by formula (3). However, unless the classification is rather crude, the difference is usually so small that it can be ignored in most statistical problems.

Example. Since the value of $\bar{x}$ for the data of Table 2 will be needed in the next two sections, its calculation by means of formula (4) is carried out here, as shown in Table 3.

Table 3

x_i	f_i	$x_i f_i$
114.5	1	114.5
124.5	4	498.0
134.5	17	2,286.5
144.5	28	4,046.0
154.5	25	3,862.5
164.5	18	2,961.0
174.5	13	2,268.5
184.5	6	1,107.0
194.5	5	972.5
204.5	2	409.0
214.5	1	214.5
Totals	120	18,740.0

$$\bar{x} = \frac{18,740}{120} = 156.2$$

Now it can be shown that the numerical value of the mean represents the point on the x-axis at which a sheet of metal in the shape of the histogram of the distribution would balance on a knife edge. Since this balancing point is usually somewhere near the middle of the base of the histogram, it follows that $\bar{x}$ usually gives a fairly good idea of where the histogram is located or centered. Thus the mean helps to describe a frequency distribution by determining where the histogram of the distribution is located along the x-axis. It is therefore often called a *measure of location.*

For the frequency distribution of Table 2, whose histogram is displayed in Fig. 2, calculations, as shown in Table 3, gave the value $\bar{x} = 156.2$. An inspection of Fig. 2 shows that the histogram there certainly ought to balance on a knife edge somewhere in the vicinity of 156. The value of $\bar{x}$ here gives one a good idea of where the histogram of the distribution is located, or centered, on the x axis.

The principal reason for classifying a set of measurements is to obtain information concerning the nature of the distribution of the variable X being studied.

Therefore, even though the data have been classified, the mean should be calculated by means of formula (3) rather than formula (4), provided the original set of measurements is still available.

Several other measures of location will be discussed in section 8.

6 THE VARIANCE

The purpose of this section is to introduce a quantity that measures the extent to which a set of measurements varies about the mean of the set. In this connection consider the two sets of measurements

$$4, 5, 6, 7, 8 \quad \text{and} \quad 2, 4, 6, 8, 10$$

and their histograms shown in Fig. 6.

The mean of each set is 6. Since the common difference between consecutive pairs of measurements in the second set is twice what it is in the first set (2 to 1) and since the range of values of the second set is twice the range of values of the first set (8 to 4), most people would agree that the second set of measurements varies twice as much about its mean as does the first set.

The simplest measure of the variation of a set of measurements is undoubtedly the *range* of the set, that is, the difference between the largest and smallest measurement of the set. However, it has certain undesirable properties that will be discussed in section 8; therefore, another more useful measure will be introduced.

Since the mean has been chosen as the desired measure of location of a set of measurements, a measure of variation should measure the extent to which the measurements deviate from their mean. For unclassified data corresponding to formula (3), the deviations are $X_1 - \overline{X}, X_2 - \overline{X}, \ldots, X_n - \overline{X}$. Values of X that are larger than $\overline{X}$ will produce positive deviations, whereas values of X smaller than $\overline{X}$ will yield negative deviations. Since only positive distances from the mean are desired, it is necessary to use the absolute values of such deviations, or possibly to use their squares. Since absolute values are difficult to work with, it is customary

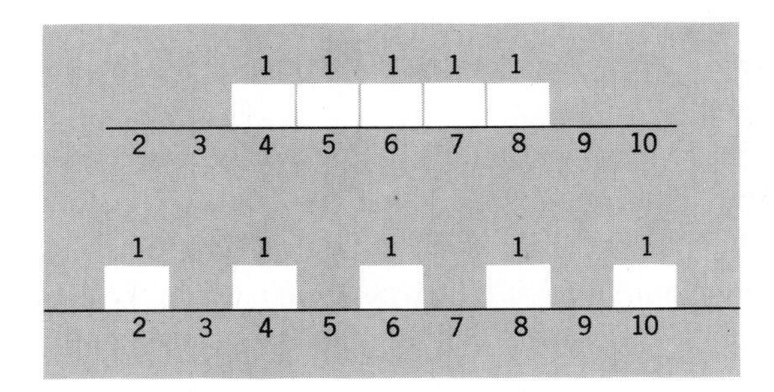

Figure 6 Two artificial distributions.

to take the squares of the deviations and average them. Thus, a measure of variation of a set of measurements about the mean of the set is given by

(5)
$$\frac{\sum_{i=1}^{n} (X_i - \overline{X})^2}{n}$$

If a set of measurements has been classified and formula (4) is being used as the measure of location, then the average of the squares of the deviations is

(6)
$$\frac{\sum_{i=1}^{k} (x_i - \overline{x})^2 f_i}{n}$$

In the solutions of statistical inference problems in later chapters, a slight modification of this measure is more useful than the measure itself. This modification consists in using the divisor $n - 1$ in place of n in these formulas. The justification for this modification will be given in Chapter 6. The resulting quantity is denoted by the special symbol s^2 and is called the *sample variance.* Thus, by definition, for unclassified data

(7)
$$s^2 = \frac{\sum_{i=1}^{n} (X_i - \overline{X})^2}{n - 1}$$

and for classified data

(8)
$$s^2 = \frac{\sum_{i=1}^{k} (x_i - \overline{x})^2 f_i}{n - 1}$$

7 THE STANDARD DEVIATION

Since the variance involves the squares of deviations, it is a number in squared units. In many problems it is desirable that quantities describing a distribution possess the same units as the original set of measurements. The mean satisfies this requirement but the variance does not; however, by taking the positive square root of the variance the desired effect can be achieved. The resulting quantity is called the standard deviation. Thus, by definition, the *standard deviation* for unclassified data is

(9)
$$s = \sqrt{\frac{\sum_{i=1}^{n} (X_i - \overline{X})^2}{n - 1}}$$

and for classified data it is

$$s = \sqrt{\frac{\sum_{i=1}^{k} (x_i - \bar{x})^2 f_i}{n - 1}} \tag{10}$$

Just as in the case of calculating the mean, one should use formula (9) rather than (10) in calculating the standard deviation, provided the original unclassified data are available. However, unless the classification is very coarse, the difference between the values of s obtained from formulas (9) and (10) will be very small and can be ignored in practice.

Example 1. The value of s for Table 2 will be needed in the next section; therefore its calculation by means of formula (10) is carried out here. This requires the value of $\bar{x}$ that was obtained in Table 3. The necessary calculations are shown in Table 4. The squares in the fourth column were found by using Table I in the appendix.

Table 4

x_i	f_i	$x_i - \bar{x}$	$(x_i - \bar{x})^2$	$(x_i - \bar{x})^2 f_i$
114.5	1	−41.7	1,738.89	1,738.89
124.5	4	−31.7	1,004.89	4,019.56
134.5	17	−21.7	470.89	8,005.13
144.5	28	−11.7	136.89	3,832.92
154.5	25	− 1.7	2.89	72.25
164.5	18	8.3	68.89	1,240.02
174.5	13	18.3	334.89	4,353.57
184.5	6	28.3	800.89	4,805.34
194.5	5	38.3	1,466.89	7,334.45
204.5	2	48.3	2,332.89	4,665.78
214.5	1	58.3	3,398.89	3,398.89
Totals	120			43,466.80

$$s = \sqrt{\frac{43,466.80}{119}} = \sqrt{365.27} = 19.1$$

Many of the modern electronic hand calculators yield the value of s without the necessity of calculating sums of squares. The calculations of Table 4 were carried out on the assumption that no such equipment is available.

For the purpose of observing how s measures variation, consider once more the two sets of measurements corresponding to Fig. 6. Since $\bar{X} = 6$ for both sets, calculation of s for the two sets by means of formula (9) will yield the values

$$s_1 = \sqrt{\frac{4 + 1 + 0 + 1 + 4}{4}} = \frac{\sqrt{10}}{2}$$

and

$$s_2 = \sqrt{\frac{16 + 4 + 0 + 4 + 16}{4}} = \sqrt{10}$$

The fact that the value of the standard deviation is twice as large for the second set as for the first set is certainly a satisfying property of the standard deviation if it is to be considered a measure of variation.

The preceding illustration, which indicates that the standard deviation increases in size as the variability of the data increases, does not give any clue to the meaning of the magnitude of the standard deviation. Thus, if the values of the standard deviation in this illustration had been $3\sqrt{10}$ and $6\sqrt{10}$, instead of $\sqrt{10}/2$ and $\sqrt{10}$, the interpretation would have been the same. This situation is similar to that occurring when two students compare scores on a test. One student may score twice as many points as the other, but this does not reveal how much absolute knowledge of the subject either student possesses.

In order to give some quantitative meaning to the size of the standard deviation, it is necessary to anticipate certain results of later work. For a set of data that has been obtained by sampling a particular type of population, called a normal population, it will be shown later that when the sample is large the interval from $\bar{x} - s$ to $\bar{x} + s$ usually includes about 68 percent of the observations and that the interval from $\bar{x} - 2s$ to $\bar{x} + 2s$ usually includes about 95 percent of the observations. A sample from a population of this type usually has a histogram that looks somewhat like the histogram of Fig. 3. As the sample increases in size, the histogram tends to approach the shape of a bell.

Example 2. As an illustration of this property, consider the data of Table 2. Table 4 yielded the value of $s = 19.1$ for those data. Since it was found earlier that $\bar{x} = 156.2$, the two intervals $(\bar{x} - s, \bar{x} + s)$ and $(\bar{x} - 2s, \bar{x} + 2s)$ become (137.1, 175.3) and (118.0, 194.4), respectively. These values have been marked off with vertical arrows on the x-axis of Fig. 7, which is the histogram for the frequency distribution of Table 2.

In computing the percentages of the data lying within each of these two intervals, it is necessary to approximate how many observations in an interval lie to the right, and to the left, of a point inside the interval. Toward this end, it is assumed that all the observations in an interval are spread uniformly along the interval. Thus, for the second interval, it is assumed that one of the four observations lies in the first quarter of the interval, another is in the second quarter of the interval, etc. With this understanding, the number of observations lying between 137.1 and 175.3 is approximately equal to $4 + 28 + 25 + 18 + 8 = 83$. The values of 4 and 8 are obtained by interpolating according to the preceding agreement on how observations are spread out. For example, 4 is obtained by realizing that the distance

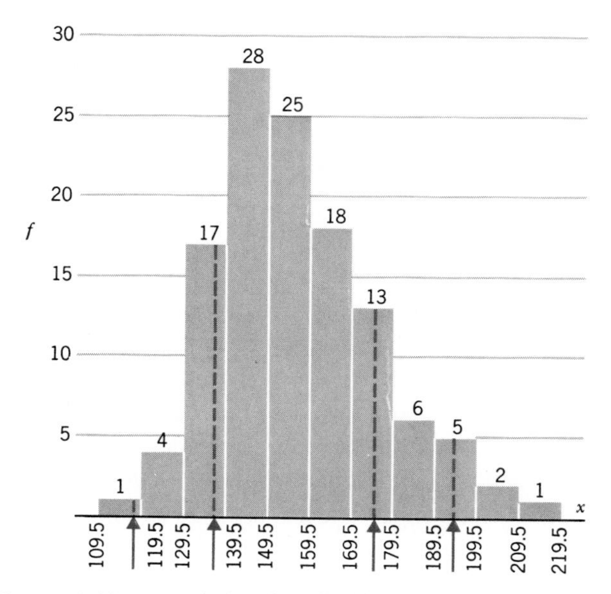

Figure 7 Histogram for the distribution of 120 weights.

from 129.5 to 137.1 is 7.6 units, whereas the distance from 129.5 to 139.5 is 10 units; therefore 76 percent of the 17 observations in this interval should be assumed to be to the left of 137.1 and 24 percent to the right. Since 24 percent of 17 is equal to 4, to the nearest integer, it follows that 4 of the 17 observations of this interval should be assumed to lie to the right of 137.1. The value of 8 in the preceding sum is obtained in a similar manner. Thus the distance from 169.5 to 175.3 is 5.8 units, whereas the distance from 169.5 to 179.5 is 10 units; hence 58 percent of the 13 measurements in this interval should be treated as being to the left of the point 175.3. Since 58 percent of 13 is 7.54, or 8 to the nearest integer, it follows that 8 of those 13 observations should be assumed to be to the left of 175.3. Now since the total number of observations is 120 and since $\frac{83}{120} = .69$, it follows that approximately 69 percent of the observations fall inside the interval $(\bar{x} - s, \bar{x} + s)$. Similar calculations will show that approximately 94 percent of the observations fall inside the interval $(\bar{x} - 2s, \bar{x} + 2s)$. These results are certainly close to the theoretical percentages of 68 and 95 for normal distributions.

For a distribution whose histogram resembles the histogram in Fig. 4 or Fig. 5, one would not expect to find the percentages for the intervals discussed to be very close to the theoretical percentages for a normal distribution, yet the percentages are often fairly close to those theoretical values.

By using the foregoing geometrical interpretation of the standard deviation it is possible to obtain a rough idea of the size of the standard deviation for familiar distributions. Consider, for example, the distribution of stature for adult males. One might guess that about 95 percent of all males would have heights somewhere between 5 feet 2 inches and 6 feet 2 inches. Since this is a 12-inch interval, and since for a normal distribution 95 percent of the observations would be expected to lie in the interval $(\bar{x} - 2s, \bar{x} + 2s)$ whose length is $4s$, one would guess that $4s = 12$ inches or that $s = 3$ inches. A crude estimate of the standard deviation of the frequency distribution of stature for adult males is therefore 3 inches. In a similar manner, one should be able to give a crude estimate of the size of the standard deviation for other familiar frequency distributions and thus acquire a feeling for the standard deviation as a measure of the variability of data.

8 OTHER DESCRIPTIVE MEASURES

It is interesting to observe in newspaper reports how different groups will employ different averages to describe the distribution of wages in their industry. Employers usually quote the mean wage to indicate the economic status of employees. Labor leaders, however, prefer to use the mode or the median as an indicator of the wage level.

The *mode* of a set of measurements is defined as the measurement with the maximum frequency, if there is one. Thus, for the set of measurements 3, 3, 4, 4, 4, 5, 5, 6, 6, 7, 8, 9, 9, the mode is 4. In certain industries there may be more laborers working at the lowest wage scale than at any of the other scales, and therefore labor leaders would naturally prefer to quote the mode in describing the distribution of wages. In most problems of this type one must know more than just a measure of location, and therefore the mode is likely to be of limited value here.

The *median* of a set of measurements is defined as the middle measurement, if there is one, after the measurements have been arranged in order of magnitude. For the set of measurements in the preceding paragraph, which is arranged in order of magnitude, the median is 5. If there is an even number of measurements, one chooses the median to be halfway between the two middle measurements. Thus, if one of the 5's were deleted from the preceding set, there would be no middle measurement and the median would become 5.5. The median is a more realistic measure to describe the wage level in certain industries than either the mode or the mean. Since the median wage is one such that half the employees receive at least this much and half receive at most this much, one usually obtains a fairly good picture of the wage level from the median. The mean has the disadvantage that if most of the wages are fairly low, but there is a small percentage of very high wages, the mean wage will indicate a deceptively high wage level.

The median would seem to be better than the mean here as an indicator of what is popularly meant by the wage level.

The median has another rather attractive property. If the variable being studied is income and incomes are listed in intervals of 500 dollars, but with all over 20,000 dollars listed as 20,000 dollars or more, it is not possible to compute the mean income because of the uncertainty of the incomes in the last interval. The median income would be unaffected by this lack of information and therefore could be calculated here.

For data that have been classified, the median is defined to be the value of x such that a vertical line through the corresponding point on the x-axis cuts the histogram into two parts having equal areas. This property is quite different from the balancing property of the mean. The calculations necessary to determine a median for classified data are based on this geometrical property. As an illustration of such calculations consider the problem for the data of Fig. 2. Since there are 120 measurements represented by that histogram, and areas are proportional to frequencies, the problem is equivalent to that of finding a point on the x-axis such that 60 of the measurements will be to the left of it and 60 to the right of it. If absolute frequencies are summed it will be found that the first four class intervals yield a total of 50 frequencies, whereas the first five intervals yield a total of 75 frequencies. The median must therefore be located in the fifth class interval, ideally halfway between the sixtieth and sixty-first measurement of the entire set. Since 10 of the 25 measurements in the fifth interval are needed to yield a total of 60 measurements, the median should be located at a point that is $\frac{10}{25}$ of the way along that interval. The interval is 10 units long and it begins at the point 149.5; therefore the median is the value

$$149.5 + \frac{10}{25}(10) = 153.5$$

Note that the median is slightly smaller than the mean for this problem. This is usually true for distributions skewed to the right. From Fig. 2 observe that there is a slight amount of this skewness here.

In view of the foregoing remarks about some of the attractive properties of the median, one might wonder why statisticians usually prefer the mean to the median. There are some computational advantages of the mean over the median; however, the principal reason for preferring the mean is that it is a much more useful and reliable measure to use in making statistical inferences. Since the ultimate objective is to solve statistical inference problems, the mean is usually preferred to the median when both measures can be found.

Another measure of location that possesses desirable properties for certain classes of problems is the *geometric mean.* It is given by the formula

$$\text{G.M.} = (X_1 X_2 \cdots X_n)^{1/n} \tag{11}$$

The geometric mean is particularly appropriate when the X's represent measurements taken over time of a quantity that is increasing at a constant rate. Thus, the geometric mean of the value of an investment at compound interest over a period of n equal time intervals will be equal to the value of the investment at the halfway point.

As an illustration, suppose that the value of an investment of $10,000 increases by 8 percent each year for 9 years. The value at the start and at the end of each of the 9 years would then be given by

$$10{,}000,\ 10{,}000(1.08),\ 10{,}000(1.08)^2,\ \ldots,\ 10{,}000(1.08)^9$$

The geometric mean of these 10 values, as given by formula (11), will be the tenth root of the product of these numbers. Thus

$$\begin{aligned} \text{G.M.} &= [10{,}000 \cdot 10{,}000(1.08) \cdot 10{,}000(1.08)^2 \cdots 10{,}000(1.08)^9]^{1/10} \\ &= [(10{,}000)^{10}(1.08)^{1+2+\cdots+9}]^{1/10} = 10{,}000(1.08)^{45/10} \\ &= 10{,}000(1.08)^{4.5} = 14{,}139 \end{aligned}$$

This is the value of the investment at the end of $4\frac{1}{2}$ years, that is, the value at the halfway point. For comparison, the arithmetic mean of the preceding values is

$$\begin{aligned} \overline{X} &= \frac{1}{10}[10{,}000 + 10{,}000(1.08) + 10{,}000(1.08)^2 + \cdots + 10{,}000(1.08)^9] \\ &= 1000[1 + (1.08) + (1.08)^2 + \cdots + (1.08)^9] \\ &= 1000[14.487] = 14{,}487 \end{aligned}$$

Thus the arithmetic mean yields a value considerably higher than the true value of the investment at the halfway point.

Several measures of variation, in addition to the standard deviation, are occasionally used. The simplest measure of the variation of a set of measurements is the range. The *range* has already been employed in the process of classifying data and is merely the difference between the largest and smallest measurements of the set. It is a popular measure in such fields as industrial quality control and meteorology. This popularity rests principally on its ease of computation and interpretation. It is a simple matter to find the largest and smallest measurements of a set, in contrast to calculating the standard deviation of the set. It is also a simple matter to explain how the range measures variation, in contrast to explaining how the standard deviation does so. However, an unfortunate property of the range is the tendency of its value to increase in size as the sample size increases. For example, one would expect the range of the weights of a sample of ten college students to be considerably smaller than the range in the weights of a sample of 100 students. It is possible to adjust the range for this growth by means of fancy formulas, but then the range loses its simplicity.

Another measure of variation that is encountered frequently is the *mean deviation.* It is defined by the formula

$$\text{M.D.} = \frac{1}{n} \sum_{i=1}^{n} |X_i - \overline{X}| \tag{12}$$

The two vertical bars enclosing $X_i - \overline{X}$ are absolute value symbols. They indicate that one should calculate the difference $X_i - \overline{X}$ and then ignore the minus sign in case the difference is a negative number. Thus one calculates all the differences of this type, ignores the minus signs, sums the resulting numbers, and divides by n. The preceding formula can be applied to classified data by rewriting it in the form

$$\text{M.D.} = \frac{1}{n} \sum_{i=1}^{k} |x_i - \bar{x}| f_i \tag{13}$$

Measures of variation can also be constructed by methods similar to that used to define the median. Instead of merely finding a value of x that divides the histogram into a lower and upper half, one could also find two other values of x, one dividing the histogram at a point such that one-fourth of the area is to the left of it and the other such that one-fourth of the area is to the right of it. These two measures, together with the median, constitute the three *quartiles* of the distribution. The smallest quartile is called the first quartile, the median is called the second quartile, and the largest quartile is called the third quartile. A simple measure of variation can be constructed from the quartiles by taking the difference between the third quartile and the first quartile. This measure, which is used in some fields, is called the *interquartile range.*

One could construct other measures of variation in a similar manner by considering deciles rather than quartiles. *Deciles* are values of x that divide the histogram into tenths. Thus one might use the difference between the ninth decile and the first decile as a measure of variation. Measures of this general type possess certain advantages over the standard deviation similar to those possessed by the median over the mean. However, the same kind of superiority of the standard deviation exists here as exists for the mean over the median when it comes to solving problems of statistical inference.

Deciles can be used for purposes other than constructing measures of variation. They are sometimes used for describing a distribution in more detail than that given by a measure of location and a measure of variation. Knowing all the deciles of a distribution would give a large amount of information concerning the nature of it.

These ideas can be extended further by introducing *percentiles,* which are values of x that divide the histogram into 100 equal area parts. Deciles and percentiles are calculated in the same manner as quartiles. Percentiles are used extensively in such fields as psychology and educational testing, for example in comparing students' results on standardized examinations with national averages. Undoubtedly some of you have had the experience of being told at what percentile you rated on a scholastic aptitude test.

The various measures of location and variation that have been described in this section were presented here to acquaint the student with the fact that special problems sometimes require special treatment; however, almost all of the theory and problems hereafter will use only the mean and the standard deviation to measure location and variation.

9 COMMENTS ON CALCULATIONS

As was stated before, one should use formulas (3) and (9) for calculating the mean and the standard deviation if the original set of data is available, even though the data have been classified into a frequency table. The purpose of the classification is primarily to observe the nature of the distribution and not to simplify the calculation of those measures. The calculations can be made very quickly using the original data and modern computers.

If the calculation of s is to be performed without the use of such computers, or if only the classified data are available, it is often easier to employ the following algebraic modification of formula (10):

$$s = \sqrt{\frac{\sum_{i=1}^{k} x_i^2 f_i - \frac{1}{n}\left(\sum_{i=1}^{k} x_i f_i\right)^2}{n-1}}$$

This formula eliminates the necessity of calculating the differences $x_i - \bar{x}$; however, if the x_i and f_i are large numbers, sums in this formula may be very large. Unless a sufficient amount of digit accuracy is carried, the difference of these sums may produce very little accuracy in the answer. Thus, judgment is required to decide whether this computing formula is better or worse than formula (10) for calculating s. Since this formula is being presented here only to acquaint students with the fact that such formulas exist, this type of decision will not be needed to solve the exercises that follow. Students may, however, use this formula in place of formula (10) if they so prefer.

10 REVIEW ILLUSTRATIONS

This section is designed to serve as a problem-solving review of some of the concepts discussed in this chapter.

Example 1. The following data represent the number of hours worked per week for 100 laborers. Using these data, (*a*) classify them into a frequency table, (*b*) draw the histogram, (*c*) calculate the mean and standard deviation, (*d*) calculate the percentage of measurements lying in the two intervals $(\bar{x} - s, \bar{x} + s)$ and $(\bar{x} - 2s, \bar{x} + 2s)$, (*e*) calculate the median and the interquartile range.

36	36	32	35	41	32	41	30	22	32	27	35	35	10	29	41	45	45	30	39
45	33	23	28	27	43	44	31	34	33	36	28	31	39	29	42	43	31	28	39
33	18	25	36	45	45	24	37	52	26	23	23	38	37	38	42	40	42	40	40
42	40	34	37	34	36	40	33	40	20	10	23	15	28	28	32	28	37	37	44
25	36	26	40	40	39	39	41	33	39	40	38	39	16	39	38	41	41	28	27

(*a*) These numbers range from 10 to 52, which is a spread of 42. Dividing by 10 and 20 gives 4.2 and 2.1, respectively; hence choose a class interval of length 3. Starting the first interval at 9.5 will give 9.5 and 12.5 as the boundaries for the first interval. The tabulation for these 100 measurements then becomes:

		x	f
9.5–12.5	//	11	2
12.5–15.5	/	14	1
15.5–18.5	//	17	2
18.5–21.5	/	20	1
21.5–24.5	~~////~~ /	23	6
24.5–27.5	~~////~~ //	26	7
27.5–30.5	~~////~~ ~~////~~ /	29	11
30.5–33.5	~~////~~ ~~////~~ //	32	12
33.5–36.5	~~////~~ ~~////~~ //	35	12
36.5–39.5	~~////~~ ~~////~~ ~~////~~ //	38	17
39.5–42.5	~~////~~ ~~////~~ ~~////~~ ////	41	19
42.5–45.5	~~////~~ ////	44	9
45.5–48.5		47	0
48.5–51.5		50	0
51.5–54.5	/	53	1

(*b*)

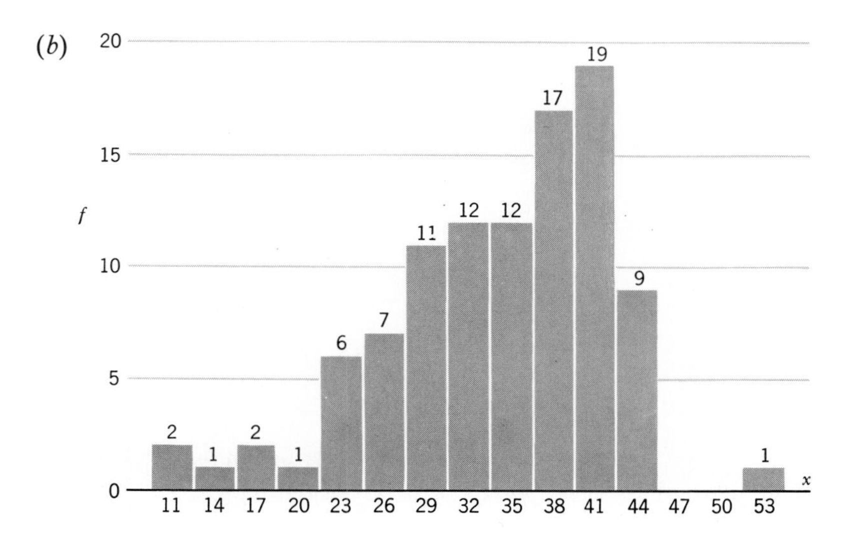

(*c*)

x_i	f_i	$x_i f_i$
11	2	22
14	1	14
17	2	34
20	1	20
23	6	138
26	7	182
29	11	319
32	12	384
35	12	420
38	17	646
41	19	779
44	9	396
47	0	
50	0	
53	1	53
Totals	100	3,407

Using formula (4),

$$\bar{x} = \frac{3{,}407}{100} = 34.07 = 34.1$$

x_i	f_i	$(x_i - \bar{x})$	$(x_i - \bar{x})^2$	$(x_i - \bar{x})^2 f_i$
11	2	−23.1	533.61	1,067.22
14	1	−20.1	404.01	404.01
17	2	−17.1	292.41	584.82
20	1	−14.1	198.81	198.81
23	6	−11.1	123.21	739.26
26	7	− 8.1	65.61	459.27
29	11	− 5.1	26.01	286.11
32	12	− 2.1	4.41	52.92
35	12	.9	.81	9.72
38	17	3.9	15.21	258.57
41	19	6.9	47.61	904.59
44	9	9.9	98.01	882.09
47	0	12.9	166.41	0
50	0	15.9	252.81	0
53	1	18.9	357.21	357.21
Totals	100			6,204.60

Using formula (10),

$$s = \sqrt{\frac{6{,}204.60}{99}} = \sqrt{62.7} = 7.9$$

(*d*) The required intervals are $(\bar{x} - s, \bar{x} + s) = (26.2, 42.0)$ and $(\bar{x} - 2s, \bar{x} + 2s) = (18.3, 49.9)$. From the histogram in (*a*), observe that the boundaries of the interval in which 26.2 lies are 24.5 and 27.5. Since the distance from 26.2 to 27.5 is 1.3 units and the distance from 24.5 to 27.5 is 3 units, the fraction $1.3/3 = .43$ of the 7 measurements inside that interval are treated as lying to the right of 26.2. But 43 percent of 7 is 3, to the nearest integer; hence 3 of those measurements should be included in the total count. Similar calculations for the interval in which 42.0 lies give the fraction $2.5/3 = .83$. To this accuracy, the total number of measurements lying inside the interval (26.2, 42.0) is therefore

$$\frac{1.3}{3}(7) + 11 + 12 + 12 + 17 + \frac{2.5}{3}(19) = 71$$

Hence, approximately 71 percent lie inside this interval. In a similar manner, the number of measurements outside the interval (18.3, 49.9) is

$$2 + 1 + \frac{2.8}{3}(2) + 1 = 6$$

Therefore approximately 94 percent lie inside the interval.

(*e*) Refined calculations give 35.5; hence 35 or 36 is a satisfactory estimate for the median. Similar calculations give: first quartile = 29.1, third quartile = 40.1, and interquartile range = 40.1 − 29.1 = 11.

Example 2. A salesman turned in the following expense account to cover his expenses for 6 trips that were made during the past month.

Trip	Days	Expenses	Expenses per day
1	2	64	32
2	8	128	16
3	$\frac{1}{2}$	40	80
4	9	108	12
5	4	80	20
6	$\frac{1}{2}$	25	50
Totals	24	445	210

The auditor felt that these expenses were excessive because the average expense per day, he says, is \$35 (= \$210 ÷ 6), and that the median, which is only \$16, is the appropriate measure. The salesman replied that the arithmetic mean is the appropriate measure and, moreover, that the median is \$26. (*a*) Explain the proper interpretation for this problem of each of the averages mentioned. (*b*) Which of them do you believe is more appropriate?

(*a*) The average daily expense per trip is \$210/6 = \$35. This average varies with the length of trips. The auditor was not justified in calling this the average daily expense. The average expense per day is \$445/24 = \$18.54. This average removes the effect of trip lengths. It is a reasonable *per diem* average.

The auditor must have calculated his median by assuming that there were 9 days of expenditures of \$12 each and 8 days of expenditures of \$16 each. If this were true and the daily expenditures were arranged in order of size, then the twelfth and thirteenth expenditures in this ranking would be \$16.

The salesman must have calculated his median by finding a value halfway between the third and fourth expense per day when they are arranged in order of size. Those values are 20 and 32; hence the halfway value would be \$26.

(*b*) Neither of these median values is a true median and neither represents a measure of average daily expenditures. To obtain the true median it is necessary to take each day's expenditure for all 24 days, arrange them in order of size, and then choose a value midway between the twelfth and thirteenth day expenditures. The most reasonable measure here is the \$18.54 average.

EXERCISES

SECTION 1

1 Toss a coin 500 times, recording the number of heads obtained in sets of 50. Calculate the cumulative proportion of heads obtained in decimal form and observe whether the resulting set of 10 proportions appears to approach $\frac{1}{2}$.

2 Roll a die 500 times, recording only the number of times that a 1 or a 6 appears in sets of 50 rolls. Calculate the cumulative proportion of such results and observe whether the proportions seem to approach $\frac{1}{3}$.

3 A manager of a supermarket is interested in giving the best possible checkout service for his store without undue costs. He is aware that the number of customers entering the store varies during the day and also from day to day during the week. For basic data he would like to know such "waiting experience" as (*i*) the fraction of customers who must wait their turn at a checkout station, (*ii*) the distribution of the number of persons in the shortest checkout queue, and (*iii*) the distribution of the amount of time each customer waits in a checkout line.

(*a*) What types of data (or variables) are involved in (*i*), (*ii*), and (*iii*) above?

(*b*) Suppose we confine our interest to one week. Describe the population or populations in (*i*), (*ii*), and (*iii*).

(*c*) Suppose Tuesday afternoon from 3 to 5 is selected for making the observations. What do you think of the representativeness of the results?

SECTION 2

4 Weights of 300 babies range from 82 ounces to 176 ounces, correct to the nearest ounce. Determine class boundaries and class marks for the first and last class intervals.

5 The thickness of 400 washers range from 0.421 to 0.563 centimeter. Determine class boundaries and class marks for the first and last class intervals.

6 Data have been obtained on 500 "wait times," that is, the time required for a customer to get service at a checkout station. The data range from 0 to 15.1 minutes. Construct a table with approximately 10 classes into which these data might be grouped.

7 If you were to study the age distribution of college students, would you consider age to be a discrete or continuous variable? What would you choose for class boundaries?

8 Draw a sample of 100 one-digit random numbers from Table XII in the appendix by taking consecutive digits from as many columns as needed. Classify these numbers into a frequency table and draw the histogram, even though the variable here is discrete. These numbers, which are discussed in Chapter 6, possess a "rectangular" distribution; hence your histogram should resemble a rectangle.

SECTION 3

9 Given the following frequency table of the diameters in feet of 56 shrubs from a common species, draw its histogram showing the class marks.

x	1	2	3	4	5	6	7	8	9	10	11	12
f	1	7	11	16	8	4	5	2	1	0	0	1

10 Given the following frequency table of the heights in centimeters of 1000 students, draw its histogram showing the class marks.

x	155–157	158–160 etc.													
f	4	8	26	53	89	146	188	181	125	92	60	22	4	1	1

11 The following data are for the traveling time to and from work in hours per day for a group of aircraft workers. Draw the histogram, assuming continuous time.

Under 1 hour	80
1 up to 2	42
2 up to 3	7
3 up to 4	4
4 up to 5	3
5 up to 6	2

12 Given the following frequency distribution of the number of children born to wives aged 40–44 during their married lives, draw its histogram.

x	0	1	2	3	4	5	6	7	8	9	10 or more
f	1230	1520	1545	962	537	301	174	108	69	51	73

13 An experiment conducted to determine the gasoline mileage of a particular type of car yielded the following data:

Miles per gallon	Percent
15 and under 16	6
16 and under 17	10
17 and under 18	16
18 and under 19	24
19 and under 20	14
20 and under 21	18
21 and under 22	12

Graph the histogram for gasoline mileage.

14 According to the *Survey of Current Business* for 1964, the incomes of U.S. families for 1962 were the following:

Income	Percent
under 2,000	6.9
2,000–2,999	6.2
3,000–3,999	8.2
4,000–4,999	9.8
5,000–5,999	10.8
6,000–7,499	16.0
7,500–9,999	18.6
10,000–14,999	14.8
15,000 and over	8.7
	100.0

(*a*) Criticize the choice of class intervals and class boundaries.

(*b*) Graph the histogram. What would the histogram look like if the frequencies in the wide intervals were scaled down in proportion to the width of the interval?

15 What type of distribution would you expect grade-point averages of college students to possess? Sketch your idea of the nature of this distribution with the proper units on the x-axis.

SECTION 5

16 For the histogram of problem 9, guess the value of $\bar{x}$. Do the same for the histograms of problems 10 and 11.

17 For the data of problem 9, calculate $\bar{x}$.

18 A brokerage firm lists 23 common stocks as ''good quality.'' Their price-earnings ratios are as follows:

11.0	13.0	10.7	40.0	23.4	22.2	10.9	9.7	11.0	20.0	26.3	15.7
10.3	16.1	11.6	22.2	25.2	9.3	13.1	17.4	12.6	16.9	44.2	

Determine the mean p-e ratio of these 23 stocks.

19 For the data of problem 10, calculate $\bar{x}$.

20 An electronics firm has two divisions, one having 15,000 employees, the other 5,000. In the first division the employees averaged 42.15 hours per week, in the second, 36.25 hours. Calculate the average hours worked per week for the entire firm.

21 A savings and loan association has the following mortgages: $20,000 at 8%, $10,000 at 7%, $40,000 at 10%, $25,000 at 9%, and $5,000 at 6%. What is the average interest rate paid on a dollar loaned on mortgages?

22 A group of 40 boys and 60 girls was tested for reading ability. The boys averaged 46 points and the girls averaged 66 points. Calculate the arithmetic mean for the combined group and comment on it.

23 The following data, as reported in *Studies in American Demography* by W. F. Willcox, give the distribution of fertility by race in the United States in 1930.

	Children per 1000 wives		
Region	White	Black	Departure of black from white
United States	632	654	+ 22
New England	654	651	− 3
Middle Atlantic	591	468	−123
East North Central	593	439	−154
West North Central	654	450	−204
South Atlantic	752	771	+ 19
East South Central	795	710	− 85
West South Central	676	642	− 34
Mountain	682	422	−260
Pacific	441	370	− 71

Explain how it is possible for blacks to have a greater average fertility than whites (654 versus 632) in the nation when they have a lower one in 8 of the 9 regions.

SECTIONS 6 and 7

24 A sample of 5 sales slips from a file gives the following amounts purchased at a particular department (rounded to the nearest dollar):

5 15 3 16 11

Compute the variance of the amounts purchased.

25 Without classifying the data, calculate (*a*) the mean and (*b*) the standard deviation for the following set of weights of 11 children: 38, 50, 37, 44, 46, 53, 48, 38, 42, 46, 42.

26 Without classifying the data, calculate (*a*) the mean and (*b*) the standard deviation for the following set of grade-point averages of 20 students: 2.4, 1.2, 1.4, 2.4, 1.0, 1.8, 1.8, 1.4, 1.8, 3.2, 2.4, 2.2, 2.4, 1.8, 3.6, 1.8, 1.2, 2.4, 1.8, 3.4.

27 For the data of problem 9, calculate s.

28 For the data of problem 10, calculate s.

29 For the histogram of problem 9, using the results in problems 17 and 27, calculate the approximate percentages of the data that lie within the intervals $\bar{x} \pm s$ and $\bar{x} \pm 2s$.

30 For the histogram of problem 10, using the results of problems 19 and 28, calculate the approximate percentages of the data that lie within the intervals $\bar{x} \pm s$ and $\bar{x} \pm 2s$.

31 Suppose you found that 5 percent of the shots on a target were a radial distance of more than three standard deviations from the center. What might you conclude about the distribution of the shots?

32 If shoe sizes of college male (or female) students are assumed to possess a normal distribution, what would you guess the standard deviation of shoe size to be if, by your knowledge of shoe sizes, you estimated a two-standard-deviation interval about the mean?

33 A purchaser of light bulbs obtains samples from two suppliers. He tests the samples in his laboratory for length of life (in hours) and obtains the following results.

Length of life	Supplier A	Supplier B
700 and under 900	12	4
900 and under 1100	14	34
1100 and under 1300	24	19
1300 and under 1500	10	3
Totals	60	60

(*a*) Which supplier's bulbs have the greater mean length of life? Comment.
(*b*) Which supplier's bulbs are more uniform in quality? Comment.

34 What can be said about a distribution if $s = 0$?

35 If the scores on a set of examination papers are changed by (*a*) adding 10 points to all scores, (*b*) increasing all scores by 10 percent, what effect will these changes have on the mean and on the standard deviation?

36 Cite some type of data for which you feel the standard deviation would tend to exaggerate the amount of variation present in the data.

37 Calculate the standard deviation for the discrete distribution consisting of the two points $x = -1$ and $x = 1$ with frequencies of 50 each and comment about the 68 percent and 95 percent interpretation of s here.

SECTION 8

38 A random sample of 5 students from a class has the following heights (in inches):

71, 72, 75, 75, 67

(*a*) What is the median?
(*b*) What is the mean?
(*c*) What is the mode?
(*d*) What is the range?
(*e*) What is the standard deviation?

39 The following are the numbers of unoccupied apartments per apartment house in a given neighborhood: 4, 3, 3, 7, 2, 6, 4, 8, 4, 3, 6, 2, 6, 9, 1, 2, 6. What is the (*a*) mean, (*b*) median, (*c*) mode, (*d*) range?

40 Cite some type of data for which you feel the median would be a more appropriate measure of location than the mean.

41 Cite some type of data for which you feel the mode would be a more appropriate measure of location than the mean.

42 At the close of each calendar year the faculty of a business school make individual forecasts of next year's economic activity, such as gross national product (GNP), automobile production, and the Dow-Jones Industrial Average for stocks. What type of average, based on these individual forecasts, do you believe should be used to represent the entire faculty's viewpoint?

43 The faculty of a business school at the close of each calendar year makes individual forecasts of GNP changes for the coming year with the following results for twenty-three participants. Determine the mean and medians of the forecasts.

+3.0 +5.2 +4.1 +1.7 +2.0 +2.8 +2.5 −1.0 −3.2 +0.5
+1.6 +0.0 +1.9 +2.4 +5.2 −3.5 +5.5 +3.2 +6.5 +2.2
+2.5 +4.5 +3.1.

44 In an industrial wage dispute, the representative of management claims that the average hourly earnings in the industry is $2.73 whereas labor claims that the prevailing wage is $2.25. Is someone clearly misrepresenting the facts? Explain.

45 A study of 241 famous authors revealed the following data on the distribution of ages at which their best book was published. (Harvey C. Lehman, *Age and Achievement,* Princeton Univ. Press, Princeton, New Jersey (1953).)

Age of author	Number of authors
20 up to 30	20
30 up to 40	73
40 up to 50	80
50 up to 60	44
60 up to 70	22
70 up to 80	2
Total	241

Compute the median and mean of the distribution.

46 The following data, taken from the 1963 edition of *Automobile Facts and Figures,* Manufacturers' Association, give the distribution by age of licensed automobile drivers in the U.S. for 1962.

Age	Number in thousands	Percentage
14–15	2	.2
16–20	60	7.4
21–29	157	19.3
30–39	219	26.9
40–49	176	21.4
50–59	118	14.6
60–69	63	7.7
70–90	21	2.5
Totals	816	100.0

(*a*) Determine the median age of drivers.
(*b*) What age determines the oldest 10 percent of drivers?

47 For the data of problem 25, calculate the mode, median, range, and mean deviation.

48 For the data of problem 9, calculate the median and range.

49 For the data of problem 9, calculate the interquartile range.

50 Jones has the 1940 and 1960 census figures on the population of Arkansas and California and wants the 1950 figures but is unable to get them. Under the circumstances he considers the arithmetic and geometric means. Show that the geometric mean is far superior to the arithmetic mean in the case of California and is about equally good for Arkansas. Explain these results. (The data are in millions of persons.)

State	1940	1950	1960
Arkansas	1.949	1.910	1.786
California	6.907	10.586	15.717

SECTION 9

51 Given the values $z_1 = 1$, $z_2 = 2$, $z_3 = 3$, $z_4 = 4$, find the following sums.

(*a*) $\sum_{i=1}^{4} z_i$ (*b*) $\sum_{i=1}^{4} 2z_i$ (*c*) $\sum_{i=1}^{4} (z_i + 2)$

(*d*) $\sum_{i=1}^{4} z_i^2$ (*e*) $\sum_{i=1}^{4} 2z_i^2$ (*f*) $\sum_{i=1}^{4} (z_i + 2)^2$

52 Show that (*a*) $\frac{1}{n} \sum_{i=1}^{n} (x_i + c) = \bar{x} + c$, (*b*) $\sum_{i=1}^{n} (x_i - \bar{x}) = 0$.

SECTION 10

53 As a review exercise, use your result from problem 8 to work the problems that were solved in the first review example of section 10.

54 As a review exercise, work the problems that were solved in the first review illustration of section 10 for the following set of weekly wages of certain laborers. Choose a class interval of length 2 units and begin the first class interval $\frac{1}{2}$ unit below the smallest measurement.

49 47 51 48 50 46 53 46 45 50 49 50 50 47 56 51 46 47 54 53 48 50 51 50 60
51 46 48 52 52 46 61 52 49 50 45 57 54 51 60 50 56 52 44 49 45 51 50 40 46
54 47 50 55 55 47 48 53 50 49 45 50 50 51 47 54 43 53 55 50 53 52 52 51 47
51 48 45 44 50 52 49 51 51 47 53 49 46 61 49 52 48 39 46 52 51 57 49 45 50

55 The price-earnings ratios of a selected set of stocks on the New York Stock Exchange were as follows:

14.6	25.4	12.2	9.3	20.3	7.6
13.4	9.6	14.7	25.3	13.7	10.1
11.9	18.9	8.5	12.3	14.8	8.6
28.2	16.2	12.8	16.2	12.1	14.0
15.4	27.5	30.4	16.2	16.5	11.6
9.7	21.1	12.9	9.8	30.4	13.3
11.7	8.7	31.6	19.5	22.4	8.3

(*a*) Classify the data into a frequency table. Choose a class interval of length 2 and begin $\frac{1}{2}$ unit below the smallest measurement.
(*b*) Draw a histogram for the frequency distribution in (*a*).
(*c*) By inspecting the histogram in (*b*), guess the value of the mean.
(*d*) Calculate the mean of the unclassified data.
(*e*) Calculate the mean of the classified data.
(*f*) By inspecting the histogram in (*b*), guess the value of the standard deviation.
(*g*) Calculate the standard deviation of the unclassified data.
(*h*) Calculate the standard deviation of the classified data.
(*i*) By inspecting the histogram in (*b*), guess the value of the three quartiles.

56 The data of Table XIII in the appendix may be used to construct various problems related to classification. For example, the values of the variable x_1 may be classified into a frequency table and the mean and the standard deviation calculated.

Chapter 3 Probability

1 INTRODUCTION

As indicated in Chapter 1, the solutions to the statistical problems posed there are given in terms of probability statements. Although probability is applied to a variety of practical situations, an understanding of the subject is made much simpler if it is applied to nonpractical situations, such as those that arise in certain games of chance. For this reason, the definition and the rules of probability are presented in terms of idealized problems, but it is assumed that the same rules may later be applied to practical statistical problems.

Before discussing probability, it is necessary to discuss experiments that can be repeated or that can be conceived of as being repeatable. Tossing a coin, reading the daily temperature on a thermometer, or counting the number of bad eggs in a carton are examples of a simple repetitive experiment. An experiment in which several rabbits are fed different rations in an attempt to determine the relative growth properties of the rations may be performed only once with those same animals; nevertheless, the experiment may be thought of as the first in an unlimited number of similar experiments, and therefore it may be considered repetitive. Selecting a sample from a population is a repetitive experiment and is, of course, the type of experiment that is of particular interest in solving statistical problems.

Consider a simple repetitive experiment such as tossing a coin twice or, equivalently, tossing two distinct coins simultaneously. In this experiment there are four possible outcomes of interest; they are denoted

HH, HT, TH, TT

The symbol HT, for example, means that a head is obtained on the first toss and a tail on the second toss. If the experiment had consisted of tossing the coin three times, there would have been eight possible outcomes of the experiment:

HHH, HHT, HTH, THH, HTT, THT, TTH, TTT

Here the three letters in a group express the outcomes of the three tosses in the given order. An experiment such as reading the temperature on a thermometer, however, has an infinite number of possible outcomes, since temperature is a continuous type of variable. In this chapter only experiments with a finite number of possible outcomes are considered. Other types are discussed in Chapter 4.

For any experiment to which probability is to be applied, it is first necessary to decide what possible outcomes of the experiment are of interest, and to make a list of all such outcomes. This list must be such that when the experiment is performed, exactly one of the outcomes will occur. In the experiment of tossing a coin three times, interest was centered on whether the coin showed a head or a tail on each of the tosses; therefore all the possible outcomes are those that were listed previously. In selecting a digit from the table of random digits in Table XII in the appendix, one might be interested in knowing which digit was obtained, in which case there are ten possible outcomes, corresponding to the digits 0, 1, . . . , 9. However, one might be interested only in knowing whether the digit was less than 3 in magnitude. Then there would be only two possible outcomes of the experiment, namely, obtaining a digit that is less than 3 or obtaining a digit that is at least as large as 3. A game of chance experiment that will be used frequently for illustrative purposes is the experiment of drawing one ball from a box of balls of different colors. Thus, suppose a box contains three red, two black, and one green ball. Then interest will be centered only on what color a drawn ball is and not on which particular ball is obtained. Here there are three possible outcomes of the experiment corresponding to the three colors. Another interesting game of chance experiment is the experiment of rolling two dice. If it is assumed that one can distinguish between the two dice and interest centers on what number of points shows on each of the dice, then there are 36 possible outcomes, because each die has six possible outcomes and these outcomes can be paired in all possible ways. Table 1 gives a list of the possible outcomes.

Table 1

11	21	31	41	51	61
12	22	32	42	52	62
13	23	33	43	53	63
14	24	34	44	54	64
15	25	35	45	55	65
16	26	36	46	56	66

The first number of each pair denotes the number that came up on one of the dice, and the second number denotes the number that came up on the other. If the two dice are not distinguishable, it is necessary to roll them one at a time rather than simultaneously.

2 SAMPLE SPACE

It is convenient, in developing the theory of probability, to visualize things geometrically and to represent each of the possible outcomes of an experiment by means of a point. Thus, in the experiment of tossing a coin three times one would use eight points. It makes no difference what points are chosen as long as one knows which point corresponds to which possible outcome. Each point is labeled with a letter or symbol to indicate the outcome that it represents. Since possible outcomes are often called *simple events,* the letter e with a subscript corresponding to the number of the outcome in the list of possible outcomes is customarily used in labeling a point. In the coin-tossing experiment, for example, one could label the eight points with $e_1, e_2, \ldots, e_8$. Thus e_1 would represent the event of obtaining HHH and e_2 that of obtaining HHT. Since a label such as HHT is self-explanatory there seems little point to introducing additional labels here. However, it is easier to write e_2 than HHT; consequently the e symbol does possess an advantage. The set of points representing the possible outcomes of an experiment is called the *sample space* for the experiment. A sample space for the coin-tossing experiment is shown in Fig. 1, in which the labeling of the points is shown by the symbols directly above the points.

A natural sample space to choose for the experiment of selecting a digit from the table of random digits consists of the 10 points on the x-axis corresponding to

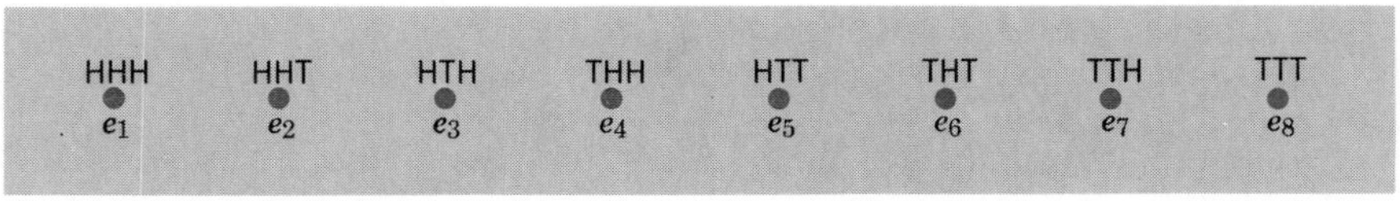

Figure 1 A sample space for a coin-tossing experiment.

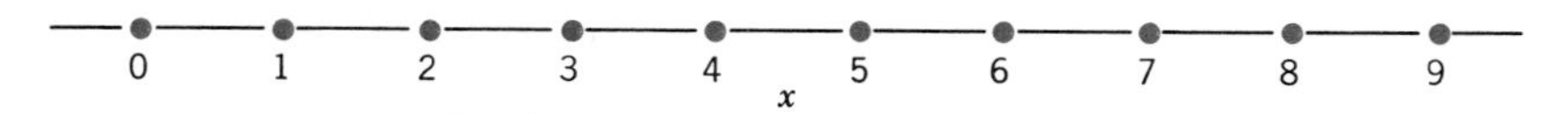

Figure 2 A sample space for a random-digit experiment.

the integers 0, 1, . . . , 9. This is shown in Fig. 2. The letter e was not used in Fig. 2 because the labeling of a point by means of its x-coordinate is about as simple as one could desire.

For the experiment of choosing a colored ball from the aforementioned box of balls, the sample space consists of three points that have been labeled in the order of the colors given, as shown in Fig. 3.

Figure 3 Sample space for colored-ball experiment.

A convenient sample space for the experiment of rolling two dice is a square array of 36 points, with six rows and six columns, attached to the elements of Table 1. This sample space will not be illustrated because Table 1 and some imagination should suffice. The symbols of Table 1 are highly descriptive of the experimental outcomes and are also very simple; there is no point in introducing a new set of symbols here.

3 THE PROBABILITY OF AN EVENT

Thus far there has been no apparent reason for introducing a geometrical representation of the outcomes of an experiment; however, later in the development of the theory the advantage of this approach will become evident.

The next step in the construction of a mathematical model for an experiment is to attach numbers to the points in the sample space that will represent the relative frequencies with which those outcomes are expected to occur. If the experiment of tossing a coin three times were repeated a large number of times and a cumulative record kept of the proportion of those experiments that produced, say, three heads, one would expect that proportion to approach $\frac{1}{8}$ because each of the eight possible outcomes would be expected to occur about equally often. Actual experiments of this kind usually show that such expectations are justified, provided the coin is well balanced and is tossed vigorously. In view of such considerations, the number $\frac{1}{8}$ would be attached to each of the points in the sample space shown in Fig. 1. The number assigned to the point labeled

e_i in a sample space is called the *probability* of the event e_i and is denoted by the symbol $P\{e_i\}$. Thus, in the coin-tossing experiment each of the events e_1, $e_2, \ldots, e_8$ possesses the probability $\frac{1}{8}$.

If the experiment of selecting a digit from the table of random digits is carried out a large number of times, each of the ten digits $0, 1, \ldots, 9$ will be obtained with approximately the same relative frequency, and therefore the experimental relative frequency for each of the digits will be close to $\frac{1}{10}$. On the basis of such experience each of the sample points in the sample space shown in Fig. 2 would be assigned the probability $\frac{1}{10}$.

The experiment of rolling two dice is treated in much the same manner as the coin-tossing experiment. Symmetry and experience suggest that each point in the sample space corresponding to Table 1 should be assigned the probability $\frac{1}{36}$.

The situation for the experiment corresponding to the sample space shown in Fig. 3 is somewhat different from the preceding ones. It is no longer true that each of the possible outcomes would be expected to occur with the same relative frequency. If the balls are well mixed in the box before each drawing and the drawn ball is always returned to the box so that the composition of the box is unchanged, one would expect to obtain a black ball twice as often as a green ball and a red ball three times as often as a green ball. This implies that in repeated sampling experiments one would expect the relative frequencies for the three colors red, black, and green to be close to $\frac{3}{6}$, $\frac{2}{6}$, and $\frac{1}{6}$, respectively. Thus, the three points e_1, e_2, and e_3 in Fig. 3 would be assigned the probabilities $\frac{3}{6}$, $\frac{2}{6}$, and $\frac{1}{6}$, respectively.

The preceding experiments illustrate how one proceeds in general to assign probabilities to the points of a sample space. If the experiment is one for which symmetry and similar considerations suggest what relative frequencies are to be expected for the various outcomes, then those expected relative frequencies are chosen as the probabilities for the corresponding points. This was the basis for the assignment of probabilities in the coin-tossing experiment, the colored-ball experiment, and the die-rolling experiment. If no such symmetry considerations are available but experience with the given experiment is available, then the relative frequencies obtained from such experience can be used for the probabilities to be assigned. The assignment of probabilities for the sample space of Fig. 2 was based partly on experience and partly on faith in the individuals who constructed Table XII. There are various methods for constructing tables of random digits, some of them being very complicated. In all such tables it is to be expected that each digit will occur about the same number of times and that there will be no discernible patterns in sequences of digits. However, since such sets of digits are often based on physical devices that are assumed to produce digits possessing such properties, it is unreasonable to expect a set of such digits to behave in this ideal manner. A good approximation is all that can be hoped for.

Since the probabilities assigned to the points of a sample space are either the expected relative frequencies based on symmetry considerations or the long-run experimental relative frequencies, probabilities must be numbers between 0 and 1 and their sum must be 1, because the sum of a complete set of relative frequencies is always 1. In the experiments related to coin tossing, colored balls, and die rolling, the probabilities obviously sum to 1 because they were constructed that way. If the probabilities for the random-digit experiment had been based entirely on the relative frequencies obtained in a long run of experiments, then those probabilities would sum to 1.

Now, in any given experimental situation, whether academic or real, it is the privilege of the statistician to assign any probabilities he desires to the possible outcomes of the experiment, provided they are numbers between 0 and 1 and provided they sum to 1. In this assignment he will be guided by the nature of the situation and his knowledge of it. It is usually quite easy to assign satisfactory probabilities to the possible outcomes of games of chance; however, this is not the case for most real-life experiments. For example, if the experiment consists of selecting an individual at random from the population of a city and interest is centered on whether the individual will die during the ensuing year, then there is no satisfactory way of assigning a probability here other than by using the experience of insurance companies. If one were interested in determining proper insurance premiums, it would be necessary to assign probabilities of death at the various ages. These are usually chosen to be the values obtained from extensive experience of insurance companies over the years. Since mortality rates have been decreasing over the years for most age groups, any mortality table based on past experience is likely to be out of date for predicting the future. Thus the probabilities assigned on the basis of past experience may not be very close to the actual relative frequencies existing today, and therefore the premiums calculated from them will not be very accurate. Fortunately for the insurance companies, premiums calculated on the basis of past experience are larger than they would be if they had been based on more up-to-date experience.

In many business situations there is very little experience on which to base the assignment of probabilities. Even for the type of situation that is encountered frequently, past experience may be out of date, as in the case of insurance rates. For new business ventures, there may be no comparable experience to assist one in the choice of probabilities. When either of such situations occurs, the assignment will have to be made on the businessman's best judgment of the chances of the various possible outcomes occurring. Once those probabilities have been assigned, mathematically they can be treated as probabilities in the same manner as probabilities that are assigned by the use of symmetry and experience for games of chance. The reliability of a mathematical model based on a set of probabilities will, of course, depend on how realistically those probabilities are assigned. The fundamental role of the statistician is to use the probabilities given him to calculate the probabilities of various contemplated actions and to assist in the interpretation of those probabilities. The businessman supplying the initial probabilities must make the ultimate decision, based on the calculated probabilities of the various possible actions and on his confidence in the accuracy of his initial probability assessments.

In view of the foregoing discussion, it follows that the probability of a simple event is to be interpreted as a theoretical, or idealized, relative frequency of the event, or possibly as an individual's measure of his betting odds that the event will occur. This does not imply that the observed relative frequency of the event will necessarily approach the assigned probability of the event for an increasingly large number of similar experiments if such experiments could be carried out; however, one hopes it will in such situations. Thus, if one has a supposedly honest die, one would hope that the observed relative frequency of, say, a 4 showing would approach the probability $\frac{1}{6}$ as an increasingly large number of rolls is made; but one would not be too upset if it did not approach $\frac{1}{6}$ because of the imperfections in any manufactured article and because of the difficulty of simulating an ideal experiment. In this connection, it should be noted that the operators of gambling houses have done well financially by assuming that dice do behave as expected. They have certainly rolled dice enough to check on such assumptions. Of course, if experience shows that a die is not behaving as expected, they will replace it very quickly with a new die.

Constructing theoretical models to explain nature is the chief function of scientists. If the models are realistic, the conclusions derived from them are likely to be realistic. A probability model is relatively easy to construct for games of chance but becomes quite difficult for business situations when there is very little experience on which to base the model. The reliability of a business probability model obviously depends heavily on the amount of knowledge one has concerning the business situation.

In the following chapters, probability will be treated largely from the relative-frequency point of view. In the last chapter, however, a considerable amount of discussion will be concerned with the estimation of probabilities and the problem of assigning probabilities when there is little pertinent experience available.

4 THE PROBABILITY OF A COMPOSITE EVENT

Now that a geometrical model has been constructed for an experiment, consisting of a set of points with labels $e_1, e_2, \ldots$ to represent all the possible outcomes and a corresponding associated set of probabilities $P\{e_1\}, P\{e_2\}, \ldots,$ the time has come to discuss the probability of composite events. The possible outcomes $e_1, e_2, \ldots$ of a sample space are called *simple events*. A *composite event* is defined to be a collection of simple events. For example, the event of obtaining exactly two heads in the coin-tossing experiment of Fig. 1 is a composite event that consists of the three simple events e_2, e_3, and e_4. Similarly, the event of obtaining a random digit smaller than 4 in the random-digit experiment of Fig. 2 is a composite event consisting of the four simple events $x = 0, 1, 2, 3$. Composite events are usually denoted by capital letters such as A, B, or C.

Now since the probabilities assigned to the simple events of Fig. 1, namely $\frac{1}{8}$, represent expected relative frequencies for their occurrence, one would expect the composite event consisting of the simple events e_2, e_3, and e_4 to occur in about three-eighths of such experiments in the long run of such experiments. Similarly, for the experiment represented by Fig. 2, one would expect the composite event consisting of the simple events $x = 0, 1, 2, 3$ to occur in the long run in about four-tenths of such experiments. In view of such expectations, and because probability is being introduced here as an idealization of relative frequency, the following definition of probability for a composite event should seem very reasonable.

(1) ***Definition.*** *The probability that a composite event A will occur is the sum of the probabilities of the simple events of which it is composed.*

As an illustration, if A is the event of obtaining two heads in tossing a coin three times, it follows from this definition and the sample space in Fig. 1 that

$$P\{A\} = P\{e_2\} + P\{e_3\} + P\{e_4\} = \frac{3}{8}$$

Similarly, if B is the event of getting a digit smaller than 4 in selecting a random digit, it follows from this definition and Fig. 2 that

$$P\{B\} = P\{0\} + P\{1\} + P\{2\} + P\{3\} = \frac{4}{10}$$

As another illustration, let C be the event of getting a red or a green ball in the experiment for which Fig. 3 is the sample space. Since C is composed of the events e_1 and e_3, it follows that

$$P\{C\} = P\{e_1\} + P\{e_3\} = \frac{4}{6}$$

Example. As illustrations for which the composite events are not quite so obvious, consider once more the experiment of rolling two dice. Table 1, with a point associated with each outcome and with the probability $\frac{1}{36}$ attached to each point, can serve as the sample space here. First, let E be the event of getting a total of 7 points on the two dice. The simple events that yield a total of 7 points are the following: 16, 25, 34, 43, 52, 61. The sum of the corresponding six probabilities of $\frac{1}{36}$ each therefore gives $P\{E\} = \frac{6}{36} = \frac{1}{6}$. Next, let F be the event of getting a total number of points that is an even number. Simple events such as 11, 13, 22, etc. satisfy the requirement of yielding an even-numbered total. The sum of two even digits, or the sum of two odd digits, will yield an even number. From Table 1 it will be observed that there are 18 points of this type; hence it follows that $P\{F\} = \frac{18}{36} = \frac{1}{2}$. Finally, let G be the event that both dice will show at least 4 points. Here the simple events such as 44, 45, 56, etc., will satisfy. Table 1 shows that there are 9 such events; therefore $P\{G\} = \frac{9}{36} = \frac{1}{4}$.

In many games-of-chance experiments the various possible outcomes are expected to occur with the same relative frequency; therefore all the points of the sample space for such experiments are assigned the same probability, namely $1/n$, where n denotes the total number of points in the sample space. This was true, for example, in the experiments of coin tossing, random digit selection, and die rolling. (It was not true, however, for the colored-ball experiment.) When the experiment is of this simple type, that is, when all the simple-event probabilities are equal, the calculation of the probability of a composite event is very easy. It consists of merely adding the probability $1/n$ as many times as there are simple events comprising the composite event. Thus, if the composite event A consists of a total of $n(A)$ simple events, the value of $P\{A\}$ can be expressed by the simple formula

$$P\{A\} = \frac{n(A)}{n} \tag{2}$$

In the experiment of rolling two dice, for example, the probability of obtaining a total of 7 points is obtained by counting the number of points in the sample space given in Table 1 that produce a 7 total, of which there are 6, and dividing this number by the total number of points, namely 36.

Although it is not often possible in real-life problems to use formula (2), it is easier to work with than the general definition (1) involving the addition of probabilities; therefore, it alone is used in the next few sections to derive basic formulas. The formulas obtained in this manner can be shown to hold equally well for the general definition and therefore are applicable to all types of problems. Since only the formulas are needed in applied problems, there will be no appreciable loss in the understanding of how to solve practical problems by following this procedure.

5 THE ADDITION RULE

Applications of probability are often concerned with a number of related events rather than with just one event. For simplicity, consider two such events, A_1 and A_2, associated with an experiment. One may be interested in knowing whether both A_1 and A_2 will occur when the experiment is performed. This joint event is denoted by the symbol (A_1 and A_2) and its probability by $P\{A_1 \text{ and } A_2\}$. On the other hand, one may be interested in knowing whether at least one of the events A_1 and A_2 will occur when the experiment is performed. This event is denoted by the symbol (A_1 or A_2) and its probability by $P\{A_1 \text{ or } A_2\}$. At least one of the two events will occur if A_1 occurs but A_2 does not, if A_2 occurs but A_1 does not, or if both A_1 and A_2 occur. Thus, the word "or" here means "or" in the sense of either one, the other, or both. The purpose of this section is to obtain a formula for $P\{A_1 \text{ or } A_2\}$.

If two events A_1 and A_2 possess the property that the occurrence of one prevents the occurrence of the other, they are called *mutually exclusive* events. For example, let A_1 be the event of getting a total of 7 in rolling two dice, and A_2 the event of getting a total of 11: then A_1 and A_2 are mutually exclusive events. For mutually exclusive events there are no outcomes that correspond to the occurrence of both A_1 and A_2; therefore the two events do not possess any points in common in the sample space. This is shown schematically in Fig. 4.

In this diagram the points lying inside the two regions labeled A_1 and A_2 represent the simple events that yield the composite events A_1 and A_2, respectively. If $n(A_1)$ denotes the number of points lying inside the region labeled A_1 and $n(A_2)$ the number lying inside the region labeled A_2, then the total number of points associated with the occurrence of either A_1 or A_2 is the sum of those two numbers; consequently, if n denotes the total number of sample points, it follows from formula (2) that

$$P\{A_1 \text{ or } A_2\} = \frac{n(A_1) + n(A_2)}{n}$$

$$= \frac{n(A_1)}{n} + \frac{n(A_2)}{n}$$

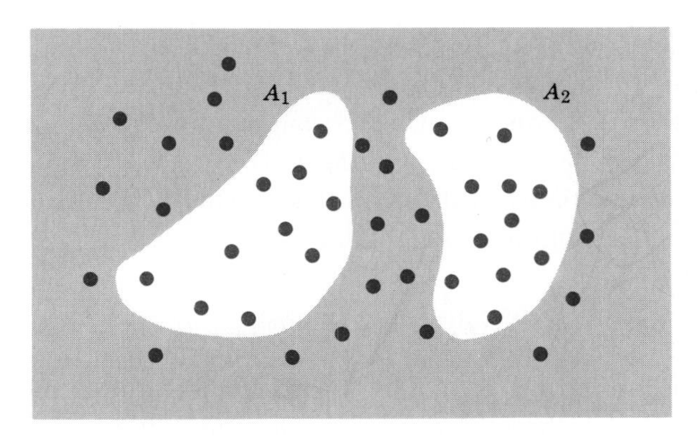

Figure 4 Sample space with two mutually exclusive events.

Since the last two fractions are precisely those defining $P\{A_1\}$ and $P\{A_2\}$, this result yields the desired addition formula, which may be expressed as follows:

(3) ***Addition Rule.*** *When A_1 and A_2 are mutually exclusive events,*

$$P\{A_1 \text{ or } A_2\} = P\{A_1\} + P\{A_2\}$$

For more than two mutually exclusive events, it is necessary merely to apply this formula as many times as required.

In the preceding illustration of rolling two dice, in which A_1 and A_2 denoted the events of getting a total of 7 and 11 points, respectively, the probability of getting either a total of 7 or a total of 11 can be obtained by means of formula (3). From Table 1 it is clear that A_1 and A_2 contain no points in common and, from counting points, that $P\{A_1\} = \frac{6}{36}$ and $P\{A_2\} = \frac{2}{36}$; therefore

$$P\{A_1 \text{ or } A_2\} = \frac{6}{36} + \frac{2}{36} = \frac{8}{36} = \frac{2}{9}$$

This result is, of course, the same as that obtained by counting the total number of points, namely 8, that yield the composite event (A_1 or A_2) and applying formula (2) directly.

As another illustration, what is the probability of getting a total of at least 10 points in rolling two dice? Let A_1, A_2, and A_3 be the events of getting a total of exactly 10 points, 11 points, and 12 points, respectively. From Table 1 it is clear that these events have no points in common and that their probabilities are given by $P\{A_1\} = \frac{3}{36}$, $P\{A_2\} = \frac{2}{36}$, and $P\{A_3\} = \frac{1}{36}$. Therefore, by formula (3), the probability that at least one of those mutually exclusive events will occur is given by

$$P\{A_1 \text{ or } A_2 \text{ or } A_3\} = \frac{3}{36} + \frac{2}{36} + \frac{1}{36} = \frac{1}{6}$$

This result also could have been obtained directly from Table 1 by counting favorable and total outcomes. Although the formula does not seem to possess any advantage here over direct counting for problems related to Table 1, it is a very useful formula for problems in which probabilities of events are available but for which tables of possible outcomes are not.

If A_2 is chosen as the event that A_1 will not occur, then, since either A_1 or A_2 must occur in an experiment, it follows that $P\{A_1 \text{ or } A_2\} = 1$. As a result, formula (3) gives

$$P\{A_1\} = 1 - P\{A_2\}$$

This formula is useful when it is easier to calculate the probability that an event will not occur, $P\{A_2\}$, than it is to calculate the probability $P\{A_1\}$ directly. For example, if A_1 is the event of obtaining a total of at least 4 points in tossing two

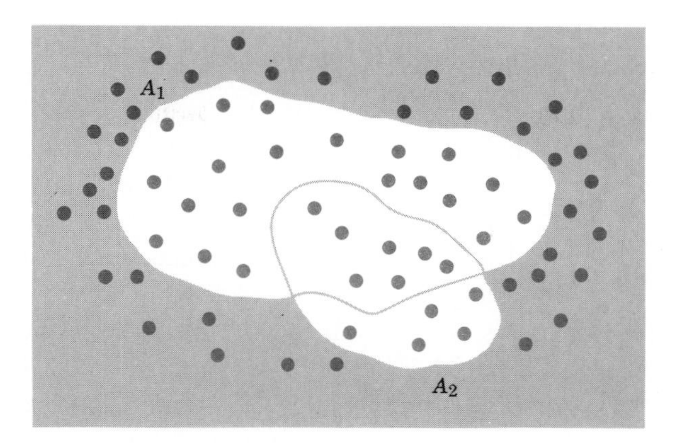

Figure 5 A sample space for two non-mutually exclusive events.

dice, it is easier to calculate the probability of obtaining a total of less than 4 points and subtract the result from 1 than it is to calculate the desired probability directly. From Table 1 the probability of obtaining less than 4 points is $\frac{3}{36}$; hence $P\{A_1\} = 1 - \frac{3}{36} = \frac{33}{36}$.

If two events A_1 and A_2 are not mutually exclusive, formula (3) is not applicable. A geometrical illustration of two non-mutually exclusive events is shown in Fig. 5. As a particular illustration related to Table 1, let A_1 be the event that at least 3 points will show on each die and let A_2 be the event that at most 4 points will show on each die. The geometry of those two regions shows that A_1 and A_2 have points in common and therefore that they are not mutually exclusive.

The correct addition formula for two non-mutually exclusive events is given by

$$P\{A_1 \text{ or } A_2\} = P\{A_1\} + P\{A_2\} - P\{A_1 \text{ and } A_2\} \tag{4}$$

The derivation of this formula is treated as an exercise in the problem sets for the benefit of those who may be interested in it. Formula (3) will suffice for all future work; therefore no illustration of how formula (4) can be applied is given here.

6 MULTIPLICATION RULE; CONDITIONAL PROBABILITY

The purpose of this section is to obtain a formula for $P\{A_1 \text{ and } A_2\}$ in terms of probabilities of single events. In order to do so, it is necessary to introduce the notion of conditional probability. Suppose one is interested in knowing whether A_2 will occur subject to the condition that A_1 is certain to occur. It is assumed here that A_1 and A_2 are not mutually exclusive events. The geometry of this problem is shown in Fig. 5.

Since A_1 must occur, the only experimental outcomes that need be considered are those corresponding to the occurrence of A_1. The sample space for this problem is therefore reduced to the simple events that comprise A_1. They are represented in Fig. 5 by the points lying inside the region labeled A_1. Among those points, the ones that also lie inside the region labeled A_2 correspond to the occurrence of both A_1 and A_2. They are the points that lie in the overlapping parts of A_1 and A_2. Let $n(A_1)$ denote the number of points lying inside A_1 and let $n(A_1$ and $A_2)$ denote the number that lie inside both A_1 and A_2. Then, from formula (2), the probability that A_2 will occur if the sample space is restricted to be the set of points inside A_1 is given by the ratio $n(A_1 \text{ and } A_2)/n(A_1)$. But this probability is what is meant by the probability that A_2 will occur subject to the restriction that A_1 must occur. If this latter probability is denoted by the new symbol $P\{A_2 \mid A_1\}$, then

$$P\{A_2 \mid A_1\} = \frac{n(A_1 \text{ and } A_2)}{n(A_1)} \tag{5}$$

Example. As an illustration of the application of this formula, a calculation will be made of the probability that the sum of the points obtained in rolling two dice is 7, if it is known that the dice show at least 3 points each. Let A_1 denote the event that two dice show at least 3 points each and let A_2 denote the event that two dice show a total of 7 points. The sample space for this problem is shown in Fig. 6; it was obtained directly from Table 1.

The points that comprise the event A_1 are all the points except those in the first two rows and the first two columns of Fig. 6. They are shown inside the white rectangle of Fig. 6. Here $n(A_1) = 16$. The points that comprise the event A_2 are the points that lie inside the white diagonal region shown in Fig. 6. The number of points that lie inside A_2 and that also lie inside

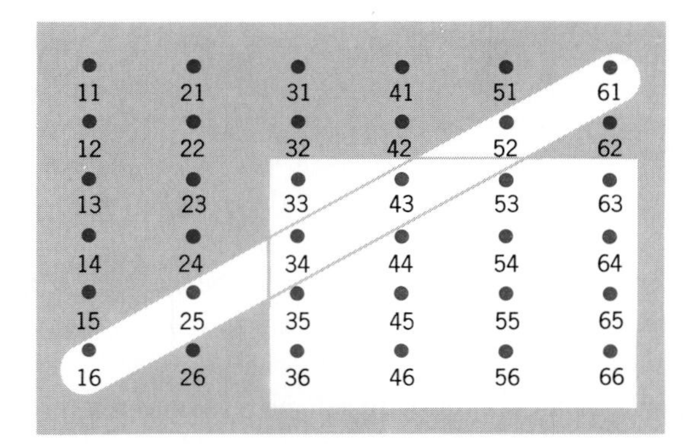

Figure 6 Sample space for a conditional probability problem.

A_1 is $n(A_1 \text{ and } A_2) = 2$. As a result, formula (5) gives

$$P\{A_2 \mid A_1\} = \frac{2}{16} = \frac{1}{8}$$

What this means in an experimental sense is that in the repeated rolling of two dice, one discards all those experimental outcomes in which either die shows a number of points less than 3. Then among the experimental outcomes that are retained one calculates the proportion of them that yields a total of 7 points. This proportion in the long run of experiments should approach $\frac{1}{8}$. It is interesting to note that the chances of getting a total of 7 points is less when one knows that each die shows at least 3 points than under ordinary rolls.

Now consider formula (5) in general terms once more. From Fig. 5 and formula (2) it is clear that

$$P\{A_1\} = \frac{n(A_1)}{n}$$

and that

$$P\{A_1 \text{ and } A_2\} = \frac{n(A_1 \text{ and } A_2)}{n}$$

Dividing the second of these two expressions by the first and canceling n gives

$$\frac{P\{A_1 \text{ and } A_2\}}{P\{A_1\}} = \frac{n(A_1 \text{ and } A_2)}{n(A_1)}$$

This result in conjunction with formula (5) yields the formula

$$P\{A_2 \mid A_1\} = \frac{P\{A_1 \text{ and } A_2\}}{P\{A_1\}} \tag{6}$$

This formula defines the *conditional probability* of A_2 given A_1, and when written in product form, yields the fundamental multiplication formula for probabilities, which may be expressed as follows:

(7) ***Multiplication Rule.*** $P\{A_1 \textit{ and } A_2\} = P\{A_1\}P\{A_2 \mid A_1\}$.

In words, this formula states that the probability that both of two events will occur is equal to the probability that the first event will occur, multiplied by the conditional probability that the second event will occur when it is known that the first event is certain to occur. Either one of the two events may be called the first event, since this is merely convenient language for discussing them, and there is no time order implied in the way they occur. Even though there is no time order implied for the two events A_1 and A_2 in the symbol $P\{A_2 \mid A_1\}$, it is customary to call this conditional probability "the probability that A_2 will occur when it is known that A_1 has occurred." Thus, if you are being dealt a five-card poker hand, someone might ask, "What is the probability that your hand will contain the ace of spades if it is known that you have received the ace of hearts?" This is merely convenient language for discussing probabilities of a poker hand that must contain the ace of hearts, and there is no implication that if the ace of spades is in the hand it was obtained after obtaining the ace of hearts. For many pairs of events, however, there is a definite time-order relationship. For example, if A_1 is the event that a high-school graduate will go to college and A_2 is the event that he will graduate from college, then A_1 must precede A_2 in time.

Example 1. As an illustration of the multiplication rule, a calculation will be made of the probability of getting two red balls in drawing two balls from a box containing three red, two black, and one green ball. It will be assumed here that the first ball drawn is not returned to the box before the second drawing is made.

Let A_1 denote the event of getting a red ball on the first drawing and A_2 that of getting a red ball on the second drawing. In order to be able to continue using formula (2) rather than the more general definition (1), it is necessary to give each ball a number and use six points in the sample space. The first three points will represent red balls, the next two the black balls, and the last one the green ball. Then, by formula (2),

$$P\{A_1\} = \frac{3}{6}$$

For the purpose of calculating $P\{A_2 \mid A_1\}$ it suffices to consider only those experimental outcomes for which A_1 has occurred. Since the first ball drawn is not returned to the box, this means considering only those experiments in which the first drawn ball was one of the three red balls. Thus, the second part of the experiment can be treated as a new single experiment in which one ball is to be drawn from a box containing two red, two black, and one green ball. As a result, $P\{A_2 \mid A_1\}$ represents the probability of getting a red ball in drawing one ball from this new box of balls. Here there are five points in the sample space, the first two representing red balls, the next two black balls, and the last one a green ball. Hence, by formula (2),

$$P\{A_2 \mid A_1\} = \frac{2}{5}$$

Application of formula (7) to these two results then gives

$$P\{A_1 \text{ and } A_2\} = \frac{3}{6} \cdot \frac{2}{5} = \frac{1}{5}$$

The advantage of using formula (7) on this problem will become apparent if one tries to solve this problem by applying formula (2) directly to the sample space that corresponds to this two-stage experiment. That sample space will consist of thirty points and will resemble the sample space shown in Fig. 6, except that the main diagonal points will be missing because they correspond to getting the same numbered colored ball on both drawings. The advantage of formula (7) is most pronounced in two-stage and multiple-stage experiments, which usually possess complicated sample spaces with a large number of points, because it reduces the calculation of probabilities to calculations for one-stage experiments only. The sample spaces for one-stage experiments are usually quite simple and much easier to visualize than those for multiple-stage experiments. Hereafter, in calculating probabilities, the techniques based on formula (7) and single-stage experiments will be used almost exclusively in order to avoid the time-consuming method based on applying formula (2) directly to the sample space of the entire experiment. It should be understood, however, that it is always possible to calculate any type of probability that may arise by working exclusively with the original sample space. A student should occasionally work a problem in this manner to test his understanding of basic concepts. It would be well for him, for example, to solve the preceding illustrative problem by constructing the sample space for that two-stage experiment.

Example 2. As another illustration of the multiplication rule, calculate the probability of getting 2 prizes in taking 2 punches on a punch board that contains 5 prizes and 20 blanks. If A_1 denotes the event of getting a prize on the first punch and A_2 the event of getting a prize on the second punch, then formula (7) gives

$$P\{A_1 \text{ and } A_2\} = \frac{5}{25} \cdot \frac{4}{24} = \frac{1}{30}$$

The value of $P\{A_2 \mid A_1\} = \frac{4}{24}$ arises from the fact that since the first punch yielded a prize there are only 4 prizes left and only 24 punches left.

Example 3. As a final illustration, consider the problem of calculating the probability of getting three red balls from the box of colored balls used in Example 1 if three balls are drawn from the box without any replacements being made. Here there are three events A_1, A_2, and A_3, corresponding to a red ball on each of the three drawings. Formula (7) can be generalized to treat three events; however, without complicating the problem further by additional notation it will be solved by taking each single stage of the experiment in order and multiplying the appropriate probabilities together. The calculations would proceed as follows:

$$P\{A_1 \text{ and } A_2 \text{ and } A_3\} = \frac{3}{6} \cdot \frac{2}{5} \cdot \frac{1}{4} = \frac{1}{20}$$

7 MULTIPLICATION RULE; INDEPENDENT EVENTS

If the events A_1 and A_2 are such that the probability that A_2 will occur does not depend upon whether or not A_1 occurs, then A_2 is said to be *independent* of A_1, and one can write

$$P\{A_2 \mid A_1\} = P\{A_2\}$$

For this case, the multiplication rule reduces to

$$P\{A_1 \text{ and } A_2\} = P\{A_1\}P\{A_2\}$$

Since the event (A_1 and A_2) is the same as the event (A_2 and A_1), A_1 and A_2 may be interchanged in (7) to give

$$P\{A_1 \text{ and } A_2\} = P\{A_2\}P\{A_1 \mid A_2\}$$

Comparing the right sides of these two formulas shows that $P\{A_1 \mid A_2\} = P\{A_1\}$. This demonstrates that A_1 is independent of A_2 when A_2 is independent of A_1. Because of this mutual independence, it is proper to say that A_1 and A_2 are independent, without specifying which is independent of the other. As a result,

When A_1 and A_2 are independent,

$$P\{A_1 \text{ and } A_2\} = P\{A_1\}P\{A_2\} \tag{8}$$

In view of this result, one can state that two events are independent if and only if the probability of their joint occurrence is equal to the product of their individual probabilities. Although it is easy to state the mathematical condition that must be satisfied if two events are to be independent, it is occasionally difficult in a real-life problem to decide whether they are independent.

In games of chance, such as roulette, it is always assumed that consecutive plays are independent events. If one were not willing to accept that assumption, then one would be forced to assume that the roulette wheel possessed a memory or that the operator of the wheel was secretly manipulating it.

To determine whether two real-life events A_1 and A_2 are independent it is necessary to ask the question: If A_1 occurs, does that change the chances that A_2 will occur from what they would be if A_1 were completely ignored? If the chances are not changed, the events are independent.

The independence of two events differs radically from the concept of two events being mutually exclusive. In real-life problems it is quite easy to determine whether two events A_1 and A_2 are mutually exclusive. All that one needs to do is ask the question: If A_1 occurs, does that make it impossible for A_2 to occur? If the answer is yes, the two events are mutually exclusive.

As an illustration of two mutually exclusive real-life events, let A_1 be the event that the price of a municipal bond will increase next month and let A_2 be the event that the estimated yield to maturity of this bond will increase next month. These events are mutually exclusive because if the price of a bond increases the yield must decrease, and therefore if A_1 occurs, A_2 cannot possibly occur. As an illustration of two independent events, let A_1 be the event that the average price of municipal bonds will increase next month and let A_2 be the event that the price of silver will increase next month. These events may be treated as independent because there is no evidence that the price of silver will increase, or decrease, if the average price of municipal bonds increases.

8 ILLUSTRATIONS

As illustrations of the application of the preceding rules of probability, consider a few simple card problems.

Two cards are drawn from an ordinary deck of 52 cards, the first drawn card being replaced before the second card is drawn.

Example 1. What is the probability that both cards will be spades? Let A_1 denote the event of getting a spade on the first draw and A_2 the event of getting a spade on the second draw. Since the first card drawn is replaced, the probability of getting a spade on the second draw should not depend upon whether or not a spade was obtained on the first draw; hence A_2 may be assumed to be independent of A_1. Formula (8) will then give

$$P\{A_1 \text{ and } A_2\} = \frac{13}{52} \cdot \frac{13}{52} = \frac{1}{16}$$

Example 2. What is the probability that the cards will be either two spades or two hearts? Let B_1 be the event of getting two spades, and B_2 the event of getting two hearts. Then, from the preceding result, it follows that

$$P\{B_1\} = P\{B_2\} = \frac{1}{16}$$

Since the events B_1 and B_2 are mutually exclusive and the problem is to calculate the probability that either B_1 or B_2 will occur, formula (3) applies; hence

$$P\{B_1 \text{ or } B_2\} = \frac{1}{16} + \frac{1}{16} = \frac{1}{8}$$

Example 3. As before, let two cards be drawn from a deck, but this time the first card drawn will not be replaced. What is the probability that both cards will be spades? Now A_2 is not independent of A_1 because if a spade is obtained on the first draw the chances of getting a spade on the second draw will be smaller than if a nonspade had been obtained on the first draw. For this problem formula (7) must be used. Here

$$P\{A_1 \text{ and } A_2\} = \frac{13}{52} \cdot \frac{12}{51} = \frac{1}{17}$$

The second factor is $\frac{12}{51}$ because there are only 51 cards after the first drawing, all of which are assumed to possess the same chance of being drawn, and there are only 12 spades left.

Example 4. As a final illustration that does not involve games of chance and that involves more than two independent events, consider the following problem. Assuming that the ratio of male children is $\frac{1}{2}$ (which is only approximately true), find the probability that in a family of 6 children (a) all the children will be of the same sex, (b) 5 of the children will be boys and 1 will be a girl.

(a) Let A_1 be the event that all the children will be boys and A_2 the event that they will all be girls. Because A_1 and A_2 are mutually exclusive events,

$$P\{A_1 \text{ or } A_2\} = P\{A_1\} + P\{A_2\}$$

Since the six individual births may be assumed to be six independent events with respect to the sex of the child, it follows by using the more general version of the multiplication formula (8) that

$$P\{A_1\} = P\{A_2\} = \left(\frac{1}{2}\right)^6$$

Hence,

$$P\{A_1 \text{ or } A_2\} = \left(\frac{1}{2}\right)^6 + \left(\frac{1}{2}\right)^6 = \frac{1}{32}$$

(b) Let A_1 be the event that the oldest child is a girl and the others are boys, A_2 the event that the second oldest is a girl and the others are boys, and similarly for events A_3, A_4, A_5, A_6. Since the event of having 5 boys and 1 girl will occur if and only if one of the six mutually exclusive events $A_1, \ldots, A_6$ occurs, it follows from (3) that

$$P\{5 \text{ boys and 1 girl}\} = P\{A_1\} + \cdots + P\{A_6\}$$

But

$$P\{A_1\} = \cdots = P\{A_6\} = \left(\frac{1}{2}\right)^6$$

Hence

$$P\{5 \text{ boys and 1 girl}\} = 6\left(\frac{1}{2}\right)^6 = \frac{3}{32}$$

Although the preceding rules of probability were derived on the assumption that all the possible outcomes of the experiment in question were expected to occur with the same relative frequency, the rules hold for more general experiments. They can even be applied to events related to experiments involving an infinite number of possible outcomes. These more general experiments are considered in Chapters 4 and 5.

9 BAYES' FORMULA

There is a certain class of important problems based on the application of formula (6) that lead to rather involved computations; therefore it is convenient to have a formula for solving such problems in a systematic manner. These problems may be illustrated by the following academic one. Suppose a box contains 2 red balls and 1 white ball and a second box contains 2 red balls and 2 white balls. One of the boxes is selected by chance and a ball is drawn from it. If the drawn ball is red, what is the probability that it came from the first box? Let A_1 denote the event of choosing the first box and let A_2 denote the event of drawing a red ball. Then the problem is to calculate the conditional probability $P\{A_1 \mid A_2\}$. This will be done by the use of formula (6) with A_1 and A_2 interchanged in that formula. Since the phrase "by chance" is understood to mean that each box has the same probability of being chosen, it follows that the probability of drawing the first box is $\frac{1}{2}$, and that of drawing the second box is the same. The calculation of the numerator term in (6) can be accomplished by using formula (7) in the order of events listed there. Thus

$$P\{A_2 \text{ and } A_1\} = P\{A_1 \text{ and } A_2\} = \frac{1}{2} \cdot \frac{2}{3} = \frac{1}{3}$$

The denominator $P\{A_2\}$ can be calculated by considering the two mutually exclusive ways in which A_2 can occur, namely, getting the first box and then a red ball or getting the second box and then a red ball. By formula (3), $P\{A_2\}$ will be given by the sum of the probabilities of those two mutually exclusive possibilities; hence

$$P\{A_2\} = \frac{1}{2} \cdot \frac{2}{3} + \frac{1}{2} \cdot \frac{2}{4} = \frac{7}{12}$$

Application of the modified version of formula (6) then yields the desired result, namely,

$$P\{A_1 \mid A_2\} = \frac{\frac{1}{3}}{\frac{7}{12}} = \frac{4}{7}$$

This problem could have been worked very easily by looking at the sample space for the experiment; however, the objective here is to work with formula (6) and attempt to obtain a formula for treating more complicated problems of the type of the present one.

The foregoing problem is a special case of problems of the following type. One is given a two-stage experiment. The first stage can be described by stating that exactly one of, say, k possible outcomes must occur when the complete experiment is performed. Those possible outcomes will be denoted by $e_1, e_2, \ldots, e_k$. In the second stage there are, say, m possible outcomes, exactly one of which must occur. These will be denoted by $o_1, o_2, \ldots, o_m$. The values of the probabilities for each of the possible outcomes $e_1, e_2, \ldots, e_k$ are given. As before, they will be denoted by $P\{e_1\}, P\{e_2\}, \ldots, P\{e_k\}$. The values of all the conditional probabilities of the type $P\{o_j \mid e_i\}$, which represents the probability that the second-stage event o_j will occur when it is known that the first-stage event e_i occurred, are also given. The problem is to calculate the probability that the first-stage event e_i occurred when it is known that the second-stage event o_j occurred. This conditional probability is written $P\{e_i \mid o_j\}$. For simplicity of notation, the calculations will be carried out for $P\{e_1 \mid o_1\}$; the calculations for any other pair are the same.

In terms of the present notation, formula (6) assumes the form

$$P\{e_1 \mid o_1\} = \frac{P\{e_1 \text{ and } o_1\}}{P\{o_1\}} \tag{9}$$

Formula (7) gives

$$P\{e_1 \text{ and } o_1\} = P\{e_1\}P\{o_1 \mid e_1\} \tag{10}$$

From the information given in this problem it will be observed that the two probabilities on the right side of (10) are known; therefore the numerator in (9) can be obtained from (10). The value of $P\{o_1\}$ in (9) can be computed by considering all the mutually exclusive ways in which o_1 can occur in conjunction with the first stage of the experiment. The second-stage event o_1 will occur if the first-stage event e_1 occurs and then o_1 occurs, or if the first-stage event e_2 occurs and then o_1 occurs, . . . , or if the first-stage event e_k occurs and then o_1 occurs. If e_1 is replaced by e with the appropriate subscript in (10), that formula can be used to calculate the probability for each of these mutually exclusive possibilities. Application of formula (3) then yields the formula

$$P\{o_1\} = P\{e_1\}P\{o_1 \mid e_1\} + P\{e_2\}P\{o_1 \mid e_2\} + \cdots + P\{e_k\}P\{o_1 \mid e_k\}$$

This result together with (10) when applied to (9) will give the desired formula, which is known as Bayes' formula. If a summation symbol is used, it becomes

(11) ***Bayes' Formula.*** $$P\{e_1 \mid o_1\} = \frac{P\{e_1\}P\{o_1 \mid e_1\}}{\sum_{i=1}^{k} P\{e_i\}P\{o_1 \mid e_i\}}$$

Returning to the problem that was solved earlier without this formula, one will observe that there were two events e_1 and e_2 in the first stage, corresponding to choosing the first or second box, and that $P\{e_1\} = P\{e_2\} = \frac{1}{2}$. The second stage also consisted of two events o_1 and o_2 corresponding to obtaining a red or a white ball. The conditional probabilities of obtaining a red ball based on what transpired at the first stage were given by $P\{o_1 \mid e_1\} = \frac{2}{3}$ and $P\{o_1 \mid e_2\} = \frac{2}{4}$. It will be observed that the substitution of these values in (11) yields the result that was obtained before.

Example. Consider now a more practical application of this formula. Suppose a test for detecting a certain rare disease has been perfected. This test is capable of discovering the disease in 97 percent of all afflicted individuals. Suppose further that when it is tried on healthy individuals, 5 percent of them are incorrectly diagnosed as having the disease. Finally, suppose that when it is tried on individuals who have certain other milder diseases, 10 percent of them are incorrectly diagnosed. It is known that the percentages of the three types of individuals being considered here in the population at large are 1 percent, 96 percent, and 3 percent, respectively. The problem is to calculate the probability that an individual, selected at random from the population at large and tested for the rare disease, actually has the disease if the test indicates he is so afflicted.

Here there are three events e_1, e_2, and e_3 in the first stage corresponding to the three types of individuals in the population. Their corresponding probabilities are $P\{e_1\} = .01$, $P\{e_2\} = .96$, and $P\{e_3\} = .03$. There are two events o_1 and o_2 in the second stage corresponding to whether the test claims that the individual has the disease or not. The conditional probabilities are given by $P\{o_1 \mid e_1\} = .97$, $P\{o_1 \mid e_2\} = .05$, and $P\{o_1 \mid e_3\} = .10$. In terms of the present notation, the problem is to calculate $P\{e_1 \mid o_1\}$. A direct application of formula (11) based on the preceding probabilities will supply the answer, namely,

$$P\{e_1|o_1\} = \frac{(.01)(.97)}{(.01)(.97) + (.96)(.05) + (.03)(.10)} = .16$$

This result may seem rather surprising because it shows that only 16 percent of the individuals whom the test indicates as having the disease actually do have it when the test is applied to the population at large. The 84 percent who were falsely diagnosed might resent the temporary mental anguish caused by their belief that they had the disease before further tests revealed the falsity of the diagnosis. They might also resent the necessity of having been required to undergo further tests when it turned out that those tests were really unnecessary. A calculation such as the preceding one might therefore cause authorities to ponder a bit before advocating mass testing.

10 COUNTING TECHNIQUES: TREES

It is convenient in solving some of the more difficult problems involving two or more stage experiments to have systematic methods for calculating the compound event probabilities that arise. A pictorial method that has proved particularly useful is based on what is known as a *probability tree.*

For the purpose of illustrating how such a tree is constructed, consider once more the first example to which formula (7) was applied. A box contains three red, two black, and one green ball and two balls are to be drawn. This is a two-stage experiment for which the various possibilities that can occur may be represented by a horizontal tree such as that shown in Fig. 7. Each stage of a multiple-stage experiment has as many branches as there are possibilities at that stage. Here there are three main branches at the first stage and three branches at each of the second stages, except for the last one, where there are only two branches: it is impossible to obtain a green ball at the second drawing if a green ball is obtained on the first drawing. The total number of terminating branches in such a tree gives the total number of possible outcomes in the compound experiment, and therefore the endpoints of those branches may be treated as the sample points of a sample space.

The probability that is attached to any branch of the tree is the conditional probability that the event listed under that branch will occur, subject to the con-

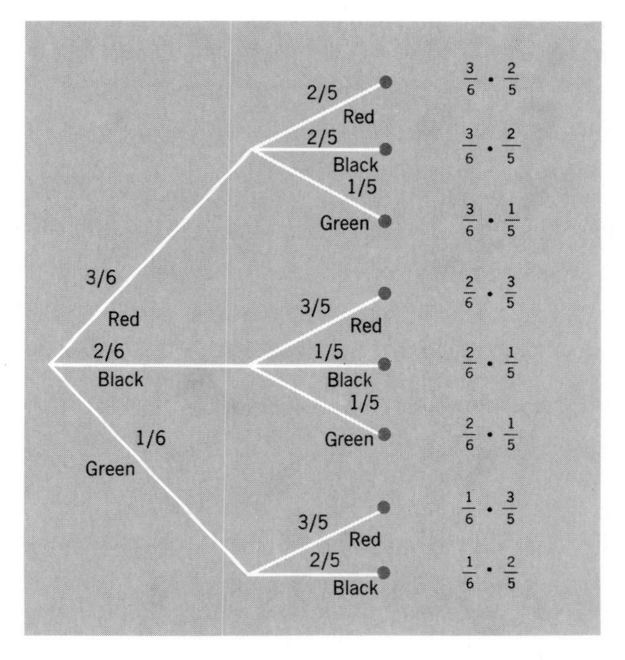

Figure 7 A probability tree.

dition that the preceding branch events all occurred. Thus, the $\frac{2}{5}$ listed above the top terminal branch is the conditional probability that a red ball will be obtained on the second drawing if a red ball was obtained on the first drawing. The probability listed at the end of a terminal branch is the probability of obtaining the sequence of events that are required to arrive at that terminal point and is obtained by multiplying the probabilities associated with the branches leading to that terminal point. Thus, the first terminal probability $\frac{3}{6} \cdot \frac{2}{5}$ is the probability of obtaining a red ball at the first drawing times the conditional probability of doing so at the second drawing. By means of this tree and its probabilities it is relatively easy to answer various probability questions.

Example 1. Probability trees yield a simple pictorial method for calculating probabilities for which Bayes' formula would normally be employed. As an example, consider the earlier problem that was used to motivate Bayes' formula, namely, the problem of calculating $P(A_1 \mid A_2)$ where A_1 is the event of selecting Box 1 and A_2 is the event of getting a red ball. The tree corresponding to this two-stage experiment is shown in Fig. 8 with the proper probabilities attached to each branch and to each of the four sample points.

Now the topmost branch corresponds to the compound event A_1 and A_2; therefore $P(A_1 \text{ and } A_2) = \frac{1}{2} \cdot \frac{2}{3}$. Furthermore, it follows from definition (1) that $P(A_2) = \frac{1}{2} \cdot \frac{2}{3} + \frac{1}{2} \cdot \frac{2}{4}$, because the event A_2 consists of the two sample points associated with the word "red." Hence, by formula (6)

$$P\{A_1 \mid A_2\} = \frac{P\{A_1 \text{ and } A_2\}}{P\{A_2\}} = \frac{\frac{1}{2} \cdot \frac{2}{3}}{\frac{2}{2} \cdot \frac{2}{3} + \frac{1}{2} \cdot \frac{2}{4}} = \frac{4}{7}$$

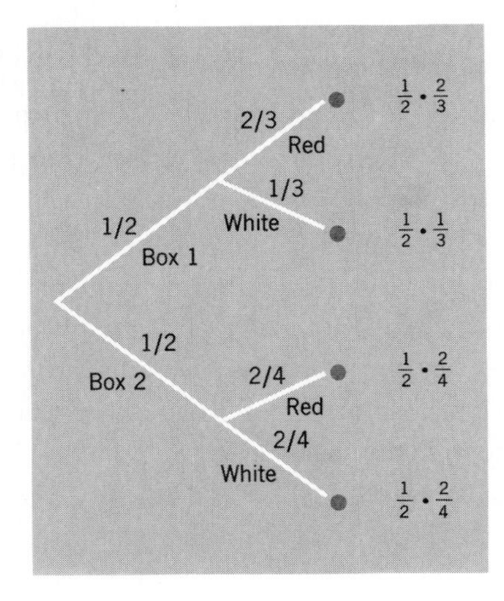

Figure 8 A tree for a Bayesian problem.

The technique is now seen to be the following one. After constructing the probability tree, select the terminal branch that corresponds to the occurrence of both A_1 and A_2. Then divide the probability associated with that terminal branch by the sum of the probabilities of all the terminal branches that are associated with the event A_2.

Example 2. The application of the tree technique to the second illustration associated with Bayes' formula proceeds as follows:

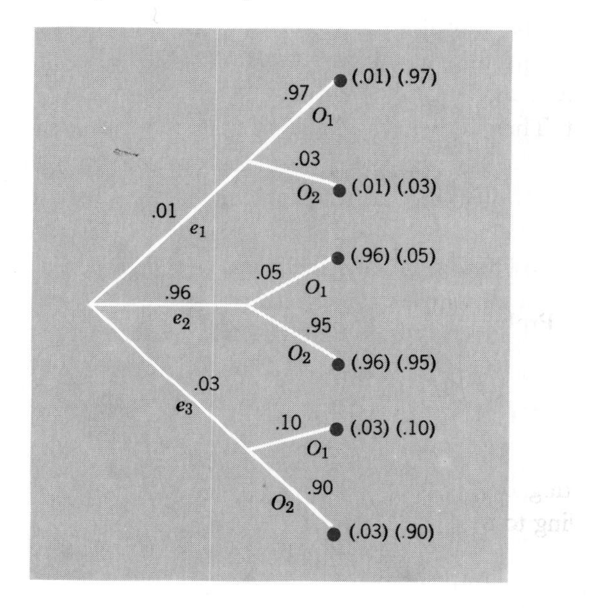

To obtain $P\{e_1 \mid o_1\}$ it now suffices to divide the probability (.01)(.97) associated with the top terminal branch by the sum of the probabilities of the terminal branches associated with the letter o_1. Thus

$$P\{e_1 \mid o_1\} = \frac{(.01)(.97)}{(.01)(.97) + (.96)(.05) + (.03)(.10)} = .16$$

Example 3. As a final illustration of the tree technique for calculating Bayes' formula probabilities, consider the following problem on accident-proneness. Suppose it is known that 10 percent of factory workers are classified as accident-prone and that the probability is .6 that such a worker will have at least one accident during a period of a year, whereas the probability is only .3 for a non–accident-prone worker. On the basis of this information, what is the probability that a worker who has had at least one accident in each of two consecutive years should be classified as accident-prone? The tree associated with this problem is the following one, where A and N denote, respectively, having had and not having had any accidents during that year:

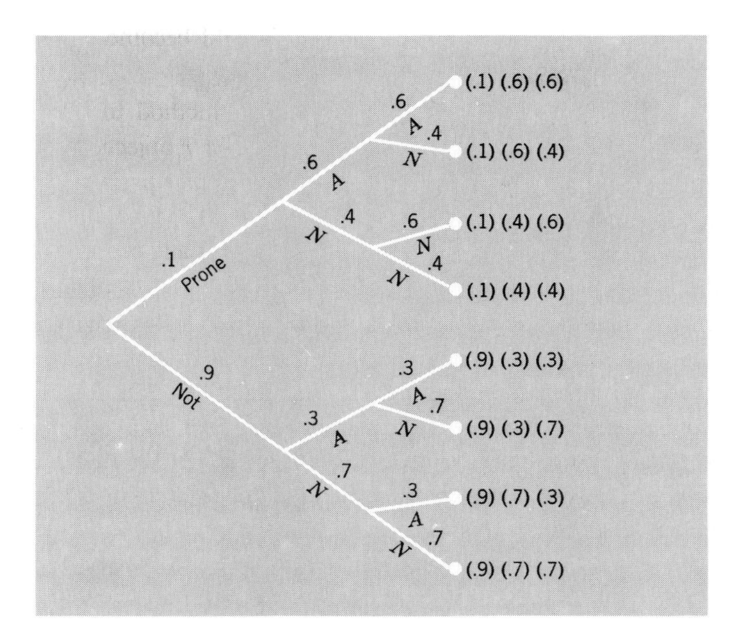

If A_1 represents being accident-prone and A_2 represents having accidents in each of two consecutive years, then the problem is to evaluate $P\{A_1 \mid A_2\}$. Here $P\{A_1 \text{ and } A_2\} = (.1)(.6)(.6)$ and $P\{A_2\} = (.1)(.6)(.6) + (.9)(.3)(.3)$; hence

$$P\{A_1 \mid A_2\} = \frac{(.1)(.6)(.6)}{(.1)(.6)(.6) + (.9)(.3)(.3)} = \frac{4}{13} = .31$$

It is clear from this result that one cannot legitimately classify a worker as accident prone on this basis, since only about $\frac{1}{3}$ of those who have had at least one accident in each of two consecutive years are of this type.

11 COUNTING TECHNIQUES: COMBINATIONS

If there are many stages to an experiment and several possibilities at each stage, the probability tree associated with the experiment would become too large to be manageable. For such problems the counting of sample points is simplified by means of algebraic formulas. Toward this objective, consider a two-stage experiment for which there are r possibilities at the first stage and s possibilities at the second stage. A tree to represent this experiment is shown in Fig. 9. Since each of the r main branches has s terminal branches attached to it, the total number of possibilities here is rs, and therefore this is the number of sample points in the sample space for this two-stage experiment. If a third stage with t possibilities were added, the total number of sample points would become *rst*. This can be extended in an obvious manner to any number of stages.

Now consider the application of this counting method to the problem of determining in how many ways it is possible to select r objects from n distinct objects. Toward this objective, first consider the particular problem of determining how many three-letter words can be formed from the five letters *a, b, c, d, e* if a letter may be used only once in a given word and if any set of three letters is called a three-letter word.

Forming a three-letter word may be thought of as a three-stage experiment in which the first stage consists of choosing the first letter of the word. In this problem there are five possibilities for the first stage but only four possibilities for the second stage, because the letter chosen at the first stage is not available at the second stage. There are only three possibilities left at the third stage; therefore, according to the preceding counting method, the total number of three-letter words that can be formed is $5 \cdot 4 \cdot 3 = 60$.

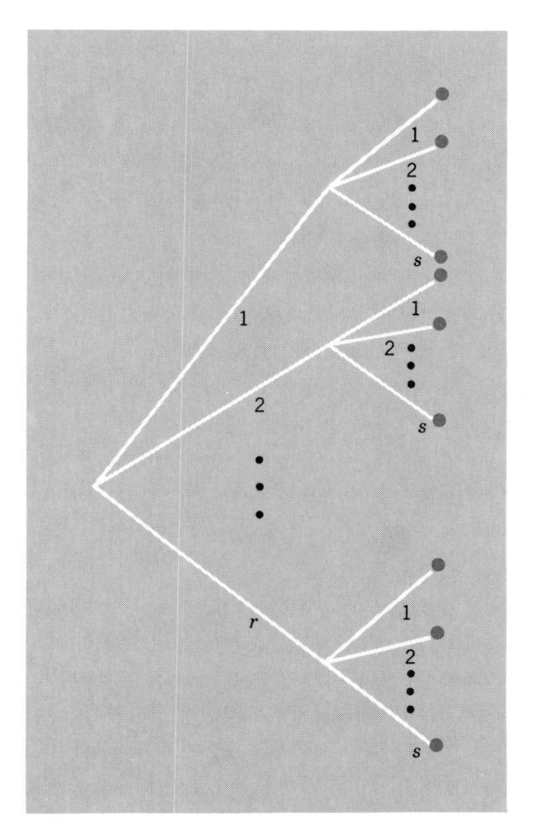

Figure 9 A tree for a two-stage experiment.

The preceding problem will be modified slightly by asking: How many committees consisting of three individuals can be selected from a group of five individuals? If the letters a, b, c, d, e are associated with the five individuals, then a three-letter word will correspond to a committee of three; however, two words using the same three letters but in a different order will correspond to the same committee. For example, *ace* and *cea* are distinct words but do not represent different committees. Since the three letters a, c, and e will produce only one committee but will give rise to $3 \cdot 2 \cdot 1 = 6$ distinct three-letter words, and this will be true for every selection of three letters, it follows that there will be only $\frac{1}{6}$ as many committees of three as there are three-letter words; hence there will be $\frac{60}{6} = 10$ such committees.

Now suppose one is given n distinct objects, say n distinct letters a, b, c, . . . , and that r of those objects are to be chosen to form r-letter words. This can be treated as an r-stage experiment in which there are n possibilities at the first stage, $n - 1$ possibilities at the second stage, etc. Then, the total number of words that can be formed is given by the formula

$$ {}_nP_r = n(n-1)(n-2) \cdot \cdot \cdot (n-r+1) \tag{12} $$

The symbol ${}_nP_r$ is called the number of permutations of n objects taken r at a time. An arrangement along a line of a set of objects is called a *permutation* of those objects; therefore, an r-letter word is a permutation of the r letters used to construct the word.

The corresponding problem of counting the number of committees consisting of r individuals that can be selected from n individuals is readily solved by means of the preceding formula. Just as in the preceding special problem, it suffices to realize that a committee is concerned only with which r letters are selected and not how they are arranged along a line. Since r distinct letters can be arranged along a line to produce $r(r-1)(r-2) \cdot \cdot \cdot 1$ distinct words but only one committee, the number of words must be divided by $r(r-1)(r-2) \cdot \cdot \cdot 1$ to give the number of distinct committees. Thus, the total number of committees that can be formed is given by the formula

$$ \binom{n}{r} = \frac{n(n-1)(n-2) \cdot \cdot \cdot (n-r+1)}{r(r-1)(r-2) \cdot \cdot \cdot 1} \tag{13} $$

The symbol $\binom{n}{r}$ is called the number of combinations of n things taken r at a time. A *combination* of r objects is merely a selection of r objects without regard to the order in which they are selected or arranged after selection. A permutation of r objects may be thought of as the outcome of a two-stage experiment in which the first stage consists in choosing a combination of r objects and the second stage in arranging that combination along a line.

A convenient symbol to use in connection with formula (13) is the *factorial* symbol, which consists of an exclamation mark after an integer, indicating that the number concerned should be multiplied by all the positive integers smaller than it. Thus $4! = 4 \cdot 3 \cdot 2 \cdot 1$ and $r! = r(r - 1)(r - 2) \cdot \cdot \cdot 1$. In order to allow r to assume the value 0, it suffices to let $r = 1$ in the relation $r!/r = (r - 1)!$ to arrive at the definition $0! = 1$. Formula (13) can therefore be written in the form

$$\binom{n}{r} = \frac{n(n-1) \cdot \cdot \cdot (n - r + 1)}{r!}$$

The numerator of this expression can also be expressed in terms of factorial notation by observing that

$$\frac{n!}{(n-r)!} = \frac{n(n-1) \cdot \cdot \cdot (n-r+1)(n-r)(n-r-1) \cdot \cdot \cdot 1}{(n-r)(n-r-1) \cdot \cdot \cdot 1}$$
$$= n(n-1) \cdot \cdot \cdot (n-r+1)$$

By using factorial symbols in this manner, formula (13) can be written in the compact form

$$\binom{n}{r} = \frac{n!}{r!(n-r)!} \tag{14}$$

Example 1. The usefulness of formula (13), or (14), for counting purposes will be illustrated in the calculation of the probability of getting five spades in a hand of five cards drawn from a deck of 52 playing cards. Here the total number of possible outcomes corresponds to the total number of five-card hands that can be formed from a deck of 52 distinct cards. Since arrangement is of no interest here, this is a combination counting problem. The total number of possible hands, according to formula (13), is given by

$$\binom{52}{5} = \frac{52 \cdot 51 \cdot 50 \cdot 49 \cdot 48}{5 \cdot 4 \cdot 3 \cdot 2 \cdot 1}$$

The number of outcomes that correspond to the occurrence of the desired event is equal to the number of ways of selecting five spades from thirteen spades. This is given by

$$\binom{13}{5} = \frac{13 \cdot 12 \cdot 11 \cdot 10 \cdot 9}{5 \cdot 4 \cdot 3 \cdot 2 \cdot 1}$$

The desired probability is given by the ratio of these two numbers; hence it is equal to

$$\frac{13 \cdot 12 \cdot 11 \cdot 10 \cdot 9}{52 \cdot 51 \cdot 50 \cdot 49 \cdot 48} = \frac{33}{66{,}640} = .0005$$

The moral seems to be that one should not expect to obtain a five-card poker hand containing only spades. Even if one settles for a hand containing five cards of the same suit, the probability is only four times as large, namely .002, which is still discouragingly small.

Example 2. As an illustration of the use of these formulas for solving a slightly less academic problem, consider the following problem related to the giving of prizes. Suppose an office staff consists of fifty individuals, of whom ten are classified as executives. Three prizes are to be given at a drawing at an office party at the end of the year. What is the probability that at most one prize will be won by an executive? For morale reasons this would be highly desirable. It follows from the addition rule that the desired probability is given by adding the probability of 0 prizes going to the executive group and 1 prize going to them. The total number of ways of giving out three prizes to fifty individuals is equal to the number of committees of three that can be selected from fifty individuals, which is given by $\binom{50}{3}$. This is, therefore, the total number of sample points in the sample space. The number of those sample points that correspond to the event of 0 prizes going to the executive group is equal to the number of ways of giving three prizes to forty (non-executive) individuals, which is given by $\binom{40}{3}$. Thus

$$P\{0\} = \frac{\binom{40}{3}}{\binom{50}{3}}$$

Now the number of ways of giving 1 prize to the executive group and 2 prizes to the others can be calculated as follows. Since only 1 executive from among the 10 available is to receive a prize, there are $\binom{10}{1}$ ways of selecting that executive. Since 2 of the 40 nonexecutive individuals are to receive prizes, there are $\binom{40}{2}$ ways of selecting those two individuals. But by the fundamental counting principal explained in the first paragraph and displayed in Fig. 9, the total number of ways of performing these two selections is given by multiplying the number of ways of performing the separate selections of this two-stage selection experiment. Therefore, the total number of ways of giving 1 prize to the executive group and 2 prizes to the nonexecutive group is given by

$$\binom{10}{1}\binom{40}{2}$$

This is the total number of sample points that correspond to the occurrence of the desired event; hence the probability that exactly 1 prize will go to the executive group and 2 prizes to the nonexecutive group is given by

$$P\{1\} = \frac{\binom{10}{1}\binom{40}{1}}{\binom{50}{3}}$$

The desired probability is therefore given by

$$P\{0\} + P\{1\} = \frac{\binom{40}{3} + \binom{10}{1}\binom{40}{2}}{\binom{50}{3}}$$

$$= \frac{\frac{40 \cdot 39 \cdot 38}{3 \cdot 2 \cdot 1} + \frac{10}{1}\frac{40 \cdot 39}{2 \cdot 1}}{\frac{50 \cdot 49 \cdot 48}{3 \cdot 2 \cdot 1}} = \frac{221}{245} = .90$$

Thus, it is fairly certain that the nonexecutive group will win at least two of the prizes. In view of this probability, one might be tempted to question the honesty of the drawing if more than one prize was won by the executives.

12 REVIEW ILLUSTRATIONS

This section considers and solves some problems that review many of the concepts and techniques introduced in this chapter.

Example 1. Each of two dice has been altered by having its one-spot changed to a two-spot. As a result, each die will contain 2 twos but no ones. The two dice are then rolled once. Assuming that the two dice can be distinguished, solve the following problems: (*a*) Construct a sample space for the experiment. (*b*) Assign probabilities to the points of the sample space. (*c*) Using definition (1) calculate the probability (*i*) of getting a total of 6 points on the two dice, (*ii*) that at least one of the dice will show a 2, (*iii*) that at least one of the dice will show a 2 if it is known that no number larger than 4 was obtained on either die. The solutions follow.

(*a*) The sample space is conveniently represented by the following 25 points, which have been labeled by indicating the outcome on each of the two dice.

22	32	42	52	62
23	33	43	53	63
24	34	44	54	64
25	35	45	55	65
26	36	46	56	66

(*b*) Since the two dice are assumed to behave in the same manner as two normal dice for which the sample space is given in Table 1 and for which each point was assigned the probability $\frac{1}{36}$, the probabilities assigned here should be in agreement with those probabilities. Hence, since the event 22 here corresponds to the composite event consisting of the simple events 11, 12, 21, and 22 for Table 1, the probability $\frac{4}{36}$ should be assigned to the point labeled 22 of the present sample space. Each of the remaining points of this space that has a 2 in its label should be assigned the probability $\frac{2}{36}$ because there are two simple events of Table 1 that produced it. All other points should be assigned the probability $\frac{1}{36}$ because they do not differ from the corresponding ones in Table 1.

(*c*) (*i*) If A denotes the event of getting a total of 6, it is seen by inspecting the sample space in (*a*) that the three points 24, 33, and 42 constitute the composite event A. Hence, applying definition (1) and using the probabilities assigned to those points in (*b*), it follows that

$$P\{A\} = \frac{2}{36} + \frac{1}{36} + \frac{2}{36} = \frac{5}{36}$$

(*ii*) If B denotes the event of getting at least one 2, it is seen that B consists of the simple events 22, 23, 24, 25, 26, 32, 42, 52, and 62. From (*b*), the point 22 was assigned the probability $\frac{4}{36}$ and the remaining points the probability $\frac{2}{36}$; consequently, since there are eight of the latter,

$$P\{B\} = \frac{4}{36} + 8\left(\frac{2}{36}\right) = \frac{5}{9}$$

(*iii*) If C denotes the event of getting at least one 2 and D denotes the event that no number larger than a 4 shows, it is seen that D consists of the simple events 22, 23, 24, 32, 33, 34, 42, 43, and 44. Of these points, only 22, 23, 24, 32, and 42 also lie in C; hence it follows from the assignment of probabilities in (*b*) that

$$P\{D\} = \frac{4}{36} + 4\left(\frac{2}{36}\right) + 4\left(\frac{1}{36}\right) = \frac{4}{9}$$

$$P\{C \text{ and } D\} = \frac{4}{36} + 4\left(\frac{2}{36}\right) = \frac{1}{3}$$

$$P\{C \mid D\} = \frac{\frac{1}{3}}{\frac{4}{9}} = \frac{3}{4}$$

Example 2. A marketing research department is studying the brand loyalty behavior of the consumers of its product, brand B. Studies indicate that consumers as a group have a rather fixed pattern of behavior with respect to choosing an alternative brand in repeated purchases. Suppose that 60 percent of brand A buyers repurchase brand A the next time they buy and that 70 percent of brand B buyers repurchase brand B the next time they buy. Assume that everyone buys either brand A or brand B and that the frequency of buying is the same for both. What is the probability that a randomly selected customer will buy brand A as his next purchase if initially brands A and B divided the market equally?

There are two types of buyers of Brand A: an A buyer who buys A again, and a B buyer who shifts to brand A. If the subscripts 1 and 2 represent the first and second purchase times, then

$$\begin{aligned} P\{A_2\} &= P\{A_1\}P\{A_2 \mid A_1\} + P\{B_1\}P\{A_2 \mid B_1\} \\ &= (.50)(.60) + (.50)(.30) = .45 \end{aligned}$$

Example 3. A box contains the following five cards: the ace of spades, the ace of clubs, the two of hearts, the two of diamonds, and the three of spades. An ace is considered as a one. Spades and clubs are black cards, whereas hearts and diamonds are red cards. Two cards are to be drawn from this box without replacing the first drawn card before the second drawing. Using the addition and multiplication formulas, calculate the probability that (a) both cards will be red, (b) the first card will be an ace and the second card will be a two, (c) both cards will the the same color, (d) one card will be a spade and the other will be a club, (e) a total of 4 on the two cards will be obtained, (f) exactly one ace will be obtained if it is known that both cards are black. The solutions follow.

(a) Applying formula (7) and considering the experiment in two stages, $P\{RR\} = \frac{2}{5} \cdot \frac{1}{4} = \frac{1}{10}$.

(b) Applying formula (7), $P\{A2\} = \frac{2}{5} \cdot \frac{2}{4} = \frac{1}{5}$.

(c) The two events RR and BB constitute the two mutually exclusive ways in which the desired event can occur; hence applying formula (3), the desired probability is given by

$$P\{RR \text{ or } BB\} = P\{RR\} + P\{BB\} = \frac{2}{5} \cdot \frac{1}{4} + \frac{3}{5} \cdot \frac{2}{4} = \frac{2}{5}$$

(*d*) The two events SC and CS will satisfy; hence

$$P\{SC \text{ or } CS\} = P\{SC\} + P\{CS\} = \frac{2}{5}\cdot\frac{1}{4} + \frac{1}{5}\cdot\frac{2}{4} = \frac{1}{5}$$

(*e*) A total of 4 will be obtained if both cards are twos, or if one card is a three and the other is a one. If a subscript is used to denote the number on a card, the events that will satisfy are the following ones: H_2D_2, D_2H_2, S_3S_1, S_1S_3, S_3C_1, C_1S_3. Since these constitute the mutually exclusive ways in which the desired event can occur and since each of these possesses the same probability, namely, $\frac{1}{5}\cdot\frac{1}{4} = \frac{1}{20}$, it follows that

$$P\{4 \text{ total}\} = \frac{6}{20} = \frac{3}{10}$$

(*f*) Let A_1 denote the event that both cards will be black and A_2 the event that exactly one ace will be obtained. Then $P\{A_2 \mid A_1\}$ is the probability needed to solve the problem. From formula (6), this requires the computation of $P\{A_1\}$ and $P\{A_1 \text{ and } A_2\}$. First, $P\{A_1\} = \frac{3}{5}\cdot\frac{2}{4} = \frac{3}{10}$. Next, both A_1 and A_2 will occur if one of the following mutually exclusive events occurs: S_1S_3, S_3S_1, C_1S_3, S_3C_1. Since each of these events has the probability $\frac{1}{5}\cdot\frac{1}{4} = \frac{1}{20}$ and there are four of them, it follows that $P\{A_1 \text{ and } A_2\} = \frac{4}{20} = \frac{1}{5}$. Hence,

$$P\{A_2 \mid A_1\} = \frac{\frac{1}{5}}{\frac{3}{10}} = \frac{2}{3}$$

Example 4. Two players are involved in a playoff in which the player who first wins two games is declared the winner. Suppose that players A and B are equally likely to win any game played. (*a*) Draw a tree showing the possible outcomes. (*b*) Calculate the probability for each of those outcomes. (*c*) If A wins the first game, what is the probability that B will still win the playoff?

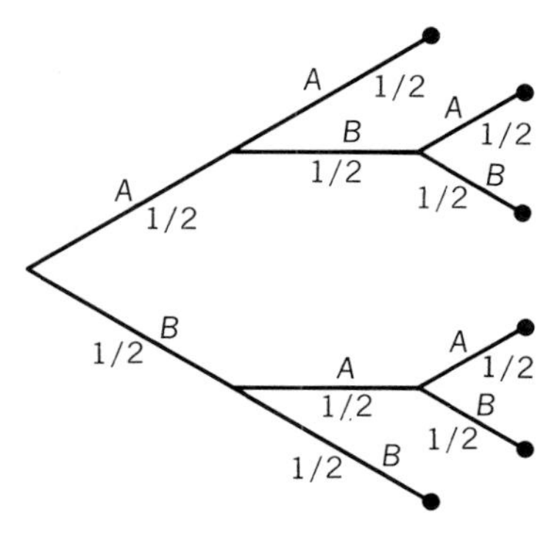

(*b*) The probabilities associated with the sample points, beginning with the highest point, are $\frac{1}{4}$, $\frac{1}{8}$, $\frac{1}{8}$, $\frac{1}{8}$, $\frac{1}{8}$, $\frac{1}{4}$.

(c) The conditional sample space needed here consists of the upper part of the tree beginning at the first branching of the main A branch. The probability of BB is then given by $\frac{1}{2} \cdot \frac{1}{2} = \frac{1}{4}$.

Example 5. Suppose that 80 percent of used-car buyers are good credit risks. Assume that the probability that a good credit risk has a charge account is .7, but that this probability is .4 for a bad credit risk. (a) Draw a probability tree for this problem. (b) Use this tree to calculate the probability that an individual with a charge account (C) is a bad credit risk (B). (c) Use this tree to calculate the probability that an applicant for credit who does not have a charge account (N) is a good credit risk (G). The solutions follow.

(a)

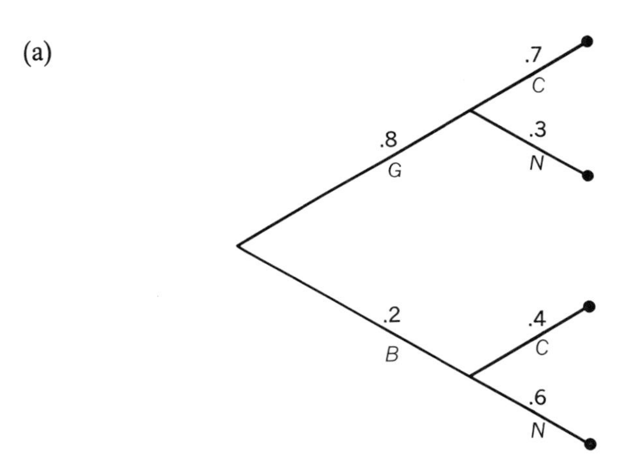

$$(b)\ \ P\{B \mid C\} = \frac{(.2)(.4)}{(.2)(.4) + (.8)(.7)} = \frac{1}{8}$$

$$(c)\ \ P\{G \mid N\} = \frac{(.8)(.3)}{(.8)(.3) + (.2)(.6)} = \frac{2}{3}$$

Example 6. A box of 15 spare parts for a certain type of machine contains 10 good parts and 5 defective ones. If 3 parts are picked at random from the box, what is the probability that (a) they will all be good parts, (b) they will all be defective, (c) 2 will be good and 1 will be defective, and (d) at least 2 will be good? The solutions will be expressed in terms of combination symbols and then evaluated.

$$(a)\ \ \frac{\binom{10}{3}}{\binom{15}{3}} = \frac{10!}{3!7!}\frac{3!12!}{15!} = \frac{10 \cdot 9 \cdot 8}{15 \cdot 14 \cdot 13} = \frac{24}{91}$$

(b) $$\frac{\binom{5}{3}}{\binom{15}{3}} = \frac{5!}{3!2!}\frac{3!12!}{15!} = \frac{5 \cdot 4 \cdot 3}{15 \cdot 14 \cdot 13} = \frac{2}{91}$$

(c) $$\frac{\binom{10}{2}\binom{5}{1}}{\binom{15}{3}} = \frac{10!}{2!8!}\frac{5!}{1!4!}\frac{3!12!}{15!} = \frac{10 \cdot 9}{2}\frac{5}{1}\frac{3 \cdot 2}{15 \cdot 14 \cdot 13} = \frac{45}{91}$$

(d) $$\frac{45}{91} + \frac{24}{91} = \frac{69}{91}$$

EXERCISES

SECTION 2

1 List all the possible outcomes if a coin is tossed 4 times.

2 A box contains 1 red, 1 black, and 1 green ball. Two balls are to be drawn from this box without replacing the first drawn ball before the second drawing. Construct a sample space for this experiment similar to Table 1.

3 A box contains 2 black and 1 white ball. Two balls are to be drawn from this box. Construct a sample space for this experiment (*a*) using 6 points, (*b*) using 3 points.

SECTION 3

4 What probabilities should be assigned to the points of the sample space corresponding to the experiment of problem 1?

5 What probabilities would you assign to the points of the sample space of problem 2? What assignment would you have made if the first ball was returned to the box before the second drawing?

6 What probabilities would you assign to the points of the two sample spaces constructed in problem 3?

7 Let e_1, e_2, and e_3 denote the events of getting a digit less than 4, getting a digit between 4 and 6 inclusive, and getting a digit larger than 6, respectively, when selecting a digit from the table of random digits.

(*a*) Construct a sample space for this experiment and assign probabilities to the points.

(*b*) Perform the experiment 200 times and calculate the experimental relative frequencies for the three events to see whether your model seems to be a realistic one.

8 Assign probabilities to the sample points of the four-point sample space representing the outcomes that can occur when two marbles are drawn from a box that contains 2 black and 2 white marbles.

SECTION 4

9 A die has its 6-spot altered to a 3 spot. When tossed, what is the probability of obtaining 4 or more dots?

10 A box contains 4 red, 3 black, 2 green, and 1 white ball. A ball is drawn from the box and then returned to the box. What is the probability that the ball will be (*a*) red, (*b*) red or black? Now simulate this experiment by means of random numbers by calling the digits 0, 1, 2, 3 red, the digits 4, 5, 6 black, the digits 7, 8 green, and the digit 9 white, and perform the experiment of selecting a digit from the table of random digits 200 times. Let A_1 and A_2 denote the events in parts (*a*) and (*b*), respectively, and keep a tabulation of the number of times A_1 and A_2 occurred. Observe whether the mathematical model assumed to hold here seems to be realistic.

11 An honest die is rolled twice. Using Table 1, calculate the probability of getting (*a*) a total of 4, (*b*) a total of less than 4, (*c*) a total that is an even number.

12 An honest coin is tossed 4 times. Using the model of problems 1 and 4, calculate the probability of getting (*a*) 4 heads, (*b*) 3 heads and 1 tail, (*c*) at least 2 heads.

13 For the experiment of rolling two honest dice, calculate the probability that (*a*) the sum of the numbers will not be 11, (*b*) neither 1 nor 2 will appear, (*c*) each die will show 3 or more points, (*d*) the numbers on the two dice will not be the same, (*e*) exactly one die will show fewer than 3 points.

SECTION 5

14 A department fills orders that vary between 0 and 10 from day to day with the following probabilities: .02, .07, .15, .20, .19, .16, .10, .06, .03, .01, and .01.
(*a*) What is the probability of fewer than 4 orders?
(*b*) What is the probability of at least 1 order?
(*c*) What is the probability of anywhere from 3 to 5 orders?

15 If you toss two dice, what is the probability that you will roll a total of 7? Either a 7 or an 11? Either a 3 or a 12?

16 For the experiment of rolling the 2 altered dice of example 1 of section 12, calculate the probability that (*a*) the sum of the numbers will be less than 6, (*b*) the number 2 will occur, (*c*) exactly 1 die will show fewer than 3 points.

17 The following table shows the length of life of wholesale grocery businesses in a metropolitan area.
(*a*) During the period studied, what is the probability that an entrant to this business will fail within 10 years?
(*b*) What is the probability he will survive at least 25 years?
(*c*) How many years would he have to survive to be among the 10 percent longest survivors?

Length of life (years)	Percent of wholesalers
0 up to 5	65
5 up to 10	16
10 up to 15	9
15 up to 25	5
25 and over	5
Total	100

SECTIONS 6 and 7

18 Two balls are to be drawn from an urn containing 2 white and 3 black balls.
(*a*) What is the probability that the first ball will be white and the second black?
(*b*) What is this probability if the first ball is replaced before the second drawing?

19 Suppose we mark two dice such that the 1-spot and 2-spot are white and the 3-, 4-, 5-, and 6-spots are black. If the two dice are tossed, what is the probability of getting at least one black die?

20 Two balls are to be drawn from an urn containing 2 white, 3 black, and 4 green balls.
(*a*) What is the probability that both balls will be green?
(*b*) What is this probability if the first ball is replaced before the second drawing?
(*c*) What is the probability that both balls will be the same color?

21 A box contains 4 coins, 3 of which are honest coins but the fourth of which has heads on both sides. If a coin is selected from the box and then is tossed 2 times, what is the probability that 2 heads will be obtained?

22 If the probability is $p = .01$ that a person thirty years of age will die within a year, find the probability that out of a group of 10 such individuals, (*a*) none, (*b*) exactly one, (*c*) not more than one, (*d*) more than one, (*e*) at least one will be dead within the year.

23 In a certain area 20 percent of the homes have a television set and 40 percent have an automatic washer. A single home is selected by chance. Do you believe that it is legitimate to compute the probability of getting a home that has both a television set and an automatic washer by (*a*) multiplying .20 by .40, (*b*) adding .20 and .40? Explain.

24 A survey is conducted to measure the readership of magazines A and B. It is found that 25 percent of the families interviewed buy A and 40 percent buy B.
(*a*) Are the events of buying magazine A and buying magazine B independent events?
(*b*) If the events are assumed to be independent, what is the probability that a family will buy at least one of the magazines?
(*c*) What will the probability be in (*b*) if the two events are mutually exclusive?

25 An electronic device will operate correctly only if all of its components operate correctly. There are 5 components in the device. Let p denote the probability that a single component will function correctly, and assume that this probability is the same for all the components. If each component operates independently

of the others, how large must p be so that the probability of the device working is at least as large as .9?

26 The following numbers were obtained from a mortality table based on 100,000 individuals.

Age	Number alive	Deaths per 1000 during that year
17	94,818	7.688
18	94,089	7.727
19	93,362	7.765
20	92,637	7.805
21	91,914	7.855

If these numbers are used to define probabilities of death for the corresponding age group and if A, B, and C denote individuals of ages 17, 19, and 21, respectively, calculate the probability that during the year (*a*) A will die and B will live, (*b*) A and B will both die, (*c*) A and B will both live, (*d*) at least one of A and B will die, (*e*) at least one of A, B, and C will die.

27 A testing organization wishes to rate a particular brand of table radios. Five radios are selected at random from the stock of radios and the brand is judged to be satisfactory if nothing is found wrong with any of the 5 radios.

(*a*) What is the probability that the brand will be rated satisfactory if 10 percent of the radios actually are defective?

(*b*) What is this probability if 20 percent are defective?

28 Three cards are to be drawn from an ordinary deck of 52 cards.

(*a*) What is the probability that all 3 cards will be spades?

(*b*) What is the probability that all 3 cards will be of the same suit?

(*c*) What is the probability that none of the 3 cards will be spades?

29 Assuming that the ratio of male children is $\frac{1}{2}$, find the probability that in a family of 6 children, (*a*) all children will be of the same sex, (*b*) the 4 oldest children will be boys and the 2 youngest will be girls, (*c*) 5 of the children will be boys and 1 will be a girl.

30 Suppose that the probability is p that the weather (sunshine or rain) is the same on any given day as it was on the preceding day. It is raining today. What is the probability that it will rain the day after tomorrow?

31 In order to function properly, an electrical device must have the two linked components, as shown in the accompanying sketch, function properly. The sketch shows that A must function and that at least one of the two independently wired B's must function.

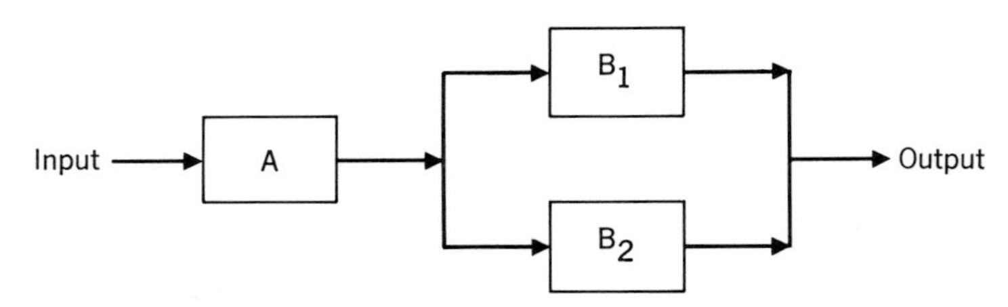

Suppose that the B components function independently of A and of each other and that the reliability (probability of functioning) of A is .9 and that of each B is .8.

(*a*) Guess the reliability of the device.

(*b*) Calculate the reliability of the device.

32 With the aid of Fig. 5, show that the addition rule for two events A_1 and A_2 that are not mutually exclusive is given by the formula

$$P\{A_1 \text{ or } A_2\} = P\{A_1\} + P\{A_2\} - P\{A_1 \text{ and } A_2\}$$

SECTION 9

33 Assume that there are equal numbers of male and female students in a high school and that the probability is $\frac{1}{5}$ that a male student will be a science major and $\frac{1}{20}$ that a female student will be one. What is the probability that (*a*) a student selected at random will be a male science student, (*b*) a student selected at random will be a science student, (*c*) a science student selected at random will be a male student?

34 Suppose that 10 percent of car owners who have an accident will have had at least one other accident during the past year. Suppose also that a driving simulator test is failed by 70 percent of such drivers but by only 20 percent of drivers who have had only one accident. If an individual car owner selected at random from those having had an accident takes the test and fails, what is the probability that he will have had additional accidents the past year?

35 Suppose a college aptitude test designed to separate high school students into promising and not-promising groups for college entrance has the following background. Among the students who made satisfactory grades in their first year at college, 80 percent passed the aptitude test. Among the students who did unsatisfactory work their first year, 40 percent passed the test. It is assumed that the test was not used for admission to college. If it is known that only 70 percent of the first-year college students do satisfactory work, what is the probability that a student who passed the test will be a satisfactory student?

36 Suppose that 70 percent of a group of individuals favors proposition A. Suppose, further, that only 80 percent of true "yes" people will respond "yes" when interviewed. The remaining 20 percent lie and respond with "no." Of the true "no" people, only 60 percent respond truthfully with a "no" and the others lie and say "yes."

(*a*) If an individual is picked at random, what is the probability that he will say "yes"?

(*b*) If a randomly chosen individual says "yes," what is the probability that he is a true "yes" individual?

37 The *Scientific American* for October 1950 posed the following problem. Suppose we have three cards. One is white on both sides, the second is white on one side and red on the other, and the third is red on both sides. The dealer shuffles the cards, takes one out, and places it flat on the table. The side showing is red. The dealer now says: ''Obviously this is not the white-white card. I will bet even money that the other side is red.'' Would you take his bet?

SECTION 10

38 To help select suitable employees for a particular job a personnel department administers an aptitude test to all applicants. To test the effectiveness of the test a sample of applicants who failed were also hired and given a fast trial at the job. It was found that of the 30 percent who passed the test, 80 percent were satisfactory, and of those who did not, only 10 percent were satisfactory. What is the probability that an applicant selected at random will prove to be satisfactory at this job? Use a tree here.

39 Millicorp is interested in the attitudes of its employees with regard to a somewhat delicate policy with which it has been toying. It has contracted with the B-attitude Research Associates to find out what its employees really think about the matter. B-attitude examines them closely by means of personal interviews. Suppose that actually 70 percent of the employees are in favor of the policy under consideration and 30 percent are opposed. Suppose also that 90 percent of those favoring the policy are willing to say so in the interview, the others stating they are opposed; and of those who actually dislike the policy, 50 percent are willing to admit it and 50 percent say they favor it. Use a probability tree to answer the following questions.

(*a*) When B-attitude turns in its report, what proportion of the employees claim they are in favor of the policy?

(*b*) Of those employees who claim they are in favor of the policy, what proportion actually is?

SECTION 11

40 The California automobile license plate contains 3 integers followed by 3 letters. Assume that all 26 letters of the alphabet and all integers (0 through 9) are used.

(*a*) How many different cars can be accommodated?

(*b*) What relative increase in the number of plates is achieved by switching from an integer to a letter?

(*c*) Suppose one of the ''words'' formed by the letters is regarded as objectionable and is therefore omitted. How many plates are ''lost''?

41 A real estate developer has 8 basic house designs. On a given street he has 5 lots.

(*a*) Suppose the community does not permit look-alike houses on a given street. How many different ways can the developer utilize his basic designs and still meet this restriction?

(*b*) If there are no restrictions, how many arrangements are possible?

42 An automobile distributer wants to stock one car of each model, color, and horsepower. If there are 5 models, 7 colors, and 2 horsepowers, how many cars are needed?

43 It is the custom in preparing ballots for elections not to list candidates alphabetically. It is generally assumed that it is very advantageous to be at the top of the list. How many different types of ballot listings are possible if there are 5 candidates running?

44 On a list of 20 individuals who volunteered to supply blood when it is needed for a transfusion, there are 15 individuals of type B blood. If 3 individuals are selected at random from the list, what is the probability that all three will be type B?

45 If a poker hand of 5 cards is drawn from a deck, what is the probability that it will contain exactly 1 ace?

46 Find the probability that a poker hand of 5 cards will contain only black cards if it is known to contain at least 3 black cards.

47 A buyer will accept a lot of 100 articles if a sample of 5 picked at random and inspected contains no defectives. What is the probability that he will accept the lot if it contains 10 defective articles?

48 If a box contains 40 good and 10 defective fuses and 10 fuses are selected, what is the probability that they will all be good? Use combination symbols here.

49 A bridge hand of 13 cards is drawn from a deck of 52 cards. Use combination symbols to calculate the probability that the hand (*a*) will contain exactly one ace, (*b*) will contain at least one ace, (*c*) will contain no spades, (*d*) will contain only spades.

50 A lady declares that by tasting a cup of tea with milk added she can tell whether the milk was placed in the cup before or after the tea was added. Eight cups of tea are mixed, four one way and four the other way. The cups are presented in random order. The lady, after tasting all eight, is asked to pick out the four in which the milk came second. She selected 3 correctly and 1 incorrectly, and hence 3 correctly and 1 incorrectly in the other four. On the assumption that she cannot discriminate, what is the probability that she will do at least this well?

SECTION 12

51 A manufactured article that cannot be used if it is defective is required to pass through two inspections before it is permitted to be packaged. Experience shows that the first inspector will miss 5 percent of the defective articles, whereas the second inspector will miss 4 percent of them. Good articles always pass inspection.

(*a*) What is the probability that a defective article will be missed by both inspectors?

(*b*) What is the probability that a defective article will be missed by one inspector but will be caught by the other inspector?

(*c*) If 10 percent of the articles turned out in the manufacturing process are defective, what percentage of the articles that are produced and pass both inspections will be defective?

52 For the third review problem of section 12, work the following problems by means of the addition and multiplication formulas. Here, however, assume that the first card is returned before the second drawing. What is the probability that (*a*) both cards will be aces, (*b*) the ace of spades is certain to be obtained, (*c*) at least one card will be an ace, (*d*) at most one card will be an ace, (*e*) a red ace will not be obtained, (*f*) the sum of the numbers on the two cards will be less than 4, (*g*) at least one ace will be obtained if it is known that neither card is a three, (*h*) the first card was an ace if it is known that the second card is not an ace?

53 Work problem 52 under the assumption that the first card drawn is not returned before the second drawing.

Chapter 4 Probability Distributions

1 INTRODUCTION

Chapter 2 was concerned with empirical frequency distributions, comprising either samples or entire populations, and their description. This chapter is concerned with theoretical frequency distributions and their properties. When a frequency distribution is based on a sample it is, in a sense, an estimate of a corresponding population distribution. If the size of the sample is large, one would expect the sample distribution to be a good approximation to the population distribution. For example, if in a study of dormitory weights one had taken a sample of 400 students and there were only 800 dormitory students on campus, one would have expected the sample and population distributions to be very similar.

In most statistical problems the sample is not large enough to determine the population distribution with much precision. However, there is usually enough information in the sample, together with information obtained from other sources, to suggest the general type of population distribution involved. For example, experience with various industrial measurements, such as piston diameters, the breaking strength of bricks, and the length of telephone conversations, shows that these variables possess distributions that are much like that shown in Fig. 3 of

Chapter 2. Thus, by combining experience and the information provided by the sample, one can often postulate the general nature of the population distribution. This postulation gives rise to probability distributions.

A probability, or theoretical, distribution is a mathematical model for an actual frequency distribution. The probability models encountered in Chapter 3 in connection with games of chance are examples of probability distributions for certain discrete variables. Thus the postulation that each of the 36 possible outcomes in rolling two dice will occur equally often in the long run yields a probability distribution for the two dice. For the experiment of weighing 120 dormitory students, the model might be a continuous probability distribution, such as the bell-shaped distribution discussed in Chapter 2 in connection with the interpretation of the standard deviation. Models for continuous variables are more difficult to explain than those for discrete variables; therefore, discrete variables are considered first.

In discussing sample distributions and their theoretical counterparts, it is customary to call a sample distribution an *empirical* distribution. The relationship between an empirical distribution and its corresponding probability, or theoretical, distribution has already been considered for a discrete variable in the illustration that produced Fig. 1 of Chapter 2. A sample of 700 was obtained by taking 700 drawings, with the drawn card replaced each time, from a set of cards consisting of 3 aces, 2 twos, and 1 three. From the results of the sampling, the relative frequencies for each type of outcome were calculated and, when expressed in decimal form, were found to be .500, .329, and .171, correct to three decimals. The corresponding theoretical distribution here is given by the three probabilities $\frac{3}{6}$, $\frac{2}{6}$, $\frac{1}{6}$. A comparison of this theoretical distribution and its empirical approximation is shown in Fig. 1 as a pair of line charts, with the white line representing theory and the blue line representing the sample. The empirical distribution is an excellent approximation to the theoretical distribution here, largely because an unusually large sample was taken.

2 RANDOM VARIABLES

In experiments of the repetitive type for which a probability model is to be constructed, one is usually interested only in a particular property of the outcome of the experiment. For example, in rolling two dice interest usually centers on the total number of points showing because that is all that really matters in the game of craps. Similarly, in taking samples of college students, interest might center on how many hours per week a student studies, or on a student's grade-point average.

Just as in Chapter 2, the quantity chosen for study in such experiments will be denoted by the letter x. Thus, in the preceding illustrations, x might represent the sum of the points on two dice, or the number of hours per week a student

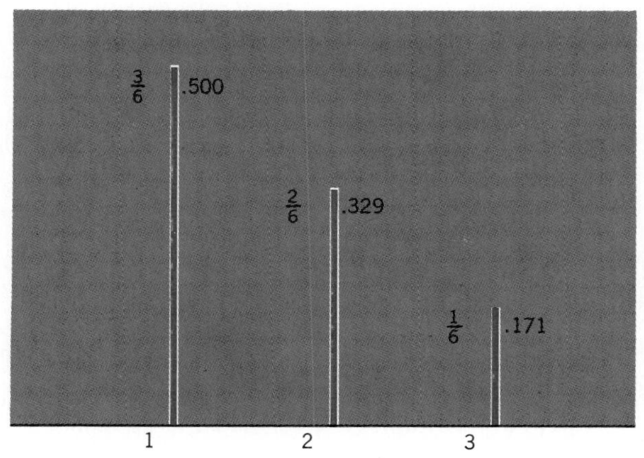

Figure 1 A theoretical distribution and its empirical approximation.

studies, or a student's grade-point average. In connection with probability distributions, the variable x is called a *random variable.*

For the purpose of seeing how a random variable is introduced in a simple experiment, return to the sample space for the coin-tossing experiment given in Fig. 1 of Chapter 3. If x is used to denote the total number of heads obtained, then each point of that sample space will possess the value of x shown directly above the corresponding point in Fig. 2. It will be observed that the random variable x can assume any one of the values 0, 1, 2, or 3, but no other values.

As another illustration, let x denote the total number of points obtained in rolling two honest dice. The sample space for this experiment is given by Table 1 of Chapter 3. This sample space has been duplicated in Fig. 3 but with the omission of the labels attached to the points. The numbers attached to the points are the values of the random variable x for this experiment. It will be observed that this random variable x can assume any one of the values 2, 3, . . . , 12.

In each of the preceding illustrations it will be observed that the value of the random variable x depends only on the particular sample point chosen. This means that x is a function of the sample points of the sample space. A formal definition is the following one.

> ***Definition.*** *A random variable is a numerical-valued function defined on a sample space.*

The word *random,* or *chance,* is used to designate variables of this type to point

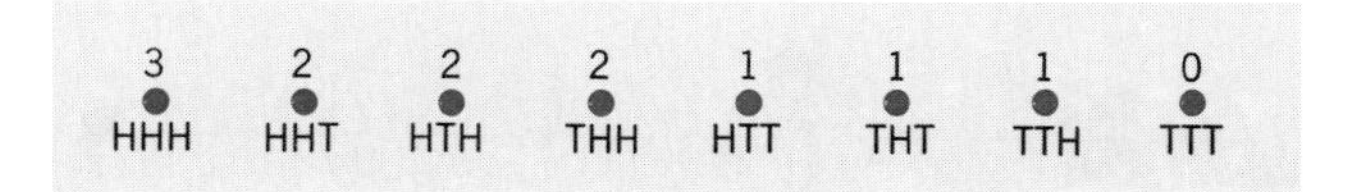

Figure 2 The values of a random variable for a coin experiment.

2	3	4	5	6	7
3	4	5	6	7	8
4	5	6	7	8	9
5	6	7	8	9	10
6	7	8	9	10	11
7	8	9	10	11	12

Figure 3 The values of a random variable for a dice experiment.

out that the value such a variable assumes in an experiment depends on the outcome of the experiment, which in turn depends on chance.

Now that interest is being centered on the values of a random variable for an experiment rather than on all the possible outcomes, a new simpler sample space can be constructed for the experiment, which can be substituted for the original sample space. Thus, in the coin-tossing experiment the only events of interest are the composite events given by $x = 0, 1, 2$, and 3. But it follows from Fig. 2 and definition (1) of Chapter 3 that the probabilities for these composite events are given by $P\{0\} = \frac{1}{8}$, $P\{1\} = \frac{3}{8}$, $P\{2\} = \frac{3}{8}$, and $P\{3\} = \frac{1}{8}$. These composite events can now be treated as simple events in a new sample space of four points, with each point associated with a value of the random variable x. The probabilities just calculated for those composite events are then assigned to the four points of the new sample space. Figure 4 shows this new sample space with its associated probabilities.

In a similar manner, a new sample space for the random variable representing the total number of points showing on two dice can be constructed by means of Fig. 3 and definition (1) of Chapter 3. Here the composite events are those corresponding to the random variable x assuming the values 2, 3, . . . , 12. The probabilities for those composite events are readily calculated by means of definition (1) of Chapter 3 to be $\frac{1}{36}, \frac{2}{36}, \frac{3}{36}, \frac{4}{36}, \frac{5}{36}, \frac{6}{36}, \frac{5}{36}, \frac{4}{36}, \frac{3}{36}, \frac{2}{36}$, and $\frac{1}{36}$, respectively. As a result a new sample space consisting of eleven points based on the values of the random variable x can be constructed as shown in Fig. 5.

After a random variable and its corresponding sample space have been introduced and the probabilities to be assigned to those sample points calculated, the

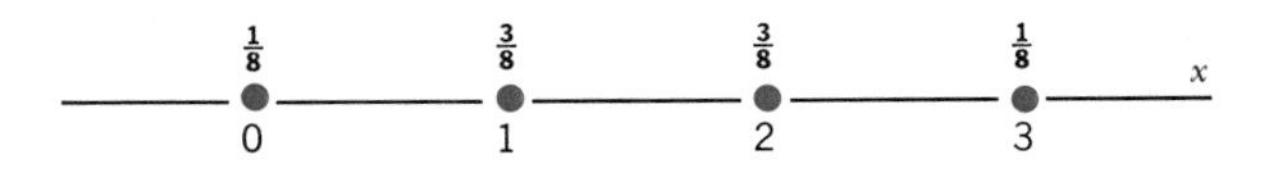

Figure 4 Sample space for a coin-tossing random variable.

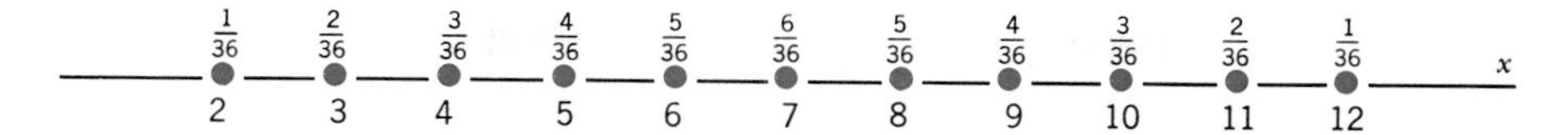

Figure 5 Sample space for a dice-rolling random variable.

desired probability distribution for the random variable has been determined. The distribution of a random variable x is always understood to be its probability distribution and not its empirical distribution. Such a distribution consists of the values that x can assume and the probabilities associated with those values. Graphs of the distributions for the random variables of Figs. 4 and 5 are given in Figs. 6 and 7.

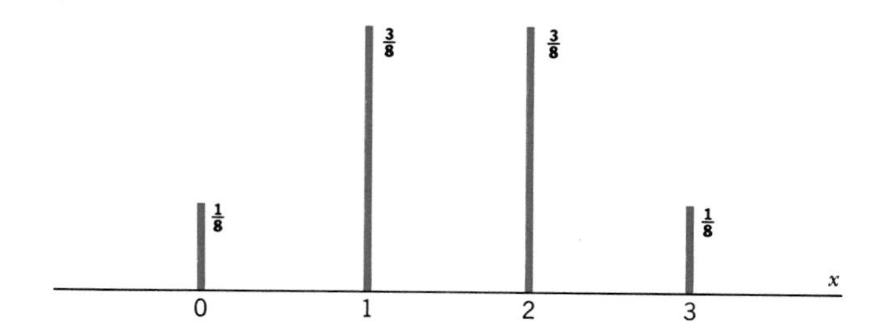

Figure 6 Distribution for a coin-tossing random variable.

A probability distribution that has been constructed in the foregoing manner is to be considered a mathematical model for a corresponding empirical distribution. Conversely, an empirical distribution of a variable x is to be considered as an approximation to a probability distribution for x.

3 PROPERTIES OF PROBABILITY DISTRIBUTIONS

The discussion of empirical distributions in Chapter 2 began with a geometrical representation by means of histograms and then it proceeded to a partial arithmetic representation by means of the mean and the standard deviation of the

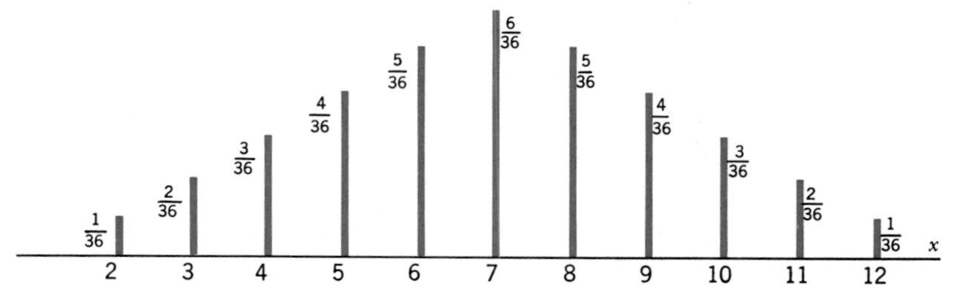

Figure 7 Distribution for a dice-rolling random variable.

distribution. The same procedure will be followed for the probability distributions that are to be used as mathematical models for empirical distributions. The only essential difference in the calculations for a probability distribution as contrasted to those for an empirical distribution is that one uses probabilities in place of observed relative frequencies. Since a probability distribution corresponds to an empirical distribution for classified data, the earlier formulas for classified data will be used.

The formula for the mean given by (4), Chapter 2, can be written in the form

$$\bar{x} = \sum_{i=1}^{k} x_i \frac{f_i}{n}$$

The corresponding theoretical mean, which is denoted by the Greek letter μ, is obtained by replacing the sample relative frequency f_i/n corresponding to the value x_i by the probability $P\{x_i\}$; hence

$$\mu = \sum_{i=1}^{k} x_i P\{x_i\} \tag{1}$$

If the sample extracted from a population is made increasingly large, it is to be expected that the relative frequency, f_i/n, of obtaining the value x_i will approach a fixed value. If so, this fixed value should be the probability $P\{x_i\}$. Thus, μ may be thought of as the value that $\bar{x}$ would be expected to approach as the sample size n becomes increasingly large. Table 1 and Fig. 1 of Chapter 2 illustrate how relative frequencies approach probability values as the sample size increases.

In a similar manner, the formula for the variance of an empirical distribution in its unmodified form and given by (6), Chapter 2, can be written

$$s^2 = \sum_{i=1}^{k} (x_i - \bar{x})^2 \frac{f_i}{n}$$

The corresponding theoretical variance, which is denoted by σ^2, is obtained by replacing f_i/n by $P\{x_i\}$ and replacing $\bar{x}$ by μ; hence

$$\sigma^2 = \sum_{i=1}^{k} (x_i - \mu)^2 P\{x_i\} \tag{2}$$

The theoretical standard deviation, σ, is the positive square root of σ^2. There is no problem here, as there was in defining s^2, about whether to divide by n or $n - 1$, because n is not involved in theoretical definitions. The value of σ^2 may be thought of as the value that s^2 would be expected to approach as the sample size n becomes increasingly large.

Example. As an illustration of the calculations involved in formulas (1) and (2), the mean and variance of the distribution given by Fig. 6 of this chapter will be computed. Since the possible values of x are $x_1 = 0$, $x_2 = 1$, $x_3 = 2$, and $x_4 = 3$, with corresponding probabilities $P\{x_1\} = \frac{1}{8}$, $P\{x_2\} = \frac{3}{8}$, $P\{x_3\} = \frac{3}{8}$, and $P\{x_4\} = \frac{1}{8}$, it follows from formula (1) that

$$\mu = 0 \cdot \frac{1}{8} + 1 \cdot \frac{3}{8} + 2 \cdot \frac{3}{8} + 3 \cdot \frac{1}{8} = \frac{3}{2} = 1.5 \tag{3}$$

Similar calculations using formula (2) yield

$$\sigma^2 = \left(0 - \frac{3}{2}\right)^2 \cdot \frac{1}{8} + \left(1 - \frac{3}{2}\right)^2 \cdot \frac{3}{8} + \left(2 - \frac{3}{2}\right)^2 \cdot \frac{3}{8} + \left(3 - \frac{3}{2}\right)^2 \cdot \frac{1}{8} \tag{4}$$

$$= \frac{9}{4} \cdot \frac{1}{8} + \frac{1}{4} \cdot \frac{3}{8} + \frac{1}{4} \cdot \frac{3}{8} + \frac{9}{4} \cdot \frac{1}{8} = \frac{3}{4}$$

The mean and variance of a probability distribution are also called the mean and variance of the random variable x whose probability distribution is used in the calculations. Thus, the values of μ and σ^2 just calculated may be called the mean and variance of the random variable x, where x denotes the number of heads obtained in tossing three coins.

It is sometimes more convenient to calculate σ^2 by means of the following formula, which is easily derived by squaring out the parentheses in (2) and summing the individual terms.

$$\sigma^2 = \sum_{i=1}^{k} x_i^2 P\{x_i\} - \mu^2 \tag{5}$$

4 EXPECTED VALUE

The mean and variance of a random variable are but two of its useful properties. They can be treated as special cases of a more general property called the *expected value.* For the purpose of describing what the phrase "expected value" means, consider a game in which you toss three honest coins and receive one dollar for each head that shows. How much money should you expect to get if you are permitted to play this game once? From Fig. 6 the probabilities associated with getting 0, 1, 2, and 3 heads in tossing three honest coins are $\frac{1}{8}$, $\frac{3}{8}$, $\frac{3}{8}$, and $\frac{1}{8}$, respectively. Intuitively, therefore, you should expect to get 0 dollars $\frac{1}{8}$ of the time, 1 dollar $\frac{3}{8}$ of the time, 2 dollars $\frac{3}{8}$ of the time, and 3 dollars $\frac{1}{8}$ of the time, if the game were played a large number of times. You should therefore expect to average the amount

$$\$0 \cdot \frac{1}{8} + \$1 \cdot \frac{3}{8} + \$2 \cdot \frac{3}{8} + \$3 \cdot \frac{1}{8} = \$1.50$$

This amount is what is commonly called the expected amount to be won if the game is played once. It will be observed from (3) that this is the same as the mean value μ. This example is an illustration of the general concept of expected value that follows.

Suppose the random variable x must assume one of the values $x_1, x_2, \ldots, x_k$ and that the probabilities associated with those values are $P\{x_1\}, P\{x_2\}, \ldots, P\{x_k\}$, where $\Sigma_{i=1}^{k} P\{x_i\} = 1$. Then the expected value of this random variable is defined to be the quantity

$$E[x] = \sum_{i=1}^{k} x_i P\{x_i\} \tag{6}$$

In the preceding illustration the random variable x was the amount of money to be won in tossing three coins, the possible values were 0, 1, 2, and 3, and the corresponding probabilities were $\frac{1}{8}$, $\frac{3}{8}$, $\frac{3}{8}$, and $\frac{1}{8}$.

A comparison of formula (1) with formula (6) shows that $E[x]$ is nothing more than the mean μ of the random variable x. Since the expected value of a random variable is the same as its mean, it would appear that there is no point in introducing this new terminology; however, expectation goes a step further. Suppose the preceding game is altered so that you win the amount $g(x_i)$ instead of x_i when the value x_i is obtained. For example, $g(x)$ might be chosen to be the function $g(x) = x^2$. Then the expected value of the game would be given by the formula

$$E[g(x)] = \sum_{i=1}^{k} g(x_i) P\{x_i\} \tag{7}$$

Thus, it is possible to talk about the expected value of any function of a random variable, rather than just of the random variable itself. As an illustration, let $g(x) = x^2$ for the preceding example; then

$$E[g(x)] = \$0 \cdot \frac{1}{8} + \$1 \cdot \frac{3}{8} + \$4 \cdot \frac{3}{8} + \$9 \cdot \frac{1}{8} = \$3$$

In view of this result, you could expect to win \$3 if the payoffs in dollars are the squares of the number of heads obtained and you are allowed to play the game once.

From definition (7) it is easily shown that the expected-value operator E possesses the following properties. (In these formulas, c is any constant.)

$$E[g(x) + c] = E[g(x)] + c \tag{8}$$

$$E[cg(x)] = cE[g(x)] \tag{9}$$

A third property that is more difficult to demonstrate is given by the formula

$$E[g(x) + h(y)] = E[g(x)] + E[h(y)] \tag{10}$$

In this last formula x and y are any two random variables and g and h are any two functions of those variables.

Example 1. For the purpose of illustrating the meanings of these formulas, suppose that $g(x) = x$, $h(y) = y^2$, $c = 2$, and that the earlier game to illustrate expectation is being played, where x and y^2 denote the amounts won in the preceding two games. Then formula (8) says that if \$2 is added to each prize of the first game, the expected amount to be won in playing it once will be \$2 more than before. Formula (9) says that if each prize of the first game is doubled, the expected amount to be won will be doubled. Formula (10) says that if the amounts to be won are combined into a single game, one can expect to win the sum of the expected values for the separate games. All of these formulas are intuitively obvious when one realizes that they represent the corresponding properties for mean values.

Example 2. As another illustration of how to calculate expected values, consider the game in which you toss a coin three times and then toss a pair of dice. You are paid \$5 for each head that shows and, in addition, as many dollars as the sum of the points showing on the two dice. How much should you expect to win? Let x represent the number of heads and y the total number of points that show. Then you should expect to win the amount $E[5x + y] = E[5x] + E[y]$. From Fig. 6 it follows that

$$E[5x] = 0 \cdot \frac{1}{8} + 5 \cdot \frac{3}{8} + 10 \cdot \frac{3}{8} + 15 \cdot \frac{1}{8} = \frac{60}{8} = 7.50$$

and from Fig. 7 it follows that

$$\begin{aligned} E[y] &= 2 \cdot \frac{1}{36} + 3 \cdot \frac{2}{36} + 4 \cdot \frac{3}{36} + 5 \cdot \frac{4}{36} + 6 \cdot \frac{5}{36} + 7 \cdot \frac{6}{36} \\ &\quad + 8 \cdot \frac{5}{36} + 9 \cdot \frac{4}{36} + 10 \cdot \frac{3}{36} + 11 \cdot \frac{2}{36} + 12 \cdot \frac{1}{36} \\ &= \frac{252}{36} = 7 \end{aligned}$$

Hence, you should expect to win \$14.50 if you are permitted to play this game once.

The variance of a probability distribution is conveniently represented by an expected value in the same manner as is the mean. This is accomplished by choosing $g(x) = (x - \mu)^2$ in (7). Then

$$E[(x - \mu)^2] = \sum_{i=1}^{k} (x_i - \mu)^2 P\{x_i\}$$

A comparison of this result with (2) shows that

$$\sigma^2 = E[(x - \mu)^2]$$

Thus, the mean and variance are seen to be special cases of the expected value of a function of a random variable.

In addition to generalizing the notions of the ordinary mean and variance of probability distributions, the expected-value operator is also a very useful tool for solving certain types of decision-making problems. Such problems will be studied rather extensively in the last chapter; nevertheless, a simple example will be explained here to indicate how E can be used to solve such problems.

Example 3. As an illustration of how the expected-value operator can be used to solve simple decision problems, consider the following problem. A car owner can't decide whether to take out a \$100-deductible collision insurance policy on his car, in addition to his liability insurance. The premium for this policy is \$80 a year. Records for his community show that the average cost to repair a car that has been responsible for an accident is \$600. Records also show that about 10 percent of drivers have an accident during the year for which they are either responsible or for which no responsibility can be assigned. If this car owner has sufficient assets so that he would not be handicapped if he had to pay out \$600 or more for car repairs, should he buy the collision policy? The information given here may be arranged in the form of a table that lists the losses that can be expected under the various possibilities. Since the \$100-deductible policy requires a payment of \$100 when an accident occurs, the table will have the following form:

	Insure	Don't insure
Accident	180	600
No accident	80	0

A decision here will be based on calculating the expected loss under the two possible choices. Let E_I and E_D denote these two expected values. Calculations give

$$E_I = 180\left(\frac{1}{10}\right) + 80\left(\frac{9}{10}\right) = 90$$

$$E_D = 600\left(\frac{1}{10}\right) + 0\left(\frac{9}{10}\right) = 60$$

If the car owner is willing to operate on the basis of what is likely to happen in the long run of experience, he will benefit financially by not buying collision insurance. This problem, of course, has been unduly simplified, because it does not, for example, consider the possibility of having more than one accident during the year or the possibility, even though it would constitute a rare event, of having a very serious accident. There are also psychological advantages to some people in possessing insurance. If the driver is a better than average driver, his probability of an accident would be less than $\frac{1}{10}$, in which case the advantage in not buying collision insurance would be even greater than \$30.

5 CONTINUOUS VARIABLES

This section is concerned with a discussion of continuous random variables and their distributions. Such variables have already been studied in Chapter 2. The essential distinction between a continuous variable and a discrete variable is that the former involves measuring, whereas the latter involves counting. The variables of Chapter 2 for which the techniques of classification of data were explained were continuous variables, whereas the variables of Chapter 3 for which the rules of probability were derived were discrete variables.

Histograms are always used to represent empirical distributions of continuous variables, whereas line charts are normally used for discrete variable distributions. In Chapter 2, however, histograms were used in Figs. 4 and 5 even though the variables there are discrete.

For the purpose of discussing how probability distributions arise in connection with continuous variables, consider a particular continuous variable, x, that represents the diameter of a steel rod obtained from the production line of a manufacturer. If the next 200 rods coming off the production line are measured, there

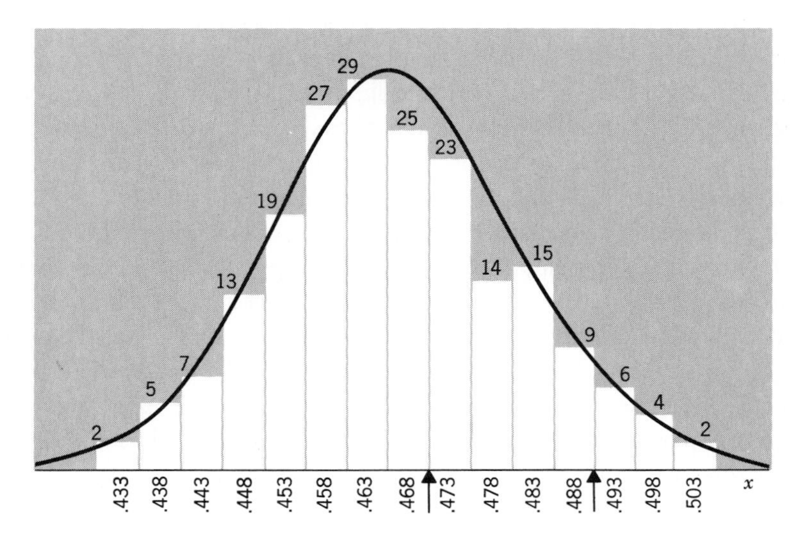

Figure 8 Distribution of the diameters of 200 steel rods, with a fitted normal curve.

will be 200 values of x with which to study the diameter variation of the production system. Classifying these 200 values of x and graphing the histogram helps to describe the distribution of diameters. The histogram of Fig. 8 illustrates the results of one experiment. If 400 rods had been measured, the resulting histogram would have been about twice as tall as that in Fig. 8. This growth in the height of a histogram as the sample size increases makes it difficult to compare histograms based on different size samples. The difficulty can be overcome by requiring that the area of the histogram always be equal to 1. This can be accomplished by choosing the proper heights for the rectangles that make up the histogram.

The advantage of using a histogram whose area is equal to 1 becomes apparent when one calculates the relative frequencies for different sets of x values. Thus, from Fig. 8, the proportion of steel rods in the sample of 200 that had a diameter in inches between .4705 and .4755 is given by the area of the rectangle for which those numbers are the boundaries. These are the boundaries for the interval whose class mark is .473. This area is, of course, $23/200 = .115$ units. Similarly, the proportion of steel rods in the sample that had a diameter in inches between .4705 and .4905 is given by the sum of the areas of the rectangles that begin at .4705 and end with .4905 and which are between the two vertical arrows shown in Fig. 8. This area is given by $(23 + 14 + 15 + 9)/200 = .305$. Although it is not customary to use areas to calculate such relative frequencies, since they are readily obtained from the observed frequencies, it is important to realize that such relative frequencies can be represented by areas of parts of the histogram, because theoretical relative frequencies will be calculated by means of areas of theoretical distributions.

With the foregoing choice of heights of rectangles to produce an area of 1, a histogram would be expected to approach a fixed histogram as the sample size is increased. For example, if a sample of 1,000 steel rods were taken from the production line and the resulting histogram with area equal to 1 were graphed, one would not expect the histogram to change its shape much if additional samples were taken, because the histogram would already be an accurate estimate of the distribution of diameters for the production process. Now, if it is assumed that x can be measured as accurately as desired, so that the class interval can be made as small as desired, then the upper boundary of the histogram would be expected to settle down and approximate a smooth curve as the sample size is increased, provided the class interval is chosen very small. Such a curve is an idealization, or model, of the relative frequency with which different values of x can be expected to occur for runs of the actual experiment. The distribution given by such a curve is a theoretical distribution for the continuous variable x.

Figure 8 shows a curve that experience has indicated should conform closely to the kind of frequency distribution expected here and that therefore represents a theoretical distribution for that variable.

Since a curve that represents a theoretical distribution is thought of as the limiting form of the histogram under continuous sampling, it must possess the essential frequency properties of the histogram. Thus the area under a theoretical distribution curve must be equal to 1 because the area under the histogram is always kept equal to 1. Further, since the area of any rectangle of a histogram is equal to the relative frequency with which x occurs between the boundaries of the corresponding class interval, the area under the theoretical distribution curve between those same boundaries represents the expected relative frequency of x occurring in that interval. If this same reasoning is extended to several neighboring intervals, it follows that the area under such a curve between any two values of x represents the expected relative frequency of x occurring inside the interval determined by those two values of x.

As an illustration, consider once more the two values $x = .4705$ and $x = .4905$ indicated by arrows in Fig. 8. The observed relative frequency for x occurring inside this interval is given by the sum of the areas of the rectangles lying inside this interval and was found to be .305. If it is assumed that the curve sketched in Fig. 8 represents a proper theoretical distribution, then the area under this curve between $x = .4705$ and $x = .4905$ represents the relative frequency with which x is expected to occur inside this interval. This area was found, by methods to be explained later, to be equal to .335.

In Chapter 3 probability was defined for discrete variables in such a manner that it was interpreted as expected relative frequency for the event in question. Although the definition and rules of probability were restricted to discrete variables for ease of explanation, the rules also apply to continuous-variable events. Consequently, in view of the discussion in the preceding paragraphs, it follows that a theoretical, or probability, distribution curve is a curve by means of which one can calculate the probability that x will lie inside any specified interval on the x-axis. Thus, in the illustration related to Fig. 8, if it is assumed that the curve there represents the theoretical distribution, one can say that when a single sample value of x is taken the probability that the value of x will lie between .4705 and .4905 is .335. In symbols this is written

$$P\{.4705 < x < .4905\} = .335$$

This kind of statement is typical of probability statements for continuous variables; the events are usually concerned with the variable x lying inside some interval, or intervals, on the x-axis. Such probabilities are given by areas under probability distribution curves.

As in the case for discrete variables, if one chooses a realistic model one can expect the probabilities calculated on the basis of that model to be close to the corresponding observed relative frequencies that are obtained when the experiment is repeated a large number of times.

Just as for discrete variables, the limiting values of $\bar{x}$ and s^2 for an empirical distribution of a continuous variable x are denoted by the letters μ and σ^2. The calculation of μ and σ^2 for a continuous-variable probability distribution such as that shown in Fig. 8 requires a knowledge of calculus and therefore cannot be considered here.

6 REVIEW ILLUSTRATIONS

Example 1. A punch board has six punches left, and it is known that there are two \$1 prizes left and four blanks. Let x denote the amount of money in dollars to be won.

(*a*) If one punch is to be taken, find the distribution of x and graph it.

(*b*) If two punches are to be taken, find the distribution of x and graph it.

(*c*) Calculate the mean and standard deviation of the distribution in part (*a*). The solutions follow.

(*a*) $x = 0, 1$

$P\{x\} = \frac{4}{6}, \frac{2}{6}$

(*b*) $x = 0, 1, 2$

$P\{x\} = \frac{4}{6} \cdot \frac{3}{5},\quad 2 \cdot \frac{2}{6} \cdot \frac{4}{5},\quad \frac{2}{6} \cdot \frac{1}{5}$

(*c*) $\mu = 0 \cdot \frac{4}{6} + 1 \cdot \frac{2}{6} = \frac{1}{3}$

$$\sigma^2 = \left(0 - \frac{1}{3}\right)^2 \cdot \frac{4}{6} + \left(1 - \frac{1}{3}\right)^2 \cdot \frac{2}{6} = \frac{2}{9}, \quad \sigma = \frac{\sqrt{2}}{3} = .47$$

If formula (5) had been used, the calculations for σ^2 would have been

$$\sigma^2 = 0^2 \cdot \frac{4}{6} + 1^2 \cdot \frac{2}{6} - \left(\frac{1}{3}\right)^2 = \frac{2}{6} - \frac{1}{9} = \frac{2}{9}$$

Example 2. A manufacturer of automobile tires has kept records of the quality of his product and obtained the following table of values based on the last six months of production.

Number of defects	0	1	2	3	4	5	≥ 6
Percentage of tires	60	22	8	5	3	2	0

Let x denote the number of defects per tire and treat the experience percentages as probabilities, converting them to the corresponding decimal proportions. Then (*a*) graph the distribution, and (*b*) calculate the mean and the standard deviation of x.

(*a*)

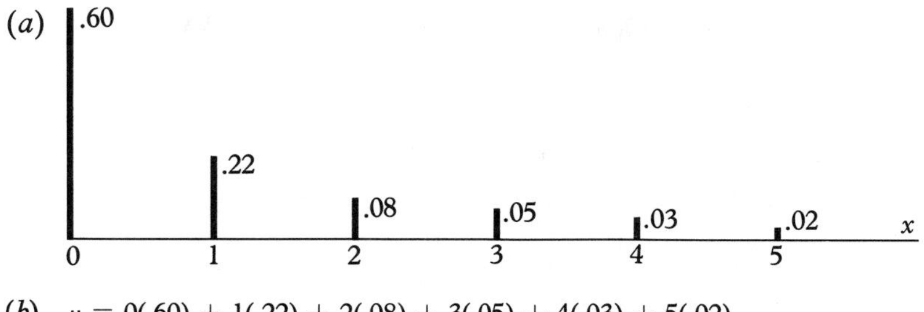

(*b*) $\mu = 0(.60) + 1(.22) + 2(.08) + 3(.05) + 4(.03) + 5(.02)$
$= .75$

$$\sigma^2 = (0 - \tfrac{3}{4})^2(.60) + (1 - \tfrac{3}{4})^2(.22) + (2 - \tfrac{3}{4})^2(.08) + (3 - \tfrac{3}{4})^2(.05) + (4 - \tfrac{3}{4})^2(.03) + (5 - \tfrac{3}{4})^2(.02) = 1.41$$

Hence, $\sigma = \sqrt{1.41} = 1.2$.

Example 3. An individual who owns the ice cream concession at a sporting event can expect to net \$600 on the sale of ice cream if the day is sunny, but only \$300 if it is cloudy, and \$100 if it rains. The respective probabilities for those events are .6, .3, and .1.

(*a*) What is his expected profit?

(*b*) If he takes out \$400 worth of insurance against rain and the insurance costs him \$90, what is his expected profit?

(*a*) $E = \$600(.6) + \$300(.3) + \$100(.1) = \$460$

(*b*) If it rains he will realize both \$100 and the insurance; hence

$$E = \$600(.6) + \$300(.3) + \$500(.1) - \$90 = \$410$$

Example 4. The demand for a particular magazine at Joe's newstand has the following probability distribution:

Number demanded, y	0	1	2	3	4	5	6
Probability, $P\{y\}$	.1	.2	.3	.1	.1	.1	.1

Joe sells this magazine for \$2.00 a copy, for which he pays \$1.00, and he is refunded \$0.10 for each unsold copy. (*a*) Calculate the mean and the standard deviation of y, the number of copies demanded. (*b*) Suppose Joe decides to stock 6 copies of the magazine. Determine the values, and their probabilities, of the random variable x that represents net profit. (*c*) Calculate the mean and the standard deviation of x by using the results in (*b*). (*d*) Suppose 3 copies are stocked. What is Joe's expected profit now? Compare with (*c*).

(*a*) $\mu = E[y] = 0(.1) + 1(.2) + 2(.3) + 3(.1) + 4(.1) + 5(.1) + 6(.1) = 2.6$

$$\begin{aligned}\sigma^2 = E[y-\mu]^2 &= (0-2.6)^2(.1) + (1-2.6)^2(.2) \\ &+ (2-2.6)^2(.3) + (3-2.6)^2(.1) \\ &+ (4-2.6)^2(.1) + (5-2.6)^2(.1) \\ &+ (6-2.6)^2(.1) \\ &= 3.24\end{aligned}$$

Hence, $\mu = 2.6$ and $\sigma = 1.8$.

(*b*)

Number demanded	0	1	2	3	4	5	6
Sales revenue	0	2.00	4.00	6.00	8.00	10.00	12.00
Refund revenue	.60	.50	.40	.30	.20	.10	0
Costs	6.00	6.00	6.00	6.00	6.00	6.00	6.00
Net profit, x	−5.40	−3.50	−1.60	.30	2.20	4.10	6.00
Probability, $P\{x\}$	.1	.2	.3	.1	.1	.1	.1

(*c*)
$$\begin{aligned}E[x] &= -5.40(.1) - 3.50(.2) - 1.60(.3) + .30(.1) + 2.20(.1) \\ &+ 4.10(.1) + 6.00(.1) \\ &= -0.46\end{aligned}$$

Therefore he will lose, on the average 46 cents on this magazine if he orders 6 copies.

$$\begin{aligned}E[x-\mu]^2 &= (-5.40 + .46)^2(.1) + (-3.50 + .46)^2(.3) \\ &+ \cdots + (4.10 + .46)^2(.1) + (6.00 + .46)^2(.1) \\ &= 11.70\end{aligned}$$

Hence, $\mu = -.46$ and $\sigma = 3.42$.

(*d*)

Number demanded	0	1	2	3	4	5	6
Sales revenue	0	2.00	4.00	6.00	6.00	6.00	6.00
Refund revenue	.30	.20	.10	0	0	0	0
Costs	3.00	3.00	3.00	3.00	3.00	3.00	3.00
Net profit, x	−2.70	−.80	1.10	3.00	3.00	3.00	3.00
Probability, $P\{x\}$	.1	.2	.3	.1	.1	.1	.1

$$\begin{aligned}E[x] &= -2.70(.1) - .80(.2) + 1.10(.3) + 3.00(.1) + 3.00(.1) \\ &+ 3.00(.1) + 3.00(.1) \\ &= 1.10\end{aligned}$$

Therefore he can expect to profit $1.10 if he orders 3 copies. A comparison with the result in (*c*) shows that ordering 3 is preferable to ordering 6.

EXERCISES

SECTION 2

1 What is the probability distribution of the variable x = number of heads when two honest coins are tossed? Graph it.

2 A box contains three cards consisting of the two, three, and four of hearts. Two cards are drawn from the box, with the first drawn card returned to the box before the second drawing. Let x denote the sum of the numbers obtained on the two cards. Use the enumeration-of-events technique to derive the distribution of the random variable x.

3 Work problem 2 under the assumption that the first card is not returned to the box.

4 In an illustration in section 8, Chapter 3, three prizes were to be given out by lot to 50 employees, 10 of whom were executives. Determine the probability distribution of the number of prizes that will be received by the executives.

SECTION 3

5 Calculate the values of μ and σ for the distribution obtained in problem 2.

6 Calculate the values of μ and σ for the distribution obtained in problem 3.

7 Calculate the values of μ and σ for the distribution of random digits, that is, for $P(x) = \frac{1}{10}$, $x = 0, 1, 2, \ldots, 9$.

8 Calculate the values of μ and σ for the distribution given by $P(-1) = \frac{3}{8}$, $P(0) = \frac{2}{8}$, $P(1) = \frac{3}{8}$. What percentage of the distribution is included in the interval $(\mu - \sigma, \mu + \sigma)$?

9 A cigarette from each of four brands is partially smoked by a blindfolded individual. After a few puffs he is required to state which brand he believes he is smoking. Let x be the random variable that represents the number of cigarettes that are correctly identified. Assume that he has no ability to identify them, which means that he names them by chance.

(*a*) Calculate the probability distribution of x.

(*b*) Graph this probability distribution.

(*c*) Calculate the mean and standard deviation of x.

SECTION 4

10 A population consists of four elements with the x-values 1, 2, 3, and 4. One element is picked at random.

(*a*) What is the expected value of x?

(*b*) What is the expected value of x^2?

(*c*) Suppose the elements are circles and the x-values are their diameters. What is the expected value of the area of such a circle?

11 The number of accidents that occur at a particular industrial plant during a workday is 0, 1, 2, or 3 with corresponding probabilities 0.94, 0.03, 0.02, 0.01. Find the expected number of accidents during a day. During 100 working days.

12 Which game would you choose if given a choice: toss two dice and receive in dollars the sum of the points showing or toss four coins and receive in dollars double the number of heads obtained?

13 A car owner wishes to sell his car and is contemplating spending \$50 to advertise it. If the probability is .5 that he will sell it at his stipulated price of \$750

without advertising and is .9 of doing so if he does advertise, should he advertise? Assume that if he does not sell it for $750 he will let a friend have it for $650.

14 Suppose we select digits at random from a page of random digits.
(*a*) Define the random variable, x.
(*b*) Find $E[x]$.
(*c*) Find $E[x - 4.5]$.
(*d*) Find $E[x^2]$.
(*e*) Find $E[(x - E[x])^2]$.
(*f*) What is the standard deviation of x?

15 An examination consisting of 10 multiple-choice questions is administered to a group of students. Five of the questions have 2 choices, 5 have 3 choices. In scoring, 10 points are given for a correct answer and 0 points for a wrong one, and answers are either right or wrong. If a student should answer the questions with the use of random numbers, given that each alternative has an equal chance, what is his expected score?

16 According to an American experience mortality table, the probability that a 30-year-old man will survive one year is 0.992, and hence that he will die within a year is 0.008. An insurance company offers to sell such a man a $1000 one-year term life insurance policy for a premium of $20. What is the company's expected gain?

SECTION 6

17 Toss 3 coins simultaneously and record the number of heads obtained. Perform this experiment 100 times and then compare your experimental relative frequencies with those given by theory in Fig. 6.

18 Suppose the probability is $\frac{1}{5}$ that a piece of property will be sold within the next year and a profit of $20,000 realized, but a loss of $4000 will occur if it is not sold by then.
(*a*) Calculate the expected profit for this selling venture.
(*b*) What probability value would make this expected profit zero?

19 In the game known as "chuck-a-luck" a player picks a number from 1 to 6. Three dice are then thrown and the player receives $1 if one die shows his number, $2 if two dice show his number, and $3 if all three dice show his number. If none of the dice shows his number he loses $1.
(*a*) What is the probability that the player will win something?
(*b*) What is his expected value in this game?

20 A box contains the following nine cards: the three, four, and five of spades, the three and four of clubs, the three and four of hearts, and the four and five of diamonds.
(*a*) If one card is to be drawn from the box and x is the random variable representing the number of black cards that will be obtained, find the distribution of x and graph it.
(*b*) If one card is to be drawn and x represents the number on the card, find the distribution of x and graph it.

(*c*) If two cards are to be drawn, with the first card being replaced before the second drawing, and x represents the number of black cards that will be obtained, find the distribution of x and graph it.

(*d*) Work part (*c*) if the first card is not replaced.

(*e*) Calculate the mean and the standard deviation of the distribution in (*b*).

Chapter 5 Some Particular Probability Distributions

In this chapter several of the most useful probability distributions for solving statistical problems will be introduced. Three of these will be discrete-variable distributions and two of them will be for continuous variables.

1 THE BINOMIAL DISTRIBUTION

Consider an experiment in which each of the possible outcomes can be classified as resulting or not resulting in the occurrence of an event A. If it results in the occurrence of A it will be classified as a success, otherwise as a failure. The word success is used here as a convenient way of describing the occurrence of an event but it does not imply that the occurrence of the event is necessarily desired. The experiment will be repeated a number of times, this number being denoted by the letter n. A random variable x will be introduced that represents the total number of successes, that is, occurrences of A, that are obtained in the n repetitions of the experiment. A random variable of this type is called a *binomial variable.*

The coin-tossing experiment that has been used so frequently can also be used here to give an example of a binomial variable. Let that experiment consist of tossing the coin once. Success will be defined as getting a head. The experiment will be repeated three times; hence $n = 3$ here. The random variable x will then represent the number of heads obtained in the three tosses, just as it did in section 2, Chapter 4. The distribution of this binomial random variable is therefore given by Fig. 6, Chapter 4.

The experiment of rolling two dice with x defined as the sum of the points on the two dice does not produce a binomial variable because one cannot classify each roll of a die as producing either a success or a failure and then have x represent the sum of the successes. There are six possibilities for each die rather than only two, as required for a binomial problem.

The experiment of drawing a ball from a box consisting of three red, two black, and one green ball is also not an experiment that leads to a binomial variable because there are three possible outcomes here rather than two. However, if one were interested only in knowing whether a red ball will be obtained, then the problem becomes a binomial distribution problem. The variable x would then represent the number of red balls obtained in performing the experiment n times. It is always understood in binomial variable problems that replacements are made before the next experiment is made when it consists of drawing objects from containers. The repetitions of the experiment must be repetitions of the original experiment in every sense.

Now look at a slightly more complicated example of a binomial random variable and how its distribution is obtained. The basic experiment will consist of rolling a die once and success will be defined as getting an ace (one-spot). The experiment will be performed three times, so that $n = 3$, and the random variable x will represent the number of aces obtained in the three rolls. To obtain the distribution of this random variable one can proceed in the same manner as for the coin-tossing experiment of section 2, Chapter 4. This consists of looking at the original sample space for the complete experiment and then reducing it to a new sample space for the random variable x by applying the definition of probability of events to the original sample space. Now for each roll of the die it is necessary merely to record whether a success or a failure occurred, where success corresponds to an ace showing. If S and F are used to represent success and failure, then there are eight points in the sample space, just as there were for three tosses of a coin in which either an H or a T must occur at each toss. These points have been represented in Table 1 by means of the letters S and F to indicate the various

Table 1

Outcome	SSS	SSF	SFS	FSS	SFF	FSF	FFS	FFF
Value of x	3	2	2	2	1	1	1	0

Table 2

Outcome	SSS	SSF	SFS	FSS	SFF	FSF	FFS	FFF
Probability	$(\frac{1}{6})^3$	$(\frac{1}{6})^2(\frac{5}{6})$	$(\frac{1}{6})^2(\frac{5}{6})$	$(\frac{1}{6})^2(\frac{5}{6})$	$(\frac{1}{6})(\frac{5}{6})^2$	$(\frac{1}{6})(\frac{5}{6})^2$	$(\frac{1}{6})(\frac{5}{6})^2$	$(\frac{5}{6})^3$

possible outcomes. The corresponding values of the random variable x have also been displayed in Table 1.

It will be observed that this table is precisely the same as for the problem of tossing a coin three times, shown in Fig. 2, Chapter 4. However, the calculation of the probabilities for the various values of x is considerably different. The probabilities for these eight possible outcomes are not equal as was the case for the coin problem. The probabilities here will be calculated by using the multiplication rule of probability. Because of the independence of the three rolls of the die, it follows, for example, that

$$P\{\text{SFS}\} = \frac{1}{6}\cdot\frac{5}{6}\cdot\frac{1}{6} = \left(\frac{1}{6}\right)^2\frac{5}{6}$$

Calculations similar to this will yield probabilities for each of the eight possible outcomes. The results of such calculations are shown in Table 2.

This sample space of eight points with its assigned probabilities can now be reduced to the sample space for the random variable x by calculating the probabilities of the composite events corresponding to the various values of the random variable. Thus, the event $x = 2$ comprises the simple events SSF, SFS, FSS; therefore its probability is obtained by adding the probabilities associated with those three points. The results of such calculations are shown in Table 3. This table gives the distribution of the desired random variable. Its graph is shown in Fig. 1, in which the probabilities have been expressed in decimal form correct to two decimals.

In view of these results, it is clear that one should not expect to get three aces when rolling three dice. Such a result will occur in the long run about once in $6^3 = 216$ experiments. It is also clear that one should not accept a wager based on even money that one will get at least one ace in rolling three dice. Such a result will occur about 42 percent of the time. If you wish to get the better of your naïve friends, give, say, 9-to-1 odds that they will not get at least two aces when rolling three dice. A fair wager would require you to give about 13-to-1 odds.

The technique employed in the two preceding illustrations can be used on any particular binomial problem that arises. One first constructs the sample space

Table 3

x	0	1	2	3
$P\{x\}$	$(\frac{5}{6})^3$	$3(\frac{1}{6})(\frac{5}{6})^2$	$3(\frac{1}{6})^2(\frac{5}{6})$	$(\frac{1}{6})^3$

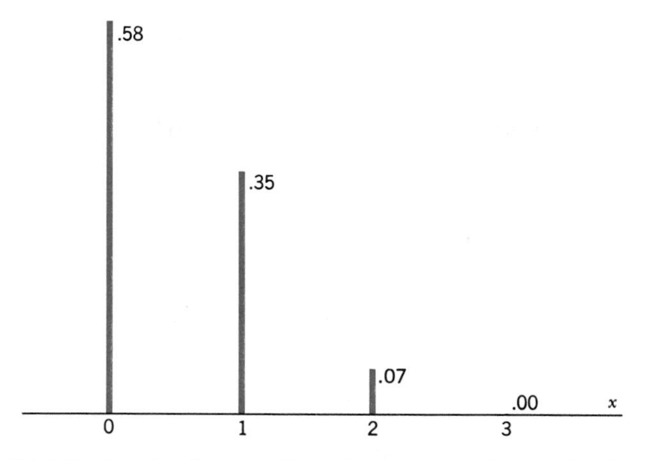

Figure 1 Distribution for the number of aces in rolling a die three times.

for the complete experiment. If there are, say, five repetitions of the basic success or failure experiment, there will be $2^5 = 32$ points in the sample space. The probability for each point is then calculated. Next, the proper value of the random variable x is associated with each point of the sample space. Then by summing the probabilities of those points that correspond to a particular value of x, the probabilities for the various values of x are obtained. These probabilities give the desired probability distribution of the binomial random variable x.

Because of the difficulty of carrying through the preceding computations each time a binomial problem arises, it is convenient to have a formula that is applicable to all such problems. For this purpose consider a general binomial problem. An experiment is to be performed for which the outcome can always be classified as either a success or a failure. The probability that it will produce a success is assumed given and is denoted by the letter p. The corresponding probability of a failure is denoted by q; hence $p + q = 1$. The experiment is to be performed n times. The number of successes that will be obtained in the n repetitions of the experiment is denoted by the letter x. The problem then is to calculate the probabilities for the various possible values of the random variable x. As stated before, these computations can be carried out systematically for any given problem, and that is what will be done here for this general problem.

There are various ways in which an experiment that is performed n times can give rise to exactly x successes and $n - x$ failures. One way is that in which all the successes occur first followed by all the failures.

$$\overbrace{SS \cdots S}^{x} \overbrace{FF \cdots F}^{n-x}$$

Another sequence is the following one in which a failure occurs first, followed by x consecutive successes, then followed by the remaining failures.

$$F\overbrace{SS\cdots S}^{x}\overbrace{FF\cdots F}^{n-x-1}$$

Because of the independence of the trials, the probability of obtaining the first of these two sequences is given by

$$\overbrace{p\cdot p\cdots p}^{x}\cdot\overbrace{q\cdot q\cdots q}^{n-x}=p^xq^{n-x}$$

The probability for the second sequence is given by

$$q\cdot\overbrace{p\cdot p\cdots p}^{x}\cdot\overbrace{q\cdot q\cdots q}^{n-x-1}=p^xq^{n-x}$$

Thus the probability for the two sequences is the same and will be the same for every sequence that satisfies the condition of having x successes and $n - x$ failures.

The number of ways in which the desired event can occur is equal to the number of different sequences of the type just displayed, those containing x letters S and $n - x$ letters F. But this is equal to the number of ways of choosing x positions out of n available positions in which to place the letter S. The remaining $n - x$ positions will automatically be assigned the letter F. The n positions may be numbered and treated like n numbered cards. The problem then is to choose x of those cards. Since only the numbers on the cards are of interest and not the order in which they are drawn, this is a combination problem. From the derivation in section 11 of Chapter 3, the number of ways of choosing x cards from n distinct cards is given by the combination formula (14) of Chapter 3, namely,

$$\binom{n}{x}=\frac{n!}{x!(n-x)!}$$

This, therefore, is the number of sequences that produce exactly x successes. As explained in section 11 of Chapter 3, the symbol $n!$ is read "n factorial." It denotes the product of all the positive integers from 1 through n. Thus, $5! = 1 \cdot 2 \cdot 3 \cdot 4 \cdot 5$. As was noted there, the value of $0!$ is defined to be 1.

Since each of these sequences represents one of the mutually exclusive ways in which the desired event can occur, and each such sequence has the same probability of occurring, namely p^xq^{n-x}, it follows that the desired probability is obtained by adding this probability as many times as there are sequences. But the number of such sequences was just found to be $\binom{n}{x}$; therefore the probability of obtaining x successes in n experiments is given by multiplying p^xq^{n-x} by $\binom{n}{x}$.

This probability, which will be denoted by $P\{x\}$ and which defines what is known as the binomial distribution, is therefore given by the following formula.

(1) ***Binomial Distribution.*** $P\{x\} = \frac{n!}{x!(n-x)!}p^x q^{n-x}$

For large values of n and x the computations involved in evaluating the quantity $n!/x!(n-x)!$ become rather heavy; consequently a table of values of this quantity for various values of n and x has been made available in the appendix as Table II. This table handles values of n from 2 through 20, and of x from 2 through $n/2$ or more. It is not necessary to list the values for larger values of x because $\binom{n}{x} = \binom{n}{n-x}$; consequently one uses this device to handle values of x larger than $n/2$.

Even with the aid of Table II, the computation of binomial probabilities often becomes heavy when the sum of several such probabilities is needed. Table III in the appendix will be useful in such situations, provided that the value of n does not exceed 10. That table lists the sum of right-tail probabilities of the form

$$P\{x \geq x_0\} = \sum_{x=x_0}^{n} \frac{n!}{x!(n-x)!}p^x q^{n-x}$$

Since the available values of p begin with .05 and proceed by steps of .05 to the value .50, it is necessary to interpolate for other values of $p \leq .50$. For values of $p > .50$, one uses $q = 1 - p$ in place of p and $n - x_0 + 1$ in place of x_0 in the table to obtain $1 - P\{x \geq x_0\}$. Subtraction of this result from 1 then gives the desired probability.

Although these two tables enable one to write down binomial probabilities rather quickly, they will not be used in the illustrative examples that follow because it seems desirable for the student to become acquainted with formula (1) by carrying out the required computations when n is fairly small. An illustration of how to use Table III, however, is given at the end of this section.

It is customary to speak of the n experiments as n independent *trials* of an experiment for which p is the probability of success in a single trial. In this language $P\{x\}$ is the probability of obtaining x successes in n independent trials of an experiment for which p is the probability of success in a single trial.

For the special case of only one trial, and hence for which x must assume the value 1 or 0, the binomial distribution is often called the *Bernoulli distribution* after a Swiss mathematician who was a pioneer in the study of probability. The binomial distribution therefore arises when one performs a sequence of n independent *Bernoulli trials* with the same p.

The frequency distribution displayed in Fig. 1 is a special case of the general binomial distribution given by (1). For that problem, $n = 3$, $p = \frac{1}{6}$, and $q = \frac{5}{6}$.

The values of $P\{x\}$ given in Fig. 1 should be checked by means of formula (1) for the purpose of becoming familiar with its use. As an illustration of such a check, if the value of $P\{3\}$ is desired for that problem, substitution of the proper values into (1) will yield

$$P\{3\} = \frac{3!}{3!0!}\left(\frac{1}{6}\right)^3\left(\frac{5}{6}\right)^0$$

Since $0! = 1$ and since by algebra any number to the zeroth power equals 1, it follows that $(\frac{5}{6})^0 = 1$, and hence that

$$P\{3\} = \left(\frac{1}{6}\right)^3$$

Although the problems used to introduce the binomial distribution were related to games of chance, the binomial distribution is very useful for solving certain types of practical problems. Such problems are solved in the next few chapters; meanwhile, a few problems, requiring only easy computations with formula (1), will be discussed.

Example 1. A manufacturer of certain parts for automobiles guarantees that a box of his parts will contain at most two defective items. If the box holds 20 parts and experience has shown that his manufacturing process produces 2 percent defective items, what is the probability that a box of his parts will satisfy the guarantee? This problem can be considered as a binomial distribution problem for which $n = 20$ and $p = .02$. A box will satisfy the guarantee if the number of defective parts is 0, 1, or 2. By means of formula (1) the probabilities of these three events are given by

$$P\{0\} = \frac{20!}{0!20!}(0.2)^0(.98)^{20} = (.98)^{20} = .668$$

$$P\{1\} = \frac{20!}{1!19!}(.02)^1(.98)^{19} = 20(.02)(.98)^{19} = .272$$

$$P\{2\} = \frac{20!}{2!18!}(.02)^2(.98)^{18} = 190(.02)^2(.98)^{18} = .053$$

Since these are mutually exclusive events, the probability that there will be at most two defective parts, written $x \leq 2$, is the sum of these probabilities; hence the desired answer is

$$P\{x \leq 2\} = .993$$

This result shows that the manufacturer's guarantee will almost always be satisfied.

Example 2. As a second illustration, suppose experience has shown that 30 percent of the used cars sold on time payment at a certain used car lot are eventually repossessed. If 10 cars are sold during a week, what is the probability that at least half of them will be repossessed? This may be treated as a binomial problem with $n = 10$, $p = .3$ and for which the desired probability is given by $P\{x \geq 5\}$. Calculations based on formula (1) give, rounded to three decimals,

$$P\{5\} = \frac{10!}{5!5!}(.3)^5(.7)^5 = .103$$

$$P\{6\} = \frac{10!}{6!4!}(.3)^6(.7)^4 = .037$$

$$P\{7\} = \frac{10!}{7!3!}(.3)^7(.7)^3 = .009$$

$$P\{8\} = \frac{10!}{8!2!}(.3)^8(.7)^2 = .001$$

$$P\{9\} = \frac{10!}{9!1!}(.3)^9(.7)^1 = .000$$

$$P\{10\} = \frac{10!}{10!0!}(.3)^{10}(.7)^0 = .000$$

The sum of these probabilities gives the desired answer:

$$P\{x \geq 5\} = .15$$

These examples illustrate the fact that fairly extensive calculations are needed to solve binomial-type problems. This is particularly true if n is large. Fortunately, in this case there exists an excellent simple approximation to the binomial distribution that can be used. This approximation will be studied in section 4.

As was stated previously, for $n \leq 20$, Table II in the appendix can be used to obtain the value of $n!/x!(n - x)!$, which should help considerably to ease the burden of computing binomial probabilities. For $n \leq 10$, Table III in the appendix can be used to calculate tail sums of binomial probabilities, and therefore, by subtraction, nontail sums also. To illustrate the use of Table III, consider the second of the two preceding examples. The Table III values needed for that example are $n = 10$, $p = .3$, and $x_0 = 5$. For those values this table gives $P\{x \geq 5\} = .1503$.

2 BINOMIAL DISTRIBUTION PROPERTIES

The discussion of empirical distributions in Chapter 2 began with a geometrical representation by means of line charts and histograms, and then it proceeded to a partial arithmetic representation by means of the mean and the standard deviation of the distribution. The same procedure will be followed for the theoretical distributions that are to be used as mathematical models for empirical distributions. Toward this objective, the calculation of the mean and the standard deviation for a binomial distribution will be discussed next.

The mean and the variance of a theoretical distribution were defined in Chapter 4 by means of formulas (1) and (2) of that chapter. They are repeated here.

$$\mu = \sum_{i=1}^{k} x_i P\{x_i\} \tag{2}$$

$$\sigma^2 = \sum_{i=1}^{k} (x_i - \mu)^2 P\{x_i\} \tag{3}$$

A variance formula that has computational advantages over the definition was also given there as formula (5) of Chapter 4. It is

$$\sigma^2 = \sum_{i=1}^{k} x_i^2 P\{x_i\} - \mu^2 \tag{4}$$

These formulas will now be employed to calculate the mean and the standard deviation of the binomial distribution given in Fig. 6, Chapter 4. When the probabilities $P\{x_i\}$ are given as fractions, it is advisable to omit the common denominator of those fractions in the calculations and then to divide the results by this denominator. The calculations for this problem are shown in Table 4.

After dividing these numbers by the denominator 8, formula (2) gives

$$\mu = \frac{12}{8} = \frac{3}{2}$$

The computation of σ^2 by means of formula (4) gives

$$\sigma^2 = \frac{24}{8} - \mu^2 = 3 - \left(\frac{3}{2}\right)^2 = \frac{3}{4}$$

Hence $\sigma = \sqrt{3}/2$.

The calculations for the binomial distribution given by Table 3 and shown in Fig. 1 will also be carried out. Here it is convenient to ignore the common denominator $6^3 = 216$ and then to divide by it at the end. The calculations for this problem are shown in Table 5.

Table 4

x	$8P\{x\}$	$8xP\{x\}$	$8x^2P\{x\}$
0	1	0	0
1	3	3	3
2	3	6	12
3	1	3	9
Totals		12	24

Table 5

x	$216P\{x\}$	$216xP\{x\}$	$216x^2P\{x\}$
0	125	0	0
1	75	75	75
2	15	30	60
3	1	3	9
		108	144

After division by 216, formulas (2) and (4) give

$$\mu = \frac{108}{216} = \frac{1}{2}$$

and

$$\sigma^2 = \frac{144}{216} - \left(\frac{1}{2}\right)^2 = \frac{5}{12}$$

As a result, $\sigma = \sqrt{5/12} = \sqrt{15}/6$.

It is possible to employ some algebraic tricks to carry out similar calculations for the general binomial distribution given by (1). Such calculations yield formulas for the mean and standard deviation that can be used for all binomial problems. Since such calculations are rather complicated, they will not be carried out here; however, the resulting formulas will be given and used for solving binomial problems. These formulas, which are very simple, are

$$\begin{aligned} \mu &= np \\ \sigma &= \sqrt{npq} \end{aligned} \tag{5}$$

The advantage of having such neat formulas becomes apparent when one applies them to the two computational problems that were just completed. For the distribution of Table 4, $n = 3$, and $p = \frac{1}{2}$; therefore formulas (5) give

$$\mu = 3 \cdot \frac{1}{2} = \frac{3}{2}$$

$$\sigma = \sqrt{3 \cdot \frac{1}{2} \cdot \frac{1}{2}} = \frac{\sqrt{3}}{2}$$

Similarly, for the distribution of Table 5, $n = 3$ and $p = \frac{1}{6}$; therefore formulas (5) give

$$\mu = 3 \cdot \frac{1}{6} = \frac{1}{2}$$

$$\sigma = \sqrt{3 \cdot \frac{1}{6} \cdot \frac{5}{6}} = \frac{\sqrt{15}}{6}$$

3 THE NORMAL DISTRIBUTION

The binomial distribution of section 2 is the most useful theoretical frequency distribution for discrete variables. The distribution that will be studied in this section is the most useful theoretical frequency distribution for continuous variables.

The histogram in Fig. 8, Chapter 4, is typical of many distributions found in nature and industry. Such distributions are quite symmetrical, die out rather quickly at the tails, and possess a shape much like that of a bell. A theoretical distribution that has proved very useful for distributions such as these, and that will presently be seen as very important in other ways also, is one that is called the *normal distribution;* but it is also called the *Gaussian distribution* after the famous mathematician Gauss who did so much research on it. The curve that has been sketched in Fig. 8, Chapter 4, is the graph of a particular normal distribution. The graph of a general normal distribution is given in Fig. 2. Although a normal distribution is defined by the equation of its curve, this equation is not used explicitly in subsequent chapters. The curve itself can be thought of as defining the distribution. For the benefit of those who are familiar with the exponential function, the algebraic definition is given by

Normal Distribution. $$f(x) = \frac{e^{-1/2((x-\mu)/\sigma)^2}}{\sqrt{2\pi}\,\sigma}$$

The value of e is approximately 2.718 and is the base for the natural logarithm, which occurs in the study of calculus.

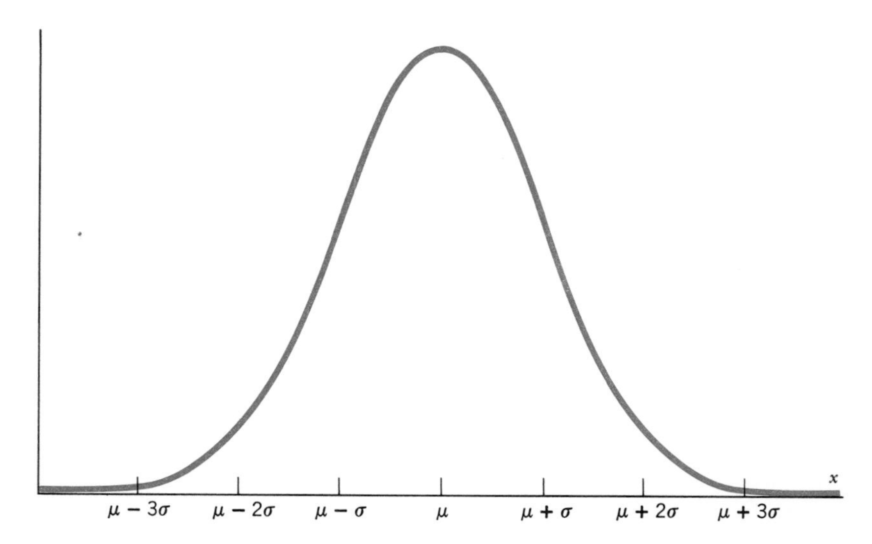

Figure 2 Typical normal distribution.

You may have heard of the normal curve in connection with the distribution of grades in large sections of certain courses. Grading on the basis of the normal curve assumes that the distribution of the mental output of students is similar to the distribution of many of their physical characteristics, which are known to be approximately normally distributed. An instructor using such a grading scheme also usually assumes that the students in his course have the proper prerequisites and study the expected number of hours. An optimist at heart! One difficulty with this system is that it assigns grades on a relative rather than an absolute basis. Thus an unusually gifted class that works hard will be assigned the same distribution of grades as a class of lazy louts. Of course, the answer to this criticism is that the students from year to year differ very little, and therefore it is highly unlikely that a large class will be made up of either brilliant workers or stupid loafers. If you have acquired a dislike for the normal curve on the basis of the grades it may have given you, realize that the normal curve is really not to blame and be charitable enough to approach its study here without prejudice. It is an exceedingly useful curve in many fields of application quite removed from school problems.

Recall from Chapter 2 that the mean of a distribution represents the point on the x-axis at which a sheet of metal in the shape of the histogram of the distribution will balance on a knife-edge. This geometrical property of the mean makes it clear that when a histogram is symmetrical about a vertical axis, the mean must be located at the symmetry point on the x-axis. This is also true for the limiting value of the mean when the size of the sample is increased indefinitely and the class interval is made increasingly small. The limiting, or theoretical, value of the mean is denoted by the Greek letter μ, just as for discrete variables. This explains why the symmetry point on the x-axis in Fig. 2 has been labeled μ. Whether a theoretical distribution is symmetrical or not, the symbol μ is used to designate the limiting, or theoretical, value of the mean.

If the standard deviation, as defined by formula (10), Chapter 2, were calculated for increasingly large samples and increasingly small class intervals, its value would also be expected to approach some value. This limiting, or theoretical, value is denoted by the Greek letter σ, just as for discrete variables.

The Greek letters μ and σ are used to represent the mean and the standard deviation of theoretical frequency distributions, corresponding to $\bar{x}$ and s for sample frequency distributions, so that there will be no confusion as to which distribution is meant when the mean and the standard deviation of a distribution are discussed. Two of the basic problems in statistics, namely, estimation and hypothesis testing, involve problems of estimating μ and σ, and testing hypotheses about them by means of their sample estimates $\bar{x}$ and s.

Suppose, now, that the limiting form of the histogram for a frequency distribution, in the sense described earlier, is a normal curve. Then it can be shown by advanced mathematical methods that σ, the limiting value of s, has the following geometrical interpretation with respect to that normal curve:

(6)
(*a*) The area under the normal curve between $\mu - \sigma$ and $\mu + \sigma$ is 68 percent of the total area, to the nearest 1 percent.
(*b*) The area under the normal curve between $\mu - 2\sigma$ and $\mu + 2\sigma$ is 95 percent of the total area, to the nearest 1 percent.
(*c*) The area under the normal curve between $\mu - 3\sigma$ and $\mu + 3\sigma$ is 99.7 percent of the total area, to the nearest .1 percent.

The axis in Fig. 2 has been marked off in units of σ, starting with the mean μ. It is clear from this sketch that there is almost no area under the curve beyond 3σ units from μ; however, the equation of the curve shows that the curve actually extends from $-\infty$ to $+\infty$.

The first two properties in (6) were used to give meaning to s in Chapter 2. Those two percentages were found there to be approximately correct for histograms whose shapes resemble a normal curve.

An interesting property of the normal curve is that its location and shape are completely determined by its values of μ and σ. The value of μ, of course, centers the curve, whereas the value of σ determines the extent of the spread. Since all normal curves representing theoretical frequency distributions have a total area of 1, as σ increases the curve must decrease in height and spread out. This is illustrated in Figs. 3 and 4, which give sketches of two normal curves with the same mean, namely 0, and standard deviations of 1 and 3, respectively. The fact that the shape of a normal curve is completely determined by its standard deviation enables one to reduce all normal curves to a standard one by a simple change of variable. For example, the curve of Fig. 3 can be made to look like the curve of Fig. 4 by changing the scale on the x-axis so that one unit on the Fig. 3 axis represents three units on the Fig. 4 axis. This corresponds, roughly, to taking

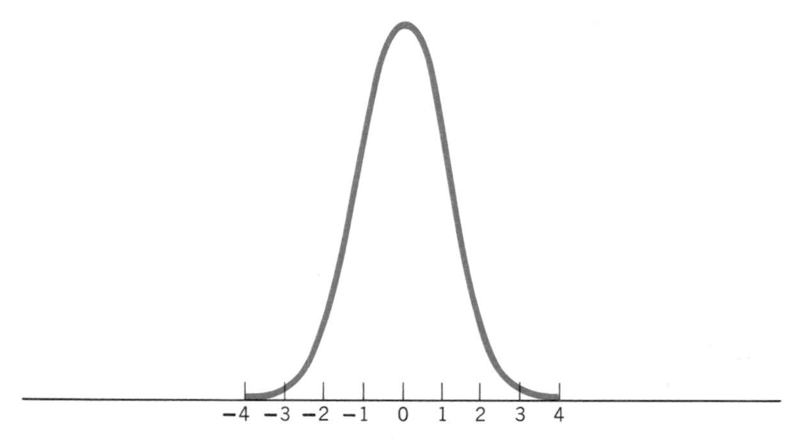

Figure 3 A normal distribution with $\mu = 0$ and $\sigma = 1$.

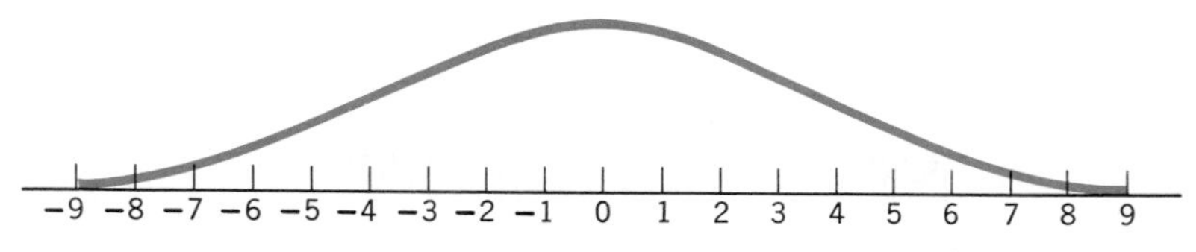

Figure 4 A normal distribution with $\mu = 0$ and $\sigma = 3$.

the curve of Fig. 3, treating it and the area beneath it as though it were made of rubber, and stretching it out to three times its width with its area preserved. Conversely, Fig. 4 could be made to go into Fig. 3 by compressing Fig. 4 to one-third its width. This assumes that the long tails of these curves are cut off and ignored. Since the simplest normal curve with which to work is the one that has its mean at 0 and whose standard deviation is 1, other normal curves are usually reduced to this *standard* one when there is need for a reduction. Now, to any point on the x-axis of a normal curve, there corresponds a point on the x-axis of the standard normal curve, and its value can be determined by stating how many standard deviations it is away from the mean point of the curve. Thus, the point $x = 6$ on Fig. 4 corresponds to the point $x = 2$ on the standard normal curve given by Fig. 3; therefore the value $x = 6$ can be obtained from Fig. 3 by stating that it is 2 standard deviations to the right of its mean. In general, if a point x on the axis of a normal curve with mean μ and standard deviation σ corresponds to a point z on the standard normal curve, then the point x is z standard deviations to the right of μ if z is positive, and z standard deviations to the left of μ if z is negative. The relationship between these corresponding points is therefore given by the formula

$$x = \mu + z\sigma$$

Or, if z is expressed in terms of x,

$$z = \frac{x - \mu}{\sigma} \tag{7}$$

This formula enables one to find the point z on the standard normal curve that corresponds to any point x on a nonstandard normal curve. Thus, the point $x = 4$ on Fig. 4 corresponds to the point $z = (4 - 0)/3 = 1\frac{1}{3}$ on Fig. 3. By this device of expressing all x-values on a normal curve in terms of corresponding values on the standard normal curve, all normal curves can be reduced to a single standard one.

In the earlier discussions in Chapter 2 about the standard deviation, it was noted that the standard deviation is unaffected by adding a constant to the values of a set of measurements, and that it is multiplied by c if each of the measurements is multiplied by c. This same property will hold for a random variable x and its theoretical standard deviation σ. That is, the variable $x - c$ will have the same standard deviation as the variable x, but the standard deviation for cx will be c times as large as for x. In view of these properties the standard deviation of $(x - \mu)/\sigma$ will be $1/\sigma$ times the standard deviation of $x - \mu$, or of x; therefore if σ is the standard deviation of x, the standard deviation of $z = (x - \mu)/\sigma$ must be 1. Since the mean of x is μ, subtracting μ from x will give a variable $x - \mu$ whose mean is 0. The mean of the variable $z = (x - \mu)/\sigma$ is therefore also 0, because multiplying a variable by a constant multiplies the mean by that constant, and multiplying 0 by $1/\sigma$ still gives 0. Thus, the variable $z = (x - \mu)/\sigma$ will possess the mean 0 and the standard deviation 1. The change of variable given by formula (7) will therefore change any variable x to one with mean 0 and standard deviation 1. This is true whether the variable x is a normal variable or not. A variable x that has been changed to the variable z by means of formula (7) is said to be measured in *standard units* after the change has been made.

Table IV in the appendix is a table for finding the area under any part of the normal curve for the variable z, that is for the normal curve that has mean 0 and standard deviation 1. If a variable z of this type is normally distributed, it is said to possess a *standard normal distribution.* The values of z in this table are given to two decimal places, with the second decimal place determining the column to use. As an illustration, suppose one wished to find the area under this standard normal curve from $z = 0$ to $z = 1.00$. The desired area is shown geometrically in Fig. 5. In Table IV one reads down the first column until the z-value 1.0 is reached, then across the entry in the column headed .00 to find .3413. This is the desired area. Table IV gives areas only from $z = 0$ to any specified positive value of z. If areas to the left of $z = 0$ are wanted, one must use symmetry and work with the corresponding right half of the curve.

Example 1. Suppose now that one wishes to find the area under part of a normal curve with mean μ and standard deviation σ. For example, suppose one wishes to find the area from $x = \mu - \sigma$ to $x = \mu + \sigma$ in Fig. 2. It follows, by symmetry, that this area is twice the area from $x = \mu$ to $x = \mu + \sigma$. From formula (7), the values $x = \mu$ and $x = \mu + \sigma$ correspond to the values $z = 0$ and $z = 1$; consequently, the area from $x = \mu$ to $x = \mu + \sigma$ is the same as that under the standard normal curve from $z = 0$ to $z = 1$, which is shown as the shaded area in Fig. 5. Since this area was found to be .3413 from Table IV, the area from $x = \mu - \sigma$ to $x = \mu + \sigma$ is twice this number, or .6826. This, of course, is the number that gave rise to the normal distribution property (a) in (6).

Example 2. As another illustration of the use of Table IV, suppose one wishes to find the area between $x_1 = 220$ and $x_2 = 280$ for x possessing a normal distribution with $\mu = 230$ and $\sigma = 20$. The desired area is the shaded area

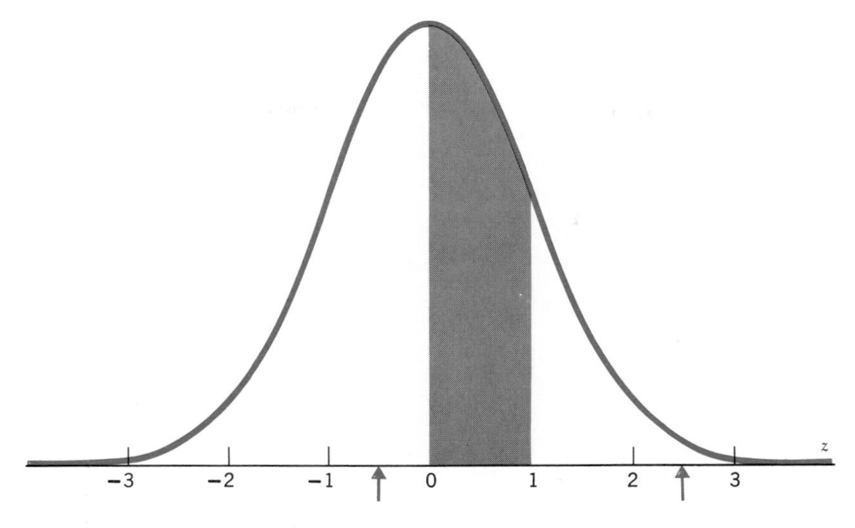

Figure 5 Standard normal distribution.

shown in Fig. 6. First it is necessary to calculate the corresponding z-values by means of (7):

$$z_1 = \frac{220 - 230}{20} = -.50$$

$$z_1 = \frac{280 - 230}{20} = 2.50$$

These two z-values are indicated by means of vertical arrows in Fig. 5.

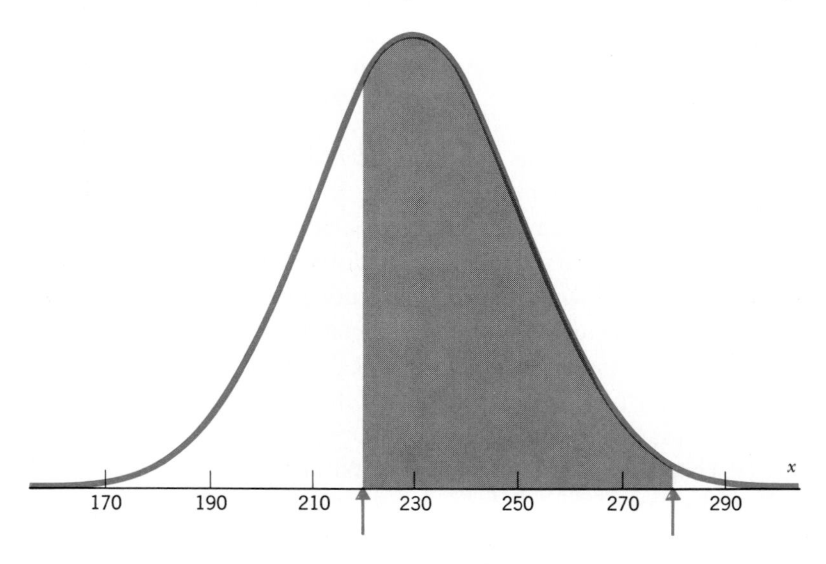

Figure 6 A particular normal distribution.

Now, the desired area is given by the area from $-.50$ to 2.50 under the standard normal curve. From Table IV, the area from $z = 0$ to $z = 2.50$ is .4938. By symmetry, the area from $z = -.50$ to $z = 0$ is the same as that from $z = 0$ to $z = .50$. The latter area is found in Table IV to be .1915; consequently, the desired area is the sum of these two areas, or .6853. Although the two normal curves in Figs. 5 and 6 would look quite different if the same x-scale were used for both, they purposely have been drawn with different x-scales so that the equivalence of corresponding areas between the arrows will be apparent.

4 NORMAL APPROXIMATION TO THE BINOMIAL DISTRIBUTION

Problems related to the binomial distribution are fairly easy to solve provided the number of trials, n, is not large. If n is large, the computations involved in using formula (1) become exceedingly lengthy; consequently, a good, simple approximation to the distribution should prove to be very useful. Such an approximation exists in the form of the proper normal distribution, with mean and standard deviation given by formulas (5). For the purpose of investigating this approximation, consider some numerical examples.

Let $n = 12$ and $p = \frac{1}{3}$ and construct the graph of the corresponding binomial distribution. By the use of formula (1), the values of $P\{x\}$ were computed, correct to three decimals, as

$$\begin{array}{lll} P\{0\} = .008 & P\{4\} = .238 & P\{8\} = .015 \\ P\{1\} = .046 & P\{5\} = .191 & P\{9\} = .003 \\ P\{2\} = .127 & P\{6\} = .111 & P\{10\} = .000 \\ P\{3\} = .212 & P\{7\} = .048 & P\{11\} = .000 \\ & & P\{12\} = .000 \end{array} \tag{8}$$

Although the graph used earlier for a binomial distribution was a line graph because of the discrete character of the variable x, this distribution will be graphed as a histogram in order to compare it more readily with normal-distribution histograms. The graph of the histogram for this distribution is shown in Fig. 7. The height of any rectangle is equal to the probability given by (8) for the corresponding class mark. Since the base length of any rectangle is 1, the area of any rectangle is equal to its height, and therefore these probabilities are also given by the areas of the corresponding rectangles. The shape of this histogram resembles somewhat that of the histogram of Fig. 8, Chapter 4, which has a normal curve fitted to it; therefore, it appears that this histogram could also be fitted fairly well by the proper normal curve.

Since a normal distribution curve is completely determined by its mean and its standard deviation, the natural normal curve to use here is the one with the

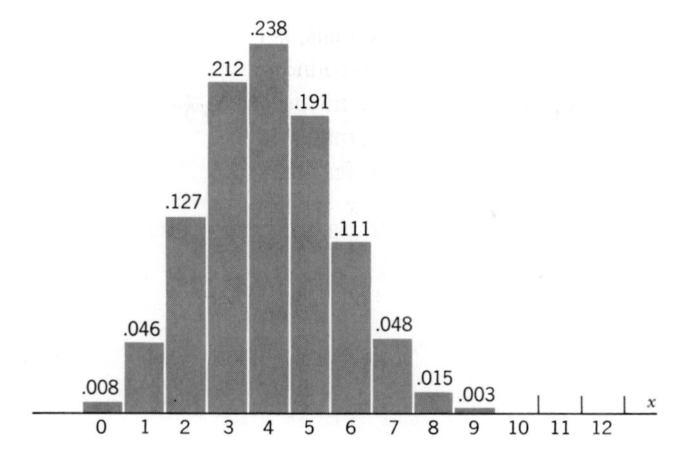

Figure 7 Binomial distribution for $p = \frac{1}{3}$ and $n = 12$.

same mean and standard deviation as the binomial distribution. From the formulas given in (5), it follows that $\mu = 12 \cdot \frac{1}{3} = 4$ and $\sigma = \sqrt{12 \cdot \frac{1}{3} \cdot \frac{2}{3}} = 1.63$. A normal curve with this mean and standard deviation was superimposed on Fig. 7 to give Fig. 8. It appears that the fit is fairly good in spite of the fact that $n = 12$ is a small value of n and advanced theory promises a good fit only for large values of n.

As a test of the accuracy of the normal-curve approximation here and as an illustration of how to use normal-curve methods for approximating binomial probabilities, consider a few problems related to Fig. 8.

Example 1. If the probability that a marksman will hit a target is $\frac{1}{3}$ and if he takes 12 shots, what is the probability that he will score at least 6 hits? The

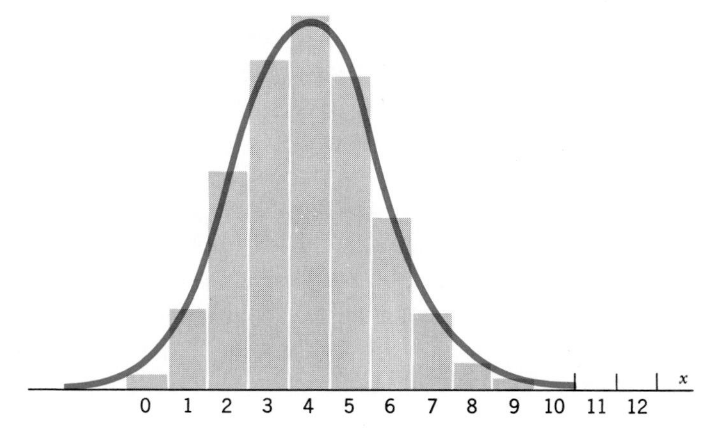

Figure 8 Binomial distribution for $p = \frac{1}{3}$ and $n = 12$, with a fitted normal curve.

exact answer, correct to three decimals, is obtained by adding the values in (8) from $x = 6$ to $x = 12$, which is found to be .177. Geometrically, this answer is the area of that part of the histogram in Fig. 8 lying to the right of $x = 5.5$. Therefore, to approximate this probability by normal-curve methods, it is merely necessary to find the area under that part of the fitted normal curve lying to the right of $x = 5.5$. Since the fitted normal curve has $\mu = 4$ and $\sigma = 1.63$, it follows that the corresponding point on the standard normal curve is

$$z = \frac{x - \mu}{\sigma} = \frac{5.5 - 4}{1.63} = 0.92$$

Now, from Table IV, the area between $z = 0$ and $z = 0.92$ is .321. Hence the area to the right of $z = 0.92$ is $.500 - .321 = .179$. This is the desired approximation to the probability of getting at least 6 hits. Since the exact answer was computed to be .177, the normal-curve approximation here is certainly good.

Example 2. To test the accuracy of normal-curve methods over a shorter interval, calculate the probability that the above marksman will score precisely six hits in twelve shots. From (8) the answer, correct to three decimals, is .111. Since this is equal to the area of the rectangle whose base runs from 5.5 to 6.5, to approximate this answer it is necessary to find the area under the fitted normal curve between $x_1 = 5.5$ and $x_2 = 6.5$. Thus, by calculating the z-values and using Table IV,

$$z_2 = \frac{6.5 - 4}{1.63} = 1.53, \qquad A_2 = .4370$$

$$z_1 = \frac{5.5 - 4}{1.63} = 0.92, \qquad A_1 = .3212$$

Subtracting these two areas gives .116, which, compared to the exact probability value of .111, is also good. From these two examples it appears that normal-curve methods give good approximations even for some situations, such as the one considered here, in which n is not very large.

Suppose, now, that the value of $p = \frac{1}{3}$ is not changed but n is allowed to increase in size. The resulting histogram, like the one in Fig. 8, will move off to the right, spread out, and decrease in height. It is difficult to inspect such a histogram and observe whether the proper normal curve would fit it well. These undesired changes in the histogram can be prevented by shifting to the corresponding variable in standard units. From (7) this means graphing the histogram for the variable

$$z = \frac{x - \mu}{\sigma}$$

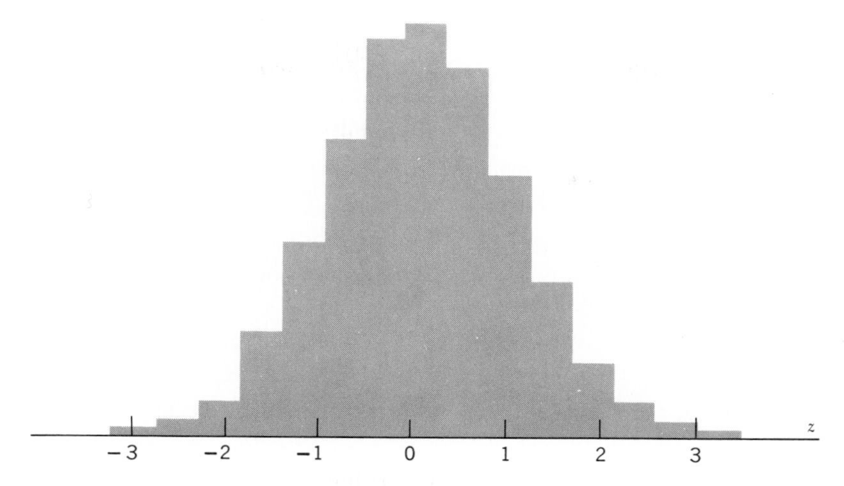

Figure 9 Binomial distribution of $(x - np)/\sqrt{npq}$ for $p = \frac{1}{3}$ and $n = 24$.

When formulas (5) are used, the standard variable z assumes the form

$$z = \frac{x - \text{np}}{\sqrt{npq}} \tag{9}$$

Since the variable z possesses a distribution with mean 0 and standard deviation 1, the histogram for z will behave itself and not go wandering off and flatten out when n becomes large, as is the case of the histogram for x.

Figures 9 and 10 show the histograms for the variable z when $p = \frac{1}{3}$ and $n = 24$ and 48, respectively. They show how rapidly the distribution of z approaches the distribution of a normal variable with mean 0 and standard deviation 1. It can be shown by advanced methods that if p is held fixed and n is allowed

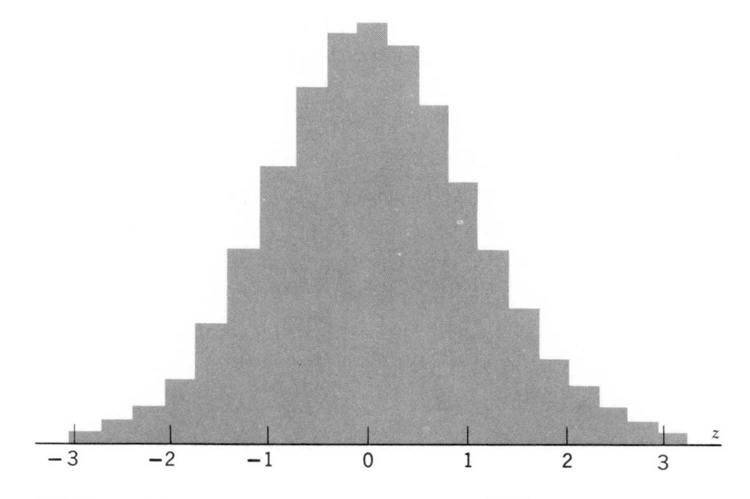

Figure 10 Binomial distribution of $(x - np)/\sqrt{npq}$ for $p = \frac{1}{3}$ and $n = 48$.

to become increasingly large, then the distribution of z will come increasingly close to the distribution of a normal variable with mean 0 and standard deviation 1. From a practical point of view, experience has shown that the approximation is fairly good as long as $np > 5$ when $p \leq \frac{1}{2}$ and $nq > 5$ when $p > \frac{1}{2}$.

The fact that the standard form of a binomial variable possesses a distribution approaching that of a standard normal variable implies that the binomial variable x possesses a histogram that can be fitted well by the proper normal curve when n is large. The proper normal curve is, of course, the one with mean and standard deviation given by formulas (5).

There are numerous occasions when it is more convenient to work with the proportion of successes, x/n, in n trials than with the actual number of successes, x. If the numerator and denominator in (9) are divided by n, then z will assume the form

$$z = \frac{\hat{p} - p}{\sqrt{\dfrac{pq}{n}}} \tag{10}$$

where the proportion of successes, x/n, is denoted by $\hat{p}$. The value of z has not changed, only the form in which it is written; consequently, z still possesses an approximate normal distribution with mean 0 and unit standard deviation when n is large. This implies that the proportion of successes, $\hat{p}$, possesses a histogram that can be fitted well by the proper normal curve when n is large. The proper normal curve is now the one with mean and standard deviation given by the formulas

$$\begin{aligned} \mu_{\hat{p}} &= p \\ \sigma_{\hat{p}} &= \sqrt{\frac{pq}{n}} \end{aligned} \tag{11}$$

Example 1. As an illustration of the use of the normal-curve approximation to the binomial distribution in the form (10), consider the following problem. Suppose a politician claims that a survey in his district showed that 60 percent of his constituents agreed with his vote on an important piece of legislation. If it is assumed temporarily that this percentage is correct, and if an impartial sample of 400 voters is taken in his district, what is the probability that the sample will yield less than 50 percent in agreement?

If it is assumed that taking a sample of 400 voters is like playing a game of chance 400 times for which the probability of success in a single game is .6, this problem can be treated as a binomial distribution problem with $p = .6$ and $n = 400$. For such a large value of n the normal curve approximation will be excellent. Using formula (10),

$$z = \frac{.5 - .6}{\sqrt{\frac{(.6)(.4)}{400}}} = -4.08$$

Now the sample proportion, $\hat{p}$, will be less than .5 provided that z is less than -4.08. The probability that $z < -4.08$ is by symmetry equal to the probability that $z > 4.08$, which is considered too small to be worth listing in Table IV. Thus, if it should happen that less than 50 percent of the sample favored the politician, his claim of 60 percent backing would certainly be discredited.

Objections may be raised, and rightfully so, that getting a sample of 400 voters is not equivalent to playing a game of chance 400 times. There are questions concerning the independence of the trials and the constancy of the probability that must be answered before one can be thoroughly happy with the binomial distribution model for this problem.

Example 2. As a second illustration of the use of formula (10), the following problem will be solved. A manufacturing company maintains the quality of its product by requiring its production manager to stop production whenever an inspector finds that more than 5 percent of the items coming off a production line are defective. If past experience shows that at most 3 percent of such items are defective under normal operating conditions, what is the probability that an inspector who examines 200 items will find that more than 5 percent are defective?

Here $n = 200$, $p = .03$, and

$$z = \frac{.05 - .03}{\sqrt{\frac{(.03)(.97)}{200}}} = 1.66$$

By Table IV, $P\{\hat{p} > .05\} = P\{z > 1.66\} = .05$; hence 5 percent of the time the inspector will claim that the quality has deteriorated more than the allowable amount, even though there has been no change in production quality.

It should not be assumed from the preceding examples that all binomial distribution problems can be treated satisfactorily by means of the normal approximation even though n may be fairly large. For example, if $n = 80$ and $p = \frac{1}{20}$, calculation of $P\{x\}$ values and a graph of those values will show that the distribution is not sufficiently symmetrical to permit a good fit by a normal curve. Since the mean is 4 here, and x cannot assume negative values, too much of the distribution is concentrated near zero for symmetry. Considerations such as these gave rise to the empirical rule for a good approximation, stated in the paragraph following (9).

5 THE POISSON DISTRIBUTION

The binomial distribution is the most useful of the well-known discrete variable distributions for solving practical problems, whereas the normal distribution is the most useful of the continuous variable distributions. One of the other discrete variable distributions that is needed quite often is the Poisson distribution, named after a Swiss mathematician who lived in the early part of the last century. It is useful for solving problems of the binomial type when n is large but p is so small that the normal-curve approximation to the binomial distribution is unsatisfactory. For example, suppose a certain machine turns out parts for an assembly line and, on the average, 1 percent of those parts are defective. If the parts are packaged in boxes of 300 each and one wishes to calculate the probability that a box will contain at most 10 defective items, then the normal approximation solution will be of questionable accuracy because the mean here is only 3 and the empirical rule given earlier requires the mean to be at least as large as 5.

The formula for the Poisson distribution that corresponds to formula (1) for the binomial distribution is the following one.

(12) ***Poisson Distribution.*** $P\{x\} = \dfrac{e^{-\mu}\mu^x}{x!}$

The possible values of the random variable x are 0, 1, 2, ... Values of $P\{x\}$ are readily calculated by means of Table XI in the appendix, which gives values of $e^{-\mu}$ corresponding to various values of μ.

This formula was obtained by considering the limiting form of (1) when n is allowed to become infinite and p approaches zero in such a manner that $\mu = np$ remains fixed. Thus, it is to be expected that the Poisson distribution will be an excellent approximation to the binomial distribution when n is very large and p is very small. For the purpose of observing the accuracy of the approximation, calculations were made of $P\{x\}$ for various values of x by means of both formula (1) and formula (12) for the special binomial problem in which $n = 64$ and $p = \frac{1}{64}$. Since $\mu = np = 64 \cdot \frac{1}{64} = 1$, the problem is to compare values of $P\{x\}$ given by the binomial formula

$$P\{x\} = \binom{64}{x}\left(\frac{1}{64}\right)^x\left(\frac{63}{64}\right)^{64-x}$$

and its Poisson approximation

$$P\{x\} = \frac{e^{-1}1^x}{x!} = \frac{e^{-1}}{x!} = \frac{.368}{x!}$$

Computations yield the results shown in Table 6, wherein values of x larger than 7 were ignored because the probabilities were less than .0001. It will be observed that the Poisson approximation is excellent. Because of the simplicity of calculating Poisson probabilities as contrasted to calculating the corresponding binomial probabilities for problems of this type, it should be clear that the Poisson distribution is very useful for treating problems in which n is large but p is very small.

In addition to serving as an approximation for certain classes of binomial distributions, the Poisson distribution can also be used to solve problems that are concerned with rare occurrences in time or space. For example, the number of typing errors per page, the number of meteorites found on an acre of desert land, the number of flaws in a yard of woolen cloth, and the number of telephone calls received at a switchboard per minute have all been found to possess a Poisson distribution, at least to a good approximation. One of the early pioneers in statistics who helped promote the use of the Poisson distribution did so by showing that the number of deaths from the kick of a horse per year per army cavalry corps in the Prussian army during the period 1875–1894 followed a Poisson distribution with high precision.

Another class of problems closely related to the preceding kind and which can often be solved with the aid of the Poisson distribution are *queueing problems.* They concern themselves with determining how many customers are likely to be waiting in line for service and how long they must wait before being served. For example, the manager of a supermarket would undoubtedly wish to know how many customers are likely to arrive at a checkout stand during a given time interval of a busy period so that he can arrange to have enough checkers available at that time. Similarly, it is important for a bank to know how many individuals are likely to arrive at a teller's window during a given period of time so that it can be prepared to serve them. If customers are required to wait too long for service, they will take their business elsewhere. The Poisson distribution has been found to be a very useful model in such problems for calculating the probabilities of various numbers of arrivals of customers seeking service.

Table 6

x	Binomial $P\{x\}$	Poisson $P\{x\}$
0	.365	.368
1	.371	.368
2	.185	.184
3	.061	.061
4	.015	.015
5	.003	.003
6	.000	.001
7	.000	.000

Example. As an illustration of the applicability of the Poisson distribution to rare events in time or space, consider the following problem. Experience has shown that the mean number of telephone calls arriving at a switchboard during one minute is 4. If the switchboard can handle a maximum of 8 calls per minute, what is the probability that it will be unable to handle all the calls that come in during a period of one minute? The desired probability can be obtained by calculating the probability of receiving exactly 9 calls, exactly 10 calls, etc., and adding those probabilities; however, it is easier to calculate the probability of getting 8 or fewer calls and then subtracting this probability from 1. Calculations give

$$
\begin{aligned}
P\{x \leq 8\} &= P\{0\} + P\{1\} + \cdots + P\{8\} \\
&= \frac{e^{-4} \cdot 4^0}{0!} + \frac{e^{-4} \cdot 4^1}{1!} + \cdots + \frac{e^{-4} \cdot 4^8}{8!} \\
&= e^{-4}\left[1 + 4 + 8 + \frac{16}{3} + \frac{32}{3} + \frac{128}{15} + \frac{256}{45} + \frac{1024}{315} + \frac{512}{315}\right] \\
&= .979
\end{aligned}
$$

Hence $P\{x > 8\} = .021$. Although this is a very small probability and therefore the chances of missing one or more calls in a given minute are small, the mean number of missed calls during an eight-hour day may be sufficiently large to give management some concern. Much would depend on whether unanswered callers will call again, and on the economic losses that can be attributed to unanswered calls.

For Poisson problems in which the value of μ is relatively small and in which one wishes to calculate the quantity $P\{x \leq k\}$, it is not necessary to calculate the value of each term as was done in the preceding example. Table VI in the appendix will yield this sum directly. From this table the desired sum for the preceding example will be found to be .979.

For the purpose of comparing the binomial distribution and its Poisson approximation when n is not large and p is not small, a graph was made for each distribution for the case when $n = 20$ and $p = \frac{1}{5}$. These graphs are shown in Fig. 11. Although both variables are discrete and their graphs should therefore be line graphs, histograms were used instead to make the degree to which a normal curve would fit the distributions more evident.

6 THE HYPERGEOMETRIC DISTRIBUTION

The binomial distribution was derived under the assumption that there were n independent trials of the same experiment. If the experiment involves selecting n individuals or objects from a finite population, the trials may not be independent.

They will be independent provided that each time an individual or object is selected, it is returned to the population, because then the population is the same at each trial. This method of selection is called *sampling with replacement.* If however, the individual or object selected at a trial is not returned to the population, the trials will not be independent and the binomial distribution model will not be applicable. This method of selection is called *sampling without replacement.* As an illustration, suppose that a shipment of 100 spark plugs contains 10 defective and 90 good plugs and that a sample of 5 plugs is taken from the shipment and tested to determine whether the shipment satisfies quality specifications. The probability of obtaining a good plug for the first of the five sample plugs is $\frac{9}{10}$; however, the probability of getting a good plug for the second of the five sample plugs is no longer $\frac{9}{10}$. Its value will depend on whether a good or a defective plug was obtained from the first sample. For problems such as this, the sample objects are not returned to the population; therefore they involve sampling without replacement. This is the type of sampling that is customarily used in industrial problems. It is particularly useful in the field of industrial quality control.

For a large finite population the error arising from assuming that p is constant from trial to trial is very small and may be ignored, in which case the binomial-distribution model is satisfactory. However, for problems in which the population is so small that a serious error will be introduced by using the binomial distribution, it is necessary to apply a more appropriate model known as the *hypergeometric distribution.* Its formula can be obtained in the following way.

Let N denote the size of the population from which a set of n individuals is to be selected. Denote by p the proportion of individuals in this finite population who possess a given property, call it A. This means that there are Np individuals

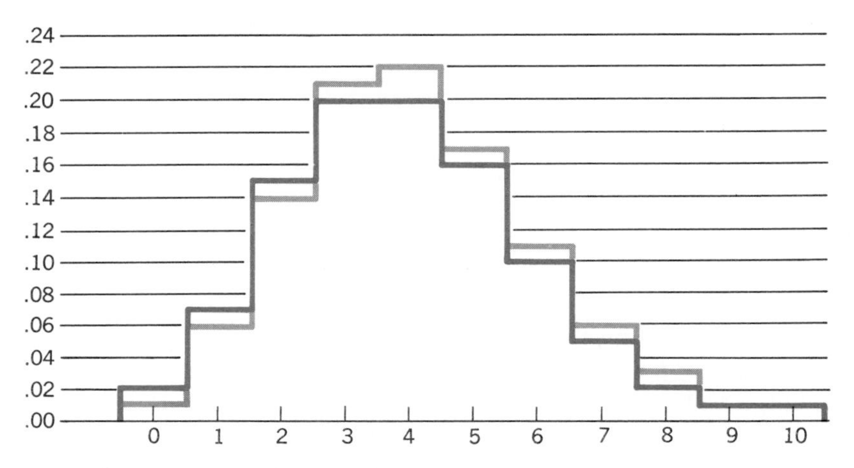

Figure 11 Graphs of the binomial (—) and Poisson (—) distributions for $n = 20$ and $p = \frac{1}{5}$.

in the population who possess property A and N–Np who do not. If x denotes the number of individuals who possess property A in the set of n individuals to be selected, then the problem is to find the probability distribution of the random variable x. Obtaining an individual with property A is equivalent to getting a success in the language of the binomial distribution.

For the purpose of calculating the probability of obtaining x such individuals, it is first necessary to calculate the total number of possible outcomes of the sampling. Since the order in which the two types of individuals are obtained is of no interest, this is a combination problem. The total number of possible outcomes is therefore equal to the number of ways of selecting n individuals from N individuals. From the techniques and notation introduced in section 11 of Chapter 3, this number is given by the value of the combination symbol $\binom{N}{n}$. Next, it is necessary to count the number of those possible outcomes that produce the desired result, which is to obtain x individuals with property A and n–x individuals without this property. Since the x individuals must come from the Np individuals having property A, the number of ways of choosing those x individuals is given by the value of the combination symbol $\binom{Np}{x}$. Similarly, since the n–x individuals must come from the N–Np individuals without property A, the number of ways of choosing those n–x individuals is given by the value of the combination symbol $\binom{N-Np}{n-x}$. The total number of ways of choosing x individuals who possess property A and n–x who do not possess the property is therefore given by the product of these two separate calculations, this is by $\binom{Np}{x}\binom{N-Np}{n-x}$. The desired probability, denoted by $P\{x\}$, is obtained by dividing the number of outcomes that produce the desired result by the total number of possible outcomes. It therefore follows from the preceding calculations that the probability of obtaining x individuals with property A may be expressed in the following manner:

(13) ***Hypergeometric Distribution.*** $$P\{x\} = \frac{\binom{Np}{x}\binom{N-Np}{n-x}}{\binom{N}{n}}$$

Calculations with this formula will show that when n is only a small percentage of N, the value of N must be quite small before there will be any appreciable difference between the values given by this formula and the binomial formula (1). As an illustration, suppose the shipment of spark plugs discussed earlier will be accepted if at most one defective plug is found in the sample of five. The problem is to calculate the probability that the shipment will be accepted. Using formula (13), this probability is given by

$$P\{x \le 1\} = \sum_{x=0}^{1} \frac{\binom{10}{x}\binom{90}{5-x}}{\binom{100}{5}}$$

$$= \frac{\binom{90}{5} + \binom{10}{1}\binom{90}{4}}{\binom{100}{5}}$$

Using formula (13) of Chapter 3 and some algebra produces the answer:

$$P\{x \le 1\} = .923$$

If the binomial formula (1) is employed, this probability assumes the value

$$P\{x \le 1\} = \sum_{x=0}^{1} \frac{5!}{x!(5-x)!}\left(\frac{1}{10}\right)^x\left(\frac{9}{10}\right)^{5-x}$$

$$= \left(\frac{9}{10}\right)^5 + 5\left(\frac{1}{10}\right)\left(\frac{9}{10}\right)^4$$

$$= \left(\frac{9}{10}\right)^4 \frac{14}{10}$$

$$= .919$$

The difference in these values is slight and is typical of the kind of differences to be expected when n is fairly small and N is fairly large.

Some extensive algebraic computations will show that the mean and the standard deviation of the hypergeometric distribution are given by the formulas

$$\mu = np$$

$$\sigma = \sqrt{npq\frac{N-n}{N-1}} \tag{14}$$

These formulas demonstrate the fact that the mean is the same whether sampling occurs with or without replacement, but that the standard deviation for sampling without replacement is smaller than that with replacement by the factor $\sqrt{\frac{N-n}{N-1}}$.

No graph of a hypergeometric distribution is included here because it would look like that of the corresponding binomial distribution unless N were very small.

7 THE EXPONENTIAL DISTRIBUTION

A continuous variable distribution that has proved very useful for solving certain classes of practical problems is the *exponential distribution.* It is given by the following formula.

(15) ***Exponential Distribution.*** $f(x) = \dfrac{e^{-x/\beta}}{\beta}, \qquad x > 0$

The parameter β is the mean of the distribution. A graph of $f(x)$ for $\beta = 1$ and $\beta = 2$ is shown in Fig. 12. The exponential distribution has been found to represent the distribution of such random variables as the length of time a customer remains in a store, the time before the next breakdown of a machine that has just been repaired, the length of life of various business firms, and the demand of a product at various price levels.

Example 1. As an illustration of the applicability of the exponential distribution, suppose that a manufacturer of television tubes has found by experience that his standard tube lasts on the average 2 years. If he guarantees his tubes to last 1 year, what proportion of his customers will be eligible for some adjustment because their tubes fail before 1 year? If x denotes the life of a tube in years and if it possesses an exponential distribution, then the probability distribution of x is given by $f(x) = \frac{1}{2}e^{-x/2}$. The problem therefore is to calculate $P\{x \leq 1\}$. This is given by the area under the graph of $f(x)$ from $x = 0$ to $x = 1$. Such areas are readily obtained by means of calculus, and for the graph of $f(x)$ whose equation is given by (15), it is given by the formula

(16) $$A(x) = 1 - e^{-x/\beta}$$

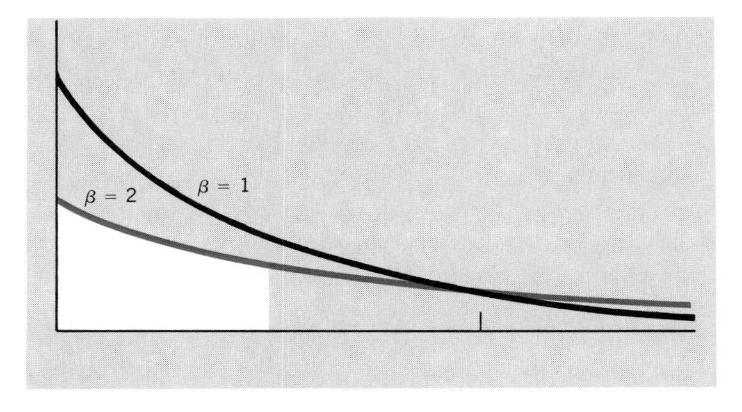

Figure 12 Exponential distributions for $\beta = 1$ and $\beta = 2$.

Here $A(x)$ denotes the area under $f(x)$ from 0 to x. Application of this formula and the table of exponentials given in Table XI of the appendix yields

$$P\{x \leq 1\} = A(1) = 1 - e^{-1/2} = .39$$

The area corresponding to this problem is shown as the white region in Fig. 12. It is clear from this result that an apparently "good" quality product, in the sense of being good on the average, may give rise to difficulties when guarantees are being set.

As indicated in the preceding section, the Poisson distribution has been found to be applicable to problems in which the random variable represents the number of occurrences of a rare event in a given time or space interval. The events are rare only in the sense that there are only a few events occurring in the given interval and the interval can be conceived of as consisting of a large number of small subinterval units. For example, if the interval is a time interval of one hour, the subintervals could be minutes or seconds. For such Poisson problems it can be shown that the random variable representing the time or distance between successive occurrences of the events possesses an exponential distribution with $\beta = 1/\mu$, where μ is the mean of the Poisson distribution. Thus, if the number of telephone calls received at a switchboard in one minute possesses a Poisson distribution with $\mu = 4$, the time between successive calls at that switchboard will possess an exponential distribution with $\beta = \frac{1}{4}$. This implies that the mean time between successive calls is 15 seconds.

Example 2. As an illustration of how the relationship between the Poisson and exponential distributions can be used in solving problems, consider the following one. Photographs by a helicopter showed that there were on the average 80 cars traveling in the fast lane of a mile stretch of a city freeway. There had been a number of accidents on that stretch in recent months attributed to "tailgating." If the distance between two cars should be at least 30 feet in this lane for safe driving, what percentage of the cars are driving too close to the car ahead for safe driving?

Assume that the cars in that mile stretch of the fast lane are distributed along that stretch with a Poisson distribution having $\mu = 80$. Experience indicates that this is often a reasonable assumption. Then the distance between two cars will possess an exponential distribution with mean $\beta = 1/\mu = 1/80$. This distance is in mile units. In units of feet, $\beta = 66$. If x denotes the distance in feet between two neighboring cars, it follows that x will possess an exponential distribution given by

$$f(x) = \frac{e^{-x/66}}{66}, \qquad x > 0$$

From (16), the probability that x will be less than 30 is given by

$$P\{x < 30\} = A(30) = 1 - e^{-30/66}$$
$$= 1 - e^{-.455} = .37$$

Hence, about 37 percent of the drivers are driving too close to the car in front of them for safe driving.

Example 3. As an illustration related to queueing theory, consider the following problem. If the average number of customers who arrive at a checkout stand during a five-minute interval is 10, what is the probability that at least a two-minute interval will occur without any customers appearing?

Since the number of arrivals may be assumed to be a Poisson variable with the mean $\mu = 10$ for a five-minute interval, it follows that $\mu = 2$ for a one-minute interval. Consequently, the mean of the related exponential distribution that gives the time x between arrivals is given by $\beta = 1/\mu = 1/2$. The associated exponential distribution is therefore

$$f(x) = 2e^{-2x}, \qquad x > 0$$

Hence,

$$A(x) = 1 - e^{-2x}$$

The desired probability is given by

$$P\{x \geq 2\} = 1 - P\{x < 2\} = 1 - A(2)$$
$$= e^{-4} = .018$$

It is therefore highly unlikely that more than two minutes will occur between successive customer arrivals.

8 REVIEW ILLUSTRATIONS

Example 1. A box contains the following nine cards: the three, four, and five of spades, the three and four of clubs, the three and four of hearts, and the four and five of diamonds. (*a*) If two cards are to be drawn with the first card being replaced before the second drawing and x represents the number of black cards that will be obtained, find the distribution of x and graph it. (*b*) Calculate the mean and the standard deviation of the distribution in (*a*) by direct calculations and by using the binomial formulas. (*c*) If the experiment in (*a*) is to be repeated 10 times instead of 2 times with the drawn card always being replaced, and x denotes the number of black cards that will be obtained, find an expression for $P\{x\}$ which gives the distribution of x.

(*d*) Use the result in (*c*) to calculate $P\{4 \leq x \leq 5\}$. (*e*) Use the normal approximation to approximate the value found in (*d*). (*f*) Calculate the approximate probability, using the normal approximation, that the proportion of black cards in 100 trials of the experiment will be less than .6.

(*a*)

Outcome	RR	RB	BR	BB
Probability	$(\frac{4}{9})^2$	$(\frac{4}{9})(\frac{5}{9})$	$(\frac{5}{9})(\frac{4}{9})$	$(\frac{5}{9})^2$
x	0	1	1	2

$\frac{16}{81}$ $\frac{40}{81}$ $\frac{25}{81}$ x

0 1 2

(*b*)

x	$81P\{x\}$	$81xP\{x\}$	$81x^2P\{x\}$
0	16	0	0
1	40	40	40
2	25	50	100
		90	140

$$\mu = \frac{90}{81} = \frac{10}{9}$$

$$\sigma^2 = \frac{140}{81} - \left(\frac{10}{9}\right)^2 = \frac{40}{81}$$

$$\sigma = \frac{\sqrt{40}}{9} = \frac{2\sqrt{10}}{9}$$

$$\mu = np = 2 \cdot \frac{5}{9} = \frac{10}{9}$$

$$\sigma = \sqrt{npq} = \sqrt{2 \cdot \frac{5}{9} \cdot \frac{4}{9}} = \frac{2\sqrt{10}}{9}$$

(*c*) $n = 10,\ p = \frac{5}{9}$,

$$P\{x\} = \frac{10!}{x!(10-x)!}\left(\frac{5}{9}\right)^x\left(\frac{4}{9}\right)^{10-x}$$

(*d*)

$$P\{4 \leq x \leq 5\} = P\{4\} + P\{5\}$$

$$= \frac{10!}{4!6!}\left(\frac{5}{9}\right)^4\left(\frac{4}{9}\right)^6 + \frac{10!}{5!5!}\left(\frac{5}{9}\right)^5\left(\frac{4}{9}\right)^5$$

$$= .39$$

(*e*)
$$\mu = np = 10 \cdot \frac{5}{9} = \frac{50}{9} = 5.56$$

$$\sigma = \sqrt{npq} = \sqrt{10 \cdot \frac{5}{9} \cdot \frac{4}{9}} = \frac{10}{9}\sqrt{2} = 1.57$$

$$z_1 = \frac{5.50 - 5.56}{1.57} = -\frac{.06}{1.57} = -.04$$

$$z_2 = \frac{3.50 - 5.56}{1.57} = -\frac{2.06}{1.57} = -1.31$$

$$A_1 = .0160, A_2 = .4049, A_2 - A_1 = .3889$$

Hence $P\{4 \leq x \leq 5\} \doteq .3889$, or .39

(*f*)
$$z = \frac{\frac{x}{n} - p}{\sqrt{\frac{pq}{n}}} = \frac{.60 - .556}{\sqrt{\frac{\frac{5}{9} \cdot \frac{4}{9}}{100}}} = \frac{.044}{.050} = .88$$

$$A = .31; \text{ hence } P\left\{\frac{x}{100} < .6\right\} \doteq .50 + .31 = .81.$$

Example 2. A restaurant can accommodate 50 customers. Experience indicates that 10 percent of those who make a reservation will not show. Suppose that the restaurant accepts 55 reservations. Let x denote the number of customers who show. (*a*) Find an expression for $P\{x\}$ that gives the distribution of x. (*b*) Calculate the mean and the standard deviation of x by means of formulas (5). (*c*) Use the normal approximation to calculate the probability that the restaurant will be able to accommodate all the customers that show.

(*a*) The 55 reservations may be treated as 55 independent trials of an experiment for which $p = \frac{9}{10}$; hence x is a binomial variable whose distribution is given by

$$P\{x\} = \frac{55!}{x!(55 - x)!}(.9)^x(.1)^{55-x}$$

(*b*) $\mu = 55(.9) = 49.5$, $\sigma = \sqrt{55(.9)(.1)} = 2.2$.

(*c*) All customers will be accommodated if at most 50 of them show; hence the problem is to calculate $P\{x \leq 50\}$. This can be obtained by finding the area under the histogram of $P\{x\}$ to the left of 50.5. The corresponding z-value is

$$z = \frac{50.5 - 49.5}{2.2} = .45$$

Hence,

$$P\{x \leq 50\} = P\{z \leq .45\} = .67$$

Example 3. A business firm consulted its dealers to determine how their sales compared with those of a year ago. A sample of 100 dealers showed that 85 of them had increased sales. If the true proportion for all its dealers is .8, (*a*) find an expression that gives the probability that at least 85 dealers out of 100 will report an increase. (*b*) Use the normal curve approximation to evaluate the probability in (*a*).

(*a*)
$$P\{x \geq 85\} = \sum_{x=85}^{100} \frac{100!}{x!(100-x)!}(.8)^x(.2)^{100-x}$$

(*b*)
$$\mu = 100(.8) = 80,$$
$$\sigma = \sqrt{100(.8)(.2)} = 4$$
$$z = \frac{84.5 - 80}{4} = 1.12$$

Hence,

$$P\{x \geq 85\} = P\{z \geq 1.12\} = .13$$

Example 4. A machine designed to drill holes in iron plates is set to drill a hole of diameter 1.02 inches in order to ensure that the hole will not be less than 1 inch in diameter. (*a*) If the standard deviation of the diameter of holes drilled by this machine is .01 inch, what percentage of the holes will be less than 1 inch? (*b*) If plates with holes wider than 1.05 inches are not usable, what percentage of the plates that contain two holes each will be rejected because at least one of the holes will be too small or too large?

(*a*) Assume that diameters are normally distributed. The problem is to calculate $P\{x < 1.00\}$, where x is this normal variable. Here

$$z = \frac{1.00 - 1.02}{.01} = -2.00$$

Hence,

$$P\{x < 1.00\} = P\{z < -2.00\} = .023$$

(*b*)
$$z = \frac{1.05 - 1.02}{.01} = 3.00$$

$$P\{x > 1.05\} = P\{z > 3.00\} = .001$$

Hence, the probability is .024 that a hole will be unsatisfactory. The probability that both of the holes will be satisfactory is $(.976)(.976) = .95$. Therefore 5 percent of the plates will be rejected.

Example 5. A seaport was visited by 180 ships during the past year. Assuming that ship arrivals are independent and that the probability of an arrival on any given day is the same for all days of the year, what is the probability that two or more ships will arrive at this port on July 4?

It appears that the Poisson distribution will be appropriate here because the event of interest, the number of ship arrivals per day, is a fairly rare event. Since $\mu = 180(\frac{1}{365}) \doteq .5$, the Poisson model to use is given by

$$P\{x\} = \frac{e^{-.5}(.5)^x}{x!}$$

Since $P\{x \geq 2\} = 1 - P\{x \leq 1\}$, it suffices to calculate $P\{0\} + P\{1\}$. This is given by

$$e^{-.5} + e^{-.5}(.5) = 1.5e^{-.5} = .91$$

Hence, $P\{x \geq 2\} = 1 - .91 = .09$.

Example 6. Suppose the probability is $\frac{1}{5}$ that a new drug will cure a certain disease. If a random sample of 30 patients with this disease are given the drug, (*a*) what is the expected number of recoveries, (*b*) what is the variance of the number of recoveries, and (*c*) what are those two quantities if the sample is taken from a single hospital that has only 60 such patients?

(*a*) Letting x represent the number of recoveries, $E[x] = np = 30 \cdot \frac{1}{5} = 6$ because x is a binomial random variable.

(*b*) $\sigma^2 = npq = 30 \cdot \frac{1}{5} \cdot \frac{4}{5} = \frac{24}{5} = 4.80$.

(*c*) $E[x] = 6$ as before but since x is now a hypergeometric variable,

$$\sigma^2 = npq\frac{N-n}{N-1} = \frac{24}{5}\frac{60-30}{59} = \frac{24}{5}\frac{30}{59} = \frac{144}{59} = 2.44$$

Thus the variance is about $\frac{1}{2}$ as large as before.

Example 7. The following is a problem involving the exponential distribution. If 5 percent of the tires purchased by a cab company last less than six months and if 20 tires are purchased, calculate the probability that at most 2 of the tires will expire in six months using (*a*) the binomial distribution, (*b*) the Poisson distribution approximation, and (*c*) the normal distribution approximation. (*d*) If the mean length of life of a tire is 2 years, use the exponential distribution to calculate the probability that the life of a tire will be less than 6 months.

$$(a)\ n = 20,\ p = \frac{1}{20},\ f(x) = \frac{20!}{x!(20-x)!}\left(\frac{1}{20}\right)^x\left(\frac{19}{20}\right)^{20-x}$$

$$P\{x \leq 2\} = f(0) + f(1) + f(2)$$

$$= \left(\frac{19}{20}\right)^{20} + 20\left(\frac{1}{20}\right)\left(\frac{19}{20}\right)^{19} + 190\left(\frac{1}{20}\right)^{2}\left(\frac{19}{20}\right)^{18}$$

$$= \left(\frac{19}{20}\right)^{20}\left[1 + \frac{20}{19} + \frac{10}{19}\right] = \left(\frac{19}{20}\right)^{20}\left(\frac{49}{19}\right) = .924$$

(*b*) $\mu = np = 20\frac{1}{20} = 1, f(x) = \frac{e^{-1} \cdot 1^{x}}{x!} = \frac{e^{-1}}{x!}$

$$P\{x \leq 2\} \doteq \frac{e^{-1}}{0!} + \frac{e^{-1}}{1!} + \frac{e^{-1}}{2!} = e^{-1}\left[1 + 1 + \frac{1}{2}\right] = \frac{5}{2}e^{-1} = .920$$

(*c*) $\sigma = \sqrt{npq} = \sqrt{20 \cdot \frac{1}{20} \cdot \frac{19}{20}} = .975, z = \frac{2.5 - 1}{.975} = 1.54$

$$P\{z \leq 1.54\} \doteq .938$$

(*d*) $A(x) = 1 - e^{-x/\mu}, \mu = 2, A(\frac{1}{2}) = 1 - e^{-1/4} = .22$

EXERCISES

SECTION 1

1 A box contains three cards consisting of the two, three, and four of hearts. Two cards are drawn from the box, with the first drawn card returned to the box before the second drawing. Let x denote the sum of the numbers obtained on the two cards. Use the enumeration-of-events technique to derive the distribution of the random variable x.

2 Verify the values given in the text in Table 3 by means of formula (1).

3 Use the method of enumeration of possible outcomes with their corresponding probabilities to derive the binomial distribution for the number of heads obtained in tossing a coin 4 times. Check your results by means of formula (1).

4 Work problem 3 for the number of aces obtained in rolling a die 4 times.

5 A coin is tossed 5 times. Using formula (1), calculate the values of $P\{x\}$, where x denotes the number of heads, and graph $P\{x\}$ as a line chart.

6 If the probability that you will win a hand of bridge is $\frac{1}{3}$ and you play 6 hands, calculate the values of $P\{x\}$, where x denotes the number of wins, by means of formula (1).

7 If a football team has an even chance of winning or losing each game played during an eight-game season (assume that no ties occur), what is the probability that this team will win at least 7 games?

SECTION 2

8 For the binomial variable for which $n = 4$ and $p = \frac{1}{4}$ calculate the mean and the standard deviation and verify your results by means of formulas (5).

9 For problem 6, calculate the mean and the standard deviation for the variable x and verify your results by means of formulas (5).

10 Given $n = 25$ and the following values of p for a binomial distribution,

$$p:\quad .1\quad .2\quad .3\quad .4\quad .5\quad .6\quad .7\quad .8\quad .9$$

(*a*) Find μ and σ^2 by means of formulas (5). (*b*) For which value of p is σ^2 largest? Smallest?

11 Derive the formula $\mu = np$ for the general binomial distribution $P\{x\}$ given by (1) by writing out the terms in $\Sigma_{x=0}^{n} xP\{x\}$, then factoring out the common factor np, then calculating $\Sigma_{x=0}^{n-1} Q(x)$, where $Q(x)$ is the same as $P\{x\}$ for $n - 1$ trials, and finally recognizing that this last sum, which has the value 1, is the other factor in $\Sigma\, xP\{x\}$. The formula seems obvious but its algebraic derivation is not.

12 Consult a more advanced statistics text to observe the algebraic technique used to derive the formula $\sigma = \sqrt{npq}$.

13 Would you expect the binomial distribution to be applicable to a calculation of the probability that the stock market will rise at least 20 of the days during the next month if you have a record for the last 5 years of the percentage of days that it did rise? Explain.

14 Explain why it would not be strictly correct to apply the binomial distribution to a calculation of the probability that it will rain at least 10 days during next January if each day in January is treated as a trial of an event and one has a record of the percentage of rainy days in January.

SECTION 3

15 Given that x is normally distributed with mean 10 and standard deviation 2, use Table IV in the appendix to calculate the probability that (*a*) $x > 12$, (*b*) $x > 11$, (*c*) $x < 9$, (*d*) $x < 9.5$, (*e*) $9 < x < 12$.

16 Assuming that stature (x) of college males is normally distributed with mean 69 inches and standard deviation 3 inches, use Table IV to calculate the probability that (*a*) $x < 65$ inches, (*b*) 65 inches $< x <$ 70 inches.

17 Suppose your score on an examination in standard units (z) is .8 and scores are assumed to be normally distributed. What percentage of the students would be expected to score higher than you?

18 The intelligence quotient (I.Q.) is approximately normally distributed with a mean of 100 and a standard deviation of 16. What is the probability that a randomly selected individual will have an I.Q. (*a*) less than 80, (*b*) greater than 140, (*c*) between 95 and 105? (*d*) What central values will include about 50 percent of individuals?

19 A food processor states on the labels of his product that the net weight is at least 16 ounces. The filling machine cannot meter the food into the cans exactly and it was found that for a given fill-setting the standard deviation was 0.2 ounces. If the processor sets up an overrun of 3 percent over the stated guaranteed amount, what fraction of this output actually fails to support his claim? Assume the fills to be normally distributed about the overrun value.

20 A high-school gym teacher announces that he grades individual athletic events by achievement relative to all his classes. If he gives 20 percent A's and if experience has shown that the mean is 4 feet 8 inches and the standard deviation is 4 inches for the high jump, how high should a student plan to jump if he expects to get an A?

21 Let x represent the weight in pounds of a king salmon caught at the mouth of a certain river and assume that x possesses a normal distribution with mean 30 and standard deviation 6. Calculate the probability that if a fisherman catches a salmon its weight will be (*a*) at least 41 pounds, (*b*) between 20 and 40 pounds, inclusive.

22 If a set of measurements is normally distributed, what percentage of them will differ from the mean by (*a*) more than half the standard deviation, (*b*) less than three-quarters of the standard deviation?

23 Suppose that the size (diameter) of a man's head is approximately normally distributed with a mean of seven inches and a standard deviation of one inch. You are in the men's hat retailing business and will stock hats in proportion to the probable size of your customer's heads. Approximately what percentage of your customers will have head sizes varying between eight and nine inches?

24 The statement was made that of one thousand 13-year-old boys, 390 have heights within 1.4 inches of the mean height, 57.3 inches. Find the value of σ here on the assumption that heights are normally distributed.

SECTION 4

25 A coin is tossed 8 times. Find the probability, both exactly by means of formula (1) and approximately by means of the normal-curve approximation, of getting (*a*) 5 heads, (*b*) at least 6 heads.

26 If the probability of your winning at pinochle is .4, find the probability, both exactly by means of formula (1) and approximately by means of the normal curve approximation, of winning 3 or more of 6 games played.

27 If 30 percent of students have defective vision, what is the probability that at least half of the members of a class of 20 students will possess defective vision? Use the normal-curve approximation.

28 If 10 percent of television picture tubes burn out before their guarantee has expired, (*a*) what is the probability that a merchant who has sold 100 such tubes will be forced to replace at least 20 of them? (*b*) What is the probability that he will replace at least 5 and not more than 15 tubes? Use the normal-curve approximation.

29 If 20 percent of the drivers in a certain city have at least 1 accident during a year's driving, what is the probability that the percentage for 300 customers of an insurance company in that city will exceed 25 percent during the next year? Use the normal-curve approximation.

30 If 10 percent of the cotter pins being manufactured are defective, what is the approximate probability that the percentage of defectives in a box of 200 will exceed 15 percent?

31 Suppose $n = 30$ and $p = \frac{1}{10}$. Calculate $P\{0\}$, and on the basis of its value argue that a good normal-curve fit to the entire distribution would not be expected here.

32 For problem 31, calculate $P\{x \leq 0\}$ by means of the normal-curve approximation.

33 Give an example of a binomial distribution for which $n > 100$ and p is not small but for which there will not be a good normal approximation.

SECTION 5

34 A study of the ownership history of 949 parcels of land on the lower east side of New York City during the period 1900–1949 provided the data below. They came from the book by Leo Grebler, *Housing Market in a Declining Area,* Columbia University Press, New York (1952). Each time a parcel changes ownership due to a foreclosure or surrender it is regarded as receiving a "transfer." The distribution of parcels by number of transfers is:

Number of transfers	Number of parcels
0	487
1	300
2	122
3	32
4	5
5	2
6	1

(*a*) Assuming the true mean transfers per parcel per 50 years is 0.700 (approximately the mean of this sample) and the Poisson is being followed, obtain the theoretical probabilities for each transfer level.

(*b*) Compute the relative frequencies observed and compare with the theoretical. Does the agreement appear good? Comment.

35 An article in a 1952 issue of the *Journal of the Royal Statistical Society* studied the accident proneness of switchmen on the South African Railways. The following table gives the distribution of the number of accidents arising in the course of a workman's employment during the first year of experience as a switchman. There were 227 switchmen of this type, all between the ages of 26 and 30.

Number of accidents, x	Number of men, f
0	121
1	85
2	19
3	1
4	0
5	0
6	1
	227

Fit a Poisson distribution to these data. Does it appear that the Poisson distribution is a reasonable model?

36 In an article published in *Biometrika* (1914), a study was made of the deaths of centenarians to determine whether such deaths were distributed randomly over time. To do so, it perused 1,000 consecutive issues of the *Utopian Seven-daily Chronicle* and obtained the following data. Here x represents the number of such deaths that occurred in one day and f represents the number of such days in the set of 1,000.

x	0	1	2	3	4	5	6	7	8
f	229	325	257	119	50	17	2	1	0

(*a*) Show that the Poisson distribution is a satisfactory model here by calculating the mean of this distribution and fitting the corresponding Poisson distribution to it.

(*b*) Calculate the variance of the data distribution and observe whether it approximately satisfies a known property of the Poisson distribution that $\sigma^2 = \mu$.

37 In the late 1930s the average number of farms per square mile in Iowa was about 4.0. A sample of farms was obtained by selecting squares of $\frac{1}{4}$-square-mile area at random and regarding those farms whose headquarters were in the sample square as belonging to the sample. If we regard the Poisson as holding in this case, use Table VI to calculate (*a*) what fraction of sample squares will contain no farm headquarters, (*b*) what fraction will contain 5 or more farms.

SECTION 6

38 In an illustration in section 11, Chapter 3, three prizes were to be given out by lot to 50 employees, 10 of whom were executives. Determine the probability distribution of the number of prizes that will be received by the executives.

39 A student preparing for an examination studies only 20 of the 25 sections that are to be covered in the examination. If the instructor selects 10 sections at random and asks one question from each selected section, what is the probability that the student will have studied at least 9 of those sections?

40 A lady declares that by tasting a cup of tea with milk added she can tell whether the milk was placed in the cup before or after the tea was added. Eight cups of tea are mixed, four in one way and four in the other way. The cups are presented in random order. The lady, after tasting all eight, is asked to pick out the four in which the milk came second. Suppose that she selected x correctly and 4-x incorrectly, and hence x correctly and 4-x incorrectly in the other four. On the assumption that she cannot discriminate, what is the probability distribution of the variable x?

41 A manufacturer of batteries has found by extensive testing that his standard automobile battery lasts on the average 2.5 years. If he guarantees his batteries to last 2 years, what fraction of his customers will be eligible for some adjustment for failure of their battery to meet the guaranteed life, assuming the exponential distribution is appropriate?

42 Following are data from a study of the lengths of lives of wholesale grocery firms in Seattle, Washington, during the period 1893 to 1951. These data came from the *Journal of Marketing* article, "Mortality of Seattle Grocery Wholesalers," published in 1953. By length of life is meant the number of years a firm stays in business. Data are based on 169 firms that started in business at some time during the period. Assuming that the average length of life of a firm in this case is 6.5 years, determine the expected fraction of firms in each of the following classes if their life lengths follow an exponential distribution. The fit appears to be quite poor; any explanation?

Length of life in years	Fraction of firms
0 up to 5	.651
5 up to 10	.160
10 up to 15	.089
15 up to 25	.047
25 and over	.053
	1.000

43 The mean life of an electric motor is 6 years. If the length of life of such a motor may be treated as an exponential variable and if the motor is guaranteed, how long should the guarantee apply in order that at most 15 percent of the motors fail before the expiration of the guarantee?

SECTION 8

44 Twelve pairs of experimental animals are placed on two different diets. The assignment of the diets to the members of each pair is made by chance. After the experiment has been completed, the difference between the weight gain of the animal on diet A and that of the animal on diet B is measured. If the difference is positive, the result is called a success. What is the probability that at least 9 successes will occur if there is no real difference in the weight-gaining properties of the two diets?

45 Breakfast food packages are listed as containing 12 ounces of cereal. The filling machine is subject to errors, with a standard deviation of .1 ounce.

(*a*) If the machine is set to fill a package with 12.1 ounces of cereal, what percentage of the packages will be at least .1 ounce short in weight?

(*b*) If the producer wishes to have at most 5 percent of his packages with a shortage of .1 ounce or more, what mean filling weight should he use?

46 An airplane can accommodate 300 passengers, of which 30 are first class and 270 are economy class. If the airline takes reservations for 30 first-class and 290 economy-class passengers and if the probability that an individual making a reservation will not show is .1, what is the probability that all passengers showing can be accommodated if first-class seats can be used for economy-class passengers?

47 A box contains four black cards numbered 1, 2, 3, and 4, three red cards numbered 2, 3, and 4, and three white cards numbered 3, 4, and 5.

(*a*) If the experiment of drawing a card from the box is repeated eight times, with the drawn card always replaced, and x denotes the number of black cards obtained, find an expression for $P\{x\}$.

(*b*) Use the result in (*a*) to calculate $P\{3 \leq x \leq 5\}$.

(*c*) Use the normal approximation to solve (*b*).

(*d*) Calculate the probability, using the normal approximation, that the proportion of black cards will be greater than $\frac{1}{2}$ if the experiment is conducted 50 times.

48 Suppose that the number of baby deliveries per day in a hospital follows a Poisson distribution and that experience over a fairly long period of time has shown that the mean number of such deliveries is 2. For what capacity should the delivery room be planned if the management wishes to be overrun no more than 3 percent of the days (about once a month)?

49 A population consists of $N = 100$ elements of which 5 possess the property of being defective. A random sample of size $n = 10$ is drawn without replacement and the number of defective elements, denoted by x, is observed. Calculate (*a*) the probability distribution for x, (*b*) the probability distribution for x if each selected element is replaced before the next selection, (*c*) the probability distribution for x if it is assumed that x possesses an approximate Poisson distribution, (*d*) the probability distribution for x if it is assumed that x possesses an approximate normal distribution. (*e*) Graph the preceding four distributions on the same graph paper, using histograms for all four distributions. (*f*) Comment about how good the various approximations to the distributions in (*a*) are. Base your comments on the graphs in (*e*).

Chapter 6 Sampling

1 INTRODUCTION

The preceding chapters have been concerned with describing frequency distributions obtained from sampling some population and with constructing probability distributions that represent the population. It has been assumed that the population distribution can be approximated with high accuracy if the sample is sufficiently large, and therefore that the sample frequency distribution satisfactorily represents the population being sampled. For example, if a manufacturer of shirts takes a large sample of neck sizes of adult males, he assumes that the distribution of those sizes for the sample will be very close to the distribution of such sizes for all adult males in the population. If this were not true, the sample could not serve as a satisfactory basis for making statistical inferences about the population being sampled.

Since the objective in sampling a population is to use the sample to draw some conclusion about the population, it is important to distinguish carefully between a sample and the population from which it was drawn. Many of the inferences to be drawn are with respect to the mean and the standard deviation of the population. The distinction then is accomplished by using the Greek letters μ and σ to represent the population mean and standard deviation and $\bar{x}$ and s to represent their sample estimates.

There are distinct advantages in relying on a sample for information about a population even if it is possible to take a complete census of the population. If the sample is properly selected it can provide timely information with remarkable accuracy at low cost, as compared to the cost of a complete census. Furthermore, there are sampling procedures, such as in industrial testing, in which the sampled items are destroyed in the testing process; therefore sampling the entire population would result in the disappearance of the population. Experience has shown that a carefully drawn sample often gives more accurate information concerning a population than will a complete census of the population because of the many errors that usually arise in taking a complete census.

Many business fields depend heavily on sampling for making decisions. For example, samples are used extensively in market surveys of consumer preferences, in TV and radio advertising effectiveness, in inspection for the quality of a product, in employee polls, in auditing, in economic indexes, and in many other activities.

If a sample is regarded as merely a portion of a population, then there are many types of samples used in various fields. A few of them are associated with the names "grab," "judgment," "volunteer," and "random." A grab sample may be obtained by grabbing items off an assembly line, or a handful of coffee beans from a bag, or the members of a ten o'clock class. In judgment sampling the sampler picks individuals he considers will represent the population well. If, for example, he were sampling college male students to determine their weight distribution, he would attempt to select a few very light and very heavy students and quite a few of moderate weight. Volunteer sampling is the type that many politicians rely upon. A volunteer sample may consist of the letters that a politician receives from his constituents, either without asking for their opinions or as the result of sending out questionnaires to them. A random sample is one that is based on probability considerations. It was defined briefly in Chapter 2. Among the various types of sampling that occur, of which the preceding ones are illustrations, only random sampling is capable of determining the precise accuracy of a sample as an estimate of the population being sampled. This property of random sampling is studied in this chapter. The next section begins with a discussion of how to draw a random sample.

2 RANDOM SAMPLING

In Chapter 2, random sampling was defined as a sampling procedure in which every member of the population has the same chance of being selected. In terms of probability this implies that the probability of any particular member being selected is $1/N$, where N denotes the number of individuals in the population. More generally, if the sample is to contain n individuals, in which case the sample is said to be of size n, the sampling is defined to be random if every combination of n individuals in the population has the same chance of being selected.

Although random sampling was advocated in Chapter 2 on the grounds that it is a method of sampling for which one can expect the sample distribution to represent the population distribution, there are more important reasons for advocating it. The most important of these is that random sampling leads to probability models for distributions. Since the conclusions to be drawn about populations by means of samples are to be based upon probabilities, samples must be selected in such a manner that the rules of probability can be applied to them. For the purpose of seeing why this is so, return to the problem discussed in the preceding chapter concerning a sample of 400 voters. There it was assumed that the sample of 400 voters could be treated as 400 independent trials of an experiment for which the probability of success in a single trial was $p = .6$. This assumption permitted the problem to be treated as a binomial distribution problem and therefore permitted the calculation of probabilities of various possible outcomes by means of the binomial distribution and its normal approximation.

If some other type of sampling had been employed to obtain the sample of 400 voters, there would be no assurance that a binomial-distribution model would be valid for calculating probabilities. For example, if a sampler started down a residential street and sampled the first 400 houses on that street he might well encounter a group of individuals whose similarity of economic status would tend to make the proportion of them favoring the candidate considerably higher, or lower, than the proportion for the entire city. One cannot, in general, make valid probability statements about the outcomes of other types of sampling methods. It is for this reason that statisticians insist that samples be randomly selected. For the present, only simple random sampling will be discussed but in a later chapter more elaborate modifications of random sampling will be introduced and described.

Random samples can be obtained by employing a game-of-chance technique. For example, if there were 10,000 voters in a city and a random sample of size 400 were to be taken, it would suffice to write the name of each voter on a slip of paper, mix the slips well in a large container, and then draw 400 slips from the container, which is thoroughly mixed after each individual drawing.

A less cumbersome and cheaper method of selecting a random sample is to employ a table of random numbers. Such a table could be constructed by writing the digits 0 to 9 on ten slips of paper, mixing them thoroughly between each drawing, replacing the drawn slip each time, and recording the digits so obtained. The resulting sequence of random digits could then be used to construct random numbers of, say, five digits each. The table of random numbers given in Table XII of the appendix could have been obtained in this fashion, but it was obtained by a more refined and foolproof method. Suppose now that the 10,000 voters are listed in a register of voters and each voter is associated with a number from 0000 to 9999. Then to obtain a sample of 400 names it is merely necessary to select 400 sets of four-digit numbers from Table XII. Since these numbers occur in columns of two digits, one would select two column elements to yield a name. There are fifty rows of numbers in each column of such numbers, therefore sixteen such columns would suffice. From the manner in which random numbers are formed, it follows that every four-digit number has the same probability of being formed at any specified place in the table; therefore, one can just as well choose the numbers systematically by reading down a column as by jumping around in the table. It could happen, of course, that one or more individuals will be selected more than once when random sampling numbers are used because these numbers were formed independently, and hence a particular four-digit number may occur more than once in the table. A few extra random numbers must then be chosen to fill out the sample.

If there had been only 7500 registered voters in the city, one would discard any four-digit number obtained from the table that is larger than 7499 because there would be no voters associated with those numbers. However, if there were only 2000 voters it would be inefficient to discard all four-digit numbers above 1999 because eighty percent of the numbers would be discarded. Instead, one could associate the even digits 2, 4, 6, 8 with the digit 0 and the odd digits 3, 5, 7, 9 with the digit 1 for the first digit of the four-digit numbers in Table XII and thereby use all of the four-digit numbers obtained. By using a little ingenuity it is possible to use random numbers to draw a random sample from any finite size population.

3 UNBIASED ESTIMATES

Since most of the statistical inferences that will be made about populations by means of samples will be concerned with such quantities as the mean and the variance, it is time to begin the study of the *accuracy* of the sample mean $\bar{x}$ and sample variance s^2 as estimates of the corresponding population parameter values μ and σ^2. This chapter will be concerned principally with studying properties of the sample mean because it is the more important of the two and is considerably easier to treat than the sample variance.

The problem of determining the accuracy of a sample estimate of a population parameter can be broken into two parts. One is concerned with determining whether the estimate is *biased.* For example, if an estimate of the average height of adult males is desired and is obtained from a random sample of college males, the value of $\bar{x}$ will tend to be larger than μ because college students tend to be taller than the rest of the older adult population. An estimate that tends to overestimate, or underestimate, a parameter value is said to be a biased estimate. A more precise definition will be given shortly.

The second part of the problem is concerned with determining how close an estimate is likely to be to the parameter it is estimating. This part of the problem will be considered in the next section.

For the purpose of discussing the property of an estimate being unbiased, consider the experiment of taking a random sample of size 10 from some population whose probability distribution has mean μ. Let $\bar{x}$ denote the mean of the sample. If this experiment were repeated a large number of times, a large number of values of $\bar{x}$ would be available for obtaining an empirical distribution of the variable $\bar{x}$. The mean of this empirical distribution would be expected to approach some fixed value as the experiment is repeated indefinitely. If the value being approached turns out to be μ, then $\bar{x}$ is said to be an unbiased estimate of μ. Stated somewhat differently, $\bar{x}$ is an unbiased estimate of μ provided that the mean of its probability distribution is μ. Since the expected value E of a random variable is the mean of its probability distribution, it follows that $\bar{x}$ will be an *unbiased estimate* of μ if, and only if, its expected value is equal to μ. This definition applies to other parameters and their corresponding estimates as well.

Now it can be shown by using the three properties of E given in Chapter 4 that

$$E[\bar{x}] = \mu \qquad \text{and} \qquad E[s^2] = \sigma^2$$

Thus, $\bar{x}$ is an unbiased estimate of μ, and s^2 is an unbiased estimate of σ^2. This latter fact is the justification for dividing $\Sigma_{i=1}^{k} (x_i - \bar{x})^2 f_i$ by $n - 1$ rather than by n in the definition of s^2 given by formula (8), Chapter 2. If n had been used in the denominator, s^2 would have been a biased estimate of σ^2 and would have produced an estimate that is smaller than σ^2 on the average.

4 THE DISTRIBUTION OF $\bar{x}$ WHEN SAMPLING A NORMAL POPULATION

In view of the fact that $\bar{x}$ and s^2 are unbiased estimates of μ and σ^2, the study of their accuracy is reduced to the study of how those estimates vary about μ and σ^2, respectively, under repeated sampling experiments. Since $\bar{x}$ is more useful than s^2 in statistical inference problems and since its sampling distribution is easier to study than that of s^2, only the distribution of $\bar{x}$ will be studied in this chapter.

There are two basic mathematical theorems on the accuracy of $\bar{x}$. Rather than merely stating them, two sampling experiments will be carried out to observe how $\bar{x}$ varies in repeated sampling and to verify that the sampling results agree with those predicted by the theorems.

Example 1. In the first experiment, samples of size 4 will be taken from the population whose probability distribution is given by Fig. 1. This distribution is a discrete approximation to the standard normal distribution. It was obtained by using Table IV to calculate the percentage of area under the standard normal curve over unit intervals, with the middle interval centered at the origin. Incidentally, the percentages obtained from Fig. 1, when the two end-interval percentages are combined with those of their neighboring intervals, give the percentages that are often used by instructors who "grade on the curve" to determine the percentage of letter grades A, B, C, D, and F to assign.

Samples of size 4 were taken from this discrete distribution by means of the random numbers found in Table XII. First, a tabulating form of the type shown in Table 1 was constructed. By this procedure all random-number pairs were divided into seven groups according to the proportions shown in Fig. 1 and associated with the class marks of Fig. 1. For example, $x = -3$ is assigned to the pair 00, which is 1 percent of all such pairs, and the value $x = -2$ is assigned to the pairs from 01 to 06 inclusive, which include 6 per cent of all random-number pairs. Four such pairs of random numbers are read from Table XII and recorded in the proper class interval to form one experiment. The results of the first three such experiments are shown in Table 1. This experiment was repeated 100 times. Next, the mean for each experiment was calculated and recorded in the last row of Table 1, labeled $\bar{x}$.

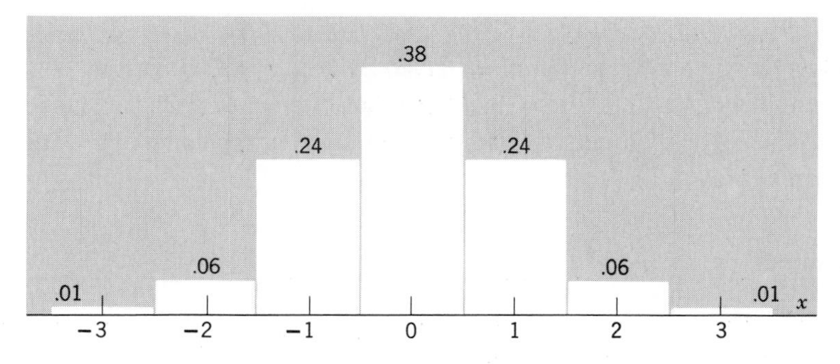

Figure 1 An approximation to the standard normal distribution.

Table 1

x	Random numbers	1	2	3	. . .
−3	00				
−2	01–06		/	/	
−1	07–30	/			
0	31–68	//	/	///	
1	69–92	/	/		
2	93–98		/		
3	99				
$\bar{x}$		0	$\frac{1}{4}$	$-\frac{1}{2}$	. . .

The values of $\bar{x}$ were next tabulated to yield the frequencies shown in Table 2, in which the third row gives the percentages in decimal fraction form of the corresponding absolute frequencies.

Finally, this frequency table was graphed as a histogram. Since total areas must be equal to 1 in comparing probability distributions, and since the class interval in this table is $\frac{1}{4}$, it is necessary to draw rectangles that are four times as high as the $f/100$ values. The resulting histogram, with area 1, is shown in Fig. 2.

On comparing Figs. 1 and 2, it is apparent that sample means based on four measurements each do not vary as much as do individual sample values. This is certainly to be expected because, for example, a large value of $\bar{x}$ would require four large values of x, and the probability of getting four large values is much smaller than the probability of getting one large value. The standard deviation of the $\bar{x}$ distribution is obviously considerably smaller than that for the x distribution. Furthermore, it is apparent that the $\bar{x}$ distribution possesses a mean that is close to 0, which is the mean of the x distribution. Finally, it appears that the distribution of $\bar{x}$, except for the difference in spread, possesses a distribution of the same approximate normal type as the x distribution.

From Fig. 1 it is clear that $\mu = 0$. Calculations based on formula (2), Chapter 4, and the probabilities given by Fig. 1 will show that $\sigma = 1.07$. If the distribution of Fig. 1 had been that of a continuous standard normal variable instead of a discrete approximation to it, the value of σ would have been exactly 1.

Table 2

$\bar{x}$	$-\frac{5}{4}$	$-\frac{4}{4}$	$-\frac{3}{4}$	$-\frac{2}{4}$	$-\frac{1}{4}$	0	$\frac{1}{4}$	$\frac{2}{4}$	$\frac{3}{4}$	$\frac{4}{4}$	$\frac{5}{4}$
f	1	4	8	11	15	19	18	12	5	4	3
$\frac{f}{100}$	.01	.04	.08	.11	.15	.19	.18	.12	.05	.04	.03

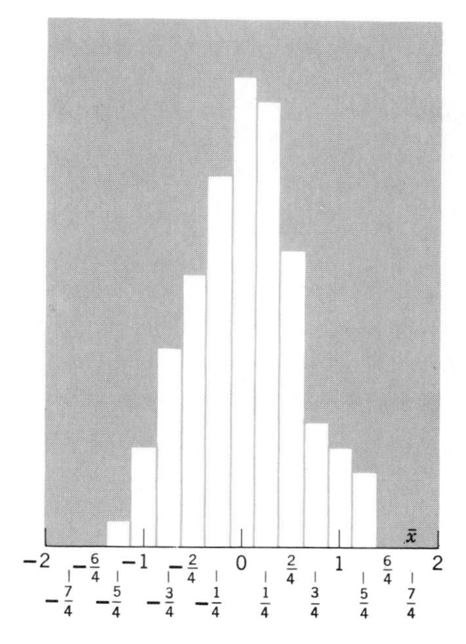

Figure 2 Distribution of $\bar{x}$ for samples of size 4 from the distribution of Figure 1.

Calculations of the mean and the standard deviation for the empirical $\bar{x}$ distribution given by Table 2 were made by means of formulas (4) and (10) of Chapter 2 and produced the values .015 and .545, respectively. Thus, the mean of the $\bar{x}$ distribution is very close to the mean, $\mu = 0$, of the x distribution, whereas the standard deviation of the $\bar{x}$ distribution is approximately one-half as large as the standard deviation, $\sigma = 1.07$, of the x distribution.

If this sampling experiment had been carried out, say, 500 times rather than just 100, irregularities such as those in Fig. 2 would disappear and it would be found that the properties of the $\bar{x}$ distribution just discussed would become increasingly apparent. Thus, it would be found that the histogram could be fitted very well with a normal curve, that the mean of the $\bar{x}$ distribution would be very close to 0, and that the standard deviation of the $\bar{x}$ distribution would have a value very close to one-half the value of the standard deviation of the x distribution. Although the samples here were taken from an approximate normal distribution as given by Fig. 1 rather than from an exact normal distribution, similar results would be obtained if the approximation were made increasingly good by choosing a very small class interval, rather than an interval of length 1 as in Fig. 1.

Fortunately, it is not necessary to carry out such repeated sampling experiments to arrive at the theoretical distribution for $\bar{x}$. By using the rules of probability and advanced mathematical methods, it is possible to derive the equation of the curve representing the distribution of $\bar{x}$ when the sampling is from the exact normal distribution rather than an approximation. This corresponds to what was done in Chapter 5 to arrive at the theoretical distribution for binomial x without performing any sampling experiments. It turns out that $\bar{x}$ will possess a normal distribution if x does, with the same mean as x but with a standard deviation that is $1/\sqrt{n}$ times the standard deviation of x. These mathematical results are expressed in the form of a theorem.

(1) ***Theorem 1.*** *If x possesses a normal distribution with mean μ and standard deviation σ, then the sample mean $\bar{x}$, based on a random sample of size n, will possess a normal distribution with mean μ and standard deviation $\sigma/\sqrt{n}$.*

The distribution of $\bar{x}$ given by this theorem is often called the sampling distribution of $\bar{x}$ because of its connection with repeated sampling experiments, even though it is derived by purely mathematical methods.

The results of the sampling experiment just completed appear to be in agreement with this theorem. The histogram of Fig. 2 looks like the type of histogram that one gets from samples from a normal population, and since $\sigma = 1$ and $n = 4$ here, its mean and standard deviation are in agreement with the theoretical values of 0 and $\sigma/\sqrt{4} = \frac{1}{2}$ given by the theorem. From this theorem one can draw the conclusion that the means of samples of size 4 from a normal population possess only one-half the variability about the mean of the population that the individual measurements do.

Example 2. As an illustration of how this theorem can be used to determine the accuracy of $\bar{x}$ as an estimate of μ when x possesses a normal distribution, consider the following problem. Let x represent the height of an individual selected at random from a population of adult males. Assume that x possesses a normal distribution with mean $\mu = 68$ inches and standard deviation $\sigma = 3$ inches. The geometry of this distribution is shown in Fig. 3. Data on male stature show that these assumptions are quite realistic. The problem to be solved is the following one: if a random sample of size $n = 25$ is taken from this population, what is the probability that the sample mean $\bar{x}$ will differ from the population mean by less than one inch?

Since $\mu = 68$, $\sigma = 3$, and $n = 25$, it follows from Theorem 1 that $\bar{x}$ will possess a normal distribution with mean 68 and standard deviation given by

$$\sigma_{\bar{x}} = \frac{\sigma}{\sqrt{n}} = \frac{3}{\sqrt{25}} = .6$$

The geometry of this distribution is shown in Fig. 4. Now, the error of estimate will be less than one inch if $\bar{x}$ falls inside the interval (67, 69). The

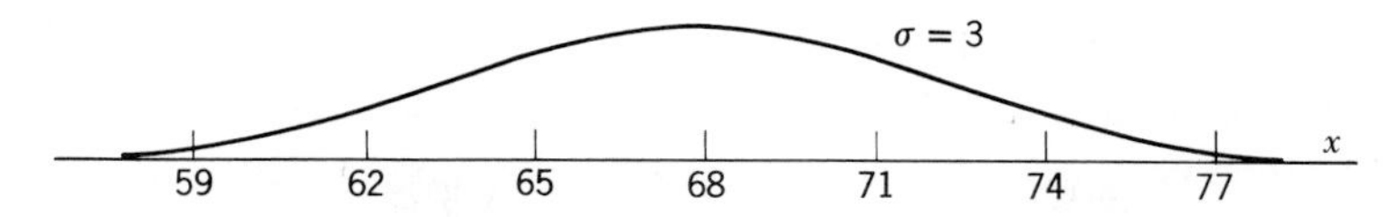

Figure 3 Normal distribution for x.

values of z corresponding to $\bar{x} = 67$ and $\bar{x} = 69$ are given by

$$z_1 = \frac{67 - 68}{.6} = -1.67 \quad \text{and} \quad z_2 = \frac{69 - 68}{.6} = 1.67$$

Since, from Table IV, the probability that z will lie between -1.67 and 1.67 is equal to .90, this value is also the probability that $\bar{x}$ will be in error by at most one inch. Thus, by taking a random sample of size 25 and using the resulting value of $\bar{x}$ to estimate the mean of the entire population, one can be quite certain that the estimate will not differ from the population mean by more than one inch.

5 THE DISTRIBUTION OF $\bar{x}$ WHEN SAMPLING A NONNORMAL POPULATION

Suppose now that the variable x does not possess a normal distribution. What then can be said about the distribution of $\bar{x}$? A number of statisticians have conducted sampling experiments with different kinds of nonnormal distributions for x to see what effect the nonnormality has on the distribution for $\bar{x}$. The surprising

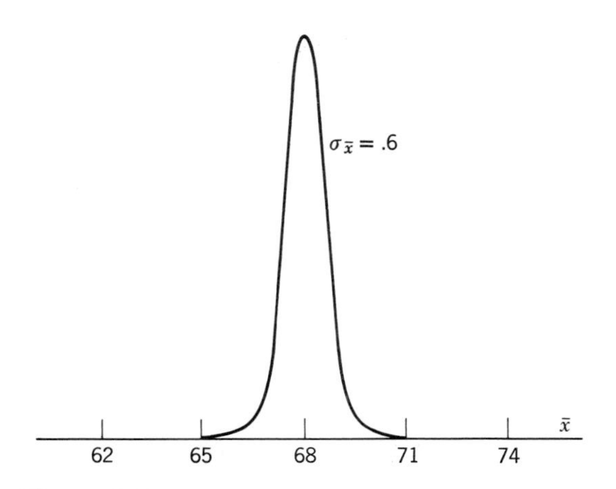

Figure 4 Normal distribution for $\bar{x}$ when $n = 25$.

result has always been that if n is larger than about 25, the distribution of $\bar{x}$ appears to be normal in spite of the population distribution chosen for x. This remarkable property of $\bar{x}$ is of much practical importance because a large share of practical problems involve samples sufficiently large to permit one to assume that $\bar{x}$ is normally distributed and thus permit the use of familiar normal-curve methods to solve problems related to means without being concerned about the nature of the population distribution.

A well-known mathematical theorem, known as a "central limit theorem," essentially states that under very mild assumptions the distribution of $\bar{x}$ will approach a normal distribution as the sample size, n, increases. This theorem can be expressed in the following manner:

(2) ***Theorem 2.*** *If x possesses a distribution with mean μ and standard deviation σ, then the sample mean $\bar{x}$, based on a random sample of size n, will have a distribution that approaches the distribution of a normal variable with mean μ and standard deviation $\sigma/\sqrt{n}$ as n becomes infinite.*

For the purpose of demonstrating that even though x is not normally distributed, $\bar{x}$ will possess an approximate normal distribution with mean μ and standard deviation $\sigma/\sqrt{n}$, provided that n is sufficiently large, the following sampling experiment was conducted. Random samples of size 10 were chosen from the population represented by the histogram in Fig. 5. The variable x here is a discrete variable that can assume only the integer values 1 to 6, with the corresponding probabilities indicated on the histogram. Theorem 2 does not require that x be a continuous variable; it may be either continuous or discrete. A histogram is used in place of a line chart merely to display the lack of normality better.

Two-digit random numbers from Table XII were divided into six groups, corresponding to the six possible values of x. The first 25 percent of those numbers, namely, all those from 00 to 24, were assigned the x-value 1. The next 25

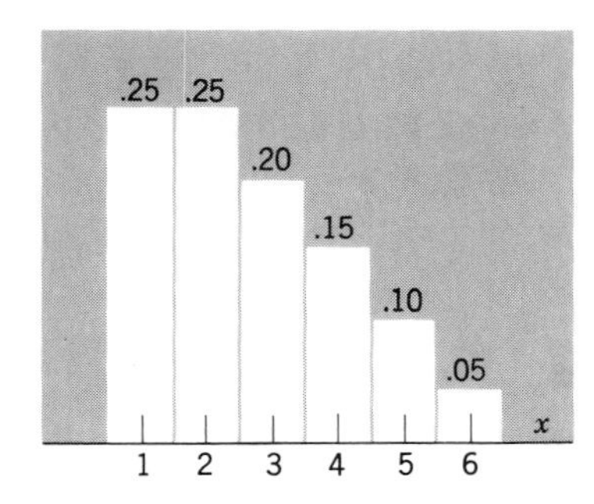

Figure 5 Population distribution for sampling experiment.

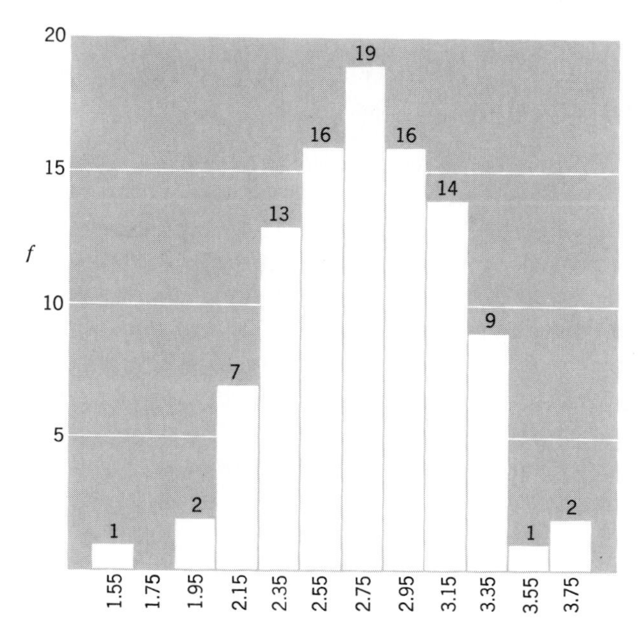

Figure 6 Histogram for 100 sample means.

percent, those from 25 to 49, were assigned the x-value 2; the next 20 percent, those from 50 to 69, the x-value 3; etc. After these assignments had been made, a column of two-digit random numbers was selected. The first 10 numbers in that column yielded the first sample of size $n = 10$ from the x population. The second set of 10 numbers in that column yielded the second sample of 10, etc. This was continued, with as many columns as needed, until 100 samples had been drawn.

In the next step of the experiment the values of $\bar{x}$ for those 100 samples were calculated. After the $\bar{x}$ values had been calculated, they were classified in a frequency table in the manner discussed in Chapter 2. The results of this classification are shown in Table 3 and Fig. 6. There was no attempt made to keep the area of the histogram equal to 1 because only the shape of the histogram is of interest here. The histogram of Fig. 6 certainly has the appearance of one that might have been obtained from sampling a normal population. Thus the claim in (2) that $\bar{x}$ should have an approximate normal distribution seems to have been fulfilled here.

Table 3

f	1	0	2	7	13	16	19	16	14	9	1	2
$\bar{x}$	1.55	1.75	1.95	2.15	2.35	2.55	2.75	2.95	3.15	3.35	3.55	3.75

Finally, the values of the mean and the standard deviation for the empirical frequency distribution of Table 3 were calculated by the methods explained in Chapter 2, with the following results:

$$\text{(3)} \qquad \begin{aligned} \text{mean of } \bar{x} \text{ distribution} &= 2.77 \\ \text{standard deviation of } \bar{x} \text{ distribution} &= .41. \end{aligned}$$

These values need to be compared with the values expected from Theorem 2. In order to make this comparison, it is necessary to know the values of μ and σ for the x population being sampled. Calculations by means of formulas (1) and (2), Chapter 4, for the probability distribution given by Fig. 5 yield the values $\mu = 2.75$ and $\sigma = 1.48$, correct to two decimal places. Since $n = 10$ in this experiement, Theorem 2 states that the approximate theoretical normal distribution for $\bar{x}$ will have as mean and standard deviation the quantities

$$\text{(4)} \qquad \begin{aligned} \mu_{\bar{x}} &= \mu = 2.75 \\ \sigma_{\bar{x}} &= \frac{\sigma}{\sqrt{n}} = \frac{1.48}{\sqrt{10}} = .47 \end{aligned}$$

The values obtained from the experiment, given in (3), appear to agree reasonably well with the theoretical values given in (4). The theoretical values in (4) are those that should be approached by the values in (3) if the sampling experiment were continued indefinitely instead of stopping after 100 trials of the experiment.

Since $n = 10$ is a small sample size and since the population distribution for x is far removed from being normal, one could hardly have expected the distribution of $\bar{x}$ to fit the theory in (2) too well, and yet it appears to do so very well.

Since $n = 25$ is sufficiently large to justify the use of Theorem 2 on the problem used to illustrate Theorem 1, it follows that it would not have been necessary to assume that stature of adult males is normally distributed in order to solve that problem.

In Chapter 5 empirical evidence was given to justify the claim that the binomial distribution can be approximated well by a normal distribution, provided that n is sufficiently large. Now Theorem 2 gives a theoretical justification for that claim, because the success ratio $\hat{p} = \dfrac{x}{n}$ can be treated as a sample mean with the sample values being 1 or 0, corresponding to success or failure, and with x denoting the sum of the sample values and n the size of the sample. Since the basic random variable here is a binomial (Bernoulli) variable based on one trial, the mean and standard deviation of it are given by $\mu = p$ and $\sigma = \sqrt{pq}$. From Theorem 2 it therefore follows that as n becomes infinite the distribution of $\hat{p}$ will approach that of a normal variable with mean p and standard deviation $\sqrt{pq}/\sqrt{n} = \sqrt{pq/n}$.

6 THE DISTRIBUTION OF THE DIFFERENCE OF TWO MEANS

Thus far only a single random variable and its properties have been studied. However, to be able to deal with a wide range of problems it is necessary to consider the sampling characteristics of two random variables. In particular, it is necessary to know what type of distribution the difference of two sample means, denoted by $\bar{x}_1 - \bar{x}_2$, will possess. For example, suppose two business machines are being compared for productivity and a record is kept of the amount of work turned out by each machine over a period of time by two equal groups of operators. If $\bar{x}_1$ and $\bar{x}_2$ denote the average amounts of work produced by the two groups, then a knowledge of the sampling distribution of the variable $\bar{x}_1 - \bar{x}_2$ is necessary in order to determine whether one machine is superior to the other.

In experimental problems of this type, the random variables $\bar{x}_1$ and $\bar{x}_2$ are obviously independent because the two groups have no influence on each other's mean productivity. Each of these variables has its own probability distribution and the probability, for example, that $\bar{x}_1$ will exceed $\bar{x}_2$ by a given amount will not depend on what value $\bar{x}_2$ assumes in the experiment. There are, however, experiments in which $\bar{x}_1$ and $\bar{x}_2$ will not be independent. For example, if the operators are not split into two comparable groups but each operator is required to alternate his use of the two types of machines from day to day for a period of two weeks, then $\bar{x}_1$ and $\bar{x}_2$ will not be independent. If a particular worker has high productivity on one machine it is likely that he will have high productivity on the other machine also. Similarly, a low producer on one machine is likely to be a low producer on the other machine as well. In general, independence will exist if the experiment consists of comparing two different groups that have been obtained by random selection.

Returning to the difference $\bar{x}_1 - \bar{x}_2$, it follows from formula (10) in Chapter 4 that

$$(5)\qquad \begin{aligned} E[\bar{x}_1 - \bar{x}_2] &= E[\bar{x}_1] - E[\bar{x}_2] \\ &= \mu_1 - \mu_2 \end{aligned}$$

Now it can be shown that the variance of the sum, or the difference, of two independent variables is equal to the sum of their individual variances; hence it follows that when $\bar{x}_1$ and $\bar{x}_2$ are independent, the variance of $\bar{x}_1 - \bar{x}_2$ is given by the formula

$$(6)\qquad \begin{aligned} \sigma_{\bar{x}_1 - \bar{x}_2}{}^2 &= \sigma_{\bar{x}_1}{}^2 + \sigma_{\bar{x}_2}{}^2 \\ &= \frac{\sigma_1^2}{n_1} + \frac{\sigma_2^2}{n_2} \end{aligned}$$

Just as was true for a single mean, the difference of two means will possess a normal distribution if the two variables x_1 and x_2 possess normal distributions.

Furthermore, if x_1 and x_2 do not possess normal distributions but n_1 and n_2 are large, the central limit theorem can be applied to show that the distribution of $\bar{x}_1 - \bar{x}_2$ will approach that of a normal variable with mean and variance given by formulas (5) and (6).

Since it was shown that $\hat{p}$, the proportion of successes in an experiment, can be treated as a sample mean to which the central limit theorem applies, it follows from the result concerning $\bar{x}_1 - \bar{x}_2$ that the difference of two independent proportions, $\hat{p}_1 - \hat{p}_2$, will possess a distribution that approaches that of a normal variable. Here the mean and variance are given by

$$E[\hat{p}_1 - \hat{p}_2] = p_1 - p_2 \tag{7}$$

and

$$\begin{aligned} \sigma_{\hat{p}_1 - \hat{p}_2}{}^2 &= \sigma_{\hat{p}_1}{}^2 + \sigma_{\hat{p}_2}{}^2 \\ &= \frac{p_1 q_1}{n_1} + \frac{p_2 q_2}{n_2} \end{aligned} \tag{8}$$

Example. As an illustration, suppose the Harris Poll and the Gallup Poll announce their findings on the popularity of a political candidate. Further, suppose Gallup says that 52% of the public favors this candidate, whereas Harris claims 48%. The sample sizes were 900 and 600, respectively. What is the probability that a difference at least this large could have arisen by chance? Since we do not know the true proportion favoring this candidate, the sample estimates will be used in place of the population values p_1 and p_2, which are equal here because the same population is being sampled by both pollsters. Then $\hat{p}_1 - \hat{p}_2$ may be treated as a normal variable with mean 0 and standard deviation

$$\sigma_{\hat{p}_1 - \hat{p}_2} = \left[\frac{(.52)(.48)}{900} + \frac{(.48)(.52)}{600}\right]^{1/2} = .026$$

The desired probability is given by $P\{| \hat{p}_1 - \hat{p}_2 | \geq .04\}$, which is equivalent to $2P\{\hat{p}_1 - \hat{p}_2 \geq .04\}$. But

$$\begin{aligned} P\{\hat{p}_1 - \hat{p}_2 \geq .04\} &= P\left\{\frac{\hat{p}_1 - \hat{p}_2}{\sigma_{\hat{p}_1 - \hat{p}_2}} \geq \frac{.04}{\sigma_{\hat{p}_1 - \hat{p}_2}}\right\} \\ &= P\left\{z \geq \frac{.04}{.026}\right\} \\ &= P\{z \geq 1.54\} \\ &= .06 \end{aligned}$$

Hence, the desired probability is .12. Thus, there is a fairly small chance that two samples of these sizes would produce a difference as large as .04. Either bad luck occurred or the sampling was not truly random.

7 THE STANDARD DEVIATION OF $\bar{X}$ WHEN SAMPLING A FINITE POPULATION

In the foregoing theory it was assumed that individuals were selected from the population by means of random sampling numbers, or that the population was so large in relation to the size of the sample that the removal of the sample had no appreciable effect on the composition of the population. Now, in real-life sampling problems this is often not the case. Most sampling plans do not permit an individual to be selected twice in a given sample; consequently, if the population is not large in relation to the size of the sample, the preceding theory will not be strictly correct. The difficulty arising in such situations can be overcome by modifying the formulas for the standard deviations of $\bar{x}$ and $\hat{p}$.

If N denotes the size of the population being sampled and n denotes the size of the sample taken without replacement, then it can be shown that the formula $\sigma_{\bar{x}} = \sigma/\sqrt{n}$ must be replaced by the formula

$$\sigma_{\bar{x}} = \frac{\sigma}{\sqrt{n}}\sqrt{\frac{N-n}{N-1}} \tag{9}$$

The value of σ is the population standard deviation corresponding to formula (2), Chapter 4.

This is similar to the formula for the standard deviation of a hypergeometric distribution, which is the distribution that applies when the random variable is the number of successes in n trials and sampling occurs without replacement. That formula is formula (14) of Chapter 5.

To see the effect of the correction factor $\sqrt{(N-n)/(N-1)}$, consider sample and population sizes for which (*a*) $n = 5$ percent of N, (*b*) $n = 10$ percent of N, (*c*) $n = 20$ percent of N. Since there is seldom any point in taking samples from populations that contain less than 100 members and since $N - 1$ will differ from N by less than 1 percent then, the foregoing correction factor may be written in the approximate form

$$\sqrt{\frac{N-n}{N}} = \sqrt{1-\frac{n}{N}} \tag{10}$$

Calculations for the three cases under consideration here yield the following values for this correction factor: (*a*) .97, (*b*) .95, and (*c*) .89. From these results, it is safe to conclude that the original formula $\sigma_{\bar{x}} = \sigma/\sqrt{n}$ will be in error by less than 10 percent unless the sample constitutes at least 20 percent of the population, and therefore that one need not worry too much about the size of the population unless the sample constitutes at least 20 percent of the population. A more conservative viewpoint would be to refrain from worrying unless the sample constitutes at least 10 percent of the population.

This same correction factor should be applied to the standard deviation of a proportion given by formula (11), Chapter 5, when the population size is small enough to justify it. Thus formula (11), Chapter 5, should be replaced by the formula

$$\sigma_{\hat{p}} = \sqrt{\frac{pq}{n}}\sqrt{\frac{N-n}{N-1}}$$

As was noted above, this is the standard deviation of the hypergeometric distribution.

8 REVIEW ILLUSTRATIONS

Example 1. Assume that scores, x, on a placement test are normally distributed with mean 160 and standard deviation 20. If a sample of size 16 is taken and the value of $\bar{x}$ computed, what is the probability that (*a*) $\bar{x}$ will exceed 165, (*b*) $\bar{x}$ will be less than 150?

$$(a)\quad \sigma_{\bar{x}} = \frac{\sigma}{\sqrt{n}} = \frac{20}{\sqrt{16}} = 5, \qquad z = \frac{165 - 160}{5} = 1,$$

$$P\{\bar{x} > 165\} = P\{z > 1\} = .16$$

$$(b)\quad z = \frac{150 - 160}{5} = -2,$$

$$P\{\bar{x} < 150\} = P\{z < -2\} = .023$$

Example 2. Let x represent the grade-point average of a randomly selected student from a certain university. It is known that the distribution of x has a mean of 2.5 and a standard deviation of .4. If a sample of 36 students is taken and the value of $\bar{x}$ calculated, what is the probability that $\bar{x}$ will (*a*) be less than 2.4, (*b*) be in the interval (2.4, 2.7).

(*a*) Although grade-point averages are not normally distributed, a sample of size 36 justifies the use of Theorem 2; hence

$$\sigma_{\bar{x}} = \frac{.4}{\sqrt{36}} = .067, \qquad z = \frac{2.4 - 2.5}{.067} = -1.50$$

$$P\{\bar{x} < 2.4\} = P\{z < -1.50\} = .07$$

$$(b)\quad z_1 = \frac{2.4 - 2.5}{.067} = -1.50, \qquad z_2 = \frac{2.7 - 2.5}{.067} = 3.00,$$

$$\begin{aligned} P\{2.4 < \bar{x} < 2.7\} &= P\{-1.50 < z < 3.00\} \\ &= P\{0 < z < 1.50\} + P\{0 < z < 3.00\} \\ &= .4332 + .4987 = .93 \end{aligned}$$

Example 3. Let x be the weight in ounces of a can of coffee of a certain brand. If the mean weight is $\mu = 16.3$ ounces and the standard deviation is $\sigma = .2$ ounce, (a) calculate the probability that the sample mean $\bar{x}$ based on a sample of size $n = 10$ will be less than 16 ounces. (b) What would this probability be for $n = 1$?

$$(a)\ \sigma_{\bar{x}} = \frac{\sigma}{\sqrt{n}} = \frac{.2}{\sqrt{10}} = .063, \qquad z = \frac{16 - 16.3}{.063} = -\frac{.3}{.063} \doteq -5$$

The probability is so small that it is not listed in the tables.

$$(b)\ z = \frac{16 - 16.3}{.2} = -\frac{.3}{.2} = -1.5, \qquad P\{z < -1.5\} = .07$$

Thus, about 7 percent of the cans will weigh less than one pound.

Example 4. A fast method of counting sets of 25 pamphlets is to weigh them. Suppose the distribution of the weight of individual pamphlets has a mean of 1 ounce and a standard deviation of .05 ounce. A weighed pile of pamphlets is counted as 25 pamphlets if it has a scale reading between 24.5 and 25.5 ounces. (a) What is the probability that a pile of 24 pamphlets will be counted as 25 by this method? (b) What is the probability that a pile of 25 pamphlets will not be counted correctly by this method?

(a) Let x_i denote the weight of the ith pamphlet. Then the weight of 24 pamphlets is given by $\Sigma_{i=1}^{24} x_i$. The problem is to calculate the probability

$$P\left\{24.5 \le \sum_{i=1}^{24} x_i \le 25.5\right\}$$

This is equivalent to

$$P\left\{\frac{24.5}{24} \le \bar{x} \le \frac{25.5}{24}\right\} = P\{1.021 \le \bar{x} \le 1.0625\}$$

Since the mean of $\bar{x}$ is 1 and its standard deviation is given by

$$\sigma_{\bar{x}} = \frac{\sigma}{\sqrt{n}} = \frac{.05}{\sqrt{24}} = .0102$$

a standard variable based on $\bar{x}$ is given by

$$z = \frac{\bar{x} - 1}{\sigma_{\bar{x}}} = \frac{\bar{x} - 1}{.0102}$$

The preceding probability can be expressed in terms of z by subtracting 1 from all elements in the inequalities and then dividing them by .0102. This gives

$$P\left\{\frac{1.021 - 1}{.0102} \le \frac{\bar{x} - 1}{.0102} \le \frac{1.0625 - 1}{.0102}\right\} = P\{2.06 \le z \le 6.13\}$$

Since the sample is sufficiently large to treat $\bar{x}$ as a normal variable, Table IV may be used to calculate this probability. It gives the value .02; hence this is the desired probability. It is highly unlikely that a set of 24 pamphlets will be counted as 25 by this weighing scheme.

(*b*) The probability that a set of 25 will be accepted is given by

$$P\left\{24.5 \le \sum_{i=1}^{25} x_i \le 25.5\right\} = P\left\{\frac{24.5}{25} \le \bar{x} \le \frac{25.5}{25}\right\}$$

$$= P\{.98 \le \bar{x} \le 1.02\}$$

Since this $\bar{x}$ has mean 1 and standard deviation

$$\sigma_{\bar{x}} = \frac{.05}{\sqrt{25}} = .01$$

the corresponding standard variable z is obtained by subtracting 1 and dividing by .01. This gives

$$P\left\{\frac{.98 - 1}{.01} \le \frac{\bar{x} - 1}{.01} \le \frac{1.02 - 1}{.01}\right\} = P\{-2 < z < 2\} = .95$$

Hence, the probability is .05 that a set of 25 pamphlets will not be counted correctly by this weighing method.

The preceding problem illustrates how the sum of a set of sample values can be treated by considering the mean of the set instead.

Example 5. Television ratings are based on samples, usually consisting of about 400 viewers. Suppose that two programs, A and B, have true ratings (percentage of households having TV sets viewing the program) of 20 and 25, respectively. A survey is taken of a random sample of 400 TV households during the showing of program A and another random sample of 400 during the showing of program B to determine the program ratings. What is the probability that the results will show program A to have a higher rating?

Here $p_1 = .20$, $p_2 = .25$, and $\hat{p}_1 - \hat{p}_2$ can be treated as a normal variable with standard deviation

$$\sigma_{\hat{p}_1 - \hat{p}_2} = \sqrt{\frac{p_1 q_1}{n_1} + \frac{p_2 q_2}{n_2}} = \sqrt{\frac{(.20)(.80)}{400} + \frac{(.25)(.75)}{400}}$$

$$= .029.$$

Since $\hat{p}_1 - \hat{p}_2$ must be at least as large as zero if A is to have a higher rating than B,

$$z = \frac{(\hat{p}_1 - \hat{p}_2) - (p_1 - p_2)}{\sigma_{\hat{p}_1 - \hat{p}_2}} = \frac{0 - (.20 - .25)}{.029} = 1.72$$

Hence, $P\{\hat{p}_1 > \hat{p}_2\} = P\{z > 1.72\} = .04$.

EXERCISES

SECTION 2

1 Suggest how to take a random sample of 100 students from the students at a university.

2 Suggest how you might set up an approximate random sampling scheme for drawing samples of (*a*) trees in a forest, (*b*) potatoes in a freight car loaded with sacks of potatoes, (*c*) children of a community under 5 years of age who have had measles. In each case indicate some variable that might be studied.

3 The number of words in a book is to be determined by selecting a sample of pages and counting the number of words on those pages. What is (*a*) the random variable, (*b*) the population?

4 Give reasons why taking every tenth name from the names under the letter A in a telephone book might or might not be considered a satisfactory random-sampling scheme for studying the income distribution of adults in a city.

5 Give an illustration of a population for which taking every twenty-fifth number in its listed order would probably yield a satisfactory approximation to random sampling for studying a particular attribute of the population.

6 Airlines often leave questionnaires in the seat pockets of their planes to obtain information from their customers regarding their services. Criticize this method of obtaining information.

7 During a prolonged debate on an important bill in the United States Senate, Senator *A* received 300 letters commending him on his stand and 100 letters reprimanding him for his stand. Senator *A* considered these letters as a fair indication of public sentiment on this bill. Comment on this.

8 A business firm sent out questionnaires to a random sample of 1000 housewives in a certain city concerning their views about paper napkins. Of these, 400 replied. Would these 400 replies be satisfactory for judging the general views of housewives on napkins?

9 How could you use random numbers to take samples of wheat in a wheat field if the wheat field is a square, each side of which is 1000 feet long, and if each sample is taken by choosing a random point in the square and harvesting the wheat inside a hoop 5 feet in diameter whose center is at the random point?

10 A survey of certain characteristics of public accounting firms in a given city is to be made. A list of certified public accountants from which to select a sample is available. For the purpose of selecting accounting firms for the sample, the *N*

accountants on the list are numbered, n random numbers from a table of random numbers are selected, and the accountants associated with those random numbers are chosen. The accounting firms to which those accountants belong are then used as the desired sample. Is this a satisfactory way of taking the sample?

11 In response to a newspaper advertisement concerning a job vacancy, six men appear at the employment office when it opens. It is likely that one of the first two or three men interviewed will be hired. How would you use Table XII to obtain an impartial ordering of the interviews of those six men?

12 In studies of the activities of employees on the job the data are frequently obtained by selecting random "instants of time" and observing an individual's activity at that instant (e.g., "working at desk," "walking," "conversing with client," "on telephone," "goofing off"). Results of such studies are helpful in understanding the manner in which various jobs are carried out but may lead to changes in arrangements of offices, general efficiency, etc. Suppose a study is to be carried out over a week and the work day is 9 to 12 and 1 to 5. Describe a method for setting up a sampling scheme from which, say, 100 "random instants" could be drawn.

SECTION 4

13 Draw a sketch of the distribution of a normal variable x and that of $\bar{x}$ if $\bar{x}$ is based on a sample of size 25 and the mean and variance of x are 10 and 25, respectively.

14 Given that x is normally distributed with mean 20 and standard deviation 4, calculate the probability that the sample mean, $\bar{x}$, based on a sample of size 64, will (*a*) exceed 21, (*b*) exceed 20.5, (*c*) lie between 19 and 21, (*d*) exceed 25, (*e*) exceed 18.

15 Given that x is normally distributed with mean 30 and standard deviation 8, calculate the probability that the sample mean, $\bar{x}$, based on a sample of size 16, will (*a*) be less than 32, (*b*) exceed 36, (*c*) exceed 28, (*d*) be less than 25, (*e*) lie between 33 and 34.

16 Show that approximately 50 percent of the items in a normally distributed population with zero mean lie in the interval from -0.6745σ to 0.6745σ. The quantity $.6745\sigma$ is called the *probable deviation.*

17 Sketch on the same piece of paper the graph of a normal curve with mean 6 and standard deviation 2 and the graph of the corresponding mean curve for a sample of size 9.

18 What would the graph of the $\bar{x}$ curve in problem 17 have looked like if the sample size had been 36?

19 Verify the values of $\bar{x}$ and s given in Section 4, Table 2, of this chapter.

20 Verify the approximate value of $\sigma = 1$ used in Section 4 of the text by deleting the decimal points in Fig. 1 and treating the resulting numbers as observed frequencies for a sample of 100 from that distribution.

SECTION 5

21 If the standard deviation of weights of first-grade children is 5 pounds, what is the probability that the mean weight of a random sample of 100 such children will differ by more than 1 pound from the mean weight for all the children?

22 For the purpose of saving time, Jones decides to find the total weight of a case of 16 packages rather than weigh each package separately. He finds the mean and standard deviation of case weights to be 24.2 and 1.2 pounds, respectively, for a large sample of such cases.
(*a*) Assuming that the cases are composed of random samples of individual packages, what is the standard deviation of weight per package?
(*b*) Suppose that it was found that the result in (*a*) did not agree with earlier results when packages were weighed individually. What explanation would you give?

23 The following quotation is from the *New York Times* for November 6, 1973, concerning the discovery of Viking artifacts on the coast of northern Newfoundland. "An acid topsoil and porous subsoil combined to destroy most of the artifacts at L'Anse Aux Meadows. But enough material was found to get a series of radiocarbon measurements. The readings all cluster around 1000, with the latest dating estimated at 1080, plus or minus 70 years." What is the probable meaning of the last sentence? Discuss the possible times of Viking habitation there on the basis of this quotation.

24 Suppose that the population of men traveling in airplanes from a large city has a distribution of weights with a mean of 160 pounds and a standard deviation of 12 pounds. What is the approximate probability that an airplane with 30 passengers will have a combined passenger weight of more than 5000 pounds?

25 Suppose that the distribution of the breaking strength of individual pieces of cotton thread has a mean of 2 pounds and a standard deviation of .4 pound. What is the probability that a cord made up of 50 such threads will support a weight of 105 pounds?

26 A marketing research firm is planning a survey of families to determine their preferences for wines. It would like to base its survey on about 500 families that drink wine, at least occasionally. Since it does not know which families drink wine, it must draw a sufficiently large sample to yield 500 wine-drinking families. In the region being studied it is believed that 40 percent of the families consume wine. On that basis it decided that a sample of 1300 families should suffice.
(*a*) What is the expected number of wine-drinking families in the sample?
(*b*) What is the probability that fewer than 490 wine-drinking families will be obtained?

SECTION 6

27 In a survey of 500 TV viewers, there were 230 women who watched an average of 24.2 hours per week and 270 men who watched an average of 23.6 hours

per week. If the standard deviations of times were 6 and 7 hours, respectively, what is the standard deviation of the sample difference in the mean times watched?

28 Two machines are producing gidgets. Machine *A* has a fault rate of 10 percent, whereas machine *B*'s rate is 8 percent. If samples of 50 are taken from the output of each machine, calculate the standard deviation of the sample difference of the two fault rates.

29 A sample of 520 of the 5,000 employees of a company is taken to determine the amount of time spent in commuting to and from work. Suppose that the standard deviation of such times is 15 for both male and female employees.

(*a*) What is the standard deviation of the difference in the sample mean $\bar{x}_M - \bar{x}_F$ if the sample contains 320 males and 200 female employees?

(*b*) What is the probability that a difference of more than 8 minutes will be observed in the sample means if $\mu_M - \mu_F = 5$?

30 The population of Scholastic Aptitude Test scores has a mean of 500 and a standard deviation of 100. What is the approximate probability that the mean score of a group of 50 students will exceed the mean score of another group of 25 students by at least 10 points?

31 The national polls of Harris and Gallup contain approximately 1000 pollees each. For an issue that has a true proportion of $p = .4$ for those favoring it, what is the standard error of the difference of the sample proportions for the two polls?

32 Two individuals are having a chat. You are told that the difference in their intelligence test scores is 60 points. Is it likely that they are a random sample from the population at large? Who might they conceivably be? Records indicate that the standard deviation of I.Q. scores is approximately 13 points.

SECTION 7

33 Work problem 21 if the total number of children is only 500.

34 Work problem 14 under the assumption that the population here consists of only $N = 500$ individuals and that Theorem (1) is still applicable. Compare your results with those of problem 14.

35 Work problem 15 under the assumption that the population here consists of only $N = 100$ individuals and that Theorem (1) is still applicable. Compare your results with those of problem 15.

SECTION 8

36 Have each member of the class perform the following experiment 10 times. From Table XII in the appendix select 10 one-digit random numbers and calculate their mean. Bring these 10 experimental means to class, where the total set of experimental means may be classified, the histogram drawn, and the mean and the standard deviation computed. These results should then be compared with theory in the same manner as in the experiment in the text. The population distribution here has $\mu = 4.5$ and $\sigma = 2.87$.

37 Perform a sampling experiment of the type used to make Theorem (2) seem plausible by taking 50 samples of size 5 from the discrete distribution given by

x	0	1	2
$P\{x\}$	.4	.2	.4

Graph the histogram of the $\bar{x}$ distribution and calculate its mean and standard deviation. Calculate the mean and standard deviation of the x distribution and compare your $\bar{x}$ results with those expected from Theorem (2). Since n is very small here and the x distribution is far from being normal, you should not expect the $\bar{x}$ distribution to look too much like a normal distribution.

Chapter 7 Estimation

1 POINT AND INTERVAL ESTIMATES

The introduction in Chapter 1 stated that one of the fundamental problems of statistics is the estimation of properties of populations. Now that probability distributions have been studied, to a limited extent at least, it is possible to discuss those properties of populations that can be estimated. The two most important population distributions that have been studied thus far are the binomial distribution and the normal distribution; therefore their properties will be investigated first.

The binomial distribution given by formula (1), Chapter 5, is completely determined by the number of trials, n, and the probability of success in a single trial, p. The symbols n and p are called the *parameters* of the distribution. The values assigned to the parameters determine the particular binomial distribution desired. Since the parameters n and p completely determine the binomial distribution, any property of a binomial distribution is also completely determined by them. Furthermore, since the number of trials, n, is almost always chosen in advance in estimation problems, the problems of estimation for binomial distributions can usually be reduced to the problem of estimating p. This is also true for the hypergeometric distribution.

The normal distribution given by Fig. 2, Chapter 5, is completely determined by the two parameters μ and σ. Problems of estimation for normal populations can therefore usually be reduced to the problems of estimating μ and σ.

The Poisson and exponential distributions given by formulas (12) and (15), Chapter 5, are completely determined by the parameters μ and β, which are their means; therefore, problems of estimation for those models are problems of estimating a mean.

There are two types of estimates of parameters in common use in statistics. One is called a point estimate and the other is called an interval estimate. A *point estimate* is the familiar kind of estimate; that is, it is a number obtained from computations on the sample values that serves as an approximation to the parameter being estimated. For example, the sample proportion $\hat{p}$ of voters favoring a certain candidate is a point estimate of the population proportion p. Similarly, the sample mean $\bar{x}$ is a point estimate of the population mean μ. An *interval estimate* for a parameter is an interval, determined by two numbers obtained from computations on the sample values, that is expected to contain the value of the parameter in its interior. The interval estimate is usually constructed in such a manner that the probability of the interval's containing the parameter can be specified. The advantage of the interval estimate is that it shows how accurately the parameter is being estimated. If the length of the interval is very small, high accuracy has been achieved. Such interval estimates are called *confidence intervals.* Both point and interval estimates are determined for binomial and normal distribution parameters in this chapter.

2 ESTIMATION OF μ

Consider the following information. A manufacturer of bricks has found from experience that the crushing strength of his bricks for a given batch is approximately normally distributed. He has also found that the mean crushing strength varies from batch to batch but the standard deviation remains fairly constant at the value $\sigma = 20$. He wishes to estimate the mean crushing strength for a new batch, so he tests a random sample of 25 bricks and finds that their mean crushing strength is $\bar{x} = 300$. With these data available, three types of estimation problems will be solved.

PROBLEM 1. How accurate is $\bar{x} = 300$ as a point estimate of the batch mean μ? To solve this problem, use is made of the theory presented in the preceding chapter. From the theory given in (2), Chapter 6, it follows that the sample mean $\bar{x}$ may be assumed to be normally distributed with mean μ and standard deviation, which is often called the *standard error of the mean,* given by

$$\sigma_{\bar{x}} = \frac{\sigma}{\sqrt{n}} = \frac{20}{\sqrt{25}} = 4$$

A sketch of this distribution is shown in Fig. 1. Since the probability is .95 that a normal variable will assume some value within two standard deviations of its mean (more accurately, 1.96 standard deviations correct to two decimals by Table IV in the appendix), it follows that the probability is .95 that $\bar{x}$ will assume some value within 8 units of μ. Since $\bar{x} = 300$ is the observed value here, the manufacturer can feel quite confident that this value differs from the population value μ by less than 8 units because in the long run, in only 5 percent of such sampling experiments will the sample value $\bar{x}$ differ by more than 8 units from μ. The magnitude of the difference $\bar{x} - \mu$ is called the *error of estimate.* In terms of this language, one can say that the probability is .95 that the error of estimate will be less than 8 units. If higher probability odds were desired, one could use, say, a three-standard deviation interval on both sides of μ and then state that the probability is .997 that the error of estimate will be less than 12 units.

Since $\bar{x}$ is based on a sample it is not possible to state how close $\bar{x}$ is to the population mean μ when μ is unknown; it is only possible to state in probability language how close $\bar{x}$ is likely to be to μ. Thus, the exact error of estimate, namely $|\bar{x} - \mu|$, is known only when μ is known. Since the problem of estimation arises only when μ is unknown, one must introduce probabilities in order to discuss the magnitude of an error of estimate in statistical problems. Even though the true value of μ is not known, one can still speak of how close $\bar{x}$ is to μ and therefore one can say something about the magnitude of the error of estimate, $|\bar{x} - \mu|$, provided the statement is couched in the proper probability language.

PROBLEM 2. Suppose the manufacturer is not satisfied with the accuracy of his estimate based on the sample of 25. How large an additional sample should he take so that he can be reasonably certain, say, with a probability of .95, that his

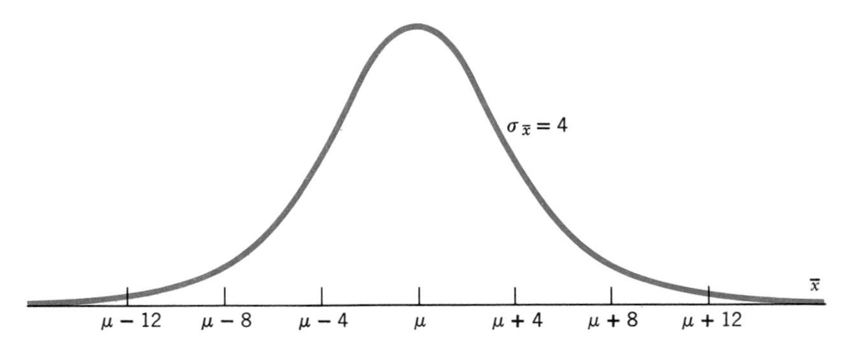

Figure 1 Distribution of $\bar{x}$ for the crushing strength of bricks.

estimate will not be in error by more than 5 units? Now, as n is increased, the normal curve for $\bar{x}$ will become taller and narrower, concentrating more and more area in the neighborhood of the mean μ. A stage will be reached when 95 percent of the area, centered at μ, is found to lie within the interval extending 5 units on both sides of μ. This value of n is the desired sample size. Since 95 percent of the central area corresponds to 1.96 standard deviations on both sides of μ, it follows that n must be such that 1.96 standard deviations for $\bar{x}$ equals 5. Thus n must satisfy the equation

$$1.96\sigma_{\bar{x}} = 5$$

Since $\sigma = 20$ here, this equation is equivalent to

$$1.96\frac{20}{\sqrt{n}} = 5$$

Solving for n gives the result $n = 61.5$. The manufacturer therefore must take an additional sample of size 37, since he has taken 25 already, in order to attain the desired accuracy of estimate.

Since 1.96 is a inconvenient number to use in equations of this type, it usually suffices to replace it by 2. The solution of the equation then becomes $n = 64$. Although this approximation does yield a difference of 2 in the answer, that is hardly a large number to worry about in a total sample of 64. Furthermore, the objective here is to learn the methods of statistics and any saving of calculating energy, which hopefully will be applied to thinking energy, is well worth the sacrifice in accuracy of computation.

If one wants something other than .95 for the probability that the error of estimate will not exceed 5, then it is necessary to replace the factor 1.96 (or 2) in the preceding equation by the proper z-value found in Table IV corresponding to the desired probability. Thus, if the probability .90 were selected, the Table IV value of z would be 1.64. This follows from the fact that 90 percent of the area of a normal curve lies within 1.64 standard deviations of the mean. The equation to be solved for n would then become

$$1.64\frac{20}{\sqrt{n}} = 5$$

The solution of this equation is $n = 43$.

Since it is bothersome to have to solve an equation like this each time, a formula that yields n more directly will be obtained. If the maximum allowable error of estimate is denoted by e and the z-value corresponding to the desired probability is denoted by z_0, then the equation that must be solved for n is given by

$$z_0\frac{\sigma}{\sqrt{n}} = e \qquad (1)$$

The solution of this equation, and therefore the desired formula, is given by

$$n = \left(\frac{z_0 \sigma}{e}\right)^2 \tag{2}$$

This formula enables one to determine how large a sample is needed in order to estimate μ to any desired degree of accuracy before a single sample has been taken, provided the value of σ is known. It is not necessary to have a preliminary sample available, as in Problem 2. If, however, one does not know σ from other sources or have a good estimate of it, then it is necessary to take a preliminary sample in order to obtain an estimate of σ that can be used in the formula for determining how large n must be.

PROBLEM 3. Consider a third type of estimation problem for this same example. What is a 95 percent confidence interval for μ based on the original sample of 25? If it is assumed that x is exactly normally distributed, it is clear from the theory given in (1), Chapter 6, or Fig. 1, that one can write

$$P\{\mu - 8 < \bar{x} < \mu + 8\} = .95 \tag{3}$$

This is an algebraic probability statement of what was stated in geometrical language in Problem 1, namely, that the probability is .95 that the point on the $\bar{x}$-axis of Fig. 1 corresponding to a sample mean $\bar{x}$ will not be more than 8 units away from the point representing the population mean μ. Now it is possible to turn this geometry, and hence the algebra, around and state that the probability is .95 that the point μ will not be more than 8 units away from the point corresponding to a sample mean $\bar{x}$. The relationship here is relative; if one point is within 8 units of a second point, the second point will be within 8 units of the first point. This reversing of the roles of the two points will now be done algebraically by the use of inequality properties.

An inequality such as $\bar{x} < \mu + 8$ can be rearranged in the same manner as an equality, except that multiplying both sides of an inequality by a negative number will reverse the inequality sign. Thus, the inequality $2 < 5$ becomes the inequality $-2 > -5$ when it is multiplied through by -1. The inequality $\bar{x} < \mu + 8$ is seen to be equivalent to the inequality $\bar{x} - 8 < \mu$ by adding -8 to both sides of the first inequality. Similarly, $\mu - 8 < \bar{x}$ is equivalent to $\mu < \bar{x} + 8$. If these two results are combined, it will be seen that the double inequality

$$\mu - 8 < \bar{x} < \mu + 8$$

is equivalent to the double inequality

$$\bar{x} - 8 < \mu < \bar{x} + 8$$

As a consequence, the probability statement (3) is equivalent to the probability statement

$$P\{\bar{x} - 8 < \mu < \bar{x} + 8\} = .95 \tag{4}$$

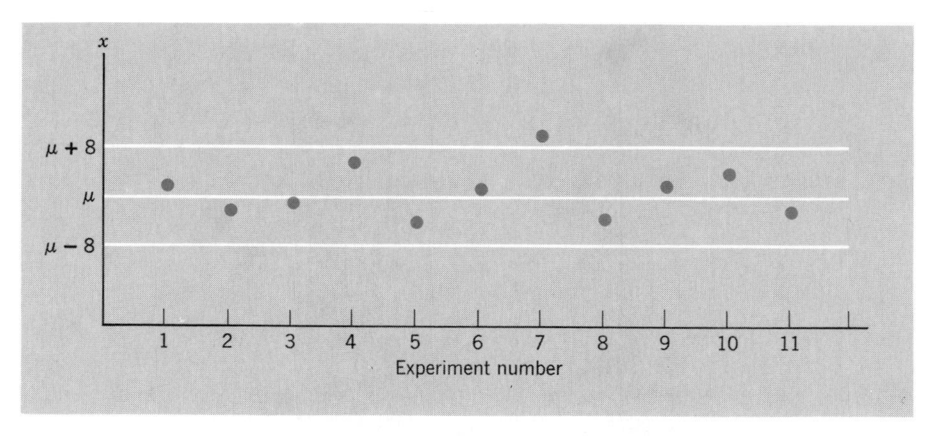

Figure 2 Repeated sampling experiments for $\bar{x}$.

In words, this says that the probability is .95 that the population mean μ will be contained inside the interval that extends from $\bar{x} - 8$ to $\bar{x} + 8$. This interval is written in the form $(\bar{x} - 8, \bar{x} + 8)$.

Although (3) and (4) are equivalent probability statements, they possess slightly different interpretations in terms of relative frequencies in repeated runs of this sampling experiment. For each such sampling experiment, a value of $\bar{x}$ is obtained. If these values of $\bar{x}$ are plotted as points, as shown in Fig. 2, then the frequency interpretation of (3) is that in such repeated sampling experiments 95 percent of the points will fall within the band shown in Fig. 2.

A frequency interpretation for (4) requires that the interval extending from $\bar{x} - 8$ to $\bar{x} + 8$, corresponding to each sampling experiment, be plotted. This has been done in Fig. 3 for the experiments that yielded Fig. 2. The frequency interpretation of (4), then, is that in such repeated sampling experiments 95 percent of the intervals will contain μ. Geometrically, it is clear from Figs. 2 and 3 that an interval in Fig. 3 will contain μ if and only if the corresponding point in Fig. 2 lies inside the band displayed there. This is very much like saying that a chalk line on the floor (μ) will be within 8 feet of you ($\bar{x}$) if and only if you are within 8 feet of the line. The advantage of the interval interpretation is that in practice one never knows what the value of μ is; otherwise there would be no point of estimating it, and therefore it is not possible to construct the band given by $\mu - 8$ and $\mu + 8$ in Fig. 2; however, it is always possible to construct the intervals given by $\bar{x} - 8$ and $\bar{x} + 8$ in Fig. 3.

Now, in practice, only one sampling experiment is conducted; therefore, only the first point and the first interval are available from Figs. 2 and 3. On the basis of this one experiment, the claim is made that the interval from $300 - 8$ to $300 + 8$, or from 292 to 308, contains the population mean μ. Using inequality symbols, this is written in the form

$$292 < \mu < 308$$

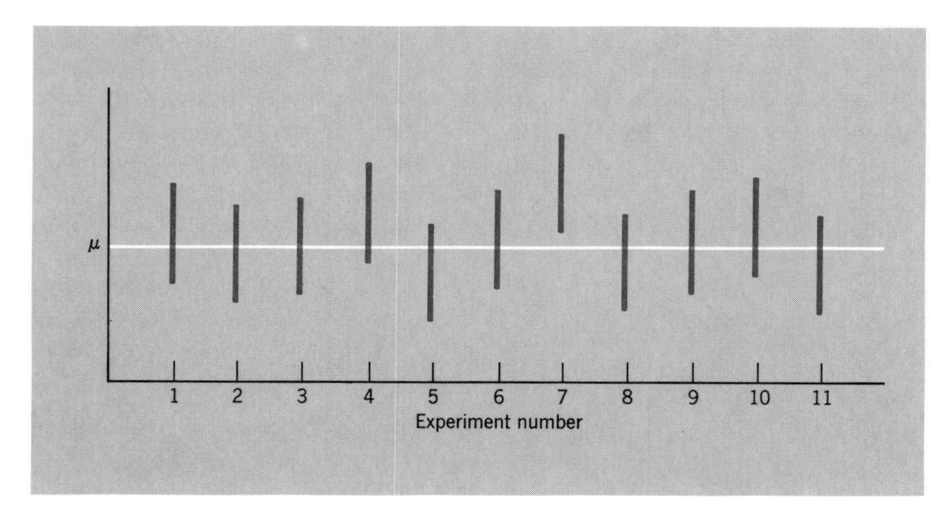

Figure 3 Intervals for repeated sampling experiments.

If for each such sampling experiment the same claim is made for the interval corresponding to that experiment, then 95 percent of such claims will be true in the long run of such experiments. In view of this property, the interval from 292 to 308 is called a 95 percent *confidence interval* for μ. The end points of the interval, namely, 292 and 308, are *confidence limits* for μ.

It should be clearly understood that one is merely betting on the correctness of the rule of procedure when applying the confidence interval technique to a given experiment. It is incorrect to make the claim that the probability is .95 that the interval from 292 to 308 will contain μ. The latter probability is either 1 or 0, depending upon whether μ does or does not lie in this fixed interval. Nontrivial probability statements are made only about variables and not about constants. It is only when one considers the variable interval from $\bar{x} - 8$ to $\bar{x} + 8$, before a numerical value of $\bar{x}$ has been inserted, that one can make probability statements such as that in (4).

The advantage of a confidence interval for μ over a point estimate of μ is that the confidence interval gives one an idea of how closely μ is being estimated, whereas the point estimate $\bar{x}$ says nothing about how good the estimate is. Thus, the confidence interval (292, 308) gives one assurance (confidence) that the true mean μ is very likely at least as large as 292 and very likely not larger than 308.

The three types of problems just solved in connection with the example introduced at the beginning of this section are the three major problems arising in the estimation of μ. They may be listed as (1) determining the accuracy of $\bar{x}$ as an estimate of μ, (2) determining the size sample needed to attain a desired accuracy of estimate of μ, and (3) determining a confidence interval for μ.

The methods for solving these problems were quite simple because it was assumed that the variable x was normally distributed and that the value of σ was known. Now, by the theory in (2), Chapter 6, it follows that it would have been safe to treat $\bar{x}$ as a normal variable even if x had not been assumed to be normally distributed, because n is large here. The problem of what to do when σ is not known is not so simple. If the sample is large it is safe to replace σ by its sample estimate s in the formulas used to solve the problems. This possibility is considered next.

Example 1. As an illustration of what to do when σ is not known but n is large, the following problem will be worked. A municipal electric company has raised its rates a substantial amount and wishes to determine what effect that will have on electric consumption. To allow customers to react to the new rates, the company waits two months after the new rates have been installed before measuring consumption. It then takes a random sample of 200 monthly bills from the set of bills about to be sent out. Suppose the results of the sampling yielded the sample values $\bar{x} = \$9.50$ and $s = \$2.10$. What is a 95 percent confidence interval for the mean of all customers? The problem is solved in the same manner as before, except that s is used in place of σ.

As before, it follows from Table IV that the probability is .95 that a standard normal variable z will satisfy the inequalities

$$-1.96 < z < 1.96$$

But if $\bar{x}$ is a normal variable, the quantity

$$\frac{\bar{x} - \mu}{\sigma_{\bar{x}}} = \frac{\bar{x} - \mu}{\sigma/\sqrt{n}} = \frac{\bar{x} - \mu}{\sigma}\sqrt{n}$$

will be a standard normal variable, and therefore the probability is .95 that it will satisfy the inequalities

$$-1.96 < \frac{\bar{x} - \mu}{\sigma}\sqrt{n} < 1.96$$

If these inequalities are solved for μ, they reduce to the following inequalities:

$$\bar{x} - 1.96\frac{\sigma}{\sqrt{n}} < \mu < \bar{x} + 1.96\frac{\sigma}{\sqrt{n}} \tag{5}$$

This result can be used as a formula for obtaining 95 percent confidence intervals for population means.

For the problem being considered, $n = 200$, $\bar{x} = 9.50$, and $s = 2.10$. Since the value of σ is not known here, it must be approximated by its sample estimate $s = 2.10$. If these values are substituted into (5), it will assume the form

$$9.50 - 1.96\frac{2.10}{\sqrt{200}} < \mu < 9.50 + 1.96\frac{2.10}{\sqrt{200}}$$

These quantities reduce to 9.21 and 9.79; hence the desired approximate 95 percent confidence interval for μ is given by

$$9.21 < \mu < 9.79$$

This is only an approximate 95 percent confidence interval because σ was replaced by its sample approximation s, and x was not assumed to be normally distributed. For a sample as large as 200, the errors arising because of these approximations will be negligible.

If one desired, say, a 90 percent confidence interval rather than a 95 percent confidence interval, it would be necessary merely to replace the number 1.96 by the number 1.64 in the preceding formulas, just as in the earlier problem of determining n. In the preceding problem the interval would then become

$$9.50 - 1.64\frac{2.10}{\sqrt{200}} < \mu < 9.50 + 1.64\frac{2.10}{\sqrt{200}}$$

If these inequalities are simplified, the desired approximate 90 percent confidence interval for μ becomes

$$9.26 < \mu < 9.74$$

Any other percentage confidence interval can be obtained in a similar manner by means of Table IV.

Example 2. The techniques employed to estimate the mean of a population can also be used to estimate the size of a finite population. The following example is a problem of this type.

In April 1946 a sample of 105 out of a total of 3,398 polling places in Greece was observed by field personnel of the Allied Mission for Observing the Greek Election (France, the United Kingdom, and the U.S.). The mean number of votes cast in the 105 polling places was 319.20 and the standard deviation was 88.35. Assuming that the behavior of voters was unaffected by the presence of observers, what are 95 percent confidence limits for the total number of votes cast in that election, that is, for the entire nation?

To solve this problem it suffices to solve the corresponding problem for the mean and then use it to obtain the desired limits.

Formula (5) gives

$$319.20 - 1.96\frac{\sigma}{\sqrt{105}} < \mu < 319.20 + 1.96\frac{\sigma}{\sqrt{105}}$$

Since σ is unknown, it will be replaced by s; hence approximate 95 percent confidence limits for μ are given by

$$319.20 - 1.96\frac{88.35}{\sqrt{105}} < \mu < 319.20 + 1.96\frac{88.35}{\sqrt{105}}$$

This reduces to

$$302.30 < \mu < 336.10$$

Since the total population of votes is given by $N\mu$, where N denotes the total number of polling places, the desired approximate 95 percent confidence interval for the total number of votes cast is given by

$$3{,}398(302.30) < N\mu < 3{,}398(336.10)$$

This reduces to

$$1{,}027{,}215 < N\mu < 1{,}142{,}068$$

In view of the preceding theory, the problems of estimation concerning the mean of a Poisson or exponential distribution are treated in the same manner as for a normal distribution, provided that n is not too small; hence they need not be treated separately here.

The methods of estimation explained in this section are called large-sample methods whenever σ is replaced by its sample estimate because they are then strictly valid only for large samples. If the sample is smaller than about 25 and the value of σ is unknown, these methods are of questionable accuracy, and therefore a more refined method is needed. A method designed to solve such small-sample problems is presented in Section 4.

3 ESTIMATION OF p

Section 2 was concerned with the estimation of the parameter μ for continuous variable distributions. This section explains how to solve similar types of problems for the parameter p associated with binomial distributions. The methods presented here are large-sample methods because they require the replacement of p by its sample estimate and also because they assume that the normal curve approximation to the binomial distribution is satisfactory.

As an example to illustrate the various types of estimation problems to be solved, consider the problem of estimating the percentage of shoppers at a market who prefer a slightly sweetened version of a brand of frozen orange juice to the regular brand. Suppose 300 shoppers are asked to taste the two versions and to give their opinions. Furthermore, suppose that 180 of them prefer the sweetened version. By using these data, the following three problems will be solved. These are the same three types of problems that were solved in the preceding section with respect to μ.

(1) What is the accuracy of the sampling proportion as an estimate of p? (2) How large a sample is needed if the probability is to be .95 that the error of estimate will not exceed .03 units? (3) What is a 95 percent confidence interval for p? All of these problems are solved in the same manner as in Section 2 because the sample size here is large enough to justify the use of normal-curve methods.

PROBLEM 1. From formula (10), Chapter 5, it follows that the sample proportion, $\hat{p}$, may be assumed to be approximately normally distributed with mean p and standard deviation

$$\sqrt{\frac{pq}{n}} = \sqrt{\frac{pq}{300}}$$

As a result, the probability is approximately .95 that $\hat{p}$ will lie within 1.96 such standard deviations of p. Thus the probability is approximately .95 that the error of estimate will be less than

$$1.96\sqrt{\frac{pq}{300}} \tag{6}$$

Since p is unknown, it must be estimated by

$$\hat{p} = \frac{x}{n} = \frac{180}{300} = .60$$

The value of (6) then assumes the approximate value

$$1.96\sqrt{\frac{(.60)(.40)}{300}} = .055$$

It can therefore be stated that the probability is approximately .95 that the sample estimate $\hat{p}$ will not differ from p by more than .055 units. This result gives a good idea of the accuracy of the sample value .60 as an estimate of p.

PROBLEM 2. To solve the problem of how large a sample is needed to attain a given accuracy of estimate for p, one uses the same reasoning as that used in section 2 for μ. This means that n must be chosen so that the proper number of standard deviations of $\hat{p}$ will equal the desired maximum error of estimate. As before, let e denote the selected maximum error of estimate and let z_0 denote the value of z corresponding to the desired probability of not exceeding this maximum error. Then, just as in (1), n must satisfy the equation

$$z_0\sqrt{\frac{pq}{n}} = e$$

Solving this equation for n yields the formula

(7) $$n = \frac{z_0^2 pq}{e^2}$$

For the particular problem being considered here, $e = .03$ and $z_0 = 1.96$. Since p is unknown, it must be estimated by the sample value $\hat{p} = .60$. If these values are substituted in (7), the value of n will be found to be approximately 1024; hence an additional sample of approximately 724 will be needed to obtain the desired accuracy of estimation. This shows that a very large sample is needed to obtain this high degree of accuracy.

PROBLEM 3. To find a confidence interval for p, one also uses the same reasoning as for μ. Since $\hat{p}$ takes the place of $\bar{x}$ in formula (5), an approximate 95 percent confidence interval for p is given by the inequalities

$$\hat{p} - 1.96\sqrt{\frac{pq}{n}} < p < \hat{p} + 1.96\sqrt{\frac{pq}{n}}$$

This formula cannot be applied in its present form because the value of p that occurs in the two radical terms is unknown. The difficulty can be overcome by replacing that value of p by $\hat{p}$. The resulting approximate confidence interval is then given by the formula

(8) $$\hat{p} - 1.96\sqrt{\frac{\hat{p}\hat{q}}{n}} < p < \hat{p} + 1.96\sqrt{\frac{\hat{p}\hat{q}}{n}}$$

Application of this formula to the present problem with $n = 300$ and $\hat{p} = .60$ gives

$$.60 - .055 < p < .60 + .055$$

These limits reduce to .545 and .655; consequently an approximate 95 percent confidence interval for p is given by the inequalities

$$.545 < p < .655$$

Repetition may be boring, yet it is worth repeating that all three solutions are based on large-sample methods. Fortunately, these methods are quite good, even for small samples, provided that $np > 5$ for $p \leq \frac{1}{2}$ or $nq > 5$ for $p > \frac{1}{2}$.

Example. A final illustration of the methods for estimating p is given because of its interest to those who enjoy politics. A well-known pollster claims that his estimate of the proportion of the voters favoring a certain presidential candidate is not in error by more than .03 units. In a close presidential race, how large a sample would he need to be certain, with a probability of .997, of being correct in his claim?

From Table IV in the appendix, 99.7 percent of the central area of a normal distribution lies within three standard deviations of the mean; therefore $z_0 = 3$ here. Since the race is very close, it may be assumed that $p = \frac{1}{2}$; hence formula (7) yields the result

$$n = \frac{9 \cdot \frac{1}{2} \cdot \frac{1}{2}}{(.03)^2} = 2500$$

A random sample of this size taken from over the country should therefore suffice to give him the desired accuracy.

The use of $p = \frac{1}{2}$ in the foregoing problem may appear to be arbitrary, particularly if the election is not really close. However, it is easy to show that pq assumes its maximum value when $p = \frac{1}{2}$, and hence that the maximum value of n in (7) occurs when $p = \frac{1}{2}$. This is done by first verifying that

$$pq = p(1 - p) = p - p^2 = \frac{1}{4} - \left(\frac{1}{2} - p\right)^2$$

Next, pq will be as large as possible, namely $\frac{1}{4}$, when the term $(\frac{1}{2} - p)^2$ that is being subtracted from $\frac{1}{4}$ has the value 0. But this will occur when $p = \frac{1}{2}$. This implies that when one is determining the size of the sample necessary for a specified accuracy of estimate, the value of n for $p = \frac{1}{2}$ will be larger than for any other value of p. As a result, the use of $p = \frac{1}{2}$ in such problems assures one that the resulting value of n is certainly large enough and possibly larger than necessary.

An interesting feature of problems like the preceding one is that, contrary to the belief of most people, the accuracy of an estimate of a proportion p does not depend on the size of the population but only on the size of the sample. Thus a sample of 2500 voters out of 50,000,000 voters is sufficient, theoretically, to determine their voting preferences with high accuracy.

For obvious economic reasons, professional pollsters do not take simple random samples. They usually combine a type of random sampling of regions with random sampling of individuals. These will be considered in a later chapter.

Unfortunately, voters do not always behave like trials in a game of chance, so that the binomial distribution model is not strictly applicable to voting problems. For example, a voter when interviewed may favor one candidate and yet a week later he may vote for another candidate, or he may not bother to vote at all. He may also misinterpret a pollster's question and therefore respond incorrectly. Experience has shown that because of uncontrolled human factors, the accuracy of an estimate of p for voters does not increase appreciably after a sample of 10,000 has been taken. It is necessary to use good sense in applying mathematical models to real life, particularly when it comes to human beings and some of their inconsistencies.

The technique for estimating the size of a population is the same as that shown in Example 2, section 2, with μ replaced by p. One first calculates limits for p and then multiplies them by N to obtain limits for Np.

The foregoing methods for estimating μ and p were based on the assumption that the population was sufficiently large to justify the use of $\sigma/\sqrt{n}$ and $\sqrt{pq/n}$ for the standard deviations of $\bar{x}$ and $\hat{p}$, respectively. When the size of the population is small it is necessary to shift to formulas (9) and (10) of Chapter 6, otherwise systematic errors will occur in the various estimation techniques that have been explained. Except for this substitution the methods for solving the three standard problems of estimation are the same.

4 A SMALL-SAMPLE METHOD FOR ESTIMATING μ

Example. As an illustration of a problem for which large sample methods would be inappropriate, suppose that an inspector wishes to run a quick check on the weight of bread produced by a bakery and takes a random sample of 15 loaves from the bakery's output. Suppose that the mean weight and standard deviation of these 15 loaves were 15.8 ounces and .3 ounce, respectively, and that the inspector wants a 90 percent confidence interval for the mean weight of the entire output. This sample is clearly too small to yield a good estimate for σ; therefore to avoid the error involved in replacing σ by s when s is based on such a small sample, a new variable called *Student's t variable* is introduced. It is defined by the formula

$$t = \frac{\bar{x} - \mu}{s}\sqrt{n} \tag{9}$$

This variable resembles the standard normal variable introduced in section 2, namely,

$$z = \frac{\bar{x} - \mu}{\sigma}\sqrt{n}$$

However, it differs from z in that it involves the sample standard deviation, s, in place of the population standard deviation, σ. Since t does not require a knowledge of σ, as is the case with z, its value can be computed from sample data, whereas the value of z cannot be computed unless σ is known. This is the reason why t can be used to solve problems without the necessity of introducing approximations to population parameters.

If a large number of sampling experiments are carried out in which a sample of size n is selected from a normal population and the value of t computed, a large number of values of t will be available for classifying into

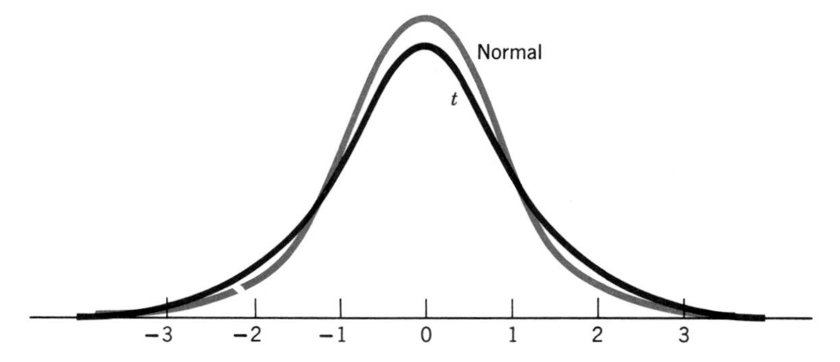

Figure 4 Standard normal distribution and a Student's t distribution.

a frequency table to obtain a good estimate of the limiting, or theoretical, distribution of t. Mathematical methods, however, yield the exact distribution. It turns out that the distribution of t depends only on the value of n, provided that the basic variable x possesses a normal distribution. Furthermore, the distribution of t is very close to the distribution of a standard normal variable z, except for very small values of n. Figure 4 shows the graph of the distribution of t for $n = 5$ and the graph of a standard normal variable z.

Table V in the appendix gives values of the variable t corresponding to what is called the number of "degrees of freedom," denoted by ν, and various probabilities. For the problem being considered here, the number of degrees of freedom is given by the formula $\nu = n - 1$. This corresponds to using the divisor $n - 1$ rather than n in defining the sample standard deviation in Chapter 2. The t distribution is used for other types of problems also in which the parameter ν is not equal to $n - 1$; otherwise this phrase would not need to be introduced here. Any column heading, such as .05, indicates the probability of t exceeding the value of t listed in that column. Because of symmetry it follows that .05 is also the probability that t will be to the left of the corresponding value of $-t$. Thus for the problem being discussed, since $n = 15$, one reads the entry in the row corresponding to 14 degrees of freedom and in the column headed .05, and finds $t = 1.761$. The probability is therefore .90 that t will satisfy the inequalities

$$-1.761 < t < 1.761 \tag{10}$$

If this is applied to (9), one can conclude that the probability is .90 that

$$-1.761 < \frac{\bar{x} - \mu}{s}\sqrt{n} < 1.761$$

These inequalities can be solved for μ in the same manner as that used to obtain formula (5). The result is

$$\bar{x} - 1.761\frac{s}{\sqrt{n}} < \mu < \bar{x} + 1.761\frac{s}{\sqrt{n}} \tag{11}$$

The desired 90 percent confidence interval for μ is obtained by substituting the sample values $n = 15$, $\bar{x} = 15.8$, and $s = .3$ into these inequalities. The result of this substitution is

$$15.8 - 1.761\frac{.3}{\sqrt{15}} < \mu < 15.8 + 1.761\frac{.3}{\sqrt{15}}$$

This simplifies to

$$15.66 < \mu < 15.94 \tag{12}$$

From the preceding analysis it would appear that the mean weight of loaves from this bakery is very likely less than one pound.

It is important to realize the distinction between this new method of finding a confidence interval for μ and the earlier large-sample method. This method does not require approximating σ by s, as is true for the large-sample method, and therefore it gives an exact rather than an approximate solution to the problem.

Formula (11) cannot be used to find a 90 percent confidence interval for μ unless the sample size is 15, because from Table V it will be found that (10) holds only for $\nu = 14$. If t_0 is used to denote the value of t found in the .05 column of Table V opposite ν degrees of freedom, then a general formula for a 90 percent confidence interval for μ is given by the inequalities

$$\bar{x} - t_0\frac{s}{\sqrt{n}} < \mu < \bar{x} + t_0\frac{s}{\sqrt{n}} \tag{13}$$

This formula is also valid for percentages other than 90 if the corresponding value of t_0 is employed. Thus a 95 percent confidence interval is obtained if one replaces t_0 by the t-value in the .025 column of Table V, which is opposite the desired degrees of freedom value.

For the purpose of comparing the old method with the new, it suffices to calculate a 90 percent confidence interval by means of formula (5) with 1.96 replaced by 1.64. If σ is estimated by s and the sample values $n = 15$, $\bar{x} = 15.8$, and $s = .3$ are substituted in (5), the desired approximate 90 percent confidence interval becomes

$$15.8 - 1.64\frac{.3}{\sqrt{15}} < \mu < 15.8 + 1.64\frac{.3}{\sqrt{15}}$$

This simplifies into

$$15.67 < \mu < 15.93$$

It will be noted that this interval is slightly narrower than that given by the small-sample method and displayed in (12). The large-sample method always gives a confidence interval that is too narrow; however, the error decreases rapidly as n grows and is hardly noticeable for n larger than 25.

Since the small-sample method based on the t distribution is an exact method, it would seem that one should always use it when σ is unknown. Unfortunately, however, the theory behind the t distribution requires one to assume that the basic variable x possesses a normal distribution; therefore, unless one can be assured that x is at least approximately normally distributed, the t distribution may not be justified.

5 A SMALL-SAMPLE METHOD FOR ESTIMATING p

The methods presented in section 3 for solving estimation problems about the parameter p of a binomial distribution required that n be sufficiently large to justify assuming that $\hat{p}$ possesses an approximately normal distribution. If the empirical rule given in Chapter 5 for determining whether n is sufficiently large to justify this assumption is not satisfied, the following small-sample method should be used.

For fixed values of n and p it is possible, by using the binomial distribution formula, to calculate a value $\hat{p}_1$ such that $P\{\hat{p} > \hat{p}_1\} = .025$ and a value $\hat{p}_2$ such that $P\{\hat{p} > \hat{p}_2\} = .975$. If this is done for n fixed but for various values of p, two curves can be constructed showing the relationship between p and $\hat{p}_1$ and between p and $\hat{p}_2$. These two curves can then be used to find a 95 percent confidence interval for p when n has this fixed value. Calculations such as these have been made for various values of n to produce Table VII in the appendix. Similar calculations could be carried out to produce tables for other confidence coefficients; however, only the 95 percent table is presented here.

Example. For the purpose of explaining how Table VII is to be used, consider a particular problem. The management of a large business firm, before proceeding with a serious study, would like to sound out its employees to determine whether they would favor starting and stopping work an hour earlier than usual. A sample of 20 employees resulted in 6 of them favoring such a change. What is a 95 percent confidence interval for the proportion of all employees who would favor it?

In Table VII the proper row to use is determined by the observed number of successes, x, located at the left margin, and the proper column to use is determined by the size of the sample, n. It is necessary to interpolate between column values for values of n that are not listed. For the particular problem being considered here, $x = 6$ and $n = 20$. From Table VII it therefore follows that a 95 percent confidence interval is given by

$$.12 < p < .54$$

This is a very wide interval and is not likely to be of much value to management in estimating what the consensus of all employees might be on this matter. It does indicate, however, that less than half of them are likely to favor earlier hours.

Suppose that large-sample methods had been used on this problem. Then, by formula (8), the desired 95 percent confidence interval for p would have been given by

$$.30 - 1.96\sqrt{\frac{(.3)(.7)}{20}} < p < .30 + 1.96\sqrt{\frac{(.3)(.7)}{20}}$$

or

$$.10 < p < .50$$

It should be noted that the large-sample technique always produces limits that are symmetrical with respect to $\hat{p}$ but that the exact method usually does not do this unless $\hat{p}$ is close to .5. As $\hat{p}$ moves away from .5, the degree of asymmetry increases.

6 ESTIMATION OF THE DIFFERENCE OF TWO MEANS AND TWO PROPORTIONS

An industrial problem that arises frequently is that of comparing two production methods and determining how much better one is than the other.

Example 1. As an illustration, suppose that a job is performed by 30 workmen using method I and that the same job is performed by 25 different workmen using method II. Interest is centered on how long it takes them to perform the job. Suppose that the mean and standard deviation times for the two groups are $\bar{x}_1 = 82$ minutes, $s_1 = 6$ minutes, and $\bar{x}_2 = 78$ minutes, $s_2 = 5$ minutes, respectively. The problem is to estimate the difference between the mean times of the two production methods, assuming that the population of workers is very large compared to the sizes of these samples.

A large-sample confidence interval for $\mu_1 - \mu_2$ can be constructed by using formula (6), Chapter 6, and the same technique as that used to construct a large-sample confidence interval for a single mean. Under the assumption that $\bar{x}_1$ and $\bar{x}_2$ are independent normal variables, a 95 percent confidence interval for $\mu_1 - \mu_2$ is given by

$$\bar{x}_1 - \bar{x}_2 - 1.96\sigma_{\bar{x}_1-\bar{x}_2} < \mu_1 - \mu_2 < \bar{x}_1 - \bar{x}_2 + 1.96\sigma_{\bar{x}_1-\bar{x}_2} \tag{14}$$

Here

$$\sigma_{\bar{x}_1-\bar{x}_2} = \sqrt{\frac{\sigma_1^2}{n_1} + \frac{\sigma_2^2}{n_2}}$$

Since σ_1^2 and σ_2^2 are unknown, they must be estimated by s_1^2 and s_2^2. Using these estimates and substituting the other sample values, the interval for this problem becomes

$$4 - 1.96\sqrt{\frac{36}{30} + \frac{25}{25}} < \mu_1 - \mu_2 < 4 + 1.96\sqrt{\frac{36}{30} + \frac{25}{25}}$$

This reduces to

$$1.1 < \mu_1 - \mu_2 < 6.9$$

Hence, management can be quite certain that the second method will save at least 1.1 minutes of time and possibly as much as 6.9 minutes.

Since the large-sample method for estimating a proportion is the same as that for the mean, one should expect to treat the difference of two proportions in the same manner as the difference of two means. On this assumption, which is justified, it follows from formula (8), Chapter 6, and from (14) that a 95 percent confidence interval for $p_1 - p_2$ is given by

$$\hat{p}_1 - \hat{p}_2 - 1.96\sigma_{\hat{p}_1-\hat{p}_2} < p_1 - p_2 < \hat{p}_1 - \hat{p}_2 + 1.96\sigma_{\hat{p}_1-\hat{p}_2} \tag{15}$$

Here

$$\sigma_{\hat{p}_1-\hat{p}_2} = \sqrt{\frac{p_1q_1}{n_1} + \frac{p_2q_2}{n_2}}$$

For the purpose of applying this formula, it is necessary to substitute the sample estimates $\hat{p}_1$ and $\hat{p}_2$ for p_1 and p_2 in the expression for $\sigma_{\hat{p}_1-\hat{p}_2}$ before numerical limits can be obtained.

Example 2. As an illustration, suppose a candy company is experimenting with two methods for making chocolates and is concerned with the percentage of broken chocolates that are produced. Suppose, further, that process I had 12 broken chocolates in a batch of 200 and process II had 24 broken chocolates in a batch of 300. What is a 95 percent confidence interval for $p_1 - p_2$? Since these are undoubtedly independent experiments and since the samples are large enough to justify using normal approximations, formula (15) may be applied, with $\hat{p}_1$ and $\hat{p}_2$ replacing p_1 and p_2 in the expression for $\sigma_{\hat{p}_1-\hat{p}_2}$. This gives

$$-.02 - 1.96\sqrt{\frac{(.06)(.94)}{200} + \frac{(.08)(.92)}{300}} < \hat{p}_1 - \hat{p}_2 < -.02 + 1.96\sqrt{\frac{(.06)(.94)}{200} + \frac{(.08)(.92)}{300}}$$

Calculations yield

$$-.065 < \hat{p}_1 - \hat{p}_2 < .025$$

The results here are too ambiguous to give the investigators much confidence in method I over method II. A considerably larger sample would be needed to shorten the confidence interval.

7 REVIEW ILLUSTRATIONS

Example 1. A gasoline additive is being tested to see whether it increases mileage. Twenty-five cars are supplied with 5 gallons of gasoline and are run until the gasoline is exhausted. At the completion of the experiment the average mileage for each car is computed. Calculations with the data of this one experiment gave a mean of $\bar{x} = 18.5$ miles per gallon and a standard deviation of $s = 2.2$ miles per gallon for the 25 cars. Experience with cars of the same kind that were used with no additives indicates that, approximately, $\mu = 18.0$ and $\sigma = 2.0$ miles per gallon. Assuming that the additive has no effect on mileage, solve the following problems. (*a*) Determine the probability accuracy of $\bar{x}$ as an estimate of μ. What is the actual accuracy? Is the sample compatible with what was to be expected by theory? (*b*) Find how large an experiment should be conducted to be certain with a probability of .95 that the estimate would not be in error by more than $\frac{1}{2}$ mile per gallon. (*c*) Find a 95 percent confidence interval for μ. Does this interval actually contain μ? (*d*) Dropping the assumption that the additive had no effect on either the mean or the variance, use Student's t variable to find a 95 percent confidence interval for μ. The solutions follow.

(*a*) Here $\mu = 18.0$, $\sigma = 2.0$, $n = 25$, and $\bar{x} = 18.5$; hence

$$1.96\sigma_{\bar{x}} = 1.96\frac{2.0}{\sqrt{25}} = .784$$

The probability is .95 that the error of estimate will not exceed .784. The actual error is $|\bar{x} - \mu| = 18.5 - 18.0 = .5$; therefore the sample value is compatible with theory.

(*b*) Since .95 gives $z_0 = 1.96$ and the maximum tolerable error is to be $e = \frac{1}{2}$, n is given by

$$n = \left[\frac{(1.96)(2.0)}{\frac{1}{2}}\right]^2 \doteq 62$$

An additional sample of 37 would suffice. Thus, approximately 37 additional cars of the same type should be run.

(*c*) A 95 percent confidence interval is given by

$$\bar{x} - 1.96\frac{\sigma}{\sqrt{n}} < \mu < \bar{x} + 1.96\frac{\sigma}{\sqrt{n}}$$

Hence, for this problem it becomes

$$18.5 - 1.96\frac{2.0}{\sqrt{25}} < \mu < 18.5 + 1.96\frac{2.0}{\sqrt{25}}$$

or

$$17.72 < \mu < 19.28$$

Since $\mu = 18.0$, it is contained inside this interval.

(*d*) Since $\nu = 24$ and a 95 percent interval is desired, $t_0 = 2.0639$; hence formula (13) gives

$$18.5 - 2.06\frac{2.2}{\sqrt{25}} < \mu < 18.5 + 2.06\frac{2.2}{\sqrt{25}}$$

or

$$17.59 < \mu < 19.41$$

This interval also contains $\mu = 18.0$; therefore it is compatible with the assumption of no additive effect.

Example 2. An article in the medical journal *Lancet* for March 2, 1974, discusses the effects of oral contraceptives on the sex ratio at birth. In a study conducted in Hungary, a record was kept of the sex of children born to mothers who had been taking the "pill" for at least 2 years before the birth of their child. There were 170 mothers of this type in the experiment. The number of male babies born to this group was 58 and therefore the number of female babies was 112. Use these data to calculate a 90 percent confidence interval for the percentage of male babies expected from mothers of this type.

Formula (8) with 1.96 replaced by 1.64 may be applied here. Substitution of the sample values $n = 170$ and $\hat{p} = \frac{58}{170} = .34$ in that formula gives

$$.34 - 1.64\sqrt{\frac{(.34)(.66)}{170}} < p < .34 + 1.64\sqrt{\frac{(.34)(.66)}{170}}$$

This reduces to

$$.28 < p < .40$$

From this study it appears that there are approximately twice as many female babies as male babies born to mothers who have been on oral contraceptives for at least 2 years before the birth of their child.

Example 3. In a simple random sample of 250 from the list of 10,000 customers of a large store, 150 had no complaints with the store's service, 70 had one or more complaints, and 30 did not reply after repeated appeals. (*a*) Estimate the number of customers who would have replied had the sample included all 10,000 customers rather than just the 250. (*b*) Ignoring the nonresponses, estimate 95 percent confidence limits for the fraction of customers with no complaints. (*c*) Considering the nonresponses, is there sufficient evidence that no matter what the nonresponders think of the store's service that a majority of the customers have no complaints?

(*a*) Let $\hat{p}_0$ be the fraction of customers in the sample responding to the inquiry, that is, $\hat{p}_0 = (150 + 70)/250$ or $\frac{220}{250} = .88$. Since there are 10,000 customers in all, the number who would respond if given the same opportunity is estimated to be $(10{,}000)\hat{p}_0 = (10{,}000)(.88) = 8800$.

(*b*) Let $\hat{p}$ = the fraction of customers in the sample "declaring no complaints"; hence $\hat{p} = \frac{150}{250} = .6$. A 95 percent confidence interval for this proportion in the population is given by

$$\hat{p} - 1.96\sigma_{\hat{p}} < p < \hat{p} + 1.96\sigma_{\hat{p}}$$

where $\sigma_{\hat{p}}$ is estimated by $[(.6)(.4)/250]^{1/2} \doteq .032$. This gives

$$\hat{p} - 1.96(.032) < p < \hat{p} + 1.96(.032)$$
$$.600 - .063 < p < .600 + .063$$
$$.537 < p < .663$$

This result ignores the possibility that some nonresponders may be noncomplainers.

(*c*) Here it is necessary to make some assumptions about the nonresponses. The worst possible assumption under the circumstances is that they all have complaints but for some reason did not care to declare them. Adding these to our overt complainers does not alter the fact that we have 95 percent confidence that p, the fraction of overt noncomplainers among both responders and nonresponders, is somewhere between 53.7 and 66.3 percent of the store's customers. Hence the nonresponders do not alter our earlier conclusion.

Example 4. Twenty patients who suffer from high blood pressure were given a drug that is supposed to lower pressure. If x denotes the amount of a patient's decrease from his normal blood pressure and if the sample of 20 produced the estimates $\bar{x} = 5$ and $s = 10$, (*a*) find a 95 percent confidence interval for μ using the t distribution. (*b*) Assuming that $\sigma = 10$ and that the 20 patients were a random sample of such patients, determine how many patients should have been sampled if the error in estimating μ is to be less than 2 units with a probability of .95.

(*a*) $n = 20$, $\bar{x} = 5$, $s = 10$; hence, formula (13) becomes

$$5 - 2.093\frac{10}{\sqrt{20}} < \mu < 5 + 2.093\frac{10}{\sqrt{20}}$$

or

$$.32 < \mu < 9.68$$

Thus, one can be quite certain of a decrease of at least .32, but this is precious little.

$$(b)\ n = \left[\frac{(1.96)(10)}{2}\right]^2 = 96$$

Example 5. In a sample of 100 small electronic components there are 2 defectives. (*a*) Use Table VII to find 95 percent confidence limits for p. (*b*) Use normal-approximation methods to find 95 percent confidence limits for p and compare them with those found in (*a*). Comment.

(*a*) Table VII gives the result

$$.00 < p < .07$$

(*b*) Here, 95 percent limits are given by

$$.02 - 1.96\sqrt{\frac{(.02)(.98)}{100}} < p < .02 + 1.96\sqrt{\frac{(.02)(.98)}{100}}$$

This reduces to

$$.02 - .027 < p < .02 + .027$$

Or, since 0 is the smallest possible value for p,

$$.00 < p < .057$$

The large-sample method is highly inaccurate for this problem.

Example 6. A purchasing agent for a company was confronted with two types machines to perform a certain operation. He was permitted to try both machines over a trial period. Twenty jobs were randomly assigned to the two machines, ten to each, with the following results: $\bar{x}_1 = 25$ hours, $s_1^2 = 135$ hours squared, $\bar{x}_2 = 17$ hours, $s_2^2 = 80$ hours squared. (*a*) Find a 95 percent confidence interval for the performance difference. (*b*) How many additional trials should be undertaken to detect a performance difference of 5 hours with 95 percent confidence?

(*a*) An approximation to $\sigma_{\bar{x}_1-\bar{x}_2}$ is given by

$$\sigma_{\bar{x}_1-\bar{x}_2} \doteq \sqrt{\frac{135}{10} + \frac{80}{10}} = 4.64$$

Hence, 95 percent confidence limits for $\mu_1 - \mu_2$ are given by

$$(25 - 17) - 1.96(4.64) < \mu_1 - \mu_2 < (25 - 17) + 1.96(4.64)$$

or

$$8 - 9.1 < \mu_1 - \mu_2 < 8 + 9.1$$

or

$$-1.1 < \mu_1 - \mu_2 < 17.1$$

(*b*) If the error of estimating $\mu_1 - \mu_2$ by means of $\bar{x}_1 - \bar{x}_2$ is to be less than 5 hours with a probability of .95, and if n denotes the equal-size sample to be taken from each group, then n must satisfy the equation

$$1.96\sigma_{\bar{x}_1-\bar{x}_2} = 1.96\sqrt{\frac{\sigma_1^2}{n} + \frac{\sigma_2^2}{n}} = 5$$

The value of $\sigma_{\bar{x}_1-\bar{x}_2}$ can be approximated by using the sample variances in the formula for this standard deviation. These approximations lead to the equation

$$1.96\sqrt{\frac{135}{n} + \frac{80}{n}} = 5$$

or

$$1.96\sqrt{\frac{215}{n}} = 5$$

The solution of this equation is $n = 33$. Hence a total sample of size 66 is needed. Since 20 trials have already been taken, it should suffice to take 46 additional trials, with 23 of them assigned to each machine. It is assumed here that the problem is one of determining the accuracy of $\bar{x}_1 - \bar{x}_2$ as an estimate of $\mu_1 - \mu_2$ and is not concerned about whether μ_1 is larger or smaller than μ_2.

EXERCISES

SECTION 2

1 Experience with workmen in a certain industry indicates that the time required for a randomly selected workman to complete a job is approximately normally distributed with a standard deviation of 12 minutes.
   (*a*) If each of a random sample of 25 workmen performed the job, how accurate is their sample mean as an estimate of the mean for all the workmen?
   (*b*) How much improvement would have resulted in the accuracy of this estimate if 100 workmen had been selected?

2 From past experience the standard deviation of the height of fifth-grade children in a school system is 2 inches.
   (*a*) If a random sample of 36 such chidren is taken, how accurate is their sample mean as an estimate of the mean for all such children?
   (*b*) What would happen to the accuracy of this estimate if the sample were made 9 times as large?

3 (*a*) In problem 1(*a*) how large a sample would one need to take if one wished to estimate the population mean to within 2 minutes, with a probability of .95 of being correct?
   (*b*) What size sample would be needed if the maximum error of estimate were to be 1 minute?

4 A gas company keeps close watch on trends in the installation of appliances in its metropolitan service area using gas. Each year it draws a random sample of customers from its files and these are interviewed to determine how many gas appliances they have. Since a customer may supply several tenants from the same meter it is possible for X_i to take on such values as 0, 1, 2, 3, ... Past experience indicates that σ for gas stoves per customer is 1.00. How large a sample is required to obtain an estimate within 5 percent accuracy with a probability of error not to exceed .01 if the mean is about 1.00?

5 A paper manufacturer determines the disposition of a shipment of pulp on the output of pulp from his own wood "digester" on the basis of the properties of paper made from the pulp. An important property is bursting strength of the paper made in a laboratory from the pulp under investigation. Suppose a sample of 4 specimens yielded a mean of 25 units and it is known from the history of such tests that the standard deviation among specimens is 5. Assuming normality of test results, what are the 95 percent confidence limits on μ for this sample?

6 The time required to complete a certain clerical task was observed on each of 25 persons who were specially trained to perform it. The mean of the 25 is 50 minutes and the standard deviation of the group is 15 minutes. Management would like to have an estimate within 3 minutes of the true mean time with a probability of error not to exceed $\frac{1}{100}$. Assuming the population is approximately normal and the 25 persons tested form a random sample of potential clerks, how large a sample is required to meet these specifications?

7 If the results of the experiment in problem 1(*a*) yielded $\bar{x} = 140$ minutes, find (*a*) 95 percent confidence limits for μ, (*b*) 90 percent confidence limits for μ.

8 If the results of the experiment in problem 2(*a*) yielded $\bar{x} = 54$ inches, find (*a*) 95 percent confidence limits for μ, (*b*) 90 percent confidence limits for μ.

9 A set of 50 experimental animals is fed a certain kind of rations for a 2-week period. Their gains in weight yielded the values $\bar{x} = 42$ ounces and $s = 5$ ounces.
(*a*) How accurate is 42 as an estimate of the population mean?
(*b*) How large a sample would you take if you wished $\bar{x}$ to differ from μ by less than 1 ounce, with a probability of .95 of being correct?
(*c*) Find 95 percent confidence limits for μ.

10 In a sample audit of a small transactions account, a careful auditing of a 50 randomly selected transactions yielded a mean error of $-\$150$ with a standard deviation of \$60. Calculate 95 percent confidence limits for μ, the true mean error of the entire account, assuming approximate normality of $\bar{x}$.

11 A new enzyme is being considered by a pharmaceutical company to observe whether it will increase the yield of one of the company's processes. The yield for a batch is defined to be the actual yield divided by a theoretical yield based on past experience. The new enzyme was tried out on 50 batches and an average yield of 1.24 was obtained with a standard deviation of .20.
(*a*) Calculate a 95 percent confidence interval for the true yield.
(*b*) On the basis of this experiment, are you certain that the true yield is above 1.00? Explain.

12 The U.S. Public Land Survey maps each state into area units of $\frac{1}{4}$ square mile each. One of the agricultural states contains 219,176 such units. Several years ago a sample of 908 units was taken and the number of farms whose headquarters were in those units was counted. The mean number of farms per unit was found to be .8722 and the standard deviation was found to be .6629. Calculate approximate 95 percent confidence limits for the total number of farms in the state at that time.

SECTION 3

13 A sample of 80 motorists showed that 20 percent had lapsed driver's licenses. How accurate is this estimate of the true percentage likely to be?

14 A manufacturer of parts believes that approximately 5 percent of his product contains flaws. If he wishes to estimate the true percentage to within $\frac{1}{2}$ percent and to be certain with a probability of .99 of being correct, how large a sample should he take? Comment.

15 A random sample of 400 citizens in a community showed that 240 favored having their water fluoridated. Use these data to find 95 percent confidence limits for the proportion of the population favoring fluoridation.

16 An advertising firm claims that its recent promotional campaign reached 30 percent of the families in a certain city. The company who hired the firm doubts this assertion and wishes to take a sample survey to get the necessary facts. How large a sample should it select in order to have 95 percent confidence that its estimate will be within 3 percentage points of the true value?

17 A telephone company wants to know how many poles should be replaced. A reasonable guess seems to be somewhere between 10 and 20 percent. A considerable saving could be made in buying and installing the poles if the company could specify its needs within 10 percent. Considering the cost of surveys on the one hand and the cost of being wrong (in its estimates) on the other, it feels that it can risk a 1-in-20 chance of being in error. How large a sample of poles should be taken?

18 A local union has a total membership of 688. A random sample of 392 was selected and asked the question, "Do union problems which do not concern your shop interest you?" Ninety-seven gave the responses "No, not at all" or "Don't know." Estimate the fraction of the 688 members who would have given similar responses if a complete census were taken. Ignoring the finiteness of the population, calculate 95 percent confidence limits for p.

19 It is common practice to use a sample of 400 homes on which to base TV program ratings. Assuming random sampling, calculate (*a*) a 70 percent confidence interval for the true program rating if the sample gave a 20 percent rating. (*b*) Work (*a*) if the program rating is only 10 percent.

20 A corporation is planning to poll its employees to determine their views on a proposed health services scheme. Management feels that it should know the true proportion of those who favor it to within four percentage points with a confidence of 95 percent. It estimates that between 60 and 80 percent will favor the plan. How large a sample is needed?

21 An estimate of the proportion of serviceable items in a surplus inventory stored under unfavorable conditions is to be obtained within $\pm.05$, with a 95 percent reliability. The total inventory consists of 10,000 items, and it is believed that the proportion of serviceable items is between .30 and .70.

(*a*) What sample size is needed to obtain an estimate with the desired accuracy?

(*b*) Can the accuracy of the sample estimate be evaluated if the judgment concerning the true proportion is not correct? Explain.

22 Obtain 80 percent confidence limits for the number of accident claims that will be paid by an insurance company during the next year if this year's experience showed that 5 percent of those carrying insurance collected claims and the company has 6000 policies.

23 To determine how many families in a community of 100,000 families qualify for the U.S. Department of Agriculture's food stamps, a random sample of 360 families was taken. It was found that 98 of those 360 families qualified. Calculate 90 percent confidence limits for the total number of families in that community that qualify for the stamps.

24 In a survey conducted by Dr. Frederick McGuire of the University of California, Irvine, College of Medicine, and reported in the *Los Angeles Times* of July 11, 1974, it was found that slightly less than 40 percent of 2,000 drivers killed in auto accidents during the preceding 5 years in Orange County were intoxicated. The California Vehicle Code definition of intoxication is based on the amount of alcohol in the blood, and the determination of whether a dead driver had been intoxicated was based on a postmortem test of his blood. On the assumption

that this five-year period is typical of Orange County driving experience and is likely to hold in the future, and on the assumption that 400 Orange County drivers will be killed next year, calculate a 95 percent confidence interval for the number of drivers that will be killed in Orange County who will be intoxicated at the time of death.

25 As reported in the *American Sociological Review* for 1955, a study was made to determine what effect sudden wealth has on an individual's desire to work. Each of 393 men was asked the question, "If by chance you inherited enough money to live comfortably without working, do you think you would work anyway or not?" The study revealed that 314 of the men answered "would keep working." On this basis calculate a 95 percent confidence interval for the true fraction of those who would give such an answer.

26 In a study made by three professors of psychiatry from the Department of Psychiatry, Rutgers Medical School, and reported in the February 1974 issue of *Archives of General Psychiatry,* the following data were published on the attitudes of United Automobile Workers who worked in a General Motors plant.

How satisfied are you with your job?	Active patients	Former patients	Classified sick	Classified well
Number satisfied	13	19	90	463
Sample size	17	26	95	481
Percentage	76	73	95	96

The two "patients" groups consisted of those workers who had been seen and diagnosed in the UAW Labor Union Clinic at Johns Hopkins Hospital. The remaining workers were classified as "sick" or "well" on the basis of their MacMillan Index, a device for attempting to measure mental health by means of interviews.

(*a*) Find a 95 percent confidence interval for the percentage of well workers who are satisfied with their job.

(*b*) Find a 90 percent confidence interval for the percentage of sick workers who are satisfied with their job.

(*c*) On the basis of these data, does it appear true that a worker's satisfaction with his job is independent of whether or not he rates high on a mental health interview, provided that he has not been a patient at the clinic?

SECTION 4

27 Given that $\bar{x} = 20$, $s = 4$, $n = 10$, with x normally distributed, use Student's t distribution to find (*a*) 95 percent confidence limits for μ, (*b*) 99 percent confidence limits for μ.

28 A sample of 15 cigarettes of a certain brand was tested for nicotine content and gave $\bar{x} = 22$ milligrams and $s = 4$ milligrams. Use Student's t distribution to find 95 percent confidence limits for μ.

29 Work problem 28 by large-sample methods and compare the results of the two methods.

30 A set of 12 experimental animals was fed a special diet for 3 weeks and produced the following gains in weight: 30, 22, 32, 26, 24, 40, 34, 36, 32, 33, 28, 30. Find 90 percent confidence limits for μ, assuming gain in weight is a normal variable.

31 Given the observations: 2, 5, 3, 8,

(*a*) compute their mean and variance.

(*b*) Suppose these observations were randomly selected from a process that we know has a normal distribution with mean μ and variance σ^2. Determine an exact interval within which you would be willing to say the true mean lies, with 95 percent "confidence."

(*c*) Suppose from previous information it was known that the process had a true variance of 7. Does this information alter your result in (*b*)?

32 Suggest how you might proceed to determine the sample size needed for estimating μ with a certain accuracy when σ is unknown by taking samples in small groups and reestimating σ as additional groups are taken.

SECTION 5

33 In a sample of 100 small castings there are 3 defectives.

(*a*) What is your estimate of the proportion of defectives in this lot?

(*b*) What is the standard deviation of this estimate?

(*c*) Use Table VII to find a 95 percent confidence interval for the proportion of defectives.

(*d*) Use a normal curve approximation to solve (*c*) and compare your two answers.

34 A lot of 700 modules for TV has been received from a supplier, the specifications requiring that 95 percent be free of defects. A sample of 30 was checked out and 4 were found to be defective. Find a 95 percent confidence interval for the proportion of defectives. Does this confidence interval include the p in the specifications?

35 A marketing department is interested in determining the size of a market for a new product that has a very limited popular appeal but for which a very high unit profit appears to be possible. A consumer survey of 1,000 families showed that 10 of them would buy the new product. The contemplated market area contains 2,000,000 families. Estimate the number of families that would buy the product. Use the survey results to find a 95 percent confidence interval for the number of families that would buy the product.

36 (*a*) If in a sample of 100 farmers none had sprayed for corn borers, what 95 percent confidence statement would you make about the true fraction?

(*b*) Suppose all the farmers in the sample had sprayed for corn borers. What 95 percent confidence interval would you then use?

37 Solve each of the inequalities in the double inequality $\hat{p} - z_0\sqrt{pq/n} < p < \hat{p} + z_0\sqrt{pq/n}$ for the variable p. This will involve the solution of a quadratic equation. Use your results to obtain a confidence interval for p that does not contain p in its limits.

SECTION 6

38 A sample of 50 workers was selected from each of two factories to study the amount of time required to complete a complicated assembly. The data gave $\bar{x}_1 = 59$ minutes and $s_1 = 10$ minutes for the north factory and $\bar{x}_2 = 55$ minutes and $s_2 = 8$ minutes for the south factory. Find a 95 percent confidence interval for the difference of mean times. Comment.

39 A sample of 300 employees was interviewed for opinions on their attitude concerning a new bonus plan being proposed by the management. The results, summarized separately for males and females, showed that 52 percent of the 180 males favored it and 55 percent of the 120 females did. Calculate an 80 percent confidence interval for the true opinion difference.

40 A sample of 400 second-grade pupils was given an achievement test in arithmetic. It was observed that the 180 girls had an average score of 80 with a standard deviation of 20, whereas the 220 boys scored an average of 82 with a standard deviation of 22. Calculate a 95 percent confidence interval for the true mean difference. Comment.

41 In the spring of 1954 the Federal Reserve System and the Research Center of the University of Michigan jointly sampled 3,000 spending units (families) in the U.S. for the purpose of studying consumer finances. A similar study was made one year later, when 3,120 units were sampled. These surveys revealed that 7.7 percent of those interviewed in 1954 intended to purchase a television set, whereas in 1955 only 5.9 percent planned on doing so. What is a 95 percent confidence interval for the true difference? Is your result consistent with the assumption of no change?

42 Jones and Smith were assigned the job of checking into the validity of vouchers issued for purchases. A sample of 100 was taken and 50 vouchers were assigned to each man. At the end of one week Jones had examined 30 and found that 5 were bogus. Smith had examined 35 and found that 12 were bogus. Because of the difference in these bogus percentages, the manager requested a 95 percent confidence interval for the true difference. Will this confidence interval justify the manager's believing that Jones and Smith differ in their procedures?

43 A company that does business through the "club" scheme in which individuals sign up to accept issues of the product (such as books, records, etc.) at certain time intervals was interested in looking into the success ratio of collections. A sample of 69,249 accounts was selected and examined for collection success. A rate of 70 percent was found. Management was skeptical of the accuracy of the result and requested that a second sample be taken. The second sample, of size 68,011, gave a success rate of 75 percent. Do you believe that management should now be satisfied? What would you tell management?

44 Jones has been watching the expense accounts of two employees, A and B, over the past 20 weeks. A and B have similar assignments and work independently of each other. Computations show that the mean and standard deviation of weekly expenditures for A are \$210 and \$10, and for B, \$215 and \$15. In order to obtain an estimate of the mean weekly difference in costs with \$2 per week accuracy and 90 percent confidence, how many weeks will Jones have to check on their accounts? Comment.

SECTION 7

45 In 1951–1952 a study was carried out on the insurance beneficiaries of the U.S Bureau of Old-Age and Survivors Insurance. A random sample of 2,533 aged/widows, roughly 1 percent of the entire population at the time, was selected and personally interviewed. On money income from all sources received during the previous year, the sample mean was \$804 with a sample standard deviation of \$589.

(*a*) Calculate 99 percent confidence limits for μ, the mean for the population.

(*b*) It was found that 15.3 percent of the aged/widows received \$1200 or more. Calculate 99 percent confidence limits for p, the fraction of the population having \$1200 or more income.

46 If you guess that between 25 percent and 75 percent of the housewives in a region own a certain type of appliance and you wish to take a sample that will, at the 95 percent confidence level, yield an estimate that will not differ by more than 6 units from the correct percentage, how large a sample should you take? Did your guess about this percentage help you cut down on the size of the sample needed? Explain.

47 A manufacturing company wishes to estimate the proportion of defective units that it produces. To do so it decides to take a sample of size n and find the proportion of defectives in the sample. It decides that its estimate must have a probability of at least .95 of being in the interval $(p - .04, p + .04)$.

(*a*) It is known from earlier experience that each lot contains at least 60 percent good units. Using this information, determine the sample size needed.

(*b*) The company has a rule that if the sample proportion of defectives is larger than .30, it will stop production. If the true proportion is .20, what is the probability of stopping production for a sample of the size found in (*a*)?

48 A company with 150,000 customers on credit wishes to determine how many of them would default if a new policy under consideration were introduced. A random sample of 200 of those customers is taken. It is found that 24 of them would have defaulted under the new policy. Calculate 95 percent confidence limits for the total number of customers who would default if the new policy were introduced. Comment.

49 The following was taken from a newspaper article: "Detectives followed 1,647 shoppers through department stores. They discovered that 7.4 percent of the women and 5.0 percent of the men were involved in shoplifting. The average value of stolen articles was estimated to be \$5.26. So reports a private police outfit." Assuming that 70 percent of the shoppers were women, determine by the use of a 90 percent confidence interval whether there is sufficient evidence here to declare a real sex difference in shoplifting.

50 As reported in the *British Journal of Sociology* for 1954, a study concerning the prestige rankings of certain occupations produced the following data. Each of 312 individuals was asked to rank the occupations civil servant and electrician in one of the categories: very high, high, average, low, and very low. These ratings were assigned the numerical values (scores) 1, 2, 3, 4, and 5.

		Occupation	
Rating	Score	Civil servant	Electrician
Very high	1	38	18
High	2	151	95
Average	3	113	194
Low	4	7	5
Very low	5	2	0
Totals		311	312

Ignoring the difference of 1 in the totals and treating the two ratings of an individual as independent (which is seldom justified), find a 95 percent confidence interval for the true mean difference. How would you have analyzed the original raw data to solve this problem?

51 The data of Table XIII in the appendix may be used to construct various estimation problems. For example, how reliable is the sample mean of the variable x_1 as an estimate of the population mean for this variable?

52 Have each member of the class find a 75 percent confidence interval for μ for a sample of size 25 from a table of one-digit random numbers (Table XII in the appendix). Use the fact that σ for this distribution is given by $\sigma = 2.87$ and that $\bar{x}$ may be treated as a normal variable. Check to see what percentage of the students' confidence intervals contain the true mean $\mu = 4.5$. About 75 percent should do so.

53 Work problem 52, but this time use a sample of size 10 and assume that the value of σ is not known. That is, use Student's t distribution to find the desired confidence interval. Check to see what percentage of the students' intervals contain μ.

Chapter 8 Testing Hypotheses

1 TWO TYPES OF ERROR

As indicated in Chapter 1, a second fundamental problem of statistics is the testing of hypotheses about populations. From the discussion on estimation in Chapter 7, it follows that the testing of hypotheses about binomial populations can usually be reduced to testing some hypothesis about the parameter p. Similarly, the testing of hypotheses about normal populations can usually be reduced to testing hypotheses about the parameters μ and σ. Since the Poisson and exponential distributions depend only on the parameters μ and β, testing problems related to those distributions are problems of testing a mean.

Examples, which are essentially hypothesis-testing problems related to binomial or normal distributions, have already been discussed. For example, the problem discussed in section 4, Chapter 5, of determining by means of a sample of 400 voters whether a politician's claim of 60 percent backing was valid is a problem of testing the hypothesis that $p = .6$ for a binomial distribution for which $n = 400$. The problem of comparing weights of dormitory and nondormitory students, introduced in Chapter 2, can be treated as a problem of testing the hypothesis that the means and standard deviations of two normal distributions are equal.

For the purpose of explaining the methods used to test a hypothesis about a population parameter, consider a particular problem.

During the last fifty years or more, archaeologists in a certain country have been attempting to classify skulls found in excavations into one of two racial groups, partly by the pottery and other utensils found with the skulls and partly by differences in skull dimensions. In particular, they have found that the mean length of all the skulls found thus far from race A is 190 millimeters, whereas the mean length of those from race B is 196 millimeters. The standard deviation of such measurements of length was found to be about the same for the two groups and approximately equal to 8 millimeters. A new excavation produced 12 skulls, which there is reason to believe belong to race A. The mean length of these skulls is $\bar{x} = 194$ millimeters. The problem is to test the hypothesis that the skulls belong to race A rather than to race B.

Since the test is to be based on the value of $\bar{x}$, it is formulated as a test of the hypothesis, denoted by H_0, that the population mean for the 12 skulls is 190, as contrasted to the alternative hypothesis, denoted by H_1, that the population mean is 196. This can be condensed as follows:

$$\begin{aligned} \mathrm{H}_0 &: \mu = 190 \\ H_1 &: \mu = 196 \end{aligned} \tag{1}$$

There are two possibilities for making the wrong decision here. If the skulls really belong to race A and on the basis of the value of $\bar{x}$ they decide to accept H_1, an incorrect decision will be made. If, however, the skulls really belong to race B and they decide to accept H_0, an incorrect decision will also be made. The first type of wrong decision is usually called a type I error, whereas the second type of wrong decision is called a type II error. These two possibilities for incorrect decisions, together with the two possibilities for correct decisions, are listed in Table 1.

Now, most people would use good sense in this particular problem and decide in favor of H_0 if $\bar{x}$ were closer to 190 than to 196 and in favor of H_1 if the reverse were true. Thus most people would accept H_1 in this problem. However, archaeol-

Table 1

	Decision	
	H_0 accepted	H_1 accepted
H_0 true	correct decision	type I error
H_1 true	type II error	correct decision

ogists who have other reasons for believing that the skulls belong to race A, such as pieces of pottery found with the skulls, would not be willing to use the halfway point between the two means as the borderline value for making decisions based on $\bar{x}$. They would undoubtedly insist that $\bar{x}$ be fairly close to the mean corresponding to H_1 before they would be willing to give up the hypothesis H_0 in favor of H_1. To study the reasonableness of using the halfway point, and other points to the right of it, for making decisions, the probabilities of making the two types of error are calculated.

For the purpose of calculating these probabilities, it is assumed that x, the length of a skull, is approximately normally distributed with standard deviation $\sigma = 8$ and with mean $\mu = 190$ if the skull is from race A, and with mean $\mu = 196$ if the skull is from race B. Then $\bar{x}$ may be assumed to be normally distributed with standard deviation

$$\sigma_{\bar{x}} = \frac{\sigma}{\sqrt{n}} = \frac{8}{\sqrt{12}} = 2.31$$

and with mean 190 if the skulls are from race A and with mean 196 if they are from race B. The graphs of the two normal curves for $\bar{x}$ corresponding to H_0 and H_1 are shown in Fig. 1.

If the halfway point, 193, is used for the borderline of decisions, then the probability of making a type I error, that is, the probability of accepting H_1 when H_0 is true, is the probability that $\bar{x} > 193$ when H_0 is true. This probability is equal to the shaded area *under the H_0 curve* to the right of $\bar{x} = 193$. Its value, which is denoted by α, was found by the methods explained in Chapter 5 to be .10. The probability of making a type II error, that is, the probability of accepting H_0 when H_1 is true, is the probability that $\bar{x} < 193$ when H_1 is true. This probability is equal to the shaded area *under the H_1 curve* to the left of $\bar{x} = 193$. Its

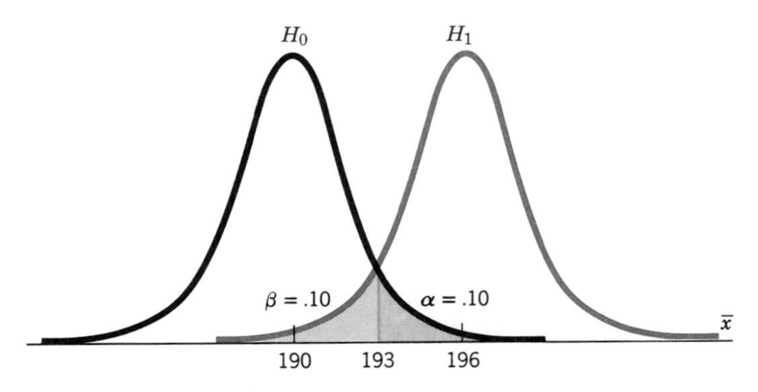

Figure 1 Distribution of $\bar{x}$ under H_0 and H_1.

value, which is denoted by β, is, by symmetry, the same as that for α; hence $\alpha = .10$ and $\beta = .10$.

If an archaeologist is fairly confident, through other sources of information, that the hypothesis H_0 is true, he will wish to make the probability of rejecting H_0 when it is actually true considerably smaller than the probability of rejecting H_1 when it is actually true. Thus he will want α to be considerably smaller than β. Now it is clear from Fig. 1 that if a point to the right of 193 were chosen for the borderline of decisions, the value of α would become smaller than .10 and the value of β would become larger than .10. Since it is not possible to decrease α without increasing β, the archaeologist will need to show some constraint in decreasing α or he will be faced with an unbearably large value of β. Suppose he decides that a value of $\alpha = .05$ will be small enough to give him the protection he desires against incorrectly rejecting H_0. This means that in only about one experiment in twenty will he incorrectly reject H_0 when it is true. With this choice agreed upon, it becomes necessary to select a value of $\bar{x}$ to the right of the halfway point such that the probability of making a type I error will be equal to $\alpha = .05$. Now, from Table IV in the appendix, it is known that 5 percent of the area of the standard normal curve lies to the right of $z = 1.64$. Since $\mu = 190$ and $\sigma_{\bar{x}} = 2.31$ here, and

$$z = \frac{\bar{x} - \mu}{\sigma_{\bar{x}}}$$

it follows that the value of $\bar{x}$ that cuts off a 5 percent right tail of the $\bar{x}$ curve is obtained by solving for $\bar{x}$ in the equation

$$1.64 = \frac{\bar{x} - 190}{2.31}$$

The solution of this equation, which is denoted by $\bar{x}_0$, is given by $\bar{x}_0 = 193.8$. Another manner of arriving at this value is to argue that it is necessary to go 1.64 standard deviations to the right of the mean of a normal distribution to obtain a value such that 5 percent of the area under the curve will be to the right of it. Thus, $\bar{x}_0$ must be given by

$$\bar{x}_0 = 190 + 1.64(2.31) = 193.8$$

Thus it follows that H_1 should be accepted here because the sample value $\bar{x} = 194$ is to the right of $\bar{x}_0 = 193.8$.

With this choice of $\bar{x}$ as the borderline value for making decisions, the value of β becomes the area under the H_1 curve to the left of $\bar{x} = 193.8$. By the methods explained in Chapter 5, the value of β will be found to be .17. Figure 2 displays these results geometrically. Although the value of β is considerably larger than the value of α here, as contrasted to using the halfway point which made $\beta = \alpha$

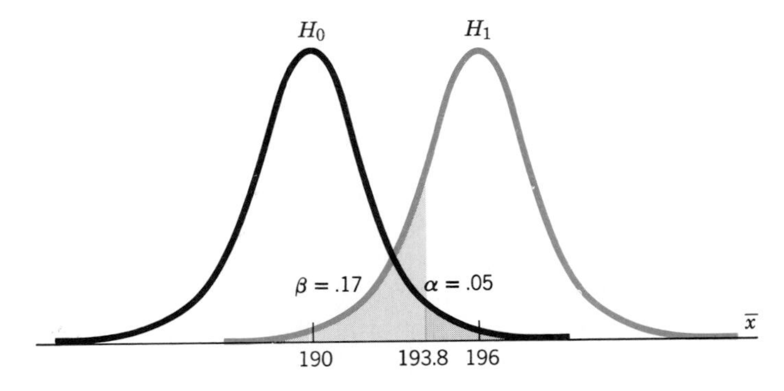

Figure 2 Distribution of $\bar{x}$ under H_0 and H_1, with selected critical region.

$= .10$, the archaeologist may consider the relative sizes of α and β to be satisfactory because he was much more concerned about making a type I error than about making a type II error. If the archaeologist should feel that the value of β is too large in relation to the value of α, all he would need to do is to decrease the value of $\bar{x}_0$ until he obtains a pair of values that are satisfactory to him in a relative sense.

The part of the $\bar{x}$-axis to the right of $\bar{x}_0$ is called the *critical region* of the test. It consists of those values of $\bar{x}$ that correspond to the rejection of H_0. The method for testing the hypothesis H_0 by means of $\bar{x}$ can be expressed very simply in terms of its critical region by stating that the hypothesis H_0 will be rejected if the sample value of $\bar{x}$ falls in the critical region of the test; otherwise H_0 will be accepted.

This method for testing the hypothesis H_0 is the method that will be used in this book for testing various hypotheses. It consists essentially of selecting a critical region for the variable being used to test the hypothesis such that if H_0 is true, the probability of the variable falling in the critical region is a fixed value α, and then agreeing to reject the hypothesis if, and only if, the sample value of the variable falls in the critical region. The experimental value of $\bar{x}$ is used only to make a decision after the critical region has been selected and is never permitted to influence the selection of the critical region.

In the foregoing problem α had the value .10 when the critical region was $\bar{x} > 193$ and the value .05 when the critical region was $\bar{x} > 193.8$. For problems of this type, the proper procedure is to choose the critical region so that the relative sizes of α and β are satisfactory; however, in many of the problems to come, this procedure would require lengthy computations and discussions of the relative importance of the two types of error involved. In order to avoid such lengthy discussions, a uniform procedure will be adopted of always choosing a critical region for which the value of α is .05. The value of $\alpha = .05$ is quite arbitrary here and some other value could have been agreed upon; however, it is the value of α most commonly used by applied statisticians. In any applied problem one can calculate the value of β and then adjust the value of α if the value of β is unsatisfactory when $\alpha = .05$. This works both ways, of course. For a very large experiment, with α fixed at .05, it might turn out that β is considerably smaller than .05. If the type I error were considered more serious than the type II error, then one would need to adjust the test to make α smaller than β, which would, of course, then make α smaller than .05.

This method of testing hypotheses requires one to choose a critical region for which $\alpha = .05$, but otherwise it does not determine how the critical region is to be chosen. Since it is desirable to make the probabilities of the two types of error as small as possible and since α is being fixed, one should choose a critical region that makes β as small as possible. Although the choice of the critical region in Fig. 2 was based on good sense, with the restriction that $\alpha = .05$, it can be shown that no other critical region with $\alpha = .05$ will have as small a value of β as the value $\beta = .17$.

Fortunately, in most simple problems an individual's good sense, or intuition, will lead him to a choice of critical region that is the best possible in the sense that it will minimize the value of β. For more difficult problems in testing hypotheses, there is a mathematical theory that enables statisticians to find best critical regions. The critical regions that have been chosen in the problems to be solved in this and later chapters are the ones obtained by using this theory whenever it applies.

2 TESTING A MEAN

The problem discussed in Section 1 is an illustration of the general problem of testing the hypothesis that the mean of a particular normal population has a certain value. That problem was rather unusual in that there was only one alternative value for the mean. In most practical problems one has no specific information about the possible alternative values of the mean in case the value being tested is not the true value. The most common situation is one in which all other values are possible. For such problems, the formulation corresponding to (1) assumes the form

$$\begin{aligned} H_0&:\mu = \mu_0 \\ H_1&:\mu \neq \mu_0 \end{aligned} \tag{2}$$

Here μ_0 denotes the particular value being tested. There are many practical problems, however, in which one is quite certain that if the mean is not equal to the value postulated under H_0 then its value must be larger than the postulated value. For such problems (2) is replaced by

$$\begin{aligned} H_0&:\mu = \mu_0 \\ H_1&:\mu > \mu_0 \end{aligned} \tag{3}$$

For problems in which one is quite certain that if the mean is not equal to μ_0 then its value must be smaller than μ_0, one would, of course, replace $\mu > \mu_0$ by $\mu < \mu_0$ in (3).

Example 1. As an illustration, suppose that a city has been purchasing brand A light bulbs for several years but is contemplating switching to brand B because of a better price. Salesmen for brand B claim that their product is just as good as brand A. Experience over several years has shown that the distribution of the length of life of light bulbs is approximately normal and that brand A bulbs have a mean life of 1180 hours, with a standard deviation of 90 hours. To test the claim of the salesmen for brand B, 100 of their bulbs, purchased from regular retail sources, were tested. This sample yielded the value $\bar{x} = 1140$. Since mean burning time is a good measure of quality, the problem now is to test the hypothesis that the mean of brand B is equal to the brand A mean against the alternative hypothesis that it has a smaller value. If the mean of brand B is denoted by μ, this test will assume the form of (3), namely,

$$H_0: \mu = 1180$$
$$H_1: \mu < 1180$$

This alternative was chosen because it was felt that if the quality of brand B bulbs were not the same as that of brand A bulbs then the brand B quality would undoubtedly be lower than the brand A quality. Salesmen are not likely to underrate their own products. Therefore, if these salesmen are telling the truth, H_0 will be true. If they are not telling the truth, their brand will be of lower quality because no salesman would be so stupid as to claim only equality when he could actually claim superiority for his product.

Now, good sense would suggest that the further $\bar{x}$ is to the left of the postulated mean of 1180, the less faith one should have in the truth of H_0 and the more faith one should have in some smaller value of μ being the true mean. Thus it is clear that the critical region should consist of small values of $\bar{x}$ and therefore of that part of the $\bar{x}$-axis to the left of some point $\bar{x}_0$. The problem therefore is to determine the point $\bar{x}_0$ so that the value of α will be .05. The technique for doing this is the same as that employed in solving the archaeologists' problem.

Since $n = 100$ here, it follows that

$$\sigma_{\bar{x}} = \frac{\sigma}{\sqrt{100}} = \frac{\sigma}{10}$$

If it is assumed that the variability of brand B bulbs is the same as that for the brand A bulbs, then $\sigma_{\bar{x}} = \frac{90}{10} = 9$.

Since the 5 percent left tail area of a standard normal curve lies to the left of the point $z = -1.64$, it follows that $\bar{x}_0$ is a point 1.64 standard deviations to the left of the mean $\mu = 1180$. The standard deviation here is $\sigma_{\bar{x}} = 9$; therefore the desired critical region is that part of the $\bar{x}$-axis to the left of

$$1180 - 1.64(9) \doteq 1165$$

These results are displayed in Fig. 3. To use algebra to obtain the value of $\bar{x}_0$, one can proceed as in the archaeologists' problem and write down the equation

$$-1.64 = \frac{\bar{x} - 1180}{9}$$

Solving for $\bar{x}$ yields the solution $\bar{x}_0 = 1165$.

Now that the critical region has been selected, one can proceed to test the hypothesis H_0. Since the sample value $\bar{x} = 1140$ falls in the critical region, the hypothesis H_0 will be rejected. It seems quite certain that a sample mean as low as 1140 could not have been obtained from a random sample of size 100 taken from a population with mean 1180. This implies that the salesmen of brand B bulbs are not justified in their claim of the same quality as brand A.

Since it is quite certain that μ is less than 1180, one should consider next the question of how much less. If a point estimate of μ were desired, then, of course, $\bar{x} = 1140$ would be selected as the estimate. One could also find a confidence interval for μ and then determine the maximum and minimum differences that are likely to exist between the two population means. Such considerations would be necessary before one could decide whether the lower price for brand B would

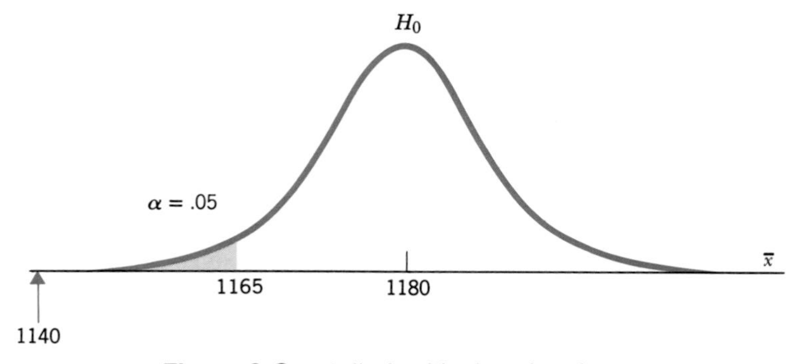

Figure 3 One-tailed critical region for $\bar{x}$.

compensate for the lower quality. Since the object of this section is to explain how to test hypotheses, these practical matters are not discussed here; however, the solutions of actual problems by statistical methods usually require such considerations.

Example 2. As an illustration for which formulation (2) would be preferred to (3), consider the following problem. Records for the last several years of applicants for a certain class of civil service positions showed that their mean score on an aptitude test was 115 and that the standard deviation of their scores was 20. An administrator is interested in knowing whether the caliber of recent applicants has changed. If he has no reason for believing that recent applicants are any better or any worse than former applicants, he should use formulation (2). This becomes

$$H_0:\mu = 115$$
$$H_1:\mu \neq 115$$

For the purpose of testing this hypothesis, the aptitude test scores of the last 50 applicants are obtained from the admissions office. Suppose that for this sample the mean is $\bar{x} = 118$.

Since the further $\bar{x}$ is from the hypothetical mean value of 115, whether to the right or the left, the less faith one would have in the truth of H_0, it is clear that the critical region here should consist of values of $\bar{x}$ out in the two tails of the $\bar{x}$ curve centered at 115. Now, for a sample as large as 50, $\bar{x}$ may be assumed to be normally distributed. Furthermore, if it is assumed that the variability of the scores for the new applicants is the same as that for former applicants, then $\sigma = 20$, and the standard deviation of $\bar{x}$ is given by

$$\sigma_{\bar{x}} = \frac{\sigma}{\sqrt{50}} = \frac{20}{\sqrt{50}} \doteq 2.8$$

Because the probability is .05 that $\bar{x}$ will assume a value more than 1.96 standard deviations away from the mean, it follows that the desired critical region of size $\alpha = .05$ should consist of the values of $\bar{x}$ out in the two tails of the $\bar{x}$ curve determined by the two values $115 - 1.96(2.8)$ and $115 + 1.96(2.8)$. These results are displayed in Fig. 4.

Since $\bar{x} = 118$ yields a point, indicated on Fig. 4 by an arrow, that does not fall into the critical region, the hypothesis H_0 will be accepted. The administration may relax in the knowledge that recent applicants are at about the same level of aptitude as former applicants.

The acceptance of a hypothesis in this manner is a practical decision matter. It does not imply that one believes that the hypothesis is precisely correct, and it

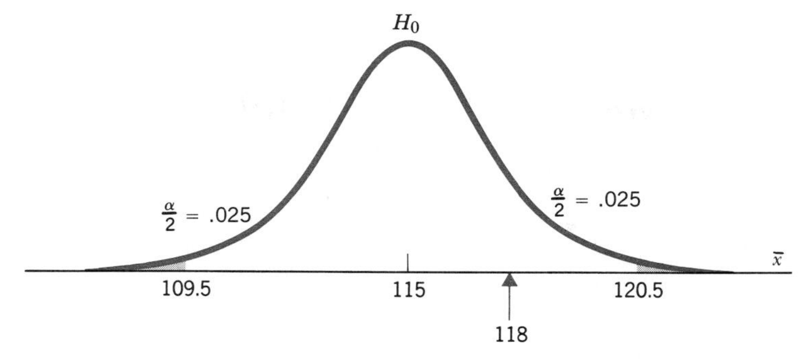

Figure 4 Two-tailed critical region for $\bar{x}$.

certainly is not a proof of the truth of the hypothesis. Rather, it implies that the sample data are compatible with the postulated value of the mean. From a practical point of view, it makes little difference whether the true mean has the postulated value or whether it has a value close to the postulated value. How close the true value of the mean must be to the postulated value in order that the hypothesis be accepted can be determined by the confidence-interval methods explained in Chapter 7. In view of these remarks, accepting a hypothesis is to be construed as admitting that the hypothesis is reasonably close to the true situation and that, from a practical point of view, one may therefore treat it as representing the true situation.

After the administrator notices that the sample mean is higher than the old mean, he will undoubtedly wish to claim that the new applicants are better than the former applicants. In view of this fact, there would be the temptation to treat this problem as one of testing $H_0{:}\mu = 115$ against $H_1{:}\mu > 115$ and to use a one-tailed critical region as in the earlier problems; however, this would be illegal because the decision as to the possible alternative values must be based on knowledge other than that given by the sample. A simple way of deciding whether to use a one-tailed or two-tailed test is to ask oneself what the logical alternative values of interest are before the sample has been taken or, what is equivalent, before the sample results have been observed.

In each of the preceding examples it was assumed that the value of σ needed to carry out the test was available from earlier experience. If that value is not available, or if there is reason to believe that the variability of the sample values is not the same as that for earlier samples, then it is necessary to replace σ by its sample estimate in the preceding tests. This approximation does not lead to any serious error if the sample is large. For small samples, a different technique is required. This will be explained in section 6. The following example is one for which the value of s would be used even if σ were available.

Example 3. A new technique for manufacturing ceramic tile is being proposed on the grounds that it produces tile of a more uniform quality without sacrificing strength. The standard process produces tile with a mean breaking strength of 42 pounds. To test the claims made for the new technique, 50 tiles were subjected to a breaking strength test. The mean was found to be 40.7 pounds and the standard deviation 3.5 pounds. The problem is to test the hypothesis that the population mean for tile of the new technique is the same as that for tile of the standard technique.

Since the new technique is believed to yield tile of more uniform quality than before, the standard deviation would be expected to be smaller than before; therefore the sample standard deviation will be used here. Now,

$$\sigma_{\bar{x}} = \frac{\sigma}{\sqrt{50}} \doteq \frac{s}{\sqrt{50}} = \frac{3.5}{\sqrt{50}} = .50$$

Since there was skepticism concerning the claim that strength would not be sacrificed, the critical region should consist of those values of $\bar{x}$ to the left of

$$\mu - 1.64\sigma_{\bar{x}} \doteq 42 - 1.64(.50) = 41.2$$

Because $40.7 < 41.2$, the hypothesis must be rejected. The new technique appears to have lowered the mean breaking strength slightly. Whether it is superior with respect to uniformity of quality would require comparing the sample standard deviation with that of the standard process. That type of problem will be considered in a later chapter.

In the preceding illustrations, each decision to accept or reject H_0 was made by looking at the geometry of the problem. An equivalent algebraic solution can be obtained by calculating the value of $z = (\bar{x} - \mu)/\sigma_{\bar{x}}$ and observing whether it is compatible with H_0. For example, in the skull problem, $z = \dfrac{194 - 190}{2.31} = 1.73$. Since the critical region was chosen to be the values of $\bar{x}$ in the 5 percent right tail area of its normal distribution, which corresponds to values of z larger than 1.64, and since $1.73 > 1.64$, it follows that H_0 should be rejected. For the problem of Example 1, $z = \dfrac{1140 - 1180}{9} = -4.44$. Here the critical region that was chosen corresponds to values of z smaller than -1.64; therefore H_0 should be rejected because $-4.44 < -1.64$. In Example 2, $z = \dfrac{118 - 115}{2.8} = 1.07$. Since a two-tailed critical region was chosen, the corresponding critical values of z consist of $|z| > 1.96$; therefore H_0 should be accepted.

The answers in the appendix are expressed in this algebraic form. They give the value of z and then the proper decision, based on this value.

The sizes of the type II errors in these problems have not been calculated because that would require considerably more discussion of the problems. This matter is considered in Section 8 for the benefit of those who are interested in knowing how large such errors are in problems such as these.

Example 4. A useful application of the idea of testing a mean arises in industrial quality-control work. Suppose that a machine is turning out a large number of parts that are used in some manufactured article and that it is important for the diameter of such a part to be very accurate. It is customary for an inspector to sample periodically from the production line to see whether the diameters are behaving properly. If 5 parts are measured every hour and their sample mean recorded, a large number of $\bar{x}$-values will be obtained after a few weeks of inspection. The mean of all these $\bar{x}$'s may be treated as the true mean of the population of diameters of parts and the standard deviation of these $\bar{x}$'s as the true value of $\sigma_{\bar{x}}$.

By treating the preceding values as true values, one can calculate the values $\mu - 3\sigma_{\bar{x}}$ and $\mu + 3\sigma_{\bar{x}}$. From normal curve properties, the probability is .997 that a sample value of $\bar{x}$ will fall between these two limits; therefore, if an $\bar{x}$-value falls outside this interval, there is good reason to believe that something has gone wrong with the machine turning out the parts. Experience has shown that a machine that is operating properly will behave very much like a random number machine in the sense that successive parts turned out behave very much like random samples from a population of parts. A three-standard deviation interval is used instead of, say, a two-standard deviation interval because about 5 percent of the $\bar{x}$-values would fall outside a two-standard deviation interval even though the machine is operating satisfactorily, and therefore the inspector would be looking for trouble too often when there is none. Furthermore, an industrial machine is only an approximation of an ideal random number machine. Experience indicates that a three-standard deviation interval is about right from a practical point of view.

After data have been gathered for a few weeks so that the limits $\mu - 3\sigma_{\bar{x}}$ and $\mu + 3\sigma_{\bar{x}}$ can be obtained, the control chart is ready to be constructed. It consists simply of a horizontal band with the horizontal axis marked off with sample numbers. A control chart of this type is shown in Fig. 5.

Each point corresponds to a sample value of $\bar{x}$ obtained from the five parts selected each hour, after the initial data-gathering period. It will be observed that the process appears to be under control. The striking advantage of a control chart is that it warns the inspector by means of probability of trouble with a machine before it has turned out a large number of bad parts, which otherwise might not be discovered until some time later when the parts were being used in assembling the article being manufactured.

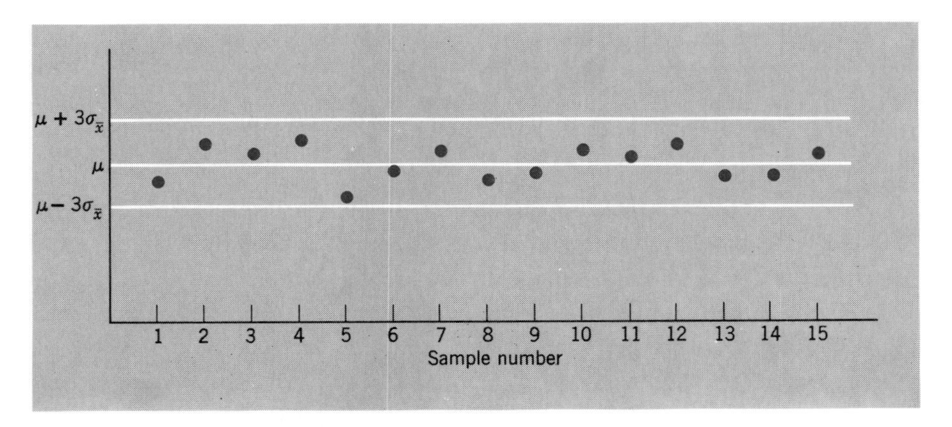

Figure 5 Control chart for the mean.

3 TESTING A PROPORTION

The large-sample normal curve methods employed to solve estimation problems for binomial p can be employed also to test hypotheses about p. As a result, the techniques for testing the hypothesis that p has a fixed value is much the same as those explained for means. As an illustration, consider the following problems.

Example 1. The executive of a large department store suspects that a competing store is attempting to undersell his store. Previously the two stores had maintained a price balance in that about half the items in one store were slightly higher than those in the other and about half were lower. To investigate this possibility he sends a "shopper" into the other store to check the prices on 200 randomly selected articles. If it is found that 120 of those articles are priced lower than in his store, is he justified in his suspicions?

This problem may be considered as a problem of testing the hypothesis

$$H_0 : p = \frac{1}{2}$$

against the hypothesis $H_1 : p \neq \frac{1}{2}$, and in which p denotes the proportion of prices that are lower in the second store. The 200 items may be treated as 200 trials of an experiment for which $p = \frac{1}{2}$ is the probability of success in a single trial. From formula (10), Chapter 5, it follows that $\hat{p} = x/n$ may be treated as a normal variable with mean $p = \frac{1}{2}$ and standard deviation given by

$$\sigma_{\hat{p}} = \sqrt{\frac{pq}{n}} = \sqrt{\frac{\frac{1}{2} \cdot \frac{1}{2}}{300}} = .035$$

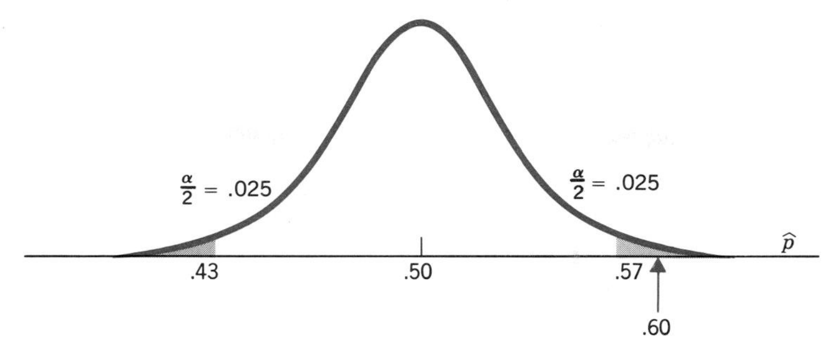

Figure 6 Two-tailed critical region for $\hat{p}$.

The problem now is much the same as the problem of testing a normal mean. The critical region here is chosen as those values of $\hat{p}$ in the two tails of the normal curve for $\hat{p}$. For $\alpha = .05$, the critical region will then consist of those values of $\hat{p}$ lying outside the interval given by

$$(p - 1.96\sigma_{\hat{p}}, p + 1.96\sigma_{\hat{p}})$$

Since $n = 200$ and $p = \frac{1}{2}$ here, computations will yield the interval (.43, .57). Figure 6 shows the approximate normal distribution for $\hat{p}$ and the critical region just determined. Since the sample value $\hat{p} = \frac{120}{200} = .60$ falls in the critical region, the hypothesis H_0 is rejected. Thus, on the basis of these data it appears that the executive is justified in his suspicions. In view of the belief that the other store had lower prices, it would have been more appropriate to use a one-tailed test here, in which case a right-tail critical region would have been selected. The conclusion would have been the same.

Example 2. As a second illustration of how to use normal curve methods to test a hypothesis about binomial p, consider the problem discussed briefly near the end of Chapter 5. A politician had claimed a 60 percent backing on a piece of legislation and a sample of 400 voters had been taken to check this claim. The question then arose as to how small the sample percentage would need to be before the claim could be rightfully refuted. This problem can be considered as a problem of testing the hypothesis

$$H_0: p = .6$$

Since the interest in this problem centers on whether $p = .6$, as against the possibility that $p < .6$, this problem is somewhat like that of the light bulbs discussed in the first section of this chapter in that the alternatives are all on one side of the hypothetical value; therefore the critical region should be under one tail of the proper normal curve. The natural alternative hypothesis here is

$$H_1: p < .6$$

The critical region of size $\alpha = .05$ for this problem should therefore be selected to be under the left 5 percent tail of the normal curve whose mean is $p = .6$ and whose standard deviation is given by

$$\sigma_{\hat{p}} = \sqrt{\frac{pq}{n}} = \sqrt{\frac{(.6)(.4)}{400}} = .0245$$

Since, from Table IV, 5 percent of the area of a standard normal curve lies to the left of $z = -1.64$, this means that the critical region should consist of all those values of $\hat{p}$ that are smaller than the value of $\hat{p}$ that is 1.64 standard deviations to the left of the mean. For this problem, the critical region therefore consists of all those values of $\hat{p}$ that are smaller than

$$p - 1.64\ \sigma_{\hat{p}} = .6 - 1.64(.0245) = .56$$

If the sample value of $\hat{p}$ turned out to be less than .56, the politician's claim would be rejected.

Another useful application of testing a binomial p arises in industrial control charts for the percentage of defective parts in mass production of parts. This application is not confined to industrial problems; it may be used wherever one has repeated operations. The technique is precisely the same as for control charts for the mean. One uses accumulated experience to obtain a good estimate of p; then one constructs the control band given by $p - 3\sqrt{pq/n}$ and $p + 3\sqrt{pq/n}$. Here n is the size sample on which each plotted proportion is based.

Example 3. As an illustration of how one would construct such a chart, consider the following problem. A record is kept during a period of ten days of the number of words mistyped by various students learning typing. During that period of time those students typed a total of approximately 20,000 words, of which 800 were mistyped. The problem is to use these data to construct a control chart for the proportion of errors made per class hour by a student who types approximately 600 words per class hour. Assuming a given student is typical, a good estimate of p is given by dividing the total number of mistyped words by the total number of typed words. This estimate is

$$p \doteq \frac{800}{20{,}000} = .04$$

Since the proportions to be plotted on the control chart are those for an hour's typing, it follows that $n = 600$ here. If these values are substituted in the formulas given in the preceding paragraph, the desired lower and upper boundaries for the control chart become

$$.04 - 3\sqrt{\frac{(.04)(.96)}{600}} \quad \text{and} \quad .04 + 3\sqrt{\frac{(.04)(.96)}{600}}$$

These simplify to .016 and .064. The chart can now be constructed in the same manner as for the mean, except that now one plots the proportion of mistyped words every hour rather than the sample mean.

4 TESTING THE DIFFERENCE OF TWO MEANS

The problem of the light bulbs that was solved in Section 2 can be modified slightly to produce a problem that is typical of many in real life. Suppose the city buying light bulbs had no experience with either brand A or brand B bulbs and wished to decide which brand to purchase, the prices being the same. It would then be necessary to test a sample of each brand, rather than just a sample of brand B as in the earlier problem. Suppose a sample of 100 bulbs from each of the two brands is tested and that the samples yield the values $\bar{x}_1 = 1160$, $s_1 = 90$, $\bar{x}_2 = 1140$, and $s_2 = 80$, in which the subscripts 1 and 2 refer to brands A and B, respectively, and the units are in hours.

Now since brand A yields a larger mean burning time than brand B, it would appear that brand A is superior to brand B; however, it might be that the reverse is true, but some bad luck with a few of the bulbs of brand B produced an unusually low sample mean. A second set of samples of 100 each might conceivably produce contrary results. The problem therefore reduces to determining whether this difference of sample means, namely $\bar{x}_1 - \bar{x}_2$, is large enough to justify the belief that brand A is superior to brand B.

In order to solve this problem it is necessary to know how the random variable $\bar{x}_1 - \bar{x}_2$ is distributed. This was discussed in section 6 of Chapter 6. There it was stated that if $\bar{x}_1$ and $\bar{x}_2$ are independent normal variables, or if they are independent and n_1 and n_2 are large enough to justify the assumption that $\bar{x}_1$ and $\bar{x}_2$ possess approximately normal distributions, then the variable $\bar{x}_1 - \bar{x}_2$ will possess a normal, or an approximately normal, distribution with parameters given by the following formulas:

$$\begin{aligned} \mu_{\bar{x}_1-\bar{x}_2} &= \mu_1 - \mu_2 \\ \sigma_{\bar{x}_1-\bar{x}_2} &= \sqrt{\frac{\sigma_1^2}{n_1} + \frac{\sigma_2^2}{n_2}} \end{aligned} \tag{4}$$

It is clear from the nature of the experiment that $\bar{x}_1$ and $\bar{x}_2$ are independent random variables and that n_1 and n_2 are large enough to justify the normality assumption. Now, from a practical point of view, it should make little difference whether one rejects the hypothesis that the brands are equally good as far as mean burning times are concerned or accepts the hypothesis that they differ in quality. From a theoretical point of view, however, it is more convenient to test the hypothesis that the brands are equally good rather than the hypothesis that they differ in quality. As a result, one sets up the hypothesis

$$H_0\colon \mu_1 = \mu_2 \tag{5}$$

An equivalent way of writing this is

$$H_0\colon \mu_1 - \mu_2 = 0$$

This type of hypothesis is known as a *null* hypothesis because it assumes that there is no difference. Very often, however, the experimenter believes that there is an appreciable difference and hopes that the sample evidence will reject the hypothesis. If the sample does reject the hypothesis, then one can claim with justification that a real difference in population means exists. If the sample does not reject the hypothesis, then there is a fair probability that the sample difference is caused by sampling variation, under the assumption that the population means are equal.

In the light of the preceding discussion, testing the hypothesis given by (5) is equivalent to testing the hypothesis that the mean of the normal variable $\bar{x}_1 - \bar{x}_2$ is 0. But this type of problem was solved in Section 2. Since the alternative hypothesis would ordinarily be chosen to be

$$H_1\colon \mu_1 \neq \mu_2$$

it follows that the same methods should be applied here as in the solution of the the aptitude test problem, which used the formulation in (2). Now, from the preceding theory, the variable $\bar{x}_1 - \bar{x}_2$ may be assumed to be normally distributed with mean 0, because $\mu_1 = \mu_2$ under H_0, and with its standard deviation given by formula (4). Unfortunately, the population values σ_1^2 and σ_2^2 are unknown; consequently, they must be approximated by their sample estimates, namely $s_1^2 = (90)^2$ and $s_2^2 = (80)^2$. Since $n_1 = n_2 = 100$ here, the approximate value of the standard deviation becomes

$$\sigma_{\bar{x}_1 - \bar{x}_2} \doteq \sqrt{\frac{8100}{100} + \frac{6400}{100}} \doteq 12$$

A critical region for which $\alpha = .05$ is chosen based on equal tail areas under the normal curve for $\bar{x}_1 - \bar{x}_2$. This means that the critical region consists of that part of the horizontal axis lying more than $1.96\sigma_{\bar{x}_1 - \bar{x}_2} (\doteq 24)$ units away from 0. Figure 7 shows geometrically the distribution of $\bar{x}_1 - \bar{x}_2$ and the selected critical region.

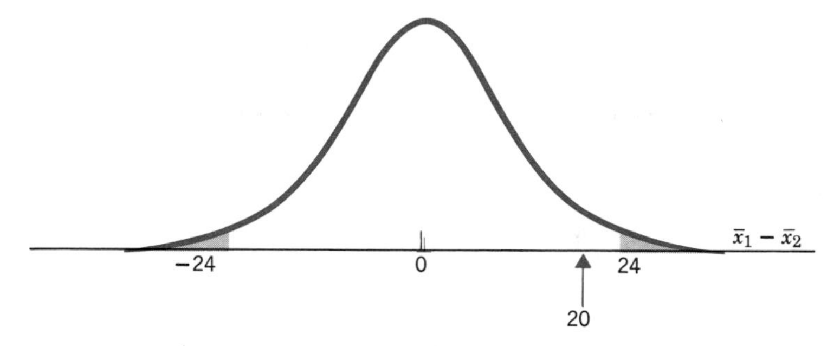

Figure 7 Two-tailed critical region for $\bar{x}_1 - \bar{x}_2$.

Since the two samples of 100 each yielded the value $\bar{x}_1 - \bar{x}_2 = 1160 - 1140 = 20$ and since 20 does not fall in the critical region for this test, the hypothesis is accepted.

It is not necessary to spend any time on testing problems concerning the mean of Poisson or exponential distributions because they can be treated satisfactorily by normal-curve methods.

Just as in the earlier problem of testing a single mean, the acceptance of a hypothesis in this manner does not imply that one believes that the hypothesis is true. It does imply, however, that one is not convinced by the sample evidence that there is an appreciable difference and that, unless further evidence is presented to the contrary, one is willing to assume that for all practical purposes there is no appreciable difference in the population means. It is a mathematical convenience to formulate a hypothesis in this manner. It would be more realistic to test whether the means differed by less than a specified amount, but the resulting theory would be much more complicated. A student should not deceive himself into believing that he has proved the hypothesis to be true just because he has agreed to accept it.

In view of the fact that sample estimates were needed as approximations for population variances, the methods used here are large-sample methods.

There are several words and phrases used in connection with testing hypotheses that should be brought to the attention of students. When a test of a hypothesis produces a sample value falling in the critical region of the test, the result is said to be *significant;* otherwise one says that the result is *not significant.* This word arises from the fact that a sample value falling in the critical region is not compatible with the hypothesis and therefore signifies that some other hypothesis is necessary. The probability of committing a type I error, which is denoted by α, is called the *significance level* of the test. For problems being solved routinely in this book, the significance level has been chosen equal to .05.

If one analyzes the technique that has been used to test the various hypotheses that have been treated thus far he will observe that it is merely a rule for making a decision. This rule is usually based on the sample value of some random variable and consists of dividing all possible values of the random variable into two groups, those that are associated with the rejection of H_0 and that form what is called the critical region of the test, and those associated with the acceptance of H_0. From this general point of view, a test of a hypothesis is merely a systematic way of making a practical decision and there is no implication made concerning the truth or falsity of the hypothesis being treated.

5 TESTING THE DIFFERENCE OF TWO PROPORTIONS

A problem of much importance and frequent occurrence in statistical work is the problem of determining whether two populations differ with respect to a certain attribute. For example, is there any difference in the percentages of smokers and nonsmokers who have heart ailments?

Problems of this type can be treated as problems of testing the hypothesis

$$H_0{:}p_1 = p_2$$

in which p_1 and p_2 are the two population proportions of the attribute. If n_1 and n_2 denote the size samples taken and $\hat{p}_1$ and $\hat{p}_2$ the resulting sample proportions obtained, then the variable to use in solving this problem is $\hat{p}_1 - \hat{p}_2$. This corresponds to using $\bar{x}_1 - \bar{x}_2$ in the problem of testing the hypotheses that $\mu_1 = \mu_2$. The methods used to solve that problem can be employed here as well because $\hat{p}_1$ and $\hat{p}_2$ may be treated as two independent normal variables. Hence, from the theory of section 6, Chapter 6, $\hat{p}_1 - \hat{p}_2$ may be considered as being approximately normally distributed with mean $p_1 - p_2$ and with standard deviation given by

$$\sigma_{\hat{p}_1 - \hat{p}_2} = \sqrt{\frac{p_1 q_1}{n_1} + \frac{p_2 q_2}{n_2}}$$

When testing the hypothesis $H_0{:}p_1 = p_2$, the mean of the distribution of $\hat{p}_1 - \hat{p}_2$ will, of course, be equal to 0.

Example. As an illustration of how to use these formulas, consider the following problem. A sample of 400 sailors was split into two equal groups by random selection. One group was given brand A pills of a seasickness preventive, and the other brand B pills. The number in each group that refrained from becoming seasick during a heavy storm was 152 and 132. Can one conclude that there is no real difference in the effectiveness of these pills?

Calculations give

$$\hat{p}_1 = \frac{152}{200} = .76, \qquad \hat{p}_2 = \frac{132}{200} = .66$$

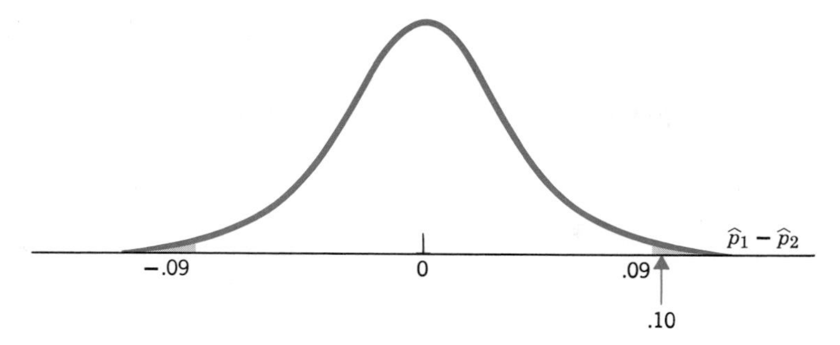

Figure 8 Two-tailed critical region for $\hat{p}_1 - \hat{p}_2$.

$$\sigma_{\hat{p}_1-\hat{p}_2} = \sqrt{\frac{p_1q_1}{200} + \frac{p_2q_2}{200}}$$

Since the values of p_1 and p_2 are unknown, they must be approximated by sample estimates. Although the values are unknown, they are assumed to be equal under the hypothesis $H_0:p_1 = p_2$. If this common value is denoted by p, then a good estimate of p is the value obtained from the sample proportion of the combined data. There were 284 of 400 sailors who were successes in this total experiment; hence p would be estimated by means of

$$\hat{p} = \frac{284}{400} = .71$$

By replacing p_1 and p_2 by p in the formula for the standard deviation and then approximating p by $\hat{p}$, one obtains

$$\sigma_{\hat{p}_1-\hat{p}_2} \doteq \sqrt{(.71)(.29)(\tfrac{1}{200} + \tfrac{1}{200})} = .045$$

The use of a two-tailed critical region of size .05 yields the critical region displayed in Fig. 8. Since $\hat{p}_1 - \hat{p}_2 = .10$ here, it falls in the critical region and therefore the hypothesis is rejected. It would appear that brand A gives somewhat better protection against seasickness than brand B, at least for sailors in stormy weather.

6 A SMALL-SAMPLE METHOD FOR ONE MEAN

In the examples of section 2 the value of σ was available from earlier experience. As was stated there, if previous experience does not supply the value of σ it is necessary to replace σ by its sample estimate, s. This is satisfactory for large samples but a serious error may result from this approximation if n is small. For such small-sample problems, one can use Student's t distribution in the same manner that normal z is used for large samples.

Example. As an illustration of a small-sample problem, consider the following problem. A new formula for the propellant of missiles is being tested to determine whether it is superior to the current formula. From past experience, the mean distance traveled by experimental missiles shot into the Pacific Ocean is 340 miles. Since the new propellant is expected to affect the variability of the distance traveled as well as the mean distance, the value of σ based on past experience will not be used here. Ten experimental missiles with the new propellant were shot into the Pacific Ocean and produced a mean distance of $\bar{x} = 360$ miles and a standard deviation of $s = 20$ miles. The problem is to test the hypothesis $H_0: \mu = 340$ against the alternative hypothesis $H_1: \mu > 340$.

It suffices to calculate the value of t given by formula (9) in Chapter 7. For this problem

$$t = \frac{\bar{x} - \mu}{s} \sqrt{n} = \frac{360 - 340}{20} \sqrt{10} = 3.16$$

From Table V in the appendix, it will be found that for $\nu = 9$ degrees of freedom, the .05-value of t is 1.833. Since the critical region is one-tailed and consists of the values of $t > 1.833$, the value $t = 3.16$ falls in the critical region and therefore H_0 is rejected. The new propellant has undoubtedly increased the mean distance traveled.

Since the derivation of Student's t distribution requires the assumption that the basic variable x is normally distributed, one must be a little careful when applying it to small samples to make certain that x possesses an approximately normal distribution. The large-sample method does not require this precaution because $\bar{x}$ is likely to be very nearly normally distributed even for moderate size samples; however the large-sample method has the more serious fault of requiring a knowledge of σ or a good estimate of it, and this is not likely to be available in small-sample practical problems.

7 A SMALL-SAMPLE METHOD FOR TWO MEANS

If the sample sizes are too small to justify replacing σ_1 and σ_2 by their sample estimates in the test of Section 4, then the appropriate Student's t test may be used. For testing the difference of two means, the theory of Student's t distribution requires one to assume that the two basic variables x_1 and x_2 possess independent normal distributions with equal standard deviations. These assumptions are considerably more restrictive than those needed for the large-sample method. If these assumptions are reasonably satisfied, then one may treat the variable

$$t = \frac{(\bar{x}_1 - \bar{x}_2) - (\mu_1 - \mu_2)}{\sqrt{(n_1 - 1)s_1^2 + (n_2 - 1)s_2^2}} \sqrt{\frac{n_1 n_2(n_1 + n_2 - 2)}{n_1 + n_2}} \tag{6}$$

as a Student's t variable with $\nu = n_1 + n_2 - 2$ degrees of freedom. This formula was obtained by replacing the denominator of the standard normal variable

$$z = \frac{(\bar{x}_1 - \bar{x}_2) - (\mu_1 - \mu_2)}{\sigma_{\bar{x}_1 - \bar{x}_2}}$$

by an appropriate sample estimate. That estimate, which uses the assumption that $\sigma_1 = \sigma_2$, and some algebra produced the expression for t. The solution is now carried out in the same manner as for testing a single mean. For example, if the problem solved in Section 4 is altered to make the sample sizes 10 each, then the value of t is

$$t = \frac{1160 - 1140}{\sqrt{9(90)^2 + 9(80)^2}} \sqrt{\frac{100(18)}{20}} = .53$$

From Table V in the appendix it will be found that the 5 percent critical value of t corresponding to $\nu = 18$ degrees of freedom for a two-tailed test is 2.10. Since the value $t = .53$ falls inside the noncritical interval, which extends from -2.10 to $+2.10$, the hypothesis is accepted. The hypothesis was accepted before for a much larger sample, therefore it would obviously be accepted here as well.

Modifications of the foregoing t test do exist for problems in which it is unreasonable to assume that the two variances are equal; however, they will not be considered here.

Matched Pairs. In some experiments designed to test whether two population means are equal, the investigator designs his experiment in such a way as to compare similar types of individuals in the hopes of making the test more efficient. For example, if an experiment is being run to determine whether two different methods of teaching arithmetic yield different results, it is often advisable to select pairs of students who have similar previous grades in arithmetic and who resemble each other in other related aspects as well. If the scores made by the ith pair of such students are denoted by x_i and y_i and there are n such pairs available, then one treats the variable $z_i = x_i - y_i$ as the ith sample value in a sample of size n of a normal variable and tests the hypothesis that the mean of this variable is zero. Thus, the test is reduced to that of testing a single normal mean. It is assumed here that after the pairings have been made, one student of each pair is selected by chance to go into the class being taught by the first method, with the other student going into the other class. Since pairing similar students is likely to eliminate much of the variability in student performance, a test based on such pairings is more likely to be able to prove that a difference exists, when it does, than is a test for which formula (6) is appropriate and which requires two sets of n randomly selected students. The preceding test is usually called a *matched-pairs test.* After the differences have been obtained, it is carried out like the test in section 6.

Example. As an illustration of a matched-pairs problem, consider the following data that give the scores made by twelve matched pairs of students in an arithmetic examination. Half the students were taught under method I and half under method II.

Method I	65	40	63	78	67	34	76	57	75	88	77	75
Method II	60	42	65	71	62	35	74	54	71	82	77	67

Taking differences and letting x_i denote the ith such difference, it follows that

i	1	2	3	4	5	6	7	8	9	10	11	12
x_i	5	−2	−2	7	5	−1	2	3	4	6	0	8

Calculations give $\bar{x} = 2.92$ and $s = 3.50$. Under the hypothesis that $\mu = 0$,

$$t = \frac{2.92 - 0}{3.50}\sqrt{12} = 2.89$$

Because there was no agreement before the experiment began that one method ought to be superior, a two-tailed test should be used here. Since $\nu = 11$, the critical value is $t_0 = 2.20$ from Table V. The value 2.89 lies in the critical region; therefore the hypothesis that the two methods are equally good is rejected. Either method I is slightly superior or the teachers using method I were doing a better job of teaching.

If the ordinary test of $H_0{:}\mu_1 = \mu_2$ had been used here, although it would be incorrect to do so because the sample values are not independent, it will be found that H_0 would be accepted.

8 THE OPERATING CHARACTERISTIC

Because of the seemingly large interval of values corresponding to the acceptance of H_0, as shown in Fig. 7 for the problem in section 4, one might be disturbed by the possibility that brand A is really better than brand B, but this fact is not being discovered by the test. In order to check this possibility, it is necessary to study the size of the type II error.

In any practical situation like this it should be possible for the individual concerned to specify the smallest difference in the population means that would be considered of practical importance to him. In the light bulb problem in section 4, for example, the purchasing agent might state that a difference smaller than 25 is too small to be important but that any larger difference is of economic importance. Now, consider the alternative hypothesis based upon this smallest important difference, namely,

$$H_1: \mu_1 - \mu_2 = 25$$

The value of β can be calculated for this H_1 in the manner of the archaeologists' problem. The distribution of $\bar{x}_1 - \bar{x}_2$ under H_1 will be the same as under H_0, except that the mean will be 25 instead of 0. The graphs of these two distributions are shown in Fig. 9. The value of β is given by the area under the H_1 curve from -24 to $+24$ because this interval is the noncritical region of the test; however, this area is practically equivalent to the area under the H_1 curve to the left of 24. Since the standard deviation here is 12 and the mean is 25, the z-value corresponding to 24 is

$$z = \frac{24 - 25}{12} = -.08$$

From Table IV in the appendix it will be found that the probability that z will lie to the left of $-.08$ is .47; consequently $\beta = .47$ here. This means that about half the time a difference of 25 in the population means will not be detected by this test.

If the difference in the population means is actually greater than 25, then, of course, the value of β will be smaller than .47. For example, if the true difference is 50, then similar calculations with the H_1 curve now centered over 50 will show that β assumes the value .015. Thus one is almost certain to detect a difference

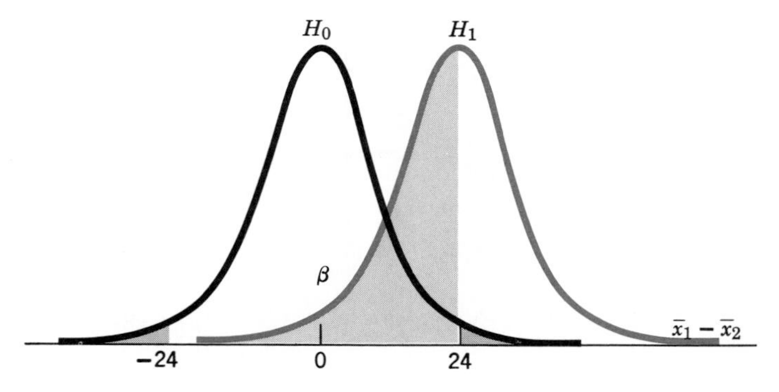

Figure 9 The distribution of $\bar{x}_1 - \bar{x}_2$ under H_0 and under H_1.

as large as 50 with this test, that is, with a sample of this size. If different values of $\mu_1 - \mu_2$ are postulated, starting with 0 and increasing in regular steps, and the value of β is calculated for each such alternative, then these values when plotted against the value of $\mu_1 - \mu_2$ will show how good the test is for detecting true differences when they exist. The values of β were calculated for this problem for steps of 10 in the value of $\mu_1 - \mu_2$. These values were graphed and a smooth curve was drawn through the resulting points, as shown in Fig. 10. The curve is symmetric; therefore, only the positive-axis half was calculated. This curve is called the *operating characteristic* of the test. It enables one to determine how good the test is for various values of $\mu_1 - \mu_2$. For example, it is clear from this graph that a difference of 10 in the population means will seldom be detected because β is very large for this value of $\mu_1 - \mu_2$. This is not a serious matter, however, if the purchasing agent is not interested in differences less than 25. A more serious matter is the relatively large value of β for $\mu_1 - \mu_2$ having a value between 25 and 35.

If the value of β for values of $\mu_1 - \mu_2$ between 25 and 35 is larger than desired, then two methods for decreasing the value of β are possible. The first method consists in choosing a larger critical region. Thus, instead of using the two $2\frac{1}{2}$ percent tails of the distribution for determining the critical region, one might choose the two 5 percent tails. This larger critical region will, of course, increase the value of α from .05 to .10. Calculations will show that the value of β for $\mu_1 - \mu_2 = 25$ will now decrease from .47 to .33. The relative importance of the two types of error here would determine how much α should be allowed to increase in order to decrease β. The second method of decreasing β consists in taking a larger sample. If α is to be fixed at a value such as .05, then the only way to decrease β is to take a larger sample. As an illustration, suppose the sample size in the problem being discussed is increased from 200 to 400. Then, assuming that the sample variances did not change, the standard deviation will be found to decrease from 12 to 8.5. The two-sided critical region is now determined by -17

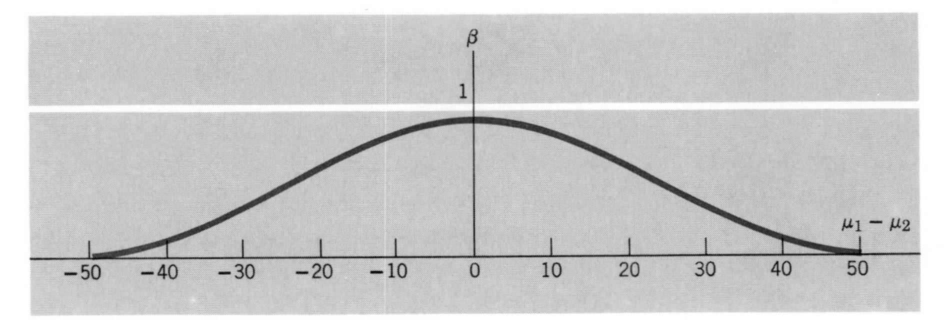

Figure 10 Operating characteristic for a two-tailed test.

and +17. Calculations of the same type as those used earlier will show that the value of β for $\mu_1 - \mu_2 = 25$ now becomes .17.

Another way of looking at the efficiency of a test is to graph $1 - \beta$, rather than β itself, for various values of the parameter being tested. Such a graph is called the *power curve* of the test and by means of it one can determine the "power" of the test, $1 - \beta$, for possible alternative values of the parameter. If the parameter has a value other than the value being tested, then $1 - \beta$ gives the probability that the alternative value of the parameter will be accepted, which is the correct decision.

To observe how the power curve changes with increasing size samples, power curves were drawn for the preceding problem for samples of sizes 200, 400, and 800. They are shown in Fig. 11. The power curve for $n = 200$ was obtained directly from the calculations made to graph Fig. 10.

If, for example, $\bar{x}_1 - \bar{x}_2 = 25$ and $n = 200$, which are the values considered earlier, then from the power curve labeled $n = 200$, the power of the test is .53. If, however, the sample size had been 400, the power would be .84. If calculations are carried to only two decimal places, the power of this test for $n = 800$ is 1.00. This last result implies that if $n = 800$ the test is almost certain to reject the hypothesis of equal means when the difference of the sample means is a large as 25.

It should be clear from the foregoing discussion of this problem that anyone who intends to take a sample for the purpose of testing some hypothesis about a mean or any other parameter should concern himself with the operating characteristic of his proposed test so that he will be able to tell whether his sample is large enough to give him the protection he desires against making various type I and

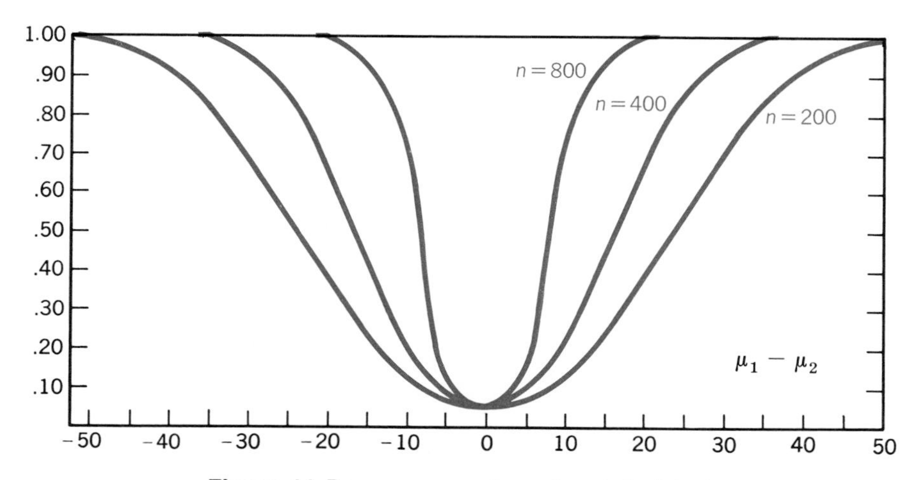

Figure 11 Power curves for a two-tailed test.

type II errors. The size sample needed to make β reasonably small for $\alpha = .05$ is often much larger than anticipated by those who apply significance tests.

In view of the preceding calculations and comments it should be apparent that the technique of testing hypotheses must be handled with care. In many problems the desired information is best given by means of a confidence interval for the parameter in question rather than in a significance test. It is only when one is required to make a definite yes-or-no decision that hypothesis testing is appropriate. In making such a decision, however, one must bear in mind that when the sample is small the hypothesis H_0 will very likely be accepted, whereas when the sample is large the reverse will be true, because hypotheses of the type H_0 that have been considered in this chapter are seldom exactly true in real-life situations.

9 REVIEW ILLUSTRATIONS

Example 1. At two similar industrial plants random samples of 30 workers each were selected and a record kept of the number of units of work turned out by them during one week. The results of that investigation yielded the values $\bar{x}_1 = 340$, $s_1 = 20$, $\bar{x}_2 = 355$, $s_2 = 30$. (*a*) According to a labor agreement workers are expected to average 350 units of work a week. Using the data from the first plant, test the hypothesis $H_0:\mu = 350$ against $H_1:\mu < 350$. (*b*) Test the hypothesis $H_0:\mu_1 = \mu_2$. (*c*) Work (*a*) and (*b*) using the t distribution.

$$(a)\quad \sigma_{\bar{x}} = \frac{\sigma}{\sqrt{30}} \doteq \frac{s}{\sqrt{30}} = \frac{20}{\sqrt{30}} = 3.65;\ \text{hence}$$

$$z = \frac{\bar{x} - \mu}{\sigma_{\bar{x}}} \doteq \frac{340 - 350}{3.65} = -2.7$$

hence reject $H_0:\mu = 350$.

$$(b)\quad \sigma_{\bar{x}_1 - \bar{x}_2} \doteq \sqrt{\frac{(20)^2}{30} + \frac{(30)^2}{30}} = 6.58;\ \text{hence}$$

$$z = \frac{\bar{x}_1 - \bar{x}_2}{\sigma_{\bar{x}_1 - \bar{x}_2}} \doteq \frac{-15}{6.58} = -2.3$$

hence reject $H_0:\mu_1 = \mu_2$.

(*c*) Reworking (*a*),

$$t = \frac{\bar{x} - \mu}{s}\sqrt{n} = \frac{340 - 350}{20}\sqrt{30} = -2.7$$

and $t_0 = 1.699$; hence reject H_0. The workers at this plant are not as productive as expected. Reworking (*b*),

$$t = \frac{\bar{x}_1 - \bar{x}_2}{\sqrt{(n_1 - 1)s_1^2 + (n_2 - 1)s_2^2}} \sqrt{\frac{n_1 n_2 (n_1 + n_2 - 2)}{n_1 + n_2}}$$

$$= \frac{-15}{\sqrt{29[20^2 + 30^2]}} \sqrt{\frac{30 \cdot 30 \cdot 58}{60}} = -2.28$$

Since 58 exceeds the table values for degrees of freedom, t is treated as a standard normal variable. Thus, $t_0 = 1.96$ for a two-sided test; consequently H_0 is rejected. It clearly suffices to use large-sample normal distribution methods here by treating s_1^2 and s_2^2 as σ_1^2 and σ_2^2.

Example 2. A manufacturer claims that at most 5 percent of his articles are defective. Two samples of 200 each were taken from shipments a month apart and inspected carefully for flaws. It was found that the first shipment contained 12 defectives and the second shipment had 20 defectives. (*a*) Combine these two sample results and test the hypothesis $H_0{:}p = .05$. (*b*) Use these sample results to test the hypothesis $H_0{:}p_1 = p_2$.

(*a*) $p = .05,\ n = 400,\ \hat{p} = \dfrac{32}{400} = .08$

$$z = \frac{\hat{p} - p}{\sqrt{\dfrac{pq}{n}}} = \frac{.08 - .05}{\sqrt{\dfrac{(.05)(.95)}{400}}} = 2.75$$

hence reject H_0. The manufacturer's claims are not substantiated.

(*b*) $\hat{p}_1 = \dfrac{12}{200} = .06,\ \hat{p}_2 = \dfrac{20}{200} = .10$

$$z = \frac{\hat{p}_1 - \hat{p}_2}{\sqrt{\hat{p}\hat{q}\left[\dfrac{1}{n_1} + \dfrac{1}{n_2}\right]}} = \frac{.06 - .10}{\sqrt{(.08)(.92)\left[\dfrac{1}{200} + \dfrac{1}{200}\right]}} = -1.47$$

hence accept H_0.

Example 3. As reported in *Newsweek* for April 27, 1974, an experiment was conducted by the Reproductive Biology Research Foundation in St. Louis to determine the effects of marijuana on sexuality. In the experiment 20 young men were selected who were in good health and who had smoked marijuana at least 4 days a week for a minimum of 6 weeks, without using any other drugs during that period. A control group of 20 young men who had never smoked marijuana was used for comparison. The measure of sexuality used was the level of the male sex hormone testosterone in the blood.

Letting the subscripts 1 and 2 correspond to the marijuana and non-marijuana groups, respectively, the experiment yielded the following testosterone-levels data:

$$\bar{x}_1 = 416, \bar{x}_2 = 742, s_1 = 152, s_2 = 130$$

(*a*) Test the hypothesis $H_0:\mu_1 = \mu_2$. (*b*) Test the same hypothesis using small-sample methods.

(*a*) Since the samples are assumed to be large enough to use large-sample methods, it suffices to calculate

$$\sigma_{\bar{x}_1-\bar{x}_2} \doteq \sqrt{\frac{(152)^2}{20} + \frac{(130)^2}{20}} = 44.7$$

Hence

$$z = \frac{416 - 742}{44.7} = -7.3$$

This value is highly significant; therefore it is quite certain that the habitual use of marijuana decreases sexual drive, as measured by the amount of the male sex hormone testosterone in the blood.

(*b*) Application of formula (6) gives

$$t = \frac{416 - 742}{\sqrt{19(152)^2 + 19(130)^2}}\sqrt{\frac{20 \cdot 20 \cdot 38}{40}} = -7.4$$

Since the critical value for t based on 38 degrees of freedom is very close to 2.02 as found in Table V, this result is highly significant. It should be observed that the small-sample method differs very little from the large-sample method for this problem. This implies that $n_1 = 20$ and $n_2 = 20$ are sufficiently large to justify the use of large-sample methods.

Example 4. The following quotation is from the *Los Angeles Times* for November 4, 1971. "More than half the 1971 American made cars tested by the state in Los Angeles exceeded California exhaust emission standards, a state Air Resources Board report showed Wednesday. And 64 percent of the 1971 foreign cars tested also failed to meet the state's limits." Facts gleaned from the report are that of the 467 cars tested, 150 were foreign, of which 64 percent failed, and 317 were domestic, of which 52 percent failed. Assuming that the survey was made properly, test the hypothesis that domestic and foreign cars do not differ in their ability to pass the emissions test.

Let $\hat{p}_1 = .64$ and $\hat{p}_2 = .52$. Then under the hypothesis that $p_1 = p_2$, the desired estimate of this common value is given by

$$\hat{p} = \frac{150(.64) + 317(.52)}{150 + 317} = \frac{261}{467} = .56$$

The corresponding large-sample estimate of $\sigma_{\hat{p}_1-\hat{p}_2}$ is therefore given by

$$\sigma_{\hat{p}_1-\hat{p}_2} \doteq \sqrt{(.56)(.44)\left[\frac{1}{150}+\frac{1}{317}\right]} = .049$$

Hence,

$$z = \frac{\hat{p}_1 - \hat{p}_2}{\sigma_{\hat{p}_1-\hat{p}_2}} \doteq \frac{.67 - .52}{.049} = 3.06$$

The hypothesis $H_0: p_1 = p_2$ must therefore be rejected. Foreign cars did not do as well as domestic cars on this test.

EXERCISES

SECTION 1

1 In a court case in which an individual is being tried for theft, what are the two types of error? Which type of error is considered more important by society?

2 Give an illustration of a hypothesis for which the type II error would be considered much more serious than the type I error.

3 Suppose you agree to reject a hypothesis if two tosses of an honest coin produce two heads. What are the sizes of the two types of error?

4 In hiring personnel, the personnel department is faced with the problem of selecting from those that apply for employment those who will turn out to be "good" employees and to avoid hiring those who will be "bad" employees. In terms of statistical hypothesis testing, the risk of rejecting a potentially good employee is the probability of a type I error and that of hiring a potentially bad employee is the probability of a type II error. In view of these risks of wrong selections and taking into consideration the costs of training new employees and the supply of applicants, what should be the relative sizes of α and β under the following conditions?
(*a*) Supply is large and training is expensive.
(*b*) Supply is small and training is inexpensive.
(*c*) Supply is small and training is expensive.

5 A coin is tossed twice. Let x denote the number of heads obtained. Consider the hypothesis $H_0: p = .5$ and the alternative $H_1: p = .7$, in which p is the probability of obtaining a head, and assume that one is going to test this hypothesis by means of the value of x. The distributions of x under H_0 and H_1 can be obtained by means of the binomial distribution formula (1) in Chapter 5. Suppose one chooses $x = 2$ as the critical region for the test. Calculate the sizes of the two types of error.

6 In problem 5 suppose $x = 0$ had been chosen as the critical region for the test. Now calculate the sizes of the two types of error and compare your results with those of problem 5. Comment about these two choices of critical region.

7 A coin is to be tossed eight times. The number of heads, x, is to be used to test the hypothesis $H_0: p = .5$ against the alternative hypothesis $H_1: p = .7$. The critical region for the test is to be chosen as the values of x that exceed 5, that is, the values 6, 7, and 8. Calculate the distribution of x under both hypotheses and represent their distributions as line charts on the same graph. Use your results to calculate the values of α and β for this test.

SECTION 2

8 Given $\bar{x} = 82$, $\sigma = 15$, and $n = 100$, test the hypothesis that $\mu = 86$.

9 Given $\bar{x} = 82$, $\sigma = 15$, and $n = 25$, test the hypothesis that $\mu = 86$.

10 A purchaser of bricks believes that the quality of the bricks is deteriorating. From past experience, the mean crushing strength of such bricks is 400 pounds, with a standard deviation of 20 pounds. A sample of 100 bricks yields a mean of 390 pounds. Test the hypothesis that the mean quality has not changed against the alternative that it has deteriorated.

11 Many years of experience with a university entrance examination in English has yielded a mean score of 64 with a standard deviation of 8. All the students from a certain city, of which there were 54, obtained a mean score of 68. Can one be quite certain that students from this city are superior in English?

12 A manufacturer of fishing line claims that his 5-pound test line will average 8 pounds test. Is he justified in his claim if a sample of size 50 yielded $\bar{x} = 8.8$ pounds and $s = 1.4$ pounds?

13 Experience with a particular type of assembly job shows that it requires, on the average, 45 seconds. A particular workman is observed over 100 cycles of this job with the following results: total time = 4,510 seconds, standard deviation of times = 2.5 seconds. Management wants to know if this workman's mean time is excessive, in which case he will be retrained. Do you believe that he should be retrained?

14 Construct a control chart for $\bar{x}$ for the following data on the blowing time of fuses, samples of 5 being taken every hour. Each set of 5 has been arranged in order of magnitude. Estimate μ by calculating the mean of all the data and estimate $\sigma_{\bar{x}}$ by first estimating σ by means of s calculated for all 60 values. State whether control seems to exist here.

42	42	19	36	42	51	60	18	15	69	64	61
65	45	24	54	51	74	60	20	30	109	91	78
75	68	80	69	57	75	72	27	39	113	93	94
78	72	81	77	59	78	95	42	62	118	109	109
87	90	81	84	78	132	138	60	84	153	112	136

SECTION 3

15 If you rolled a die 240 times and obtained 50 sixes, would you decide that the die favored sixes?

16 Past experience has shown that 40 percent of students fail a university entrance examination in English. If 50 out of 110 students from a certain city failed, would one be justified in concluding that the students from this city are inferior in English?

17 A biologist has mixed a spray designed to kill 50 percent of a certain type of insect. If a spraying of 200 such insects killed 120 of them, would you conclude his mixture was satisfactory?

18 What is the difference between saying that a coin is honest for all practical purposes and saying that it is honest in a mathematical sense?

19 Explain why you might hesitate from practical considerations to reject the hypothesis that a coin is honest if you tossed the coin 1000 times and obtained 535 heads.

20 Examine the following article which appeared in a newspaper.

LOUDER COMMERCIALS

> Are commercials really louder than regular programs or does it just seem that way?
>
> The H. H. Scott Company, maker of a fine line of noise measuring equipment and hi-fidelity components, recently ran a series of tests to determine the answer to this common question. The company monitored 40 programs broadcast by three Boston television channels.
>
> The tests determined that on 68 percent of the programs the commercials were louder than the rest of the show. In the case of two programs, the commercials were twice as noisy. Both sponsors were producers of detergents.

Test the hypothesis $H_0: p = \frac{1}{2}$ against $H_1: p \neq \frac{1}{2}$, where p represents the proportion of programs in which the commercials were louder. Could you have chosen $H_1: p > \frac{1}{2}$?

21 A seed company advertises that its grass seed, although expensive, is much more viable than its competitor's and therefore is a better buy. As a tester in a consumer testing organization you take a sample of 100 seeds from a bag and subject them to a viability test. The seed company guarantees a minimum of 96 percent viability. Your sample yields 90 percent viability. What do you conclude?

22 A sample of 400 items from those filed by a clerk contained 12 incorrectly filed items. If an error rate of 2 percent or less is satisfactory, can you conclude that this clerk's work is unsatisfactory?

23 It is assumed in an airplane-parts manufacturing plant that the weekly turnover of employees should not be more than 50 employees of the total of 2500 employed. This is a 2 percent turnover. If terminations exceed this rate, the situation is serious enough to warrant investigation. One week, 60 employees terminated employment.

(*a*) Is this a large enough number to warrant investigation?

(*b*) Accepting an $\alpha = .05$, what number would be critical?

24 Fitmaster Sweater Corp. bought 10,000 sweaters from a new supplier who, although granting a lower price than the previous supplier, agreed to meet the same quality specifications. When the shipment arrived Fitmaster selected a

sample of 100 sweaters for a detailed examination and found 10 to be rejects. Specifications require that the defective rate be 5 percent or less. Has the supplier met its commitment? (Assume Fitmaster is willing to accept a 5 percent risk for a false accusation.)

25 The following data were obtained for a daily percentage of defective parts for a production averaging 1000 parts a day. Construct a control chart and indicate whether control seems to exist here. The data are for the percentage (not proportion) of defectives and are to be read a row at a time. Estimate p by calculating the mean of these sample values.

2.2	2.3	2.1	1.7	3.8	2.5	2.0	1.6	1.4	2.6	1.5	2.8	2.9	2.6	2.5
2.6	3.2	4.6	3.3	3.0	3.1	4.3	1.8	2.6	2.1	2.2	1.8	2.4	2.4	1.6
1.7	1.6	2.8	3.2	1.8	2.6	3.6	4.2							

SECTION 4

26 Given two random samples of size 100 each from two normal populations with sample values $\bar{x}_1 = 20$, $\bar{x}_2 = 22$, $s_1 = 5$, $s_2 = 6$, test the hypothesis that $\mu_1 = \mu_2$.

27 Two sets of 50 elementary school children were taught to read by two different methods. After instruction was over, a reading test gave the following results: $\bar{x}_1 = 73.4$, $\bar{x}_2 = 70.3$, $s_1 = 8$, $s_2 = 10$. Test the hypothesis that $\mu_1 = \mu_2$.

28 In an industrial experiment a job was performed by 40 workmen according to method I and by 50 workmen according to method II. The results of the experiment yielded the following data on the length of time required to complete the job: $\bar{x}_1 = 54$ minutes, $\bar{x}_2 = 57$ minutes, $s_1 = 6$ minutes, $s_2 = 8$ minutes. Test the hypothesis that $\mu_1 = \mu_2$.

29 A test of two filing procedures, A and B, was carried out to determine which is superior. A sample of 60 clerks was selected and randomly assigned to two groups of 30 each. After proper training in the procedure assigned, the results of the experiment in terms of the average number of filings per person were: $\bar{x}_A = 30$, $\bar{x}_B = 27$, $s_A = 4$, $s_B = 3$. Test the hypothesis of no real difference in mean times.

30 In a survey consisting of a sample of 500 TV viewers, the 230 women of the sample watched TV 24.2 hours during one week, whereas the 270 men watched 23.6 hours during that period. If the standard deviations of viewing time in the two groups were 12 and 14 hours, is there any significant difference in TV watching time between the sexes?

31 In a study of the effects of heavy doses of vitamin C on colds, as reported in the *Canadian Medical Association Journal* for November 1972, a record was kept of the number of colds caught by each individual in the experiment during the experimental period. The following table of values gives part of that record in a slightly altered form.

	Number of individuals	Mean number of colds	Standard deviation of number of colds
Vitamin group	407	1.38	1.23
Placebo group	411	1.48	1.14

Use these data to test the hypothesis that vitamin C doses had no effect on the mean number of colds caught by individuals from the population of this experiment.

SECTION 5

32 In a poll taken among college students, 46 of 200 fraternity men favored a certain proposition, whereas 51 of 300 nonfraternity men favored it. Is there a real difference of opinion on this proposition?

33 In a poll of the television audience in a city, 60 out of 200 men disliked a certain program, whereas 75 out of 300 women disliked it. Is there a real difference of opinion here?

34 In one section of a city, 64 out of 480 taxpayers were delinquent with their tax payments, whereas in another section, 42 out of 500 were delinquent. Test to see if the delinquency rate is the same for those two sections of the city.

35 On the basis of their replies to a questionnaire, students are classified as "inner directed" or "other directed." A sample of 120 males and 170 females showed that 58 percent of the males and 73 percent of the females were other directed. Is this difference significant?

36 A poll in the gubernatorial contest in California during the summer of 1962 showed Nixon was favored by 47.5 percent and Brown by 52.5 percent of those who had decided on their candidate and also intended to vote. If the sample was based on 900 interviews (assumed taken at random over the state) would you say Nixon was definitely trailing Brown?

37 Two classes of the same size are taught by the same instructor. One class is a regular class, whereas the other one is in an adjoining classroom and receives the instructor's lecture by closed-circuit television. Discuss possible factors that might prevent a valid statistical comparison to be made of the two learning methods.

38 A marketing survey on brand preferences is carried out in two successive years on independently drawn samples of 400 housewives each. The preferences for brand "X" were 33 percent and 29 percent, respectively. Has there been a shift in the preferences in the population?

39 A test of 100 youths and 200 adults showed that 50 of the youths and 60 of the adults were careless drivers. Use these data to test the claim that the youth percentage of careless drivers is larger than the adult percentage by 10 percentage points against the alternative that it exceeds this amount.

40 An experiment was conducted by a group of doctors at the University of Pennsylvania, as reported in the *Los Angeles Times* for January 3, 1973, on the side effects of several common antianxiety medications. The study involved 166 patients with mild to moderate symptoms of anxiety and tension. They were divided into four groups by random selection and were asked to report any side effects such as dizziness, drowsiness, confusion, and nausea. The results of the experiment are given in the following table.

Medication	Aspirin	Compoz	Placebo	Libritabs
Number of patients	38	39	35	36
Number having side effects	3	18	4	14

It should be apparent that Compoz and aspirin differ with respect to side effects, nevertheless, to make certain that this is so, test the hypothesis that there is no difference between the aspirin and Compoz proportions of side effects.

This study revealed that neither aspirin nor Compoz was effective in relieving anxiety, however the drug Libritabs, which is often prescribed for such relief, was effective in doing so. This feature of the study involves other data and is not considered here.

SECTION 6

41 Given that x is normally distributed and given the sample values $\bar{x} = 42$, $s = 5$, $n = 20$, test the hypothesis that $\mu = 44$ using the t distribution.

42 Four markets were selected to test the sales of a new product. The sales per month were 3000, 2600, 3500, and 2900. Management has decided that their break-even volume of this product is 3100 per month per market area. Does this evidence indicate that they should not plan on distributing this product nationally?

43 An accountant for a firm has been challenged by the auditors on his miscellaneous expenditures account, which contains 1500 entries and totals $1,500,000. They claim he has understated its true status by at least 10 percent. The accountant disagrees, stating that some errors exist but that they are both positive and negative and that the total account is correct. To settle the matter, a random sample of 10 entries is examined and is found to contain the following errors in dollars:

$$-200,\ -100,\ -150,\ +100,\ -300,\ -250,\ -100,\ +50,\ -150,\ -100.$$

The mean is $-\$120$ and the standard deviation is $123.

The auditors claim they are correct because the sample indicates an error of 12 percent. The accountant claims, however, that these errors can be attributed to sampling variation. Who is correct?

SECTION 7

44 Given two random samples of sizes 10 and 12 from two independent normal populations with $\bar{x}_1 = 20$, $\bar{x}_2 = 24$, $s_1 = 5$, $s_2 = 6$, test by means of the t distribution hypothesis that $\mu_1 = \mu_2$, assuming that $\sigma_1 = \sigma_2$.

45 The following data give the weight gains of 20 rats, of which half received their protein from raw peanuts and the other half received their protein from roasted peanuts. Test by means of the t distribution to see whether roasting the peanuts had an effect on their protein value.

Raw	61	60	56	63	56	63	59	56	44	61
Roasted	55	54	47	59	51	61	57	54	62	58

46 As reported in *Management Science* in 1955, a study was made to compare the effectiveness of two communication networks with respect to the time required to complete certain tasks. Two five-man groups were given a task to perform, where the completion depended on the passage of written messages among the members of the group. One network, called the "all-channel" network, permits members of the group to send messages to any other member of the group. A "wheel" network permits only one man to send messages to all the others, while the others are permitted to send messages to this "hub" man only. The efficiency of a network is measured by the time (in minutes) required to complete the task. The experiment produced the following results.

Network	Number of groups	Average time for completion	Standard deviation of time
All-channel	20	24.38	4.82
Wheel	15	19.12	3.09

Test the hypothesis that there is no difference in the mean times for the two networks.

47 The purchasing agent for a corporation is considering which of two brands of batteries to purchase. He buys 9 batteries of brand A and 9 batteries of brand B and installs them in regular operation. Recording the length of life of each battery produced the following statistics, in units of a month:

$$\bar{x}_A = 25, \quad s_A = 4, \qquad \bar{x}_B = 28, \quad s_B = 3$$

Does the purchasing agent have enough evidence to determine whether brand B is superior to brand A?

48 The following data give the corrosion effects in various soils for coated and uncoated steel pipe. Taking differences of pairs of values, test by the t distribution the hypothesis that the mean of such differences is zero.

Uncoated	42	37	61	74	55	57	44	55	37	70	52	55
Coated	39	43	43	52	52	59	40	45	47	62	40	27

49 The following data give the gains in weight of eight pairs of experimental animals, matched with respect to various growth factors. One of each pair, selected by

chance, was given a supplement of vitamin B_{12} to its regular diet. Carry out a significance test on these differences to determine whether B_{12} was beneficial.

With B_{12}	1.60	1.68	1.75	1.64	1.75	1.79	1.78	1.77
Without B_{12}	1.56	1.52	1.52	1.49	1.59	1.56	1.60	1.56
Difference	.04	.16	.23	.15	.16	.23	.18	.21

50 A company selling machinery to a plant claims that a new expensive machine it has developed will double the output of the old type of machines being used. The plant installs one of the new machines and runs it along with the old ones for six consecutive weeks, obtaining the following output results (in rounded-off units of a thousand).

Average of old machines	3	2	3	4	5	4
Output of new machines	4	4	8	6	8	6

Is management justified in stating that the new machine has not lived up to its claims?

SECTION 8

51 Calculate the values of β when $p = 0, .1, .3, .5, .7, .9$, and 1 for the test of problem 5. Graph these values of β against p and draw a smooth curve through them to obtain the operating characteristic for that test. Discuss how good this test is for discovering alternative values of p.

52 For the test of problem 7, calculate the value of β for $p = 0, .5, .6$, and 1 and sketch the operating characteristic for the test based on these four values and the value found previously for $p = .7$. Comment on the usefulness of this test.

53 A consumer testing organization rates electric toasters by means of a sample of 4 chosen from regular commercial channels. If no defectives are found the brand is regarded as satisfactory.

(*a*) What is the probability of a satisfactory rating when 10 percent are actually defective?

(*b*) What is the probability in (*a*) if 20 percent are effective?

(*c*) Compute probabilities of accepting a brand with the above acceptance rule for various proportions of defectives.

(*d*) Use the results in (*c*) to construct a graph of this decision rule. What role can this graph play in decision making?

SECTION 9

54 A true-false examination consists of 100 questions of which 50 are true and 50 are false. Scoring is one point for each correct answer and zero for each incorrect answer. A student takes this examination and answers all questions.

(*a*) What is his expected score if he is totally ignorant and tosses a coin to decide each answer?

(*b*) At what total score level would you conclude that he had some knowledge? Explain.

55 A study of naval officers showed that 14 out of 31 leading admirals in the Navy had come from the upper 25 percent of their class at Annapolis. Can one conclude from this that scholastic success at Annapolis is helpful in becoming a leading admiral?

56 A private public opinion poll was employed by Republican headquarters to measure the popularity of Goldwater in the states of New York and Pennsylvania. Random samples of 5,000 voters in New York and 4,000 in Pennsylvania were questioned concerning their voting intentions. In New York 49 percent favored Goldwater whereas 51 percent did so in Pennsylvania. Is this a significant difference?

57 The following article occurred in a British newspaper.

> **High Living Said to Produce Girls More than Boys**
>
> London: Wealthy well-fed parents are more likely to give birth to a daughter than a son, Dr. Sprenger said today. Sprenger said a check on Debrett's, the Who's Who of British nobility, showed that the aristocracy listed there had 57 percent daughters. There seems little doubt, he said, that the nutritional state of parents has some influence on the sex ratio of the family.

How many members of the aristocracy would be needed before this percentage would differ significantly from 50 percent?

58 Two cities are separated by a river. The chamber of commerce in city *A* claims that the average family income of that city is $1,000 more than in city *B*. The chamber of commerce in city *B* disputes this claim and employs you to obtain a sample of 100 from each city's families to invalidate the claim. The sample gives $\bar{x}_A = 13{,}000$, $s_A = 2{,}000$, $\bar{x}_B = 12{,}500$, $s_B = 3{,}000$. What conclusions do you draw from these data relevant to the claim?

59 As reported in the *American Sociological Review* for 1951, a study was made to determine the attitudes of individuals 18 years of age and older concerning mental illness. Respondents were requested to classify as true or false the statement, "Most mental illness is inherited." To determine whether there were pronounced differences between younger and older respondents, the data for male respondents were broken down as follows:

Response	Males 18–24: percent	Males 25 and over: percent
False	75	52
True	18	29
Don't know	7	19
Totals	100	100
Number interviewed	260	188

Are you justified in claiming that older males are more likely to attribute mental illness to inheritance than younger males? Consider false percentages only.

60 Take a sample of size 25 from a table of one-digit random numbers (Table XII in the appendix) and test the hypothesis that $\mu = 4.5$. Use the fact that $\sigma = 2.87$ for this distribution and that $\bar{x}$ may be treated as a normal variable. Bring your result to class. Approximately 95 percent of the class should accept this hypothesis because it is true.

61 Work problem 60, choosing $\alpha = .20$, using a sample of size 10 and assuming that the value of σ is not known; that is, use Student's t test here. Compare the results of the students with expectation.

62 Have each member of the class draw 10 pairs of one-digit random numbers (Table XII in the appendix). The second number should be subtracted from the first number for each pair. Each student should bring to class the sum of the squares of these differences as well as the sum of the differences. By combining the class results, an estimate of σ_z^2, in which $z = x - y$, can be obtained and compared to the value expected from the formula $\sigma_{x-y}^2 = \sigma_x^2 + \sigma_y^2 = (2.87)^2 + (2.87)^2 = 16.5$. Here x and y denote, respectively, the first and second random digits of each pair. From earlier work it was found that $\mu_x = 4.5$ and $\sigma_x = 2.87$.

63 Select two sets of 20 one-digit random numbers and test the hypothesis (which is true here) that the two population means are equal. Work first by large-sample methods, using the fact that $\sigma = 2.87$; then work by means of Student's t distribution.

64 The data of Table XIII in the appendix may be used to construct various hypothesis-testing problems. For example, the data for the variable x_1 may be split into two parts, say, the first half and the second half, and a test made to observe whether the two parts possess the same theoretical means. Since these data came from different sections of the county, these means need not be the same.

Chapter 9 Correlation

1 LINEAR CORRELATION

The statistical methods presented thus far have all been concerned with a single variable x and its probability distribution. In particular, the preceding two chapters have been concerned with the estimation of, and the testing of hypotheses about, the parameters of binomial and normal variable probability distributions. Many of the problems in statistical work, however, involve several variables. This chapter is devoted to explaining one of the techniques for dealing with data associated with two or more variables. The emphasis is on two variables, but the methods can be extended to deal with more than two.

In some problems the several variables are studied simultaneously to see how they are interrelated; in others there is one particular variable of interest, and the remaining variables are studied for their possible aid in throwing light on this particular variable. These two classes of problems are usually associated with the names *correlation* and *regression*, respectively. Correlation methods are discussed in this chapter and regression methods in Chapter 10.

A correlation problem arises when an individual asks himself whether there is any relationship between a pair of variables that interests him. For example, is there any relationship between smoking and heart ailments, between stock prices and the rate of inflation, between the age of skilled workers and their productivity?

For the purpose of illustrating how one proceeds to study the relationship between two variables, consider the data of Table 1, which consists of the scores that 20 salesmen made on a test designed to measure their aptitude for sales work and their sales productivity over a period of time. The test score is denoted by x and the sales productivity by y. The choice of which variable to call x and which to call y is arbitrary here.

The investigation of the relationship of two variables such as these usually begins with an attempt to discover the appropriate form of the relationship by graphing the data as points in the x,y-plane. Such a graph is called a *scatter diagram.* By means of it, one can quickly discern whether there is any pronounced relationship and, if so, whether the relationship may be treated as approximately linear. The scatter diagram for the 20 points obtained from the data of Table 1 is shown in Fig. 1.

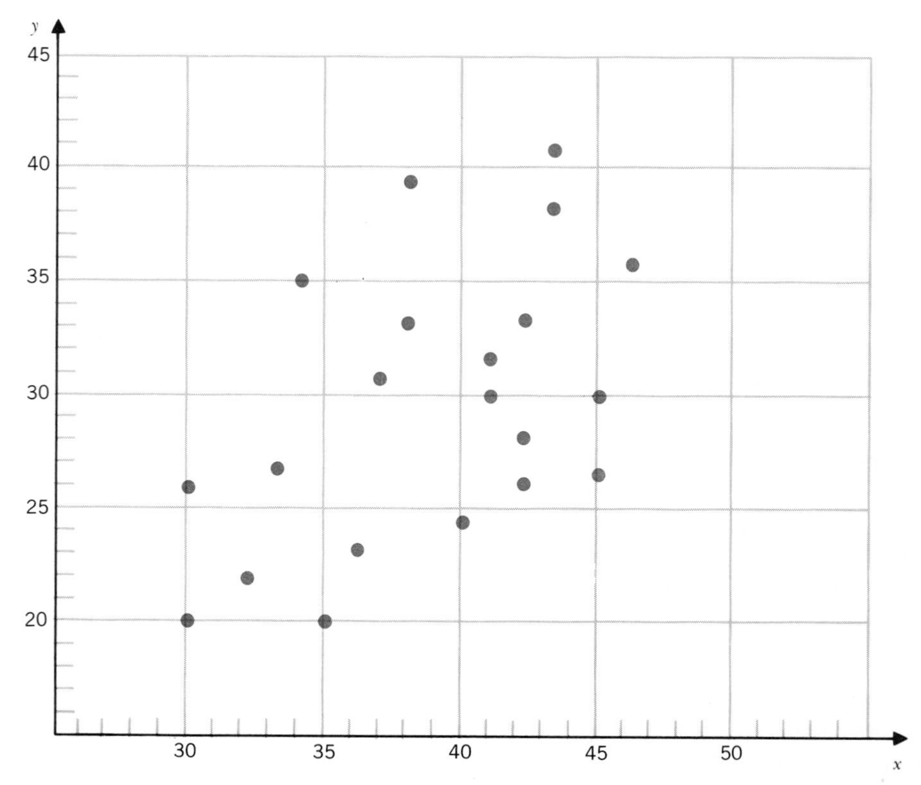

Figure 1 Scatter diagram for aptitude and productivity scores.

An inspection of this scatter diagram shows that there is a tendency for small values of x to be associated with small values of y and for large values of x to be associated with large values of y. Furthermore, roughly speaking, the general trend of the scatter is that of a straight line. In determining the nature of a trend, one looks to see whether there is any pronounced tendency for the points to be scattered on both sides of some smooth curve with a few waves or whether they appear to be scattered on both sides of a straight line. It would appear here that a straight line would do about as well as some mildly undulating curve. For variables such as these, it would be desirable to be able to measure in some sense the degree to which the variables are linearly related. For the purpose of devising such a measure, consider the properties that would be desirable.

A measure of relationship should certainly be independent of the choice of origin for the variables. The fact that the scatter diagram of Fig. 1 was plotted with the axes conveniently chosen to pass through the point (25, 15) implies that

Table 1

x	y	x	y
41	32	38	29
35	20	38	33
34	35	46	36
40	24	36	23
33	27	32	22
42	28	43	38
37	31	42	26
42	33	30	20
30	26	41	30
43	41	45	30

the relationship was admitted to be independent of the choice of origin. This property can be realized by using the deviations of the variables from their mean rather than the variables themselves. This was done in defining the standard deviation in Chapter 2. Thus one uses the variables $x_i - \bar{x}$ and $y_i - \bar{y}$ in place of the variables x_i and y_i in constructing the desired measure of relationship. The notation x_i, y_i denotes the ith pair of numbers in Table 1.

A measure of relationship should also be independent of the scale of measurement used for x and y. Thus, if the x and y scores of Table 1 were doubled in order to make them look more like conventional examination scores, with a maximum of 100, the relationship between the variables should be unaffected. Similarly, if one were interested in studying the relationship between the stature of husbands and wives, one would not want the measure of the relationship to depend upon whether stature is measured in centimeters or inches. This property can be realized by dividing x and y by quantities that possess the same units as

x and y. For reasons that will be appreciated soon, the quantities that will be chosen here are s_x and s_y, the two sample standard deviations. Both properties are therefore realized if the measure of relationship is constructed by using the variables x_i and y_i in the forms $u_i = (x_i - \bar{x})/s_x$ and $v_i = (y_i - \bar{y})/s_y$. This merely means that the x's and y's should be measured in sample standard units. This corresponds to doing for samples what was done in (7), Chapter 5, to measure a variable in theoretical standard units.

The scatter diagram of the points (u_i, v_i) for the data of Table 1 is shown in Fig. 2. It will be observed that most of the points are located in the first and third quadrants and that the points in those quadrants tend to have larger coordinates, in magnitude, than those in the second and fourth quadrants. A simple measure of this property of the scatter is the sum $\Sigma_{i=1}^{n} u_i v_i$. The terms of the sum contributed

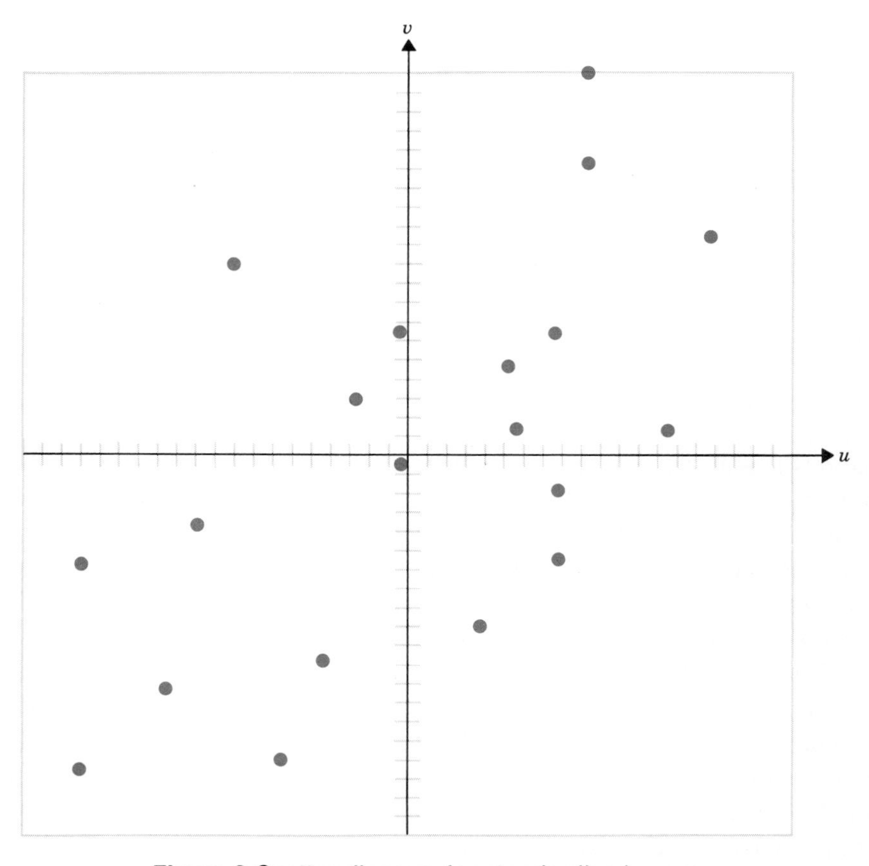

Figure 2 Scatter diagram for standardized scores.

by points in the first and third quadrants will be positive, but those corresponding to points in the second and fourth quadrants will be negative. A large positive value of this sum would therefore seem to indicate a strong linear trend in the scatter diagram. This is not strictly true, however, for if the number of points were doubled without changing the nature of the scatter the value of the sum would be approximately doubled. It is therefore necessary to divide the sum by n, the number of points, before using it as a measure of relationship. There are theoretical reasons for preferring division by $n - 1$ rather than n here, just as in the case of defining the sample variance. The resulting sum $\Sigma u_i v_i/(n - 1)$ is the desired measure of relationship. It is called the *correlation coefficient* and is denoted by the letter r; hence, in terms of the original measurements, r is defined by the following formula:

(1) ***Correlation Coefficient.*** $$r = \sum_{i=1}^{n} \left(\frac{x_i - \bar{x}}{s_x}\right)\left(\frac{y_i - \bar{y}}{s_y}\right) \Big/ (n - 1)$$

Although the value of r can be calculated directly from this definition, a better form for computational purposes is obtained by multiplying out factors, inserting computing forms for s_x and s_y, and using some algebra, with the result

(2) $$r = \frac{n\Sigma xy - \Sigma x \Sigma y}{\sqrt{[n\Sigma x^2 - (\Sigma x)^2][n\Sigma y^2 - (\Sigma y)^2]}}$$

The sums in this formula are written in abbreviated notation. For example Σxy is the sum $\Sigma_{i=1}^{n} x_i y_i$, that is, the sum of the products of all x,y pairs. As in the case of calculating the variance, large values of these sums may lead to inaccurate differences unless the calculating equipment has high capacity, in which case formula (1) should be used.

Calculations with the data of Table 1 using (2) will show that $r = .61$ for those data. In order to interpret this value of r and to discover what values of r are likely to be obtained for various types of relationships between x and y, a number of different scatter diagrams have been plotted and the corresponding values of r computed in Fig. 3. Diagrams (a)-(d) correspond to increasing degrees, or strength, of linear relationship. Diagram (e) was obtained from diagram (c) by looking at its scatter from the reverse side of the page through a strong light. This illustrates the fact that the absolute value of r measures the strength of the relationship but that the sign of r is positive if y tends to increase with increasing x and is negative if y tends to decrease with increasing x. Diagram (f) illustrates a scatter in which x and y are closely related but in which the relationship is not linear. This illustration points out the fact that r is a useful measure of the strength of the relationship between two variables only when the variables are linearly related.

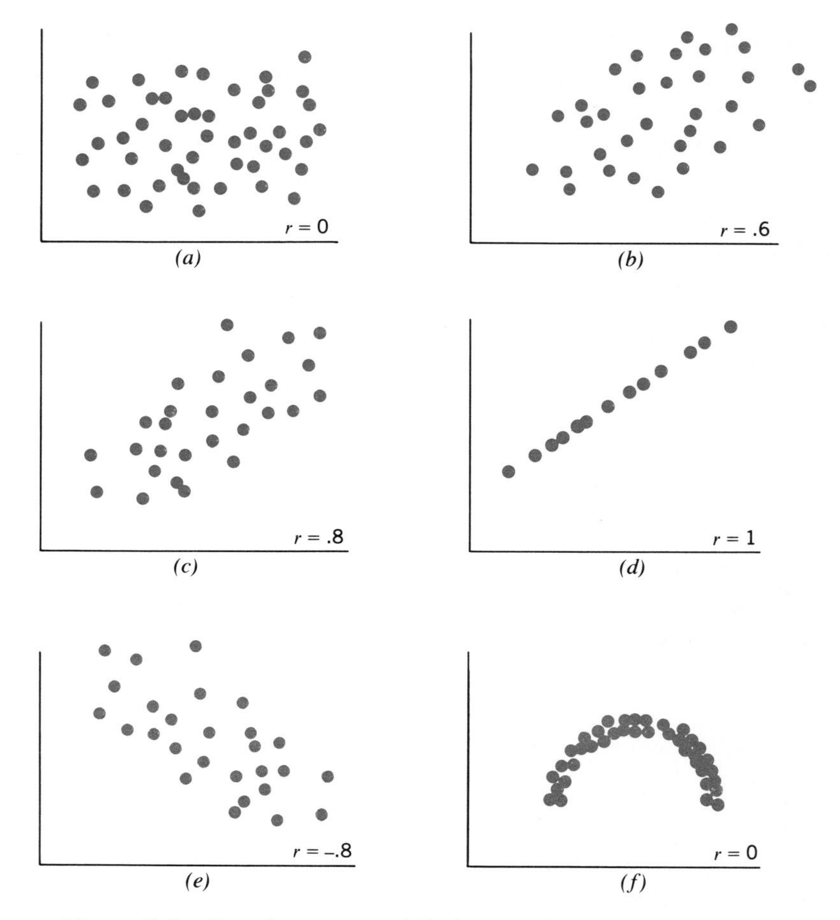

Figure 3 Scatter diagrams and their associated values of r.

The diagrams of Fig. 3, together with the associated values of r, make plausible two properties of r, namely, that the value of r must satisfy the inequalities

$$-1 \leq r \leq 1$$

and that the value of r will be equal to plus 1 or minus 1 if, and only if, all the points of the scatter lie on a straight line. These properties of r can be demonstrated to be correct by mathematical methods.

To give meaning to the value of $r = .61$ that was obtained for the correlation between aptitude and productivity scores for salesmen, it is instructive to compare this correlation value with that for pairs of familiar variables.

Several studies on the correlation between high school and college grade-point averages have yielded values of r in the neighborhood of .50 to .60. These surprisingly low values can be explained partly by the fact that there is large variation in the quality of, and the grading in, high schools with the result that the high school grade-point average is not as reliable for predicting college success as might have been expected. Other studies, however, have revealed that there is a fairly strong correlation between the grades made in various college subjects. This is also true for the scores made on certain aptitude tests. For example, one study yielded the value of $r = .74$ for the correlation between verbal aptitude and mathematical aptitude. A value as high as this would not be expected, however, for a pair of subjects such as art and mathematics.

An extensive study made in England many years ago yielded a value of $r = .51$ for the correlation between the heights of fathers and the heights of their sons. A high correlation that should come as no surprise is the value of $r = .80$ that was obtained for the correlation between the age of a husband and the age of his wife, among English couples.

These examples should help to give some feeling for how large a correlation coefficient is likely to be in any given problem that involves familiar pairs of variables.

2 INTERPRETATION OF r; SPURIOUS CORRELATION

It is possible to give a quantitative interpretation to the magnitude of r by associating it with corresponding problems of regression that are discussed in the next chapter; however, here it will be treated largely from a qualitative point of view. Thus, if the sample is sufficiently large to make r reliable, a value of r close to 0 will lead to the conclusion that the variables are not linearly related, whereas a value close to 1 in magnitude will show that they are strongly linearly related. An intermediate value in the neighborhood of .5 would represent a fairly weak, but possibly useful, linear relationship between the variables. From this point of view the value of $r = .61$ obtained for the data of Table 1 indicates that the aptitude test does possess merit in the sense that a high score on it is more likely to be obtained by a good salesman than by a poor one. If it were necessary to choose between two potential salesmen who appeared to be equally promising on all counts, except that one of them had a higher aptitude test score than the other, it would be a good bet to select the one with the higher score.

The correlation coefficient is most useful from this qualitative point of view for discovering whether a pair of variables are possibly linearly related. If they are found to be so related, then further steps can be taken to capitalize on the relationship. These ideas will be pursued further in the next chapter.

The interpretation of a correlation coefficient as a measure of the strength of the linear relationship between two variables is a purely mathematical interpretation and is completely devoid of any cause or effect implications. The fact that two variables tend to increase or decrease together does not imply that one has any direct or indirect effect on the other. Both may be influenced by other variables in such a manner as to give rise to a strong mathematical relationship. For example, over a period of years the correlation coefficient between teachers' salaries and the consumption of liquor turned out to be .98. During this period of time there was a steady rise in wages and salaries of all types and a general upward trend of good times. Under such conditions, teachers' salaries would also increase. Moreover, the general upward trend in wages and buying power, together with the increase in population, would be reflected in increased total purchases of liquor. Thus the high correlation merely reflects the common effect of the upward trend of the two variables. A high correlation such as this is often called "spurious" correlation because it is caused by the effect of some common variable, here time, rather than by a direct linear relationship. Correlation coefficients must be handled with care if they are to give sensible information concerning relationships between pairs of variables. Success with them requires familiarity with the field of application as well as with their mathematical properties.

3 RELIABILITY OF r

The value of r obtained for Table 1 may be thought of as the first sample value of a sequence of sample values $r_1, r_2, r_3, \ldots$ that would be obtained if repeated sets of similar data were obtained. Such sets of data are thought of as having been obtained from drawing random samples of size $n = 20$ from a population of salesmen. If the values of r were classified into a frequency table and the resulting histogram sketched, a good approximation would be obtained to the limiting, or theoretical, frequency distribution of r. As in the case of other variables, such as $\bar{x}$ and $\bar{x}_1 - \bar{x}_2$, the limiting distribution can be derived by mathematical methods if the proper assumptions concerning the variables x and y are made. If, for example, it is assumed that x and y are independent normal variables, in which case they are necessarily uncorrelated, the derivation shows that the desired sampling distribution of r depends only on n. Figure 4 shows the nature of this distribution when $n = 10$ and $n = 20$. By means of these distributions one can determine whether a sample value of r is large enough, numerically, to refute the possible claim that x and y are actually uncorrelated variables, that is, whether the theoretical correlation coefficient ρ, of which r is a sample estimate, has the value 0. Table VIII in the appendix gives critical values of r for these distributions, that is, values such that the probability is less than α that a sample value of r will exceed the critical value. The values of α selected are .05, .025, and .005. If there is no reason to believe that the correlation between x and y is positive, or negative, provided it is not zero, then a two-tailed test should be used. Thus,

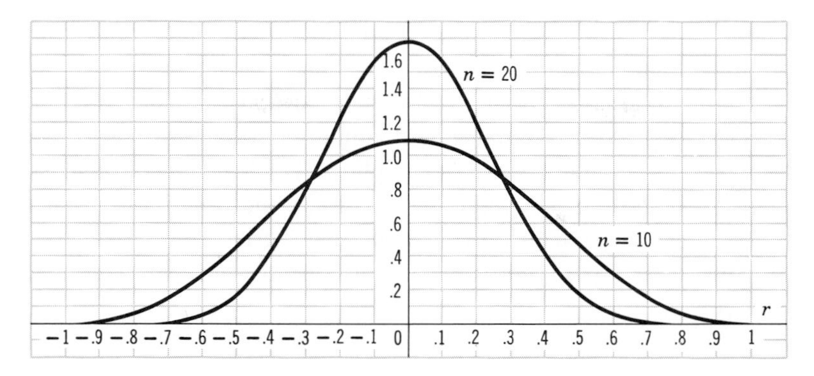

Figure 4 The distribution of r for $\rho = 0$ when n = 10 and n = 20.

the critical value of r corresponding to $\alpha = .025$ yields a two-tailed test with an .05 significance level. From Fig. 4 it will be observed, for example, that the .025 critical value for $n = 20$ should be somewhere between .4 and .5, which is in agreement with the Table VIII value of .444. For a sample size 20, therefore, the hypothesis of no correlation between x and y would be accepted unless the magnitude of r exceeded .444.

Since $n = 20$ and $r = .61$ for the data of Table 1, it follows from the Table VIII value of .378 that a hypothesis of no correlation between aptitude and productivity would be rejected here. The .05 column should be used in this problem because the alternative to zero correlation would certainly be positive correlation, and therefore only the right-tail critical region should be selected.

From Table VIII it should be apparent that a large sample is required before the claim that no correlation exists between two variables can be refuted, unless, of course, the value of r is very large. The moral here seems to be that the correlation coefficient is an interesting and often useful tool for studying the interrelationships between variables but that it must be handled with care if it is to give reliable information.

4 SERIAL CORRELATION (AUTOCORRELATION)

The statistical methods that have been presented in the preceding chapters are based on the assumption that the samples employed were obtained by random sampling. This would certainly not be true, for example, if one were interested in studying the distribution of stock prices and selected 500 consecutive daily stock prices, because there is a high correlation between neighboring prices. However, in a study of the productivity of a class of skilled workmen, it is unlikely that data taken over a few years' time will differ appreciably from data obtained by sampling at a single time point.

In view of the necessity that samples be random before standard statistical techniques can be applied to them, it is desirable to have a method for checking on the randomness of data obtained by sampling. A technique that has been found very useful for doing this is based on correlation principles and in particular on what is known as the serial correlation coefficient. Serial correlation is also used to solve many other types of problems related to time data; therefore it is well worth studying.

Let $x_1, x_2, \ldots, x_n$ be a set of observations taken over time and suppose there is reason to believe that consecutive values of x may be positively correlated. For example, the x's might represent the daily Dow-Jones stock averages for thirty consecutive days, or the mean daily temperatures in a given city for a month. Then the lack of randomness in such a sequence of values could be demonstrated by showing that there exists a positive correlation between consecutive pairs of values. This is accomplished by calculating the correlation between the following pairs of x- and y-values, where y_i is chosen to be the value x_{i+1}:

(3)

x	x_1	x_2	$\cdots$	x_i	$\cdots$	x_{n-1}
y	x_2	x_3	$\cdots$	x_{i+1}	$\cdots$	x_n

The correlation coefficient of the values given in this table is called the *serial correlation coefficient* of order 1, or of lag 1. The name *autocorrelation* is also commonly used; however, in the study of time series the latter name is technically reserved for the quantity $E[x_i x_{i+1}]$ or its estimate. The term "lag" refers to the shift in the x-values to obtain the corresponding y-values. Here there is a shift of one time unit. If, for example, one were studying the randomness of consecutive daily counts of automobile accidents in a given city, it is likely that a lack of randomness would arise from the fact that there is more traffic on certain days of the week than on others, and therefore a serial correlation coefficient with lag 7 would be a more appropriate tool than a correlation coefficient based on lag 1.

If n is reasonably large, say $n > 15$, there is a simple method to test for randomness. It is based on a modification of the method for testing the hypothesis $H_0{:}\rho = 0$ that was explained in the preceding section. First, one looks up the critical value of r in Table VIII for $N = n + 2$, where n is the number of terms in the series and N takes the place of n in Table VIII. Let this value be denoted by r_0. Then one calculates the two values

$$\pm r_0 - \frac{1}{n - 1} \tag{4}$$

These are the two critical values for the test based on the serial correlation coefficient. They are only approximations to the correct critical values, but become increasingly good as n increases in size. Since there is little point in testing for randomness in a short series, the preceding approximations should be adequate.

If one were using a lag k serial correlation coefficient for making the test, it would be necessary to replace n by $n - k + 1$ in the preceding formulas.

The serial correlation coefficient has many other uses besides that of testing for a lack of randomness in a series of sample values taken over time. One such application is that of determining whether there exist cyclical properties in a sequence of such observations. For example, it is claimed that sunspot activity follows an eleven-year cycle. This could be verified by calculating the serial correlation coefficient with lag 11 of measurements of this activity, or of some variable, such as radio static, that may be affected by it, and seeing whether it differed significantly from zero. The serial correlation coefficient is often used to determine whether there exist business cycles in the time series of economic variables. This application will be considered in the chapter on time series.

Example. As an illustration of how serial correlation is applied to test for randomness, consider the data of Table 2, which gives the annual precipitation in inches for Los Angeles from 1908 to 1973. This table should be read a row at a time. Calculations give

$$\sum_{i=1}^{65} x_i = 933, \quad \sum_{i=1}^{65} x_{i+1} = 936, \quad \sum_{i=1}^{65} x_i^2 = 16{,}011$$

$$\sum_{i=1}^{65} x_{i+1}^2 = 16{,}104, \quad \sum_{i=1}^{65} x_i x_{i+1} = 13{,}102$$

and $r = -.13$. It is not necessary to calculate these various sums if an electronic calculator is available that yields the value of r directly.

Table 2

14	24	5	18	10	17	23	17	23	8
17	9	11	20	15	6	8	9	19	19
9	8	13	19	11	19	15	14	18	18
27	12	20	31	7	23	17	13	16	4
8	11	7	14	25	4	14	12	14	13
17	6	10	6	15	12	8	27	13	24
8	26	17	9	7	17				

Because there is no reason for believing that the correlation will be positive, or negative, if it is not zero, a two-tailed test should be used. Since the total number of elements in the series is $n = 66$, it is necessary to find the .025 critical value in Table VIII corresponding to $N = 68$. Interpolation gives .240. From formula (4) it follows that the critical values are given by

$$\pm .240 - \frac{1}{65}$$

Hence, the critical values are .225 and $-.255$. Since the sample value of $r = -.13$ lies between these two critical values, the hypothesis that $\rho = 0$ is accepted.

The preceding result implies that if the rainfall is above average one year there is no tendency for it to be above or below the average the following year. Expressed another way, the rainfall during a given year is not influenced by how much it rained the preceding year. Incidentally, there have been claims made that the amount of rainfall is influenced by sunspot activity and therefore that it possesses an eleven-year cycle. This could be tested by calculating the serial correlation coefficient of order 11 for Table 2. Since problems of cycle determination will be discussed in Chapter 16, this possibility will not be pursued here; however, the interested student might do this.

5 REVIEW ILLUSTRATIONS

Example 1. An instructor believes that an examination of the "true-false" type is just as effective in measuring knowledge as is the "problem-discussion" type. For the purpose of testing this belief, he gave his 30 students an examination including both types of questions. The results are given in the following table, where x denotes the true-false score and y the problem-discussion score.

x	y	x	y	x	y
52	56	68	73	56	52
64	65	100	97	76	73
100	94	68	67	76	70
92	87	60	70	84	86
64	83	76	72	64	70
88	90	100	99	76	82
84	80	48	68	80	85
96	100	72	70	92	83
84	79	72	75	56	61
64	61	96	97	60	63

(*a*) Calculate the correlation coefficient. (*b*) Test the hypothesis that $p = 0$. (*c*) Is the instructor justified in his belief?

(*a*) $\Sigma x = 2268$, $\Sigma y = 2308$, $\Sigma x^2 = 178{,}256$, $\Sigma y^2 = 182{,}544$, $\Sigma xy = 179{,}724$,

$$r = \frac{30(179{,}724) - (2268)(2308)}{\sqrt{[30(178{,}256) - (2268)^2][30(182{,}544) - (2308)^2]}} = .90$$

(*b*) Use a one-tailed test because positive correlation is to be expected. From Table VIII, $r_0 = .306$ for $n = 30$. Since .90 is much larger than this, the hypothesis of $H_0\!:\rho = 0$ is rejected.

(*c*) Yes, because .9 is a very strong correlation and the instructor might not get a better correlation than this if he gave a second test of the same type as the first test and correlated the scores. The correlation between the scores made on the odd-numbered and the even-numbered questions of a test often does not exceed .90.

Example 2. The following data give the relationship between the production (in 10^6 boxes) and the price (dollars/box) of Florida grapefruit over a period of 20 years. Assuming that these data do not depend on time, (*a*) plot the scatter diagram, (*b*) guess the value of r, (*c*) calculate the value of r, (*d*) test the hypothesis that $\rho = 0$.

Production	Price	Production	Price
32.0	1.27	38.3	.57
29.0	.63	37.4	.89
33.0	.26	31.1	.98
30.2	.67	35.2	1.04
24.2	1.79	30.5	1.05
33.2	.94	31.6	.96
36.0	.52	35.0	.67
32.5	.76	30.0	1.24
42.0	.49	26.3	2.24
34.8	.63	31.9	1.47

(*a*)

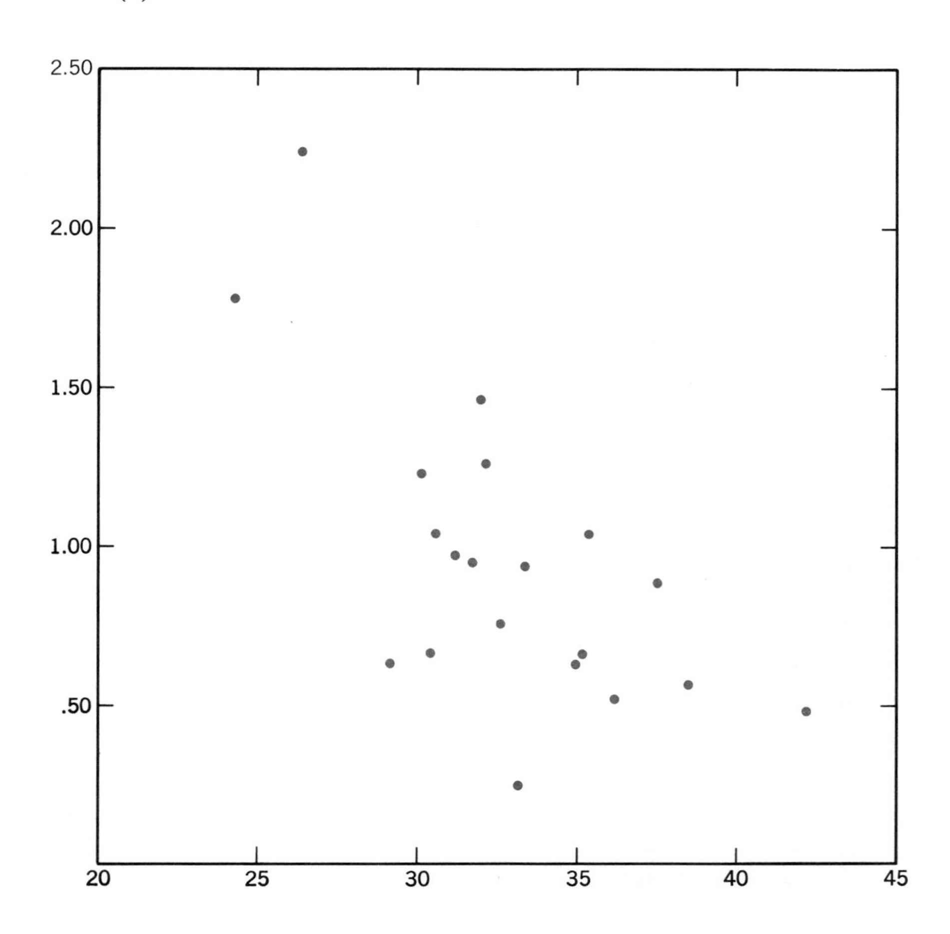

(*b*) Guess $r = -.6$.

(*c*) $\Sigma x = 654.2$, $\Sigma y = 19.07$, $\Sigma x^2 = 21{,}716.22$, $\Sigma y^2 = 22.4471$, $\Sigma xy = 599.193$,

$$r = \frac{20(599.193) - (654.2)(19.07)}{\sqrt{[20(21{,}716.22) - (654.2)^2][20(22.4471) - (19.07)^2]}} = -.67$$

(*d*) Use a one-tailed test because negative correlation is to be expected. From Table VIII, $r_0 = .378$ for $r = 20$. Since $r = -.67$ exceeds this critical value numerically, the hypothesis that $\rho = 0$ is rejected.

Example 3. The following data give the monthly call-money rates for two consecutive years. The serial correlation coefficient will be used to determine the degree of dependence between a current month's rate and the next month's rate.

4.6, 2.4, 3.9, 5.1, 5.6, 2.8, 3.5, 3.8, 10.8, 7.6, 4.9, 6.9,
5.8, 2.9, 6.0, 4.2, 2.4, 3.1, 2.5, 2.0, 2.3, 2.7, 5.2, 5.5

(*a*) Plot x_{i+1} against x_i. Does there appear to be a relationship? Guess the value of r. (*b*) Calculate the value of r and test for significance.

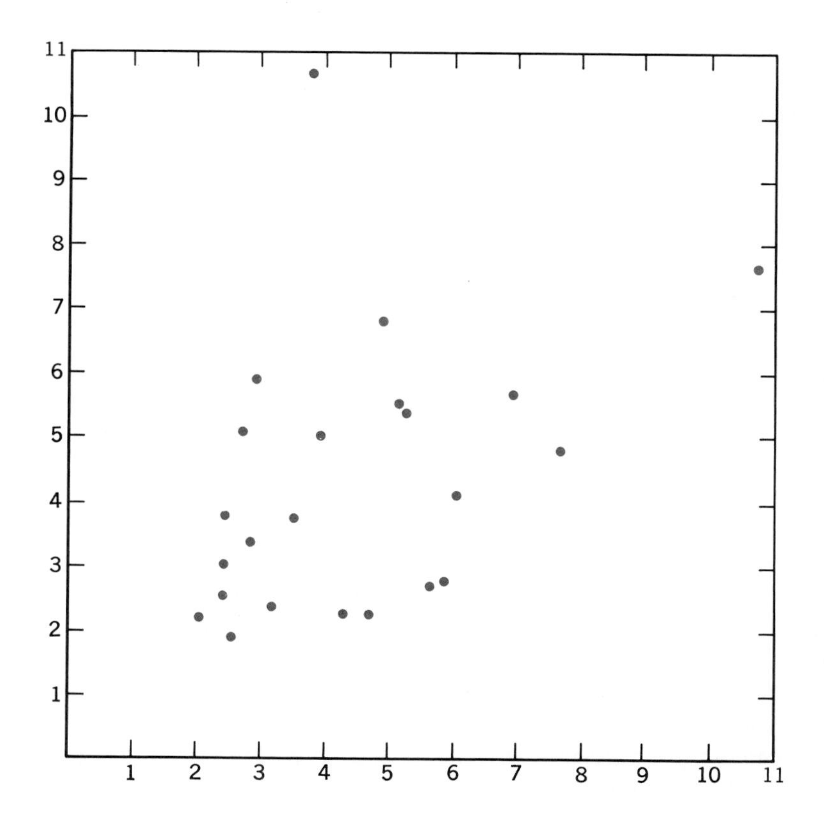

A very weak relationship. Guess $r = .2$.

(*b*) $r = .12$. Here $N = n + 2 = 26$. If there is no reason for expecting a positive, or negative, correlation, provided there is some correlation, a two-tailed test should be used. From Table VIII the critical value of r is approximately .33; hence

$$\pm r_0 - \frac{1}{n-1} = \pm .33 - \frac{1}{23}$$

This gives $-.29$ and .37 for the critical values of the serial correlation coefficient. Since .12 lies inside this acceptance region, the hypothesis of $\rho = 0$ is accepted.

EXERCISES

SECTION 1

1 By plotting the points and observing the scatter, what guesses would you make for the value of the correlation coefficient in each of the following three cases?

(*a*)		(*b*)		(*c*)	
x	*y*	*x*	*y*	*x*	*y*
5	20	48	41	16	9
11	31	2	21	11	21
13	37	67	68	8	25
22	50	99	4	3	31
16	46	71	87	4	34
8	22	66	92	17	5
20	40	43	83	4	33
25	45	27	85	11	18
7	32	39	12	12	16
17	27	36	31	8	23

2 What would you guess the value of r to be for the following pairs of variables: (*a*) the number of man-hours of work and the number of units of a product produced in a given industry, (*b*) the size of a city and the crime rate, (*c*) the cost per unit of producing an article and the number of units produced?

3 Guess the value of r for the following pairs of variables: (*a*) mathematics and foreign language grades, (*b*) consumption of butter and the price of butter, (*c*) amount of rain in the spring and the mean temperature.

4 For the following data on the heights (x) and weights (y) of 12 college students, (*a*) plot the scatter diagram, (*b*) guess the value of r, (*c*) calculate the value of r.

x	65	73	70	68	66	69	75	70	64	72	65	71
y	124	184	161	164	140	154	210	164	126	172	133	150

5 Calculate the value of r for the following data on intelligence test scores and grade-point averages, after first plotting the scatter diagram and guessing the value of r.

I.T.	295	152	214	171	131	178	225	141	116	173	230	195	174	236	198	217	143	135	146	227
G.P.A.	3.4	1.6	1.2	1.0	2.0	1.6	2.0	1.4	1.0	3.6	3.6	1.0	2.8	2.8	1.8	2.0	1.2	2.4	2.2	2.4

6 The following pairs of numbers are the scores made by 30 students on a verbal aptitude test and a mathematical aptitude test. Here x represents the verbal test score and y the mathematics test score.

x	y	x	y	x	y
77	90	80	80	42	60
57	71	56	60	92	85
93	79	66	58	64	80
66	62	74	60	71	80
86	84	79	71	87	92
62	66	54	48	54	57
74	81	73	76	65	72
70	67	72	66	70	70
68	62	63	60	83	71
69	75	83	93	87	94

(*a*) Plot the scatter diagram and observe whether the relationship may be treated as linear.
(*b*) Guess the value of r.

7 Calculate the value of r for the data of problem 6. Compare it with your guess in problem 6(*b*).

8 The following data represent a sample from a study made on "absence proneness." They consist of the number of absences of 20 employees for a six-month period (x) and a second six-month period (y).

x	7	6	6	2	3	6	6	1	1	1	1	1	0	1	0	0	2	0	0	6
y	5	3	8	1	4	6	3	0	2	2	1	1	3	1	2	3	1	4	2	3

(*a*) Plot the scatter diagram.
(*b*) Calculate the value of r. Comment.

SECTION 2

9 What interpretation would you give if told that the correlation between the number of automobile accidents per year and the age of the driver is $r = -.60$ if only drivers with at least one accident are considered?

10 Explain why it would not be surprising to find a fairly high correlation between the density of traffic on Wall Street and the height of the tide in Maine if observations were taken every hour from 6:00 A.M. to 10:00 P.M. and high tide occurred at 7:00 A.M. Plot a scatter diagram of relative values of these two variables for each hour of the day to assist you in the explanation.

11 A survey has been made on 50 firms that sell both as retailers and wholesalers. A correlation analysis revealed that the correlation coefficient between wholesale sales and gross sales was 0.7 and the correlation between retail sales and wholesale sales was approximately zero. Are the results consistent with what you might expect? Explain.

12 The following data were given as support of the theory that high-protein diets reduce the fertility of people. Are data of this sort adequate to establish a "cause-and-effect" relationship between these two variables? Discuss the problem of drawing such an inference by correlation analysis.

Country	Birth rate (per 1000)	Protein in diet (grams)
Formosa	45.6	4.7
Malay States	39.7	7.5
India	33.0	8.7
Japan	27.0	9.7
Yugoslavia	25.9	11.2
Greece	23.5	15.2
Italy	23.4	15.2
Bulgaria	22.2	16.8
Germany	20.0	37.3
Ireland	19.1	46.7
Denmark	18.3	59.1
Australia	18.0	59.9
United States	17.9	61.4
Sweden	15.0	62.6

13 For the data of problem 5, delete those items for which the intelligence test score is less than 150 and more than 225. Now calculate the value of r and compare with the value obtained for problem 5. What does this comparison seem to indicate?

14 What would be the effect on the value of r for the correlation between height and weight of males of all ages if only males in the 20–25 age group were sampled? In answering this question, it is helpful to observe the effect this restriction would have on the type of scatter diagram expected for these two variables.

SECTION 3

15 Test the hypothesis that $\rho = 0$ if a sample of size 25 gave $r = .35$.

16 A random sample of 52 overdue accounts receivable was drawn, and the correlation coefficient was calculated to examine a possible relationship between account size and length of time overdue. The value of r was found to be 0.6. Test the hypothesis that $\rho = 0$ in the population.

17 Test the hypothesis that $\rho = 0$ for the data of problem 8.
18 Test the hypothesis that $\rho = 0$ for the data of problem 5.
19 Test the hypothesis that $\rho = 0$ for the data of problems 6 and 7. Comment.

SECTION 4

20 Following are the first 30 digits on line 1 of Table XII in the appendix. Calculate the serial correlation coefficient for lag 1 and test whether it accepts the hypothesis $H_0:\rho = 0$, which it should, assuming that large-sample methods are satisfactory.

3, 1, 7, 5, 1, 5, 7, 2, 6, 0, 6, 8, 9, 8, 0,
0, 5, 3, 3, 9, 1, 5, 4, 7, 0, 4, 8, 3, 5, 5

21 Suppose you are interested in determining the relationship between wholesale and retail prices of some commodity category (food, say) and you are particularly interested in the nature of the time lag between changes in wholesale prices and their consequences on retail prices. Outline the procedure you think would be appropriate for such a study.
22 The following numbers are the prices of American railroad stocks from 1912 to 1936. Test for randomness by means of the serial correlation coefficient with lag 1. Comment on the result. The prices are 93, 93, 83, 73, 83, 81, 62, 66, 61, 60, 58, 68, 65, 81, 91, 98, 115, 131, 124, 91, 32, 25, 42, 32, 41.

SECTION 5

23 The following data appeared in *Forbes Magazine* in 1966. They were gathered to observe whether a company's book value per share is related to its net working capital per share.
(*a*) The value of the correlation coefficient here is .30. Assuming that these companies are typical, does this value of r justify one in claiming that the two variables are related?
(*b*) Plot the points. What companies appear to be inconsistent with the others?

Company	Book value/share	Working capital/share
Bath Iron Works	$68	$49
Bond Stores	34	24
Howard Stores	25	20
Neiman Brothers	28	13
Schick Electric	10	8
Belding Hemingway	24	15
Budd Company	33	8
Checker Motors	25	5
Cone Mills	36	15
Jones and Laughlin	80	7
Montgomery Ward	52	24
Republic Steel	57	3
Woolworth	38	4

24 One method for obtaining preharvest estimates of the yield of a crop is to ask the local producers to predict their individual crop yields by inspecting their fields. During the 1940s the preharvest yields of sugar beets in Colorado were obtained in this manner and then compared with the actual yields as determined by processing plant reports. The following data relating the predicted and actual yields were obtained from the 1950 *Proceedings of the American Society of Sugar Beet Technologists.* The units here are tons per acre.

Year	1941	1942	1943	1944	1945	1946	1947	1948	1949
Preharvest	14.8	13.9	19.9	12.6	13.3	12.6	13.7	13.1	14.7
Actual	14.8	12.0	12.8	12.6	12.2	13.0	15.3	13.7	16.5

(*a*) Plot the data and guess the value of r.
(*b*) Calculate the value of r.
(*c*) Test the hypothesis that $\rho = 0$ against the alternative that $\rho > 0$.
(*d*) Take the differences between predicted and actual yields and plot them against actual yields.
(*e*) Calculate the value of r for this pair of variables.
(*f*) Explain what this value of r indicates about the possible biases of producers in estimating their yields.

25 The data of Table XIII in the appendix may be used to construct various correlation problems. For example, one might calculate the correlation coefficient between the variables x_2 and x_4 and then determine whether it will refute the hypothesis $H_0: \rho = 0$.

26 The following pairs of numbers were obtained by choosing three random digits from Table XII and then adding the third digit to each of the first two to yield a pair labeled x and y.
(*a*) Plot the scatter diagram. Does the relationship appear to be linear?
(*b*) Calculate the value of r using formula (2).
(*c*) Test the hypothesis that $\rho = 0$.

x	8	8	4	10	11	1	13	12	6	14	7	8	6	13	8	4	7	8	5	14	5	10	8	13	17
y	7	8	11	14	13	3	8	8	3	8	3	11	5	11	5	10	11	3	14	13	2	10	17	10	15

x	14	9	7	10	13	9	5	13	12	5	9	6	12	14	0	10	10	17	9	13	12	9	1	12	12
y	7	11	10	1	12	2	5	13	14	14	15	11	10	15	5	13	3	8	11	9	10	14	5	14	12

27 Form 30 pairs of numbers by choosing four random digits from Table XII and adding the last two to each of the first two to yield a pair labeled x and y. Work the problems of exercise 26. Do you believe that the value of ρ here is close to $\frac{1}{2}$? If not, what would you guess its value to be?

Chapter 10 Regression

1 LINEAR REGRESSION

In the preceding chapter it was stated that correlation methods are used when one is interested in studying how two or more variables are interrelated. It often happens, however, that one studies the relationship between the variables in the hope that any relationship that is found can be used to assist in making estimates or predictions of a particular variable. Thus, for the two variables of Table 1, Chapter 9, the relationship that is indicated by the value of $r = .61$ could be used in predicting a salesman's success from a knowledge of his score on an aptitude test. The correlation coefficient is merely concerned with determining whether two variables are linearly related and if so how strong the relationship is. It is not capable of solving prediction problems. Methods that have been designed to handle prediction problems are known as regression methods. This chapter discusses the simplest of such methods.

The role that correlation often plays in regression may be illustrated by the following problem. Suppose a weather forecaster in Chicago wishes to predict the amount of precipitation that Chicago will receive seven days after the prediction is made. For the purpose of constructing a function for making such predictions he would undoubtedly select such variables as the amount of precipitation during the preceding week at various weather stations throughout the United States, and the barometric pressure and wind velocity at those stations. By inspecting past records on such variables he would be able to calculate the correlation coefficient between each such variable and the amount of precipitation. Variables that produced a very small value of r would be eliminated from further consideration. Among the remaining variables, the calculation of correlation coefficients between pairs of them might reveal some extremely high values, indicating that one member of such a pair was so strongly related to the other that it would supply no information that was not already available in the other. Thus, by such correlation calculations additional variables might be eliminated. The weather forecaster would now have a reduced set of variables with which to construct a regression function for making predictions.

The preceding illustration is typical of many problems that arise in the social sciences in which there is a great deal of uncertainty as to which variables are worth studying in casting light upon a selected variable. The correlation coefficient serves as a very useful exploratory tool in helping to solve such problems.

For the purpose of explaining regression methods, consider first a problem that involves only two variables. In particular, consider the problem of predicting the productivity score of a salesman by means of his aptitude test score, based on the data of Table 1 of Chapter 9. These data are repeated here in Table 1 and their graph is given in Fig. 1. From this graph it appears that x and y are approximately linearly related for this range of x-values. A straight line will therefore be fitted to this set of points for the purpose of using it to predict the value of y from the value of x. Such a line has been fitted in Fig. 1. Now, for any given value of x, say $x = 43$, the predicted value of y is chosen as the distance up to the line directly above the value of x. Reading across to the y-axis, it will be observed that the predicted value of y for $x = 43$ is approximately 33. Corresponding to $x = 43$ there are two observed values of y, namely, 38 and 41. They certainly differ considerably from the predicted value of 33.

Table 1

x	y	x	y
41	32	38	29
35	20	38	33
34	35	46	36
40	24	36	23
33	27	32	22
42	28	43	38
37	31	42	26
42	33	30	20
30	26	41	30
43	41	45	30

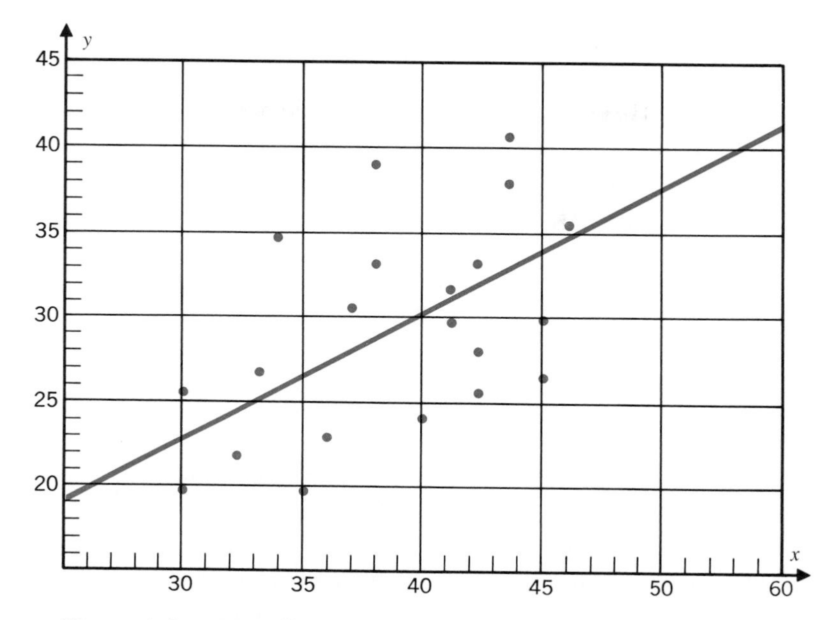

Figure 1 A scatter diagram for aptitude and productivity scores.

Suppose now that it is assumed that the relationship between *mean* productivity and aptitude scores is strictly linear over this range of x-values. This implies that if this experiment had been repeated a large number of times under the same conditions and if the y-values corresponding to each of the x-values had been averaged separately, then those averages would yield a set of points lying almost precisely on a straight line. The larger the number of such repetitions, the greater the expected precision. This assumption states essentially that there is a theoretical straight line that expresses the linear relationship between the theoretical mean value of y and the corresponding value of x.

If one accepts the linearity assumption, then one would expect the sample straight-line value of approximately 33 to be closer to the theoretical line value for $x = 43$ than either of the observed values of 38 and 41, because one would expect the sample straight line, which is based on all 20 experimental points, to be more stable than one or two observed points. In view of this reasoning, one would predict the theoretical line value corresponding to $x = 43$ to be the corresponding y-value on the sample straight line. Similar predictions could be made for the other values of x in Fig. 1. Furthermore, if one were interested in an intermediate value of x, the sample straight-line value could be used as the predicted value of y for that value of x. Since it is being assumed that the relationship is linear only for this range of x-values, it may not be legitimate to use the sample straight line to predict y-values for x-values outside this range.

2 THE METHOD OF LEAST SQUARES

In view of the preceding discussion, the problem of linear prediction reduces to the problem of fitting a straight line to a set of points. Now, the equation of a straight line can be written in the form

$$y = a + bx \tag{1}$$

in which a and b are parameters determining the line. For example, the equations

$$y = 2 + 3x$$

and

$$y = 4 - 2x$$

determine the two straight lines graphed in Fig. 2. The parameter a determines where the line cuts the y-axis. Thus the two lines in Fig. 2 cut the y-axis at 2 and 4, respectively. The parameter b determines the slope of the line. The slope 3 for the first line means that the line rises 3 units vertically for every positive horizontal unit change. A negative value such as -2 means that the line drops 2 units for every positive horizontal change of 1 unit.

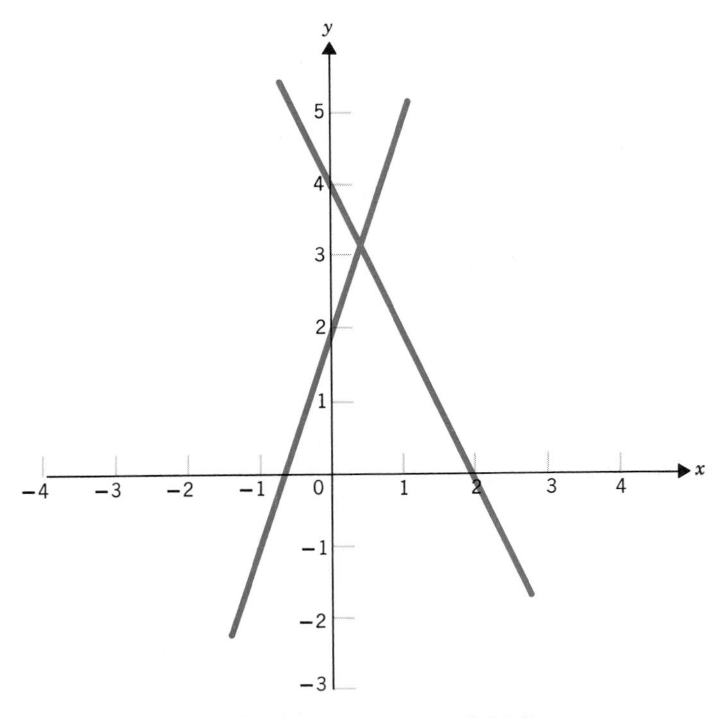

Figure 2 Graphs of two straight lines.

Since the problem is to determine the values of the parameters a and b so that the line will fit a set of points well, the problem is essentially one of estimating the parameters a and b in some efficient manner. Although there are numerous methods for performing the estimation of such parameters, the best known for regression problems is the *method of least squares.* This method will be explained next.

Since the desired line is to be used for prediction purposes, it is reasonable to require that the line be such that it will make the errors of prediction small. By the error of prediction is meant the difference between an observed value of y and the corresponding straight-line value of y. For example, the error of prediction in Fig. 1 for $x = 34$ is approximately equal to $35 - 26 = 9$. The various errors of prediction for all 20 values of x are shown in Fig. 3 by means of the vertical line segments that connect the points to the line.

The points lying above the line yield positive errors and the points lying below the line yield negative errors; therefore, it will not do to require that the sum of the errors be as small as possible because a poor-fitting line could have a very small such sum. This difficulty can be avoided by requiring that the sum of the absolute values of the errors be as small as possible. However, sums of absolute values are not convenient to work with; consequently, the difficulty is overcome by requiring that the sum of the squares of the errors be as small as possible. The values of the parameters a and b that minimize the sum of the

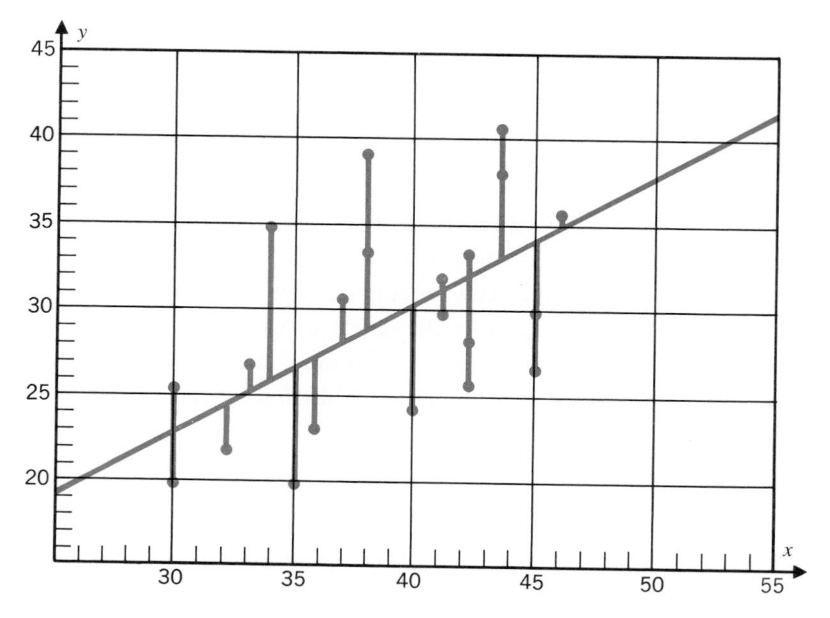

Figure 3 Errors of prediction.

squares of the errors determine what is known as the *best-fitting straight line in the sense of least squares.* It is clear from inspecting Figs. 1 and 2 that by varying a and b properly one should be able to find the equation of a line that fits the points of Fig. 1 well. The problem, however, is one of finding a best-fitting line in some systematic rational way, and this is where the principle of least squares enters.

The problem of determining the least-squares values of a and b in (1) requires more mathematics than is expected for this book; therefore, it is not solved here. The results of applying the proper mathematical methods, however, are written down. It turns out that these least-squares values may be obtained by solving the following two equations for a and b:

$$an + b \sum_{i=1}^{n} x_i = \sum_{i=1}^{n} y_i$$

$$a \sum_{i=1}^{n} x_i + b \sum_{i=1}^{n} x_i^2 = \sum_{i=1}^{n} x_i y_i$$

It is assumed here that there are n pairs of values of x and y, such as those in Table 1 where $n = 20$, and that x_i and y_i denote the ith pair of values. These equations are traditionally called the *normal equations* of least squares.

These equations can be solved for a and b to obtain explicit formulas for a and b. The resulting values when substituted in (1) yield the equation of the least-squares line. This line is called the *regression line of y on x*. The results of this procedure may be summarized as follows:

(2) ***Regression Line.*** *$\hat{y} = a + bx$, where $a = \bar{y} - b\bar{x}$ and*

$$b = \frac{n \sum_{i=1}^{n} x_i y_i - \sum_{i=1}^{n} x_i \sum_{i=1}^{n} y_i}{n \sum_{i=1}^{n} x_i^2 - \left(\sum_{i=1}^{n} x_i\right)^2}$$

The symbol $\hat{y}$ is used to represent an ordinate on the least-squares line, to distinguish it from an observed value of y. A pioneer in the field of applied statistics gave the least-squares line the name regression line in connection with some studies he was making on estimating the extent to which the stature of sons of tall parents reverts, or regresses, toward the mean stature of the population.

Table 2 illustrates the computational procedure for the data of Table 1. Computations based on using the formulas in (2) gave

$$\hat{y} = 1.01 + .734x \tag{3}$$

This is the equation of the line that was graphed earlier in Fig. 1. Incidentally, some of the new electronic hand calculators produce the regression equation directly without the necessity of calculating any sums or sums of squares. The computations in Table 2 were carried out on the assumption that such calculators are not available.

Table 2

x_i	y_i	x_iy_i	x_i^2
41	32	1312	1681
35	20	700	1225
34	35	1190	1156
40	24	960	1600
33	27	891	1089
42	28	1176	1764
37	31	1147	1369
42	33	1386	1764
30	26	780	900
43	41	1763	1849
38	29	1102	1444
38	33	1254	1444
46	36	1656	2116
36	23	828	1296
32	22	704	1024
43	38	1634	1849
42	26	1092	1764
30	20	600	900
41	30	1230	1681
45	30	1350	2025
Totals: 768	584	22,755	29,940

$$b = \frac{20(22{,}755) - (768)(584)}{20(29{,}940) - (768)(768)} = .734$$

$$a = \frac{584}{20} - .734\left(\frac{768}{20}\right) = 1.01$$

3 THE RELATION TO CORRELATION

After a regression line has been determined, it is of interest to know how useful the line is for predicting y from x. A natural measure of the usefulness is the ratio of the sum of squares of the errors of prediction based on the regression line to the sum of the squares of those errors when no attempt is made to fit a regression line, that is, when the relationship between x and y is ignored. In the latter case one would use $\bar{y}$ as the predicted value of y for all values of x. This ratio is therefore given by

$$\frac{\sum_{i=1}^{n} (y_i - \hat{y}_i)^2}{\sum_{i=1}^{n} (y_i - \bar{y})^2}$$

Unfortunately, this ratio will equal 0 when there is perfect prediction and will equal 1 when the predicted values are all equal, and hence equal to $\bar{y}$, in which case the regression line is worthless. Since it is conventional to have 0 correspond to a worthless line and to have 1 correspond to a line that predicts perfectly, a preferable form for this measure of usefulness is obtained by subtracting the preceding ratio from 1. The desired measure is therefore given by

$$1 - \frac{\sum_{i=1}^{n} (y_i - \hat{y}_i)^2}{\sum_{i=1}^{n} (y_i - \bar{y})^2}$$

For easier interpretation, it is convenient to write this expression in the form

$$\frac{\sum_{i=1}^{n} (y_i - \bar{y})^2 - \sum_{i=1}^{n} (y_i - \hat{y}_i)^2}{\sum_{i=1}^{n} (y_i - \bar{y})^2}$$

In this form the numerator represents the amount the sum of the squares of the errors of prediction have been reduced by using the regression line for prediction; hence this expression gives the percentage (as a decimal) that the sum of the squares of the errors of prediction have been reduced using the regression line for prediction.

By using the formulas given in (2) and performing some rather lengthy algebraic manipulations, it can be shown that the preceding measure of usefulness is equal to r^2, where r is the correlation coefficient as defined in the preceding chapter. Thus, the square of the correlation coefficient may serve as a measure of how useful a regression line is for predicting y from x. In this connection, the value of r^2 is often called the *coefficient of determination.*

As an illustration, consider the application of this measure to the regression problem of the preceding section. In Chapter 9, the value of the correlation coefficient for the data of that problem was found to be .61; hence $r^2 = .37$. This implies that the sum of the squares of the errors of prediction have been reduced about 37 percent by using the regression line for prediction.

Since r^2 will be considerably smaller than r, unless the value of r is close to 1, it takes a fairly large value of r before one can expect the relation between x and y to be of practical value. In the preceding illustration, for example, the value $r = .61$ appeared to represent a fairly strong relationship; however the value $r^2 = .37$ as a measure of how useful the regression line is for predicting sales success on the basis of the salesman's aptitude test score deflates the value of r considerably as a useful quantitative measure of the relationship between those two variables.

4 A PROBABILITY MODEL

For the purpose of making probability statements about the reliability of a regression line, it is necessary to treat it as a sample estimate of a corresponding theoretical line. Since the sample line is of the form (1), the equation of this theoretical line will be written

$$y = \alpha + \beta x \tag{4}$$

To simplify the discussion, assume that in a regression experiment there is only one observed value of y corresponding to each value of x. This is not true of the experiment that produced the data of Table 1 but it could be made to apply to that experiment if a single y were selected at random for each of those x's for which more than one y occurred. Assume, furthermore, that the experiment can be repeated a large number of times with those same x-values. Figure 4 illustrates the results that were obtained when an experiment of this type, involving seven values of x, was repeated eight times. The eight values of y corresponding to each fixed value of x in this diagram may be thought of as eight sample values of a random variable y associated with that value of x. Each of those y variables will have its own distribution with its own mean and variance. These distributions occur along the vertical lines directly above the corresponding x-values as shown in Fig. 4 and as indicated by the distribution of points on those lines.

For the purpose of obtaining a useful probability model for the general type of random variables introduced in the preceding paragraph, assume that there are

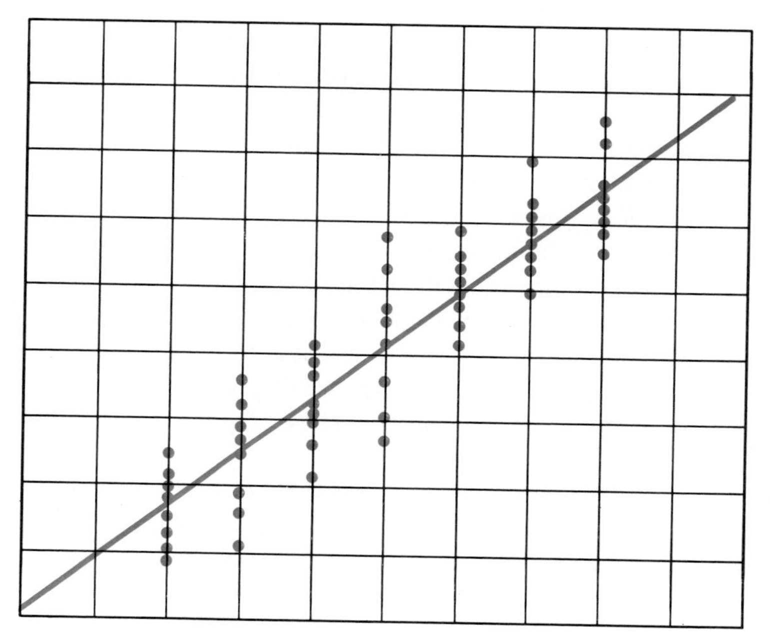

Figure 4 The results of eight replications of an experiment.

n pairs of values of x and y in the experiment and let $x_1, x_2, \ldots, x_n$ denote the x-values. Then assume that the random variables $y_1, y_2, \ldots, y_n$ corresponding to those values of x are independent variables with their means lying on a straight line and all possessing the same variance. The equation of this straight line will be written in the form (4). The line shown in Fig. 4 represents such a line.

The assumption that the means of the y_i lie on a straight line of the form (4) can be expressed algebraically by stating that

$$E[y_i] = \alpha + \beta x_i, \qquad i = 1, 2, \ldots, n \tag{5}$$

The assumption that the variables y_i possess the same variance, which will be denoted by σ_ε^2, may be expressed by writing

$$V(y_i) = \sigma_\varepsilon^2, \qquad i = 1, 2, \ldots, n \tag{6}$$

The preceding assumptions will be used in the following sections to obtain methods for determining the reliability of a regression line.

The preceding model assumes that in repeated experiments the same set of x-values will be obtained each time. This means that one can choose a set of x-values in advance of experimentation and then take a single random observation of y at each of those x-values to produce one experiment. Successive experiments are then repetitions of this first experiment. If, however, the first experiment consists of taking a random sample of n individuals and recording their associated x- and y-values, in which case the x-values are not chosen in advance of experimentation, then in repeated experiments those same x-values must be chosen and a single random observation of y taken at each of those x-values to produce a repetition of the first experiment. This would be the situation for the salesman problem of Section 1, because the data were obtained from a sample of n salesmen and no attempt was made to choose the x-values.

Many regression problems are the type in which the x-values are chosen in advance. For example, if a manufacturer of steel rods wishes to study the effect of changing the carbon content on the tensile strength of his rods, he would undoubtedly wish to run experiments at equally spaced carbon content values over the normal range of such values. Or, if a market analyst wishes to determine the effect of price on the purchases of coffee, he would probably wish to choose a set of equally spaced prices over a reasonable range of prices and conduct experiments on that basis at various markets.

Regression methods are very versatile because they can be applied to problems in which the x's are chosen in advance and also to problems in which the x's are obtained by random selection. In the latter case, probability statements require repetitions of the experiment to have the same set of x's as in the first random experiment.

Correlation methods, however, require that both x and y be random variables. They cannot be applied to problems in which the x's are chosen in advance, because the value of r will usually depend heavily upon the choice of x-values. In this connection it is proper to use r^2 as a measure of how useful a regression line is for prediction purposes even though the x's have been selected in advance; however, it is incorrect to treat the square root of r^2 as an ordinary correlation coefficient and use it as a measure of how strongly x and y are linearly related. It is, of course, also incorrect then to treat r as an estimate of ρ and use it to test the hypothesis that $\rho = 0$.

5 THE STANDARD ERROR OF ESTIMATE

A useful measure of the accuracy of prediction when using a regression line for making predictions is obtained by calculating the mean of the sum of squares of the errors of prediction. This mean is given by the expression

$$\frac{\sum_{i=1}^{n} (y_i - \hat{y}_i)^2}{n} \tag{7}$$

For problems related to determining the reliability of a regression line it is better to divide the sum of squares of the errors by $n - 2$ rather than by n, just as it was considered better to divide by $n - 1$ rather than by n in defining the ordinary sample variance. The same type of arguments apply here as applied there. It can be shown that when n is replaced by $n - 2$ in (7), that expression will be an unbiased estimate of the error variance σ_ε^2 defined in (6).

The square root of the resulting expression is denoted by s_e and is called the *standard error of estimate.* It is given by

$$\textbf{\textit{Standard Error of Estimate.}} \quad s_e = \sqrt{\frac{\sum_{i=1}^{n} (y_i - \hat{y}_i)^2}{n - 2}} \tag{8}$$

As an illustration of how the standard error of estimate may be used to help determine the usefulness of a regression line, consider its application to the data of Table 1. The predicted values $\hat{y}_i$ must first be obtained by using equation (3). Although some of the x's have more than one corresponding y-value, this merely means that some of the y's will have the same predicted value. The probability model introduced in the preceding section implies that each x has only one y-value. However, if the various pairs of x,y-values are treated as though the x's were distinct, the theory of that section is still applicable. The calculations needed to obtain s_e are shown in Table 3.

Table 3

x_i	y_i	$\hat{y}_i$	$y_i - \hat{y}_i$	$(y_i - \hat{y}_i)^2$
41	32	31.1	.9	.81
35	20	26.7	−6.7	44.89
34	35	26.0	9.0	81.00
40	24	30.4	−6.4	40.96
33	27	25.2	1.8	3.24
42	28	31.8	−3.8	14.44
37	31	28.2	2.8	7.84
42	33	31.8	1.2	1.44
30	26	23.0	3.0	9.00
43	41	32.6	8.4	70.56
38	29	28.9	.1	.01
38	33	28.9	4.1	16.81
46	36	34.8	1.2	1.44
36	23	27.4	−4.4	19.36
32	22	24.5	−2.5	6.25
43	38	32.6	5.4	29.16
42	26	31.8	−5.8	33.64
30	20	23.0	−3.0	9.00
41	30	31.1	−1.1	1.21
45	30	34.0	−4.0	16.00
				Total: 407.06

$$s_e = \sqrt{\frac{407.06}{18}} = 4.76$$

If the predicted values $\hat{y}_i$ are not needed for other purposes, it is simpler to calculate s_e by means of the following formula, which can be shown to be equivalent to formula (8).

$$s_e = \sqrt{\frac{\Sigma y_i^2 - a\Sigma y_i - b\Sigma x_i y_i}{n - 2}} \tag{9}$$

It should be noted that all the quantities needed in this formula, except Σy_i^2, are available from the computations needed to obtain the equation of the regression line.

To illustrate the use of this formula, the value of Σy_i^2 was calculated for the data of Table 3 and was found to have the value 17,704. Using this value and the necessary values of Table 2, formula (9) gives

$$s_e = \sqrt{\frac{17{,}704 - 1.01(584) - .734(22{,}755)}{18}} = 4.78$$

This result differs slightly from the value obtained using the definition of s_e because of the inaccuracy in $a = 1.01$ and $b = .734$ due to rounding errors.

In addition to the assumptions made in the preceding section, assume that the errors of prediction are normally distributed. Then it is possible to make probability statements concerning the errors of prediction. For example, one can state that approximately 95 percent of the errors of prediction will be smaller than $1.96s_e$ in magnitude. The approximation arises because $1.96\sigma_\varepsilon$ has been replaced by its sample estimate $1.96s_e$ and because only the sample regression line is available; therefore this is a large-sample statement.

Assuming that the preceding assumptions are applicable to the data of Table 1, the value of $1.96s_e$ will be calculated for those data and then used to give probability information concerning errors of prediction. Since $s_e = 4.76$, it follows that $1.96s_e = 9.3$; therefore it is to be expected that approximately 95 percent of the points of Fig. 1 will lie within the band obtained by drawing two lines parallel to the regression line, one being 9.3 units above it and one being 9.3 units below it. These two lines are shown in Fig. 5.

It should be observed that 19 of the 20 sample points lie inside the band, which is in good agreement with expectation, even though this is not a large sample. It seems clear from the fairly large value of the standard error of estimate here that an aptitude test score alone is of limited value in predicting sales success. This observation is in agreement with the earlier comment that the value of $r^2 = .37$ is not sufficiently large to justify a lot of confidence in the regression line as a predictor of sales productivity.

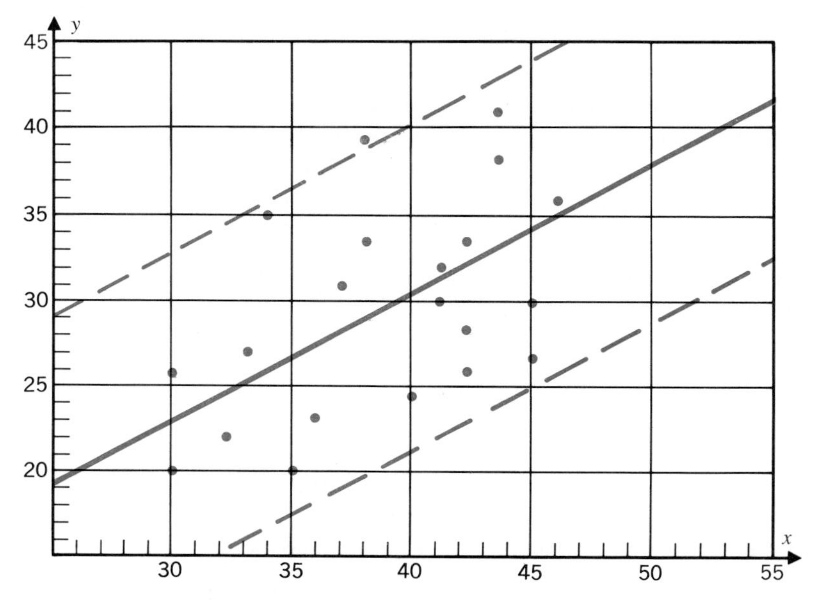

Figure 5 An approximate 95 percent prediction band.

This geometrical manner of looking at the prediction problem is a useful one for giving the experimenter a rough idea of what y-values he is likely to obtain if he performs experiments at other x-values.

If the experimenter is interested in predicting y for only a single specified value of x, then a more refined technique is needed to determine the accuracy of the prediction. That technique will not be considered here because it is of questionable validity in most business data problems.

The problem of predicting a y-value for an x-value beyond the range of observed x-values is a considerably more difficult problem than that of predicting for a value of x inside the interval of observations. Prediction beyond the range of observations is called *extrapolation*, whereas prediction inside the range is called *interpolation*. The difficulty with extrapolation is that the assumptions necessary to justify it are seldom realized in real-life situations.

Thus, it is highly unlikely that a salesman who made an extremely high score on an aptitude test would have a correspondingly extremely high productivity score. In many problems it is also unrealistic to assume that the variance of y for values of x beyond the observed range of x-values is the same as for y's corresponding to values of x within the observed range. Extrapolation is a legitimate technique only when the experimenter has valid reasons for believing that his model holds beyond the range of available observations.

A kind word needs to be said for the usefulness of straight-line regression models in realistic situations. If the scatter diagram for a set of data indicates that the relationship between x and y is not linear, it may still be possible to use the linear model if one can find a function of x and also a function of y such that the relationship between those two functional values is linear. There are many known relationships in science that are not linear but which can be made linear by taking the proper functions of x and y. For example, the functions of $y = ax^b$ and $y = ab^x$ can be converted into linear relationships by taking the logarithm of both sides and then defining new variables in terms of the logarithms of the old variables. This technique of considering the relationship between functions of x and y rather than between x and y extends the range of applicability of linear methods considerably.

6 THE RELIABILITY OF COEFFICIENTS

The probability model of section 4 enables one to make probability statements concerning the accuracy of a and b as estimates of α and β. First, consider the problem of determining how good a is as an estimate of α.

Using the expected-value operator, it can be shown that $E[a] = \alpha$; consequently, a is an unbiased estimate of α. Furthermore, the normality assumption made in section 5 enables one to show mathematically that a is a normal variable with mean α and with variance given by the formula

$$\sigma_a^2 = \sigma_\varepsilon^2\left[\frac{1}{n} + \frac{\bar{x}^2}{\Sigma(x_i - \bar{x})^2}\right] \tag{10}$$

Thus, the normal-variable techniques used in the chapter on estimation may be employed here to determine the accuracy of a as an estimate of α.

As an illustration of how one applies the preceding theory to a large-sample problem, consider once more the data of Table 1 and ignore the fact that this is really not a very large sample. First, since a is a normal variable with mean α and variance σ_a^2 given by formula (10), it follows that a 95 percent confidence interval for α is given by

$$a - 1.96\sigma_a < \alpha < a + 1.96\sigma_a$$

Substituting the value of σ_a given by formula (10) and using the computations of Table 2 that led to (3) gives

$$1.01 - 1.96\sigma_\varepsilon(.368) < \alpha < 1.01 + 1.96\sigma_\varepsilon(.368)$$

Since σ_ε is unknown, it is necessary to replace it by its sample estimate, s_e. As a result, an approximate 95 percent confidence interval for α is given by

$$1.01 - 1.96(4.76)(.368) < \alpha < 1.01 + 1.96(4.76)(.368)$$

Calculations reduce this to

$$-2.42 < \alpha < 4.44$$

Next, consider the problem of determining how good b is as an estimate of β. Here also it can be shown by using the expectation operator that $E[b] = \beta$ so that b is an unbiased estimate of β. Using the same techniques as for a, it can be shown mathematically that b is a normal variable with mean β and variance given by the formula

$$\sigma_b^2 = \sigma_\varepsilon^2\left[\frac{1}{\Sigma(x_i - \bar{x})^2}\right] \tag{11}$$

Thus, the estimation chapter techniques may be employed on this problem also.

As an illustration, consider the problem of finding a 95 percent confidence interval for the slope coefficient β for the data of Table 1. In the same manner as for α, a 95 percent confidence interval for β is given by

$$b - 1.96\sigma_b < \beta < b + 1.96\sigma_b$$

Substituting the value of σ_b given by formula (11) and using previous computations yields the following approximate 95 percent confidence interval:

$$.734 - 1.96\frac{4.76}{\sqrt{448.8}} < \beta < .734 + 1.96\frac{4.76}{\sqrt{448.8}}$$

Calculations reduce this to

$$.29 < \beta < 1.17$$

It should be observed that σ_ε was replaced by its estimate s_e, just as was done in finding an approximate confidence interval for α.

The preceding results show that neither α nor β is being estimated with much accuracy in this problem. The large variation in the y's together with the fairly small sample size preclude high accuracy of estimation.

Problems of testing hypotheses concerning α and β can be carried out in the usual manner by the techniques of Chapter 8 for normal variables, provided that the sample is sufficiently large and the assumptions of sections 4 and 5 are reasonably satisfied.

The preceding methods of sections 5 and 6 are large-sample methods. Methods for treating small samples are discussed next.

7 SMALL-SAMPLE METHODS

The statistical inference problems that were solved in sections 5 and 6 were solved by the use of large-sample methods. For example, in finding confidence intervals for α and β it was necessary to replace σ_ε by its sample estimate, s_e. However, under the probability model assumptions of section 4, together with the normality assumption made in section 5, it can be shown that Student's t distribution is applicable to these problems. Thus, it can be shown that the two variables

$$t = \frac{a - \alpha}{s_a} \quad \text{and} \quad t = \frac{b - \beta}{s_b} \tag{12}$$

both possess Student's t distribution with $n - 2$ degrees of freedom. In these formulas s_a is the sample estimate of σ_a obtained by replacing σ_ε by s_e in formula (10), and s_b is the sample estimate of σ_b obtained by replacing σ_ε by s_e in formula (11).

For the purpose of comparing small- and large-sample results, the two problems solved in section 6 will be solved by means of formulas (12). Since $n - 2 = 18$ here, the value of t needed for a 95 percent confidence interval is found in Table V to be 2.10. The first of the preceding formulas will therefore produce the following 95 percent confidence interval for α:

$$a - 2.10s_a < \alpha < a + 2.10s_a$$

Substituting the numerical values obtained previously gives

$$1.01 - 2.10(1.75) < \alpha < 1.01 + 2.10(1.75)$$

or

$$-2.66 < \alpha < 4.68$$

The large-sample method produced the interval

$$-2.42 < \alpha < 4.44$$

which differs very little from the exact small-sample interval.

From the second formula in (12), a 95 percent confidence interval for β is given by

$$b - 2.10s_b < \beta < b + 2.10s_b$$

Substituting the previously obtained numerical values gives

$$.734 - 2.10(.225) < \beta < .734 + 2.10(.225)$$

or

$$.26 < \beta < 1.21$$

The large-sample method produced the interval $.29 < \beta < 1.17$, which differs very little from the small-sample result. In both cases, however, the large-sample method yields an interval that is too narrow. If the sample size had been considerably smaller than 20, the large-sample intervals would have been found to be much too narrow when compared to the exact small-sample intervals.

8 THE REGRESSION FALLACY

The name "regression," which is commonly given to the least-squares line, is also associated with an error frequently made in the interpretation of observations taken at two different periods of time. This error, which is called the *regression fallacy,* is best explained by means of illustrations.

Suppose a teacher studies the scores of his students in a given course on their first two hour tests. If he selects, say, the top five papers from the first test and calculates the mean score for those five students on both tests, he will very likely discover that their second test mean is lower than their first test mean. Similarly, if he calculates the mean score on both tests for the five students having the lowest scores on the first test, he will undoubtedly find their mean has risen. Thus he might conclude that the good students are slipping, whereas the poor students are improving. The explanation lies partly in the reaction of students to test scores but also partly in the natural variability of students' test scores. Even if students did not vary from one test to another in their study habits and in their total relative knowledge of the subject, the inaccuracy of a test to measure this knowledge would cause considerable variation in a student's performance from test to test. As a result, some of the top five students on the first test may be there because of fortuitous circumstances; however, the second test is likely to bring them down to their natural level, thereby dragging the mean of this group of five students down with them. The same type of reasoning applies to the lowest five students on the first test to account for their improved mean.

There is undoubtedly a psychological factor operating here also to accentuate the "regression" of high and low scores toward the mean of the entire group. Students who did poorly on the first test would be expected to study considerably harder than before and thus raise their scores on the second test. Since early success often leads to overconfidence, the students making the highest scores on the first test might be expected to ease up slightly and thus lower their mean score on the second test. Nevertheless, the natural variation of test scores alone will suffice to produce a fair amount of regression toward the group mean.

The regression fallacy often occurs in the interpretation of business data. For example, in comparing the profits made by a group of similar business firms for two consecutive years, there might be a temptation to claim that the firms with high profits are becoming less efficient, whereas those with low profits are becoming more efficient because the mean profit for each of those two extreme groups would tend to shift toward the mean of the entire group. The firms with high profits the first year were high in the list either because they are normally highly efficient or because they were fortunate that year but are normally of lower efficiency. The latter group would be expected to show lower profits the second year and thus decrease the mean for the first year's high-profit group. The same type of reasoning would explain the apparent increased efficiency of the first year's low-profit group.

The original study of the relationship between the stature of fathers and sons, which gave rise to the name regression, is another illustration of this type of possible misinterpretation. It was found that the tallest group of the total group of men being studied had sons whose mean height was lower than that of the fathers. It was also found that the shortest group had sons whose mean height was higher than that of the fathers. As in the other illustrations, the explanation lies in the natural variation of subgroups of a population. Since many tall men come from families whose parents are of average size, such tall men are likely to have sons who are shorter than they are; consequently, when a group of tall men is selected, the sons of such men would not be expected to be quite so tall as the fathers. There are, of course, factors such as the tendency of tall men to marry taller-than-average women and the steady increase in stature from generation to generation to dampen the above regression tendency somewhat.

9 LIMITATIONS

The problem of predicting values of y by means of linear regression and then determining the accuracy of such predictions is a difficult problem, because the mathematical assumptions that are required to arrive at a solution are rather stringent and may not be satisfied in real-life situations. Unless an investigator has valid reasons for believing that those assumptions are reasonably satisfied, he should treat the results of his regression analysis as being only rough approximations to the truth.

Many regression problems in the social sciences involve data taken over time. For such data consecutive values of y may not be independent. For example, if a study is made to see whether the retail price index can be predicted for one month from today by means of a regression line that is determined from the retail prices during the last twelve months, it will undoubtedly be found that the retail price indexes for consecutive months are not independent. There is, of course, the added difficulty here that this is an extrapolation problem.

In summary, an investigator who uses the preceding regression techniques should be aware of the assumptions on which they are based, and realize that those techniques constitute a powerful tool that must be used with judgment.

This chapter has been concerned with the problem of predicting a variable y by means of a linear function of a related variable x. A more sophisticated problem is that of predicting the variable y by means of several related variables. The methods for solving that type of problem are similar to those introduced in this chapter. They are studied in the next chapter.

10 REVIEW ILLUSTRATIONS

Example 1. The following data were taken from the census tracts of Los Angeles County. They were obtained by selecting a set of house values (x) ranging from \$15,000 to \$49,000 and then taking a single random sample of income levels (y) associated with each house value. The units are in thousands of dollars.

House value (x)	Income (y)	House value (x)	Income (y)
16	6.6	34	17.2
18	7.4	36	15.8
20	10.7	38	15.1
22	12.3	40	18.4
24	11.3	42	19.2
26	12.9	44	16.3
28	14.5	46	16.4
30	12.2	48	22.7
32	15.0		

(a) Plot the scatter diagram and observe whether the relationship may be treated as being linear. (b) Find the equation of the least-squares line. (c) Calculate the value of the coefficient of determination and use it to comment on the usefulness of a knowledge of the value of houses in a tract for estimating income level. (d) Calculate the standard error of estimate by means of formula (9). (e) Graph the regression line; construct a $2s_e$-band about it; determine the percentage of points of the scatter diagram that lie inside this band.

(a)

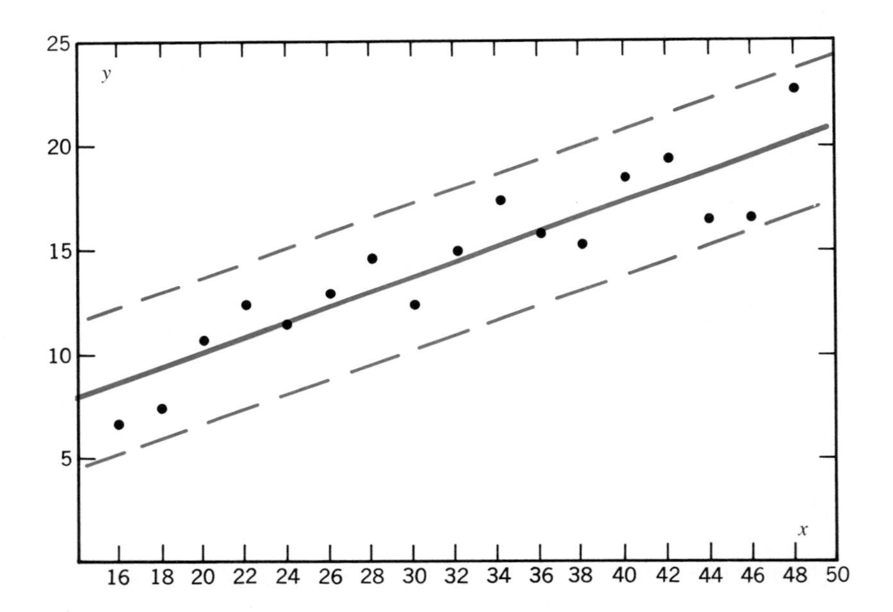

(b) $\Sigma x = 544$, $\Sigma x^2 = 19{,}040$, $\Sigma y = 244$, $\Sigma y^2 = 3772.92$, $\Sigma xy = 8413$,

$$b = \frac{17(8413) - (544)(244)}{17(19{,}040) - (544)(544)} = \frac{10{,}285}{27{,}744} = .3707$$

$$a = \frac{244}{17} - .3707\left(\frac{544}{17}\right) = 2.49$$

$$\hat{y} = 2.49 + .371x$$

(c) $$r = \frac{17(8413) - (544)(244)}{\sqrt{[17(19{,}040) - (544)(544)][17(3772.92) - (244)(244)]}}$$

From the calculations in (b) this becomes

$$r = \frac{10{,}285}{\sqrt{27{,}744[17(3772.92) - (244)(244)]}}$$

$$= \frac{10{,}285}{\sqrt{(27{,}744)(4603.64)}} = .91$$

Hence, $r^2 = .83$. This high value shows that the regression line can be very useful for estimating the income levels in tracts on the basis of their house values.

(d) $$s_e = \sqrt{\frac{3772.94 - 2.49(244) - .3707(8413)}{15}} = 1.76$$

(e) $2s_e = 3.5$: the band is shown on the scatter diagram of (a) by means of broken lines. All the points lie inside this band.

Example 2. The following data give the amount of water applied in inches, x, and the yield of hay in tons per acre, y, on an experimental farm.

Water (x)	12	18	24	30	36	42	48
Yield (y)	5.27	5.68	6.25	7.21	8.02	8.71	8.42

(a) Plot the scatter diagram and observe whether the relationship may be treated as being linear. (b) Find the equation of the least-squares line and graph it on the scatter diagram of (a). (c) Calculate the standard error of estimate. Comment about using it as in Fig. 5. (d) Calculate a 90 percent confidence interval for β. (e) Comment about the validity of using this line for extrapolation.

(a)

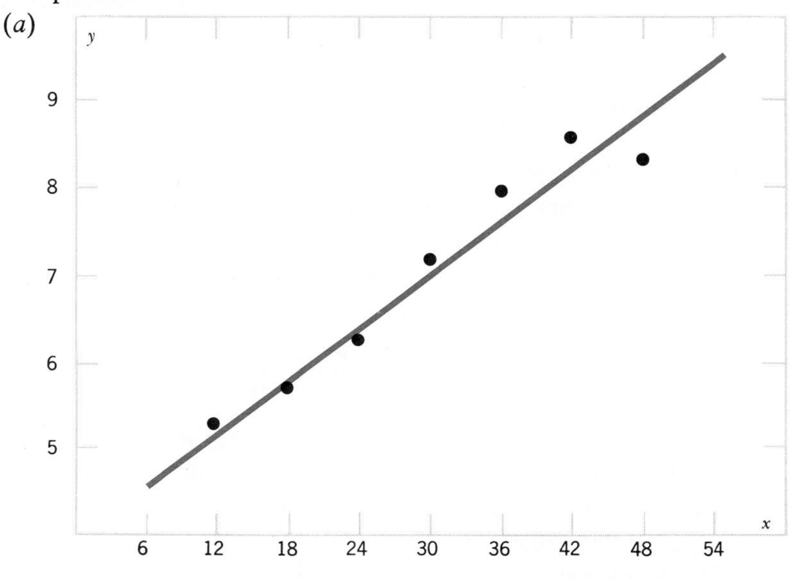

Figure 1. Hay yield as a function of amount of irrigation.

(*b*)

x_i	y_i	x_iy_i	x_i^2
12	5.27	63.24	144
18	5.68	102.24	324
24	6.25	150.00	576
30	7.21	216.30	900
36	8.02	288.72	1296
42	8.71	365.82	1764
48	8.42	404.16	2304
210	49.56	1,590.48	7308

$$b = \frac{7(1{,}590.48) - (210)(49.56)}{7(7{,}308) - (210)(210)} = .103$$

$$a = \frac{49.56}{7} - .103\left(\frac{210}{7}\right) = 3.99$$

$$\hat{y} = 3.99 + .103x$$

(*c*) $$s_e = \sqrt{\frac{362.16 - 3.99(49.56) - .103(1{,}590.48)}{5}} = .345$$

The sample is much too small to apply the Fig. 5 technique.

(*d*) $s_b = \dfrac{s_e}{\sqrt{\Sigma(x_i - \bar{x})^2}} = \dfrac{.345}{\sqrt{1008}} = .011$; hence, since $t_0 = 2.015$,

$$.103 - 2.015(.011) < \beta < .103 + 2.015(.011)$$
$$.081 < \beta < .125$$

(*e*) Adding considerably more water would probably decrease the amount of hay rather than increase it according to the apparent linear relationship shown in (*a*); therefore extrapolation here would be completely unjustified.

Example 3. A music store that specializes in selling records at a discount wishes to predict the first year's sales of new releases by means of the sales after one month in order that it may take advantage of quantity discounts in purchasing. A record was kept during the past year with the following results, wherein x denotes the first month's sales and y the annual sales.

x	5	15	7	12	7	5	8	6	7	10
y	15	50	20	34	21	12	24	16	18	40

(*a*) Plot these pairs of points. (*b*) Find the equation of the regression line. (*c*) If a record sells 10 units the first month, what is your estimate of annual sales? Why does this differ considerably from the pair (10,40) actually found? (*d*) Find 95 percent confidence limits for the true slope of the regression line. Comment.

(*a*)

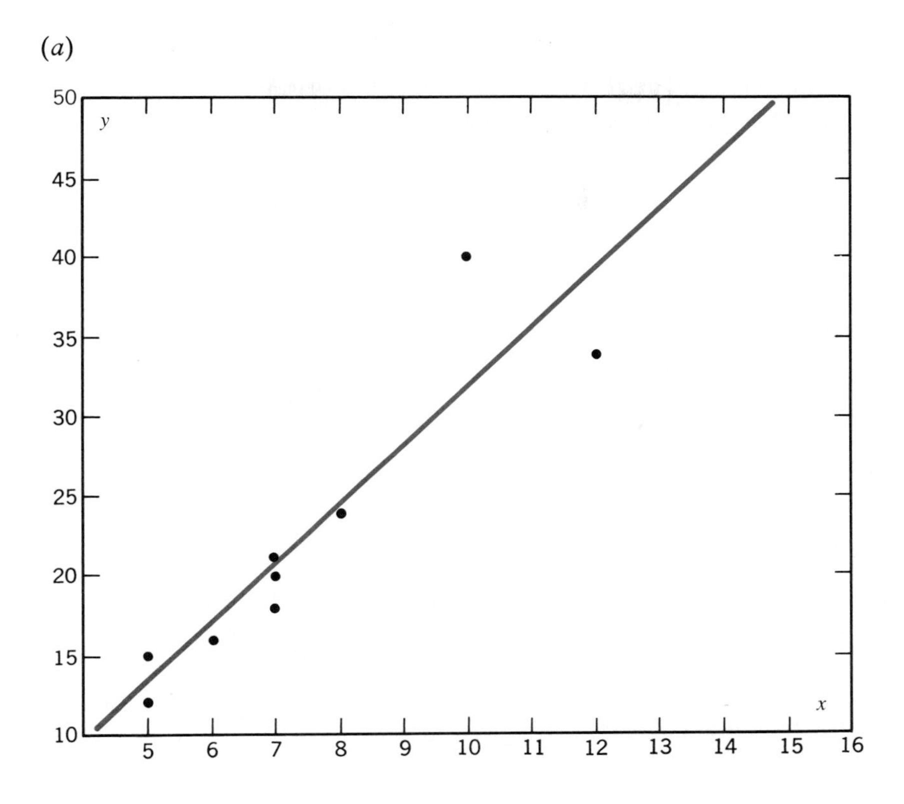

(*b*) $\Sigma x = 82, \quad \Sigma y = 250, \quad \Sigma x^2 = 766, \quad \Sigma y^2 = 7622, \quad \Sigma xy = 2394,$

$$b = \frac{10(2394) - (82)(250)}{10(766) - (82)(82)} = 3.675$$

$$a = \frac{250}{10} - 3.675\left(\frac{82}{10}\right) = -5.14$$

$$\hat{y} = -5.14 + 3.675x$$

(*c*) $\hat{y} = 31.6$. This sample point is out of line with most of the others. Furthermore, the prediction for a large x is more variable than for a small x.

(*d*) $s_e = \sqrt{\dfrac{7622 - (-5.14)(250) - 3.675(2394)}{8}} = 3.7, \quad \Sigma(x - \bar{x})^2 = 93.6;$

hence

$$3.675 - 2.306\frac{3.7}{\sqrt{93.6}} < \beta < 3.675 + 2.306\frac{3.7}{\sqrt{93.6}}$$

or

$$2.80 < \beta < 4.56$$

The estimate of β is based on such few sample values that β cannot be estimated with much accuracy. Furthermore, it is not clear that the y-values are independent; therefore this result may not be justified.

EXERCISES

SECTION 1

1 Graph the line whose equation is $y = .3x - 1$ for values of x between 0 and 5.

2 Graph the line whose equation is $y = -.5x + 1.5$ for values of x between -2 and 4.

3 The regression line for estimating the yearly family expenditure on food by means of yearly income, in dollars, is given by $y = 200 + .20x$.

(*a*) What is the average expenditure for families with incomes of 2000 dollars, 5000 dollars?

(*b*) Why would you hesitate to use this formula for incomes of 0 dollars?

4 Specimens of several brands of mayonnaise were analyzed for fat content by a rapid method and by the standard method used by the *Association of Official Agricultural Chemists.* Denoting the rapid method by x and the standard method by y, the following data were obtained:

x	80.5	30.3	25.2	77.4	48.1	35.7	18.6
y	79.3	30.4	26.0	77.9	47.5	35.5	16.7

(*a*) Plot these points and fit an approximate regression line to them by using a ruler and good judgment.

(*b*) If the rapid method yielded a value of (*i*) 50 percent, (*ii*) 70 percent, what corresponding values would your regression line give for the standard method?

5 As reported in a 1954 *Agriculture Information Bulletin* (132) entitled "Food Consumption in the United States," a study was made to determine the relationship between food expenditures and income of families receiving less than \$7,500 per year. This study produced the following table of values.

Annual income	Number of families	Average income	Average weekly food costs
under 1,000	53	610	13.0
1,000–1,999	204	1,555	15.7
2,000–2,999	410	2,505	20.9
3,000–3,999	351	3,485	24.2
4,000–4,999	167	4,421	27.9
5,000–7,500	154	5,861	29.0
over 7,500	72	11,766	40.2
Total:	1411		

(a) Let x denote the average income in thousands of dollars and y the weekly food costs in dollars. Plot the points corresponding to these seven pairs of values.

(b) Fit a freehand regression line to those points.

(c) Use your line in (b) to estimate the weekly food costs for a family with an annual income of $4,500.

(d) Suppose the equation of this fitted line is $\hat{y} = 12.3 + 3.2x$. What would your estimate then be for $x =$ $4,500?

(e) Comment about the adequacy of using a straight line to estimate average food costs.

SECTION 2

6 Find the equation of the regression line for the following data. Graph the line on a scatter diagram.

x	1	2	3	4	5
y	2	5	3	8	7

7 The city of Santa Monica, California, contains 709 blocks about which the U.S. Bureau of the Census has published basic housing and demographic data. A random sample of 14 of those blocks was selected and the number of households residing in them was obtained from the 1950 and the 1960 censuses. This study produced the following data, in which x represents the 1950 count and y the 1960 count.

x	34	25	21	51	44	27	33	31	29	0	19	30	60	39
y	37	24	22	94	100	27	61	14	55	6	22	54	51	48

(a) Calculate the regression equation for predicting y from x.

(b) In 1950 the mean number of households per block was 38.9. Use this value and your equation in (a) to predict the 1960 mean value.

(c) The actual 1960 mean value was 49.5. This represents about a 27 percent increase over 1950. What percent increase does your equation in (a) give? Comment.

8 The following data are for tensile strength and hardness of die-cast aluminum. Find the equation of the regression line for estimating tensile strength (y) from hardness (x) and graph it on the scatter diagram.

y	293	349	368	301	340	308	354	313	322	334	377	247
x	53	70	84	55	78	64	71	53	82	67	70	56

9 The following data are the scores made by students on an entrance examination (x) and a final examination (y).

(a) Find the equation of the regression line.

(b) Graph the line on the scatter diagram.

x	129	179	347	328	286	256	477	430	327	245	286	326
y	370	361	405	302	496	323	374	332	435	165	375	466

10 The following data were taken from a study made by E. S. Olcott in a 1955 issue of the *Journal of Operations Research* on the flow of traffic through a vehicular tunnel. Observations were taken at 5-minute intervals.

5-minute interval (P.M.)	Average density (vehicles/mile) x_i	Average speed (mi/hr) y_i
4:30–4:35	69	16.8
4:35–4:40	88	14.8
4:40–4:45	65	19.1
4:45–4:50	84	14.9
5:10–5:15	63	21.6
5:30–5:35	53	25.7
6:05–6:10	82	16.8
6:10–6:15	53	25.1
6:15–6:20	70	19.7
6:20–6:25	34	31.8

(*a*) Plot the data on a scatter diagram. Use x as the independent variable. Does the relationship appear to be linear?

(*b*) Fit a regression line to the data. Interpret the resulting regression function for changes in vehicular density.

11 Catalogs listing text books were examined to discover the relationship between the cost of a book and the number of pages it contains. This perusal yielded the following data for ten books.

Pages	750	550	280	580	450	350	680	500	700	300
Price	12.00	10.00	5.00	10.00	7.00	7.00	11.00	9.00	13.00	5.00

Computations yielded the following results where x denotes the number of pages and y the price.

$$\Sigma x = 5{,}140, \quad \Sigma y = 89, \quad \Sigma(x - \bar{x})^2 = 255{,}240$$

$$\Sigma(y - \bar{y})^2 = 70.9, \quad \Sigma(x - \bar{x})(y - \bar{y}) = 4{,}134$$

(*a*) Use these results to find the equation of the regression line.

(*b*) What do you estimate the price should be for a 400-page book?

(*c*) What price increase would you expect for a book if it was decided to increase the size of the book by 100 pages?

12 The following data are total production of Florida grapefruit of all types and the average on-tree price by crop year for the period 1940 to 1964.

Crop year	Production (10^6 boxes)	Price (dollars/box)	Crop year	Production (10^6 boxes)	Price (dollars/box)
1940–41	24.6	0.33	1955–56	38.3	0.57
1941–42	19.2	0.63	1956–57	37.4	0.89
1942–43	27.3	0.92	1957–58	31.1	0.98
1943–44	31.0	1.31	1958–59	35.2	1.04
1944–45	22.3	1.70	1959–60	30.5	1.05
1945–46	32.0	1.27	1960–61	31.6	0.96
1946–47	29.0	0.63	1961–62	35.0	0.67
1947–48	33.0	0.26	1962–63	30.0	1.24
1948–49	30.2	0.67	1963–64	26.3	2.24
1949–50	24.2	1.79	1964–65	31.9	1.47
1950–51	33.2	0.94			
1951–52	36.0	0.52			
1952–53	32.5	0.76			
1953–54	42.0	0.49			
1954–55	34.8	0.63			

(*a*) Plot the data in a scatter diagram with price against production. Identify each dot with crop year. Is there good reason to reject the years 1940–1941 and 1941–1942? Does there seem to be a linear relationship between price and production?

(*b*) Fit a linear regression to the data excluding the first two crop years. In 1965–1966 production was 34.9 million boxes and the price \$1.36. Compare with your predicted price.

13 The data in problem 12 are the amounts produced and the price per box of Florida grapefruit for the period 1940–1965. Data such as these that are collected over time are called *time series data* and are analyzed by methods designed for such series. Such methods are discussed in Chapter 16. They may be analyzed by simple regression methods provided they satisfy the assumptions made in section 4 of this chapter. These assumptions require, for example, that time not be a relevant factor in the relationship. One method of checking on this property is to examine the errors of prediction by plotting them as a function of time and observing whether they appear to be independent of time. Do this with the data of problem 12, using the answers obtained for that problem. Comment.

SECTION 3

14 Calculate the value of r^2 for example 3 of section 10 and use it to comment on the usefulness of that regression line for prediction purposes.

15 Calculate the value of r^2 for problem 8 and use it to comment on how useful a measure of hardness is for predicting the tensile strength of aluminum.

16 Use the answer to problem 5, Chapter 9, to calculate the value of r^2. From this value, may you conclude that a student's intelligence score is highly useful for predicting his grade point average?

17 Using the data of problems 6 and 7, Chapter 9, (*a*) calculate the equation of the regression line, (*b*) calculate the value of s_e, (*c*) comment on the value of knowing a student's verbal test score for estimating his aptitude for mathematics.

SECTION 5

18 Calculate the standard error of estimate in problem 6.

19 Calculate the standard error of estimate in problem 8 and find what percentage of the errors exceed it. What percentage of the errors exceed twice the standard error of estimate?

20 Explain why the standard error of estimate calculated in problem 18 and also in problem 19 would not be reliable for predicting the sizes of future errors of estimate.

21 A restaurant manager is concerned about the large number of individuals who make reservations but do not show. Since he has limited space and therefore can accept only a limited number of reservations, he wants to know the relation between the number of reservations and the number of showing guests. He believes that the relationship is linear. For the purpose of checking on this belief he has collected data for 27 nights. Letting x denote the number of reservations and y the number who showed, he obtained the following statistics:

$$\bar{x} = 40, \quad \bar{y} = 33, \quad \Sigma(x - \bar{x})^2 = 4{,}000$$
$$\Sigma(y - \bar{y})^2 = 3{,}250, \quad \Sigma(x - \bar{x})(y - \bar{y}) = 3{,}170$$

(*a*) Find the equation of the regression line.
(*b*) If 50 reservations are made, how many can be expected to show?
(*c*) Calculate the standard error of estimate.
(*d*) If the restaurant can accomodate at most 50 guests, do you believe it would be safe to accept 60 reservations? Explain.

22 As reported in *Science* for May 28, 1971, Bache et al., in studying the relationship of mercury and methylmercury salts to age in lake trout obtained the following data on 27 fish of various ages netted from Cayuga Lake in Ithaca, New York. The age of a fish is denoted by x and the total amount of mercury and methylmercury salts is denoted by y.

x	y	x	y	x	y
1	.24	4	.44	7	.44
1	.28	4	.41	8	.60
1	.19	4	.44	8	.59
2	.25	5	.43	8	.47
2	.26	6	.46	9	.53
2	.31	6	.55	11	.58
3	.38	6	.50	12	.62
3	.45	7	.40	12	.66
3	.28	7	.46	12	.44

(*a*) Plot the scatter diagram and comment about any apparent relationship.
(*b*) Calculate the equation of the regression line, omitting the last pair of observed values, which appear to be out of line with the rest. Possibly an incorrect age or an error of recording occurred.
(*c*) Calculate the standard error of estimate and determine how many points differ from their regression values by more than $2s_e$. What happens to the point that was omitted?

23 A marketing experiment was carried out by the importer of a foreign automobile to help determine a satisfactory pricing policy. A sample of 25 agencies was selected from the more than 500 agencies throughout the country. They were considered to be quite equal in sales potential. They were assigned certain sticker prices and were required to hold prescribed limits on them. Prices ranged from \$2500 to \$3000 per standard equipped car. After three months, the number of sales by agency and other statistical results were as follows:

Average price, $\bar{x}$ = \$2750
Average sales, $\bar{y} = 360$ per dealer
Regression coefficient, $b = -.40$
Standard error of estimate $s_e = 20$

(*a*) Find the equation of the regression line.
(*b*) Is there reasonable evidence that price would have an effect on the volume of sales for all 500 agencies?
(*c*) What change in sales does this experiment suggest would result if the price of an automobile were reduced by \$100?

SECTION 6

24 In a regression problem, $n = 18$, $\Sigma(x - \bar{x})^2 = 90$, $\Sigma(y - \bar{y})^2 = 600$, $\Sigma(x - \bar{x})(y - \bar{y}) = 180$, $\bar{x} = 20$, $\bar{y} = 30$.
(*a*) Find the equation of the regression line.
(*b*) Test the hypothesis that $\beta = 1.5$ against the alternative that $\beta > 1.5$, using large-sample methods.

25 The following data resulted from selecting one house value (y) at random from each of the associated income levels (x), taken from the census tracts of Los Angeles County.

Income (x)	House value (y)	Income (x)	House value (y)
6.5	18.1	13.5	28.6
7.5	22.2	14.5	30.2
8.5	22.6	15.5	37.2
9.5	18.8	16.5	43.5
10.5	21.9	17.5	33.4
11.5	23.8	18.5	40.3
12.5	25.4	19.5	42.7

(*a*) Plot the data.

(*b*) Calculate the equation of the regression line.
(*c*) Find an 80 percent confidence interval for β, using large-sample methods.

SECTION 7

26 Given $n = 8$, $b = 2$, $s_e = .2$, and $\Sigma(x_i - \bar{x})^2 = 50$, find 95 percent confidence limits for β.

27 Find 95 percent confidence limits for β in problem 8.

28 Test the hypothesis that $\beta = 1.0$ in problem 6.

29 Test the hypothesis that $\beta = 2.5$ in problem 8.

30 By inspecting formula (11), determine how you would choose ten values of x between $x = 0$ and $x = 10$ at which to take observations on y so as to make the confidence interval for β as short as possible.

31 Metal bars are known to elongate when heated. For temperatures that are not excessively high the elongation is a linear function of the temperature. The following data give the results of an experiment performed on 10 bars. The experimenter suspects that in addition to the usual errors of measurement, the measuring instrument was badly adjusted for the present series and that this resulted in a constant bias in the values of y. In these data, x (heat) and y (elongation) are scaled variables such that the relationship between x and y should be $y = 1.01x$.

x	100	200	300	400	500	600	700	800	900	1000
y	106	202	308	410	504	610	710	812	912	1012

(*a*) Test the hypothesis that $\beta = 1.01$. Comment.
(*b*) Test the hypothesis that $\alpha = 0$. Comment.

32 Moisture content of sweet corn is regarded as a good index for maturity in the canning stage. The most accurate method for determining moisture content is by means of a vacuum oven, a rather lengthy and cumbersome method. A rather quick and simple method has been developed that measures the moisture content electrically. To test the new method against the standard vacuum oven method, tests were run on corn of various levels of moisture content, a part of the data being presented below for 10 specimens, using both methods on each.

Vacuum, y:	78.2	77.3	72.8	72.7	71.2	68.4	68.7	67.2	66.4	66.4
Electric, x:	75.9	75.1	71.6	72.4	71.0	69.2	69.3	67.1	67.8	67.4
Difference, d:	−2.3	−2.2	−1.2	−0.3	−0.2	+0.8	+0.6	−0.1	+1.4	+1.0

(*a*) Compare the performance of the two methods by calculating the mean differences between the electric and vacuum methods and testing the hypothesis that the mean difference in the population is zero. Use the t test of Chapter 8.
(*b*) Compare the two methods by finding the regression function of the differences, d, on the readings of the electric method, x. Test the hypothesis that β, the regression coefficient in the population, is zero.
(*c*) Do results in (*a*) and (*b*) agree? Discuss.

SECTION 8

33 Explain the regression fallacy in the statement that track stars seem to go downhill after establishing a record.

34 Give an illustration of data for which the regression fallacy might easily be made.

35 Jones examined 100 firms in a given industry in 1965 and 1975. Ranking the 100 firms according to their earnings each year and calculating the equation of the regression line of 1975 rankings (y) against 1965 rankings (x), he found that the regression coefficient was less than 1. Explain.

SECTION 10

36 The average daily grades and the final examination grades for ten students in a class in calculus are given in the table below:

x = the average daily grade
y = the grade on the final examination

Student	x	y	Student	x	y
1	86	71	6	96	94
2	93	76	7	80	71
3	73	61	8	70	60
4	66	52	9	95	85
5	88	75	10	63	55

(*a*) Find the line of regression of y on x.
(*b*) Find s_e, the standard error of estimate.
(*c*) If $x = 90$, what estimate would you give for y?
(*d*) Find 95 percent confidence limits for β.

37 Girshick and Haavelmo, in an article published in *Econometrica* in 1947, made an analysis of the demand for food in the U.S. for the years 1922 to 1941. One feature of their analysis involved studying the relation between disposable income (x) and investment per capita (y), both adjusted for the cost of living. The data used in this part of the study are the following.

Year	Income	Invest	Year	Income	Invest
1922	87.4	92.9	1932	75.1	52.4
1923	97.6	142.9	1933	76.9	40.5
1924	96.7	100.0	1934	84.6	64.3
1925	98.2	123.8	1935	90.6	78.6
1926	99.8	111.9	1936	103.1	114.3
1927	100.5	121.4	1937	105.1	121.4
1928	103.2	107.1	1938	96.4	78.6
1929	107.8	142.9	1939	104.4	109.5
1930	96.6	92.9	1940	110.7	128.6
1931	88.9	97.6	1941	127.1	238.1

Assume that this relationship is independent of time.
(*a*) Calculate the regression line of investment as a function of income.

(*b*) Calculate the standard error of estimate.

(*c*) Calculate a 90 percent confidence interval for β using large-sample methods. Comment.

38 Using the results of the computations of problem 37, calculate a 90 percent confidence interval for β using small-sample methods. Compare your result with that of problem 37.

39 Use formula (2) of Chapter 9, together with the definitions of s_x and s_y, to show that the slope coefficient b as defined in (2) can be written in the form

$$b = r\frac{s_y}{s_x}$$

40 Given the following data, work parts (*a*) through (*d*) for the first of the 2 review problems of section 10.

x	1	9	6	4	5	7	7	1	2	8	6	3
y	15	75	55	42	33	45	55	17	32	80	48	45

41 Given the following data, (*a*) plot the points, (*b*) find the equation of the least-squares line, (*c*) graph the line on the graph containing the plotted points, (*d*) calculate the errors of prediction, (*e*) predict the value of y for $x = 16.5$, (*f*) calculate the standard error of estimate, (*g*) draw two lines parallel to the least-squares line to form a band within which 68 percent of the points might be expected to lie under a normal distribution assumption and comment, (*h*) test the hypothesis that $\beta = .5$, (*i*) find 90 percent confidence limits for β.

x	5	6	7	8	9	10	11	12	13	14	15	16	17
y	7.6	9.5	9.3	10.3	11.1	12.1	13.3	12.7	13.0	13.8	14.6	14.6	14.7

42 The data of Table XIII in the appendix may be used to construct various regression problems. For example, what is the regression of the variable x_1 on the variable x_2, and how reliable is it?

Chapter 11 Multiple Regression

1 MULTIPLE LINEAR REGRESSION

Most practical problems of prediction involve more than one related variable by means of which to make the prediction. For example, if a college wishes to develop a regression formula for admissions, based on predicting academic success of an entering freshman, it would certainly wish to incorporate in the regression function such variables as the high school grade-point average, the score on an aptitude test, and the rating made on an interview. Or, in forecasting the demand for electricity ten years from now in a large city, the city planners would need to consider a large number of pertinent variables, including such factors as industrial growth, the dynamics of future housing demands, the movement of industry to suburban areas, etc.

Methods for dealing with problems of predicting one variable by means of several other variables, rather than by means of just one other variable, are similar to those for one variable. For example, if one were to predict the variable y by means of a linear function of the two variables x_1 and x_2, the problem would become one of finding the best-fitting plane, in the sense of least squares, to a scatter diagram of points in three dimensions. The geometry of such a problem is illustrated in Fig. 1.

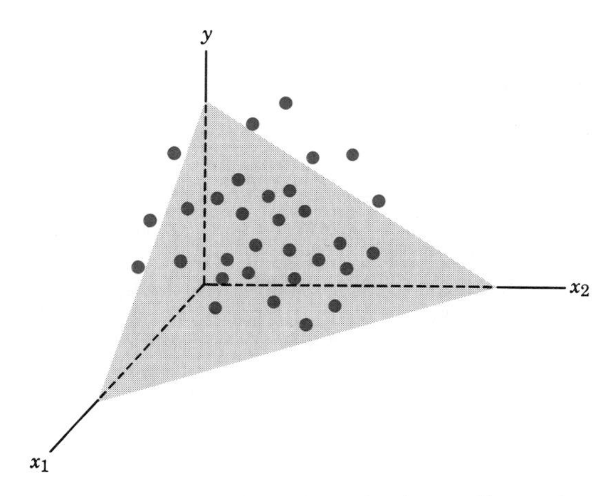

Figure 1 Regression plane in three dimensions.

Since the equation of any plane in three-dimensional space determined by the variables y, x_1, and x_2 can be written in the form

$$y = a_0 + a_1x_1 + a_2x_2 \tag{1}$$

the problem is one of estimating the three parameters a_0, a_1, and a_2 by the method of least squares. The errors of prediction here are the perpendicular distances of the points from the plane. Minimizing the sum of the squares of those errors is accomplished by the same mathematical techniques as for simple linear regression. If there are n points available, it turns out that the least-squares values of a_0, a_1, and a_2 are obtained by solving the following set of three linear equations, where it is assumed that the sums are taken over all n values of the variables.

$$\begin{aligned} a_0n + a_1\Sigma x_1 + a_2\Sigma x_2 &= \Sigma y \\ a_0\Sigma x_1 + a_1\Sigma x_1^2 + a_2\Sigma x_1x_2 &= \Sigma x_1y \\ a_0\Sigma x_2 + a_1\Sigma x_1x_2 + a_2\Sigma x_2^2 &= \Sigma x_2y \end{aligned} \tag{2}$$

This result generalizes for additional variables. Thus, if one has four independent variables, x_1, x_2, x_3, and x_4, by means of which to predict y, there are five equations in the five unknowns a_0, a_1, a_2, a_3, and a_4 to solve.

One of the attractive features of the least-squares approach in regression problems is the simple manner of writing down the equations that need to be solved to obtain estimates of the regression equation parameters. For example, equations (2) can be written down by carrying out the following operations. First sum both sides of equation (1) to obtain the first least-squares equation. Next, multiply each term of equation (1) by x_1 and sum both sides to obtain the second least-squares equation. Finally, multiply each term of equation (1) by x_2 and sum both sides to obtain the third equation.

If there are five independent variables instead of two in the regression equation, one continues this procedure by multiplying by x_3 and summing, then by x_4 and summing, and finally by x_5 and summing. There are then six least-squares equations to be solved to obtain estimates of the six unknown regression parameters.

With the ready availability of high-speed computing facilities these days, it is possible to incorporate a large number of variables in a regression function and thereby increase one's ability to predict the value of one variable as a function of various other related variables. It is not uncommon now to use twenty or more variables, when previously the computations involved in such an undertaking were prohibitively heavy. One of the striking advantages of a regression function that incorporates most of the important variables that are related to a given variable y is that it enables one to study how one of those variables affects y when all the other variables are held fixed. Thus, it permits an intimate analysis of the various relationships, which is not possible when all variables are changing simultaneously in their natural setting. For example, suppose the variable x_2 in (1) is assigned a fixed value. Then this regression function states that on the average, the value of y will increase by the amount a_1 if x_1 is allowed to increase by one unit. This is only an average relationship and is based on assumptions similar to those made in simple linear regression.

Example. As an illustration of how such interpretations are made, consider the regression problem associated with the data of Table 1. These data represent the scores made by 20 employees of a company on two psychological tests and their rating scores that management had previously decided upon. The regression problem is to determine whether an employee's success, as measured by his rating score, can be predicted on the basis of his scores on the two psychological tests.

Table 1

Test scores		Rating score	Test scores		Rating score
x_1	x_2	y	x_1	x_2	y
77	76	84	35	51	59
83	71	85	35	10	48
64	96	95	54	74	73
30	2	55	43	10	64
59	96	95	33	52	66
53	55	73	32	27	53
77	73	89	54	16	57
37	16	65	28	21	59
15	15	47	54	19	69
28	51	71	45	6	57

The least-squares equations are

$$\begin{aligned} 20a_0 + 936a_1 + 837a_2 &= 1{,}364 \\ 936a_0 + 50{,}380a_1 + 46{,}283a_2 &= 68{,}012 \\ 837a_0 + 46{,}283a_1 + 53{,}493a_2 &= 64{,}845 \end{aligned}$$

The solution of these equations is given by $a_0 = 41.2$, $a_1 = .308$, and $a_2 = .302$; therefore, the equation of the linear regression function is

$$y = 41.2 + .308x_1 + .302x_2 \tag{3}$$

Since a regression function is assumed to represent only the mean value of y corresponding to any chosen values of the independent variables, the interpretation of the regression coefficients must be in terms of mean values. Thus, if one considers only those employees who made, say, a score of 70 on the first test, then equation (3) states that on the average, management's rating of an employee will increase by the amount .302 for each unit increase on his second test score. For example, if one employee scores 10 points more than another employee on the second test and they have the same first test score, one would expect the first employee to score about 3 points higher than the second employee on management's rating. This interpretation does not depend upon what particular value is assigned to x_1. A similar interpretation can be made concerning the coefficient of x_1 if the value of x_2 is held fixed.

Suppose that the value of x_2 had not been introduced in this problem and the simple linear regression of y on x_1 had been computed. Would there have been any noticeable change in the coefficient of x_1, and would there have been any appreciable loss in the accuracy of prediction? To answer these questions, the equation of the simple linear regression of y on x_1 was calculated and found to be

$$y = 38.47 + .635x_1 \tag{4}$$

The constant term a_0 has changed very little; however, the coefficient of x_1 has about doubled in value. This should not be too surprising because if x_1 and x_2 had the same value, the contribution of the sum of the x_1 and x_2 terms in (3) would then about equal the contribution of x_1 alone in (4). This, however, raises the question as to whether x_1 alone might not be just as good as x_1 and x_2 combined in predicting y. To settle that question, it suffices to calculate the standard error of estimate for each of the regression functions (3) and (4). Calculations yield the values 5.6 and 9.2. It is clear from these values that using x_1 alone is not nearly as good as using both x_1 and x_2. The errors of prediction will be considerably smaller when using x_1 and x_2 than when using just x_1.

2 COMPUTER TECHNIQUES

With high-speed computing facilities now readily available to investigators, it is possible for them to incorporate a large number of variables in a regression problem. However, the advantage of a regression function that contains a large number of variables over one that contains a small number of the really meaningful variables is often slight. Furthermore, the difficulty and the additional costs involved in applying the larger function may give the advantage to the smaller function. The problem, therefore, arises as to how one should choose a few important variables out of a large set of variables for the purpose of constructing a linear regression function for predicting one of those variables, y, in terms of the selected variables.

One of the methods for doing this is called *stepwise regression* and proceeds as follows. First, the linear regression equation of y on each of the independent variables is obtained and the standard error of estimate calculated for each of those regressions. The variable yielding the smallest standard error of estimate is the first variable to be selected. Next, the multiple regression equation of y based on two independent variables, one of which must be the variable just selected, is obtained for each of the remaining independent variables. As before, the standard error of estimate is calculated for each regression function and the variable that produces the smallest such value is selected as the second variable to be retained. This procedure is continued, each time with a regression function containing all the previously selected variables and one of the remaining variables, until the desired number of variables has been obtained, or until the standard error of estimate has not decreased in value appreciably. No improvement is likely to result by incorporating additional variables in the regression function if there is no appreciable decrease in the value of the standard error of estimate at any stage.

A regression function that contains a large number of independent variables must be treated with much delicacy because there may be some fairly strong correlations among subsets of those variables, in which case the regression equation may be unreliable. The stepwise regression technique is not likely to select independent variables that are highly correlated in the early stages but will select the best of even highly correlated variables if continued too far. Many of the variables that are the natural ones to select in business regression problems are likely to be correlated; therefore, investigators should not let their enthusiasm for high-speed computers and lots of variables carry them away from reality and into the trap of unjustified interpretations.

3 NONLINEAR REGRESSION

The discussion in Chapter 9 pointed out the possibility of changing many nonlinear relationships between two variables x and y into linear ones by choosing the proper functions of x and y. Although this is always possible theoretically, the proper functions may be exceedingly complicated or one may not be able to determine what the proper functions are. It is then often more satisfactory to work with nonlinear relationships between x and y. The simplest curve to use as a regression model if a straight line will not suffice is a parabola. Its equation can be written in the form

$$y = a_0 + a_1x + a_2x^2 \tag{5}$$

The problem of fitting a curve of this type to a set of points in the x,y-plane by least squares is similar to that of fitting a multiple regression equation to a set of points in three dimensions. As a matter of fact, the results that were obtained in section 1 can be applied directly to this problem. It is necessary merely to choose $x_1 = x$ and $x_2 = x^2$ in equation (1) to obtain equation (5). As a result the least-squares equations for (5) are obtained by making these same substitutions in the least-squares equations given in (2). Thus, the equations whose solutions give the desired estimates for a_0, a_1, and a_2 in fitting the parabola (5) to a set of points are the following:

$$\begin{aligned} a_0n + a_1\Sigma x + a_2\Sigma x^2 &= \Sigma y \\ a_0\Sigma x + a_1\Sigma x^2 + a_2\Sigma x^3 &= \Sigma xy \\ a_0\Sigma x^2 + a_1\Sigma x^3 + a_2\Sigma x^4 &= \Sigma x^2y \end{aligned} \tag{6}$$

For the purpose of observing to what extent a parabola fits a set of points better than a straight line, it suffices to calculate the errors of prediction for both fitted curves and compare them. The standard error of estimate may also be calculated for both curves and compared; however, the standard error of estimate for a parabola uses the divisor $n - 3$ in place of $n - 2$ in formula (8), Chapter 10. If there is very little difference between the two calculated standard errors of estimate, not much is gained by using a parabola rather than a straight line for the regression curve.

These methods can be extended to polynomial curves of higher degree. All that one needs to do is to replace x_3 by x^3, x_4 by x^4, etc., in the more general forms of (1) and (2) to obtain the desired degree polynomial regression curve equation and its corresponding least-squares equations for estimating the coefficients.

Example. As an illustration of the procedure for fitting a parabola to a set of points, consider the data of Table 2, which gives the record over a period of 18 weeks of the number of units, x, produced in a factory and the estimated cost in dollars per unit, y, at that production level.

Table 2

x	y	x	y
242	107	200	120
255	108	210	122
261	102	237	114
268	103	284	103
275	105	293	102
282	100	298	100
222	113	302	104
214	118	306	103
230	112	313	105

The scatter diagram for these data is shown in Fig. 2. It seems clear that a straight line will produce a poor fit but that a parabola might do quite well. A curvilinear type of relation might be expected here because a factory that is designed to produce most efficiently at a given rate of output will be less efficient if it produces at a lower rate than the ideal, but because of overtime costs, the same is true if it attempts to produce at a higher rate than its designed capacity.

The calculations become simpler if, say, 200 is subtracted from all the x-values and 100 is subtracted from all the y-values. Let $X = x - 200$ and $Y = y - 100$, and let the equation of the parabola be written

$$Y = a_0 + a_1X + a_2X^2$$

In terms of the original data variables this is equivalent to

$$y - 100 = a_0 + a_1(x - 200) + a_2(x - 200)^2 \tag{7}$$

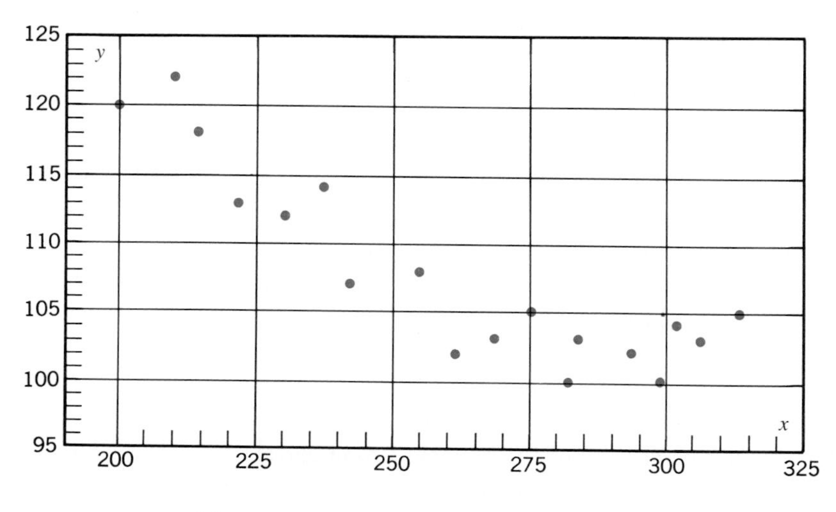

Figure 2 A scatter diagram for Table 2.

The least-squares parabola for the original data can be obtained from this equation by multiplying out the parentheses and collecting terms.

The least-squares equations (6) that resulted from the modified data are

$$18a_0 + 1{,}092a_1 + 88{,}250a_2 = 141$$

$$1{,}092a_0 + 88{,}250a_1 + 7{,}880{,}538a_2 = 4{,}800$$

$$88{,}250a_0 + 7{,}880{,}538a_1 + 741{,}676{,}598a_2 = 305{,}608$$

These are equivalent to

$$a_0 + 60.667a_1 + 4{,}902.778a_2 = 7.833$$

$$a_0 + 80.815a_1 + 7{,}216.610a_2 = 4.396$$

$$a_0 + 89.298a_1 + 8{,}404.267a_2 = 3.463$$

Subtracting the first equation from the remaining equations gives

$$20.148a_1 + 2{,}313{,}832a_2 = -3.437$$

$$28.631a_1 + 3{,}501.489a_2 = -4.370$$

or, equivalently,

$$a_1 + 114.84a_2 = -.1701$$

$$a_1 + 122.30a_2 = -.1526$$

Solving this pair of equations and using the answers to obtain a_0 from an earlier equation gives

$$a_2 = .002346, \qquad a_1 = -.4395, \qquad a_0 = 22.976$$

The result of substituting these values in (7) and collecting terms is

$$y = 304.9 - 1.378x + .00235x^2$$

The graph of this equation fitted to the scatter diagram of Fig. 2 is shown in Fig. 3. The fit appears to be very good.

For a problem in which it is not so obvious that a parabola gives a better fit than a straight line, one can calculate the standard error of estimate for both regressions and observe whether there is any appreciable gain in using a parabola rather than a straight line. In doing so, formula (9) of the next section should be employed.

As indicated in the preceding section, one must proceed with caution when interpreting a linear multiple regression function that is based on several independent variables because of the possible strong correlation effects. There is the additional difficulty in such problems that the linearity assumption may not be satisfied. If there is good evidence, theoretical or empirical, for believing that the

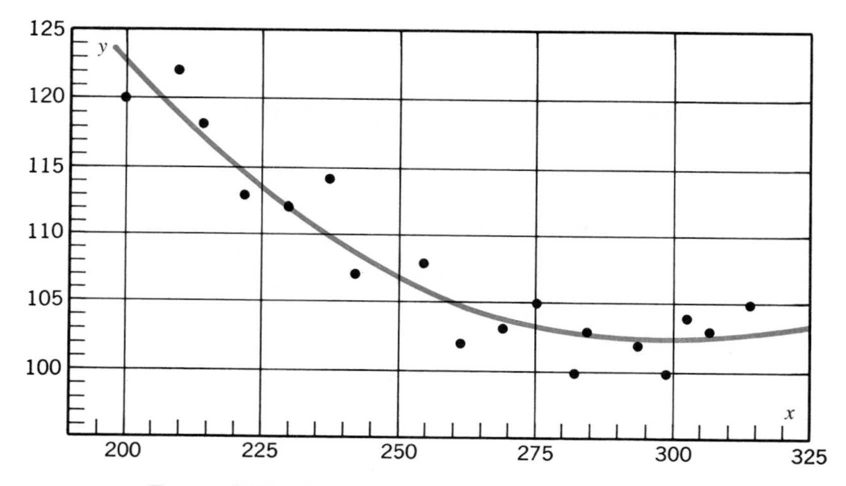

Figure 3 The least squares parabola for Table 2.

means of the y's do not lie on a plane, one can construct a more general regression function by introducing second-degree terms among the independent variables. For example, the linear regression function (1) might be replaced by the nonlinear regression function

$$y = a_0 + a_1x_1 + a_2x_2 + a_3x_1^2 + a_4x_1x_2 + a_5x_2^2 \tag{8}$$

The technique for finding the least-squares equations is the same as before. It is merely necessary to treat the variables x_1^2, x_1x_2, and x_2^2 as three new variables, say, x_3, x_4, and x_5, and write down the corresponding six least-squares equations. Since including second-degree terms rapidly increases the number of terms in a regression function, this technique is seldom used unless the number of independent variables is very small.

4 RELIABILITY

Formulas for determining the reliability of a multiple, or a nonlinear, regression function are similar to those for simple linear regression. The standard error of estimate is modified slightly to compensate for the number of coefficients that need to be estimated. It is defined as follows:

$$s_e = \sqrt{\frac{\sum_{i=1}^{n} (y_i - \hat{y}_i)^2}{n - k}} \tag{9}$$

where k is the number of coefficients, including the constant term, in the regression formula. For example, k is 3 for formulas (1) and (5), and it is 6 for formula (8). Under the same kind of assumptions as for simple linear regression, the standard error of estimate may be used to indicate what size errors can be expected when predicting a value of y by means of the regression function.

The formulas for determining the reliability of the regression coefficients as estimates of corresponding theoretical coefficients are considerably more complicated than those for simple linear regression. Some of the computing programs for multiple regression print out information that can be used to solve those reliability problems. In particular, they enable one to test if any specified coefficient has a zero value, in which case the corresponding variable is of little use in the regression formula. Problems of this type are not studied here.

In summary, the techniques employed in simple linear regression can also be used in multiple and nonlinear regression but with more trepidation because the assumptions needed to justify them are less likely to be satisfied in actual problems.

5 REVIEW ILLUSTRATIONS

Example 1. Given the following data, (a) find the equation of its least-squares plane, (b) calculate the errors of prediction, (c) calculate the standard error of estimate, (d) omit the variable x_2 and then find the equation of the least-squares line, (e) calculate the standard error of estimate for the line obtained in (d) and compare with the value obtained in (c). Was much gained by introducing the variable x_2 in addition to x_1?

x_1	1	7	6	2	5	2	8	3	7	6
x_2	4	6	8	0	1	5	8	8	3	0
y	11	21	23	7	13	18	30	18	21	20

(a) Calculations give $\Sigma x_1 = 47$, $\Sigma x_2 = 43$, $\Sigma y = 182$, $\Sigma x_1^2 = 277$, $\Sigma x_2^2 = 279$, $\Sigma x_1 x_2 = 218$, $\Sigma x_1 y = 972$, $\Sigma x_2 y = 904$; hence the following equations need to be solved:

$$a_0 10 + a_1 47 + a_2 43 = 182$$

$$a_0 47 + a_2 277 + a_2 218 = 972$$

$$a_0 43 + a_1 218 + a_2 279 = 904$$

The solution of these equations is $a_0 = 5.48$, $a_1 = 1.81$, $a_2 = .98$; hence the equation of the regression plane is

$$\hat{y} = 5.48 + 1.81x_1 + .98x_2$$

(*b*)

y_i	11	21	23	7	13	18	30	18	21	20
$\hat{y}_i$	11.2	24.0	24.2	9.1	15.5	14.0	27.8	18.8	21.1	16.3
e_i	−.2	−3.0	−1.2	−2.1	−2.5	4.0	2.2	−.8	−.1	3.7

(*c*) $\Sigma e_1^2 = 56.32$; hence $s_e = 2.84$.
(*d*) From (*a*), $\bar{y} = 18.2$, $b = 2.08$; hence $\hat{y} = 8.42 + 2.08x$.

(*e*)

y_i	11	21	23	7	13	18	30	18	21	20
$\hat{y}_i$	10.5	23.0	20.9	12.6	18.8	12.6	25.1	14.7	23.0	20.9
e_i	.5	−2.0	2.1	−5.6	−5.8	5.4	4.9	3.3	−2.0	−.9

Here $\Sigma e^2 = 142.5$, $s_e = \sqrt{142.5/8} = 4.22$. Hence it follows from (*c*) where $s_e = 2.84$ that x_2 was very useful because it reduced the value of s_e by about $\frac{1}{3}$.

Example 2. Given the following data, (*a*) plot the points, (*b*) find the equation of the least-squares line, (*c*) graph the line on the graph containing the plotted points, (*d*) calculate the errors of prediction, (*e*) predict the value of y for $x = 16.5$, (*f*) calculate the standard error of estimate, (*g*) fit a parabola to the points, (*h*) calculate the standard error of estimate for the fitted parabola and compare with the result in (*f*).

x	5	6	7	8	9	10	11	12	13	14	15	16	17
y	7.6	9.5	9.3	10.3	11.1	12.1	13.3	12.7	13.0	13.8	14.6	14.6	14.7

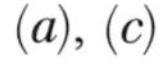

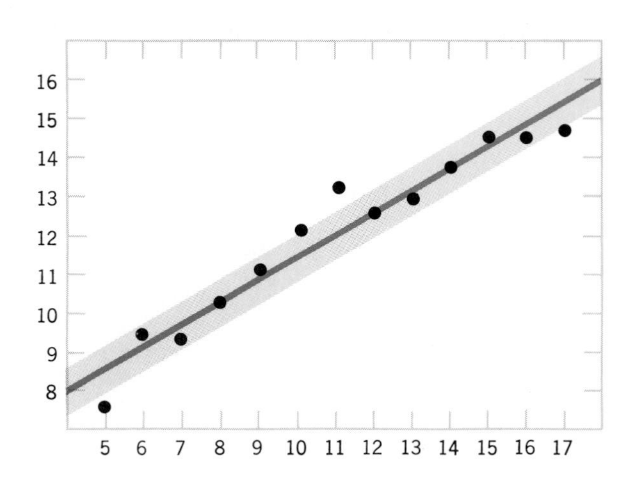

(*b*)

x_i	y_i	x_i^2	x_iy_i
5	7.6	25	38.0
6	9.5	36	57.0
7	9.3	49	65.1
8	10.3	64	82.4
9	11.1	81	99.9
10	12.1	100	121.0
11	13.3	121	146.3
12	12.7	144	152.4
13	13.0	169	169.0
14	13.8	196	193.2
15	14.6	225	219.0
16	14.6	256	233.6
17	14.7	289	249.9
143	156.6	1755	1826.8

$$\bar{x} = \frac{143}{13} = 11$$

$$\bar{y} = \frac{156.6}{13} = 12.05$$

$$b = \frac{1826.8 - 143(12.05)}{1755 - 143(11)} = .570$$

$$\hat{y} = 5.78 + .570x$$

(*d*)

y_i	7.6	9.5	9.3	10.3	11.1	12.1	13.3	12.7	13.0	13.8	14.6	14.6	14.7
$\hat{y}_i$	8.6	9.2	9.8	10.3	10.9	11.5	12.0	12.6	13.2	13.8	14.3	14.9	15.5
e_i	−1.0	.3	−.5	.0	.2	.6	1.3	.1	−.2	.0	.3	−.3	−.8

Each value of $\hat{y}_i$ was obtained from the preceding one by adding .57 to it. The first value was obtained by substituting $x = 5$ in the equation.

(*e*) $\hat{y} = 15.2$ for $x = 16.5$.

(*f*) From part (*d*), $\Sigma e^2 = 4.30$; hence $s_e = \sqrt{4.30/11} = .63$.

(*g*)

x	y	x^3	x^4	x^2y
5	7.6	125	625	190.0
6	9.5	216	1,296	342.0
7	9.3	343	2,401	455.7
8	10.3	512	4,096	659.2
9	11.1	729	6,561	899.1
10	12.1	1,000	10,000	1,210.0
11	13.3	1,331	14,641	1,609.3
12	12.7	1,728	20,736	1,828.8
13	13.0	2,197	28,561	2,197.0
14	13.8	2,744	38,416	2,704.8
15	14.6	3,375	50,625	3,285.0
16	14.6	4,096	65,536	3,737.6
17	14.7	4,913	83,521	4,248.3
		23,309	327,015	23,366.8

$$a_0 13 + a_1 143 + a_2 1755 = 156.6$$
$$a_0 143 + a_1 1755 + a_2 23{,}309 = 1826.8$$
$$a_0 1755 + a_1 23{,}309 + a_2 327{,}015 = 23{,}366.8$$

or

$$a_0 + a_1 11 + a_2 135 = 12.046$$
$$a_0 + a_1 12.273 + a_2 163 = 12.775$$
$$a_0 + a_1 13.281 + a_2 186.33 = 13.314$$

or

$$a_1 1.273 + a_2 28 = .729$$
$$a_1 2.281 + a_2 51.33 = 1.268$$

or

$$a_1 + a_2 21.995 = .573$$
$$a_1 + a_2 22.505 = .556$$

or

$$a_2 = -.033, \qquad a_1 = 1.299, \qquad a_0 = 2.212$$

Hence

$$y = 2.21 + 1.30x - .033x^2$$

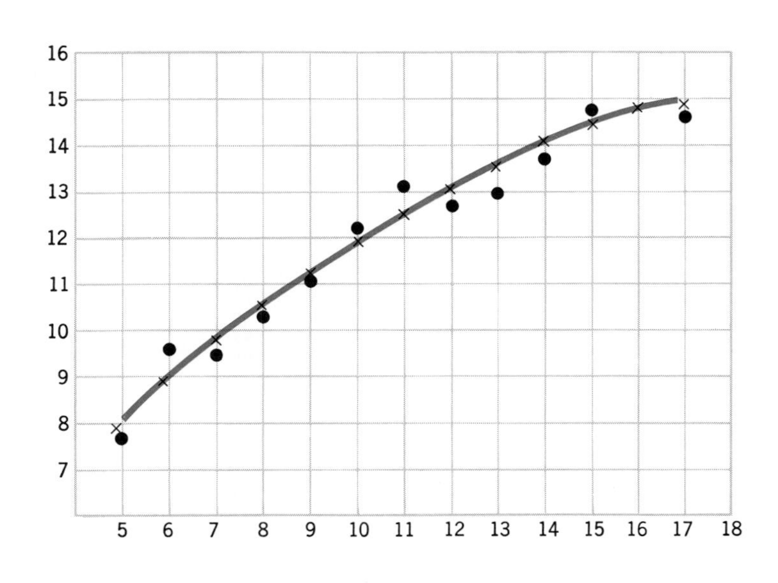

(*h*)

y_i	7.6	9.5	9.3	10.3	11.1	12.1	13.3	12.7	13.0	13.8	14.6	14.6	14.7
$\hat{y}_i$	7.9	8.8	9.7	10.5	11.2	11.9	12.5	13.1	13.5	13.9	14.3	14.6	14.8
e_i	−.3	.7	−.4	−.2	−.1	.2	.8	−.4	−.5	−.1	.3	.0	−.1

Here $s_e = \sqrt{\Sigma e^2/10} = .45$. Since $s_e = .63$ for the straight line, the parabola gives a better fit. The standard error has been decreased about $\frac{1}{3}$.

EXERCISES

SECTION 1

1 A study was made several years ago of the relative importance of the spring rainfall and the spring temperature on the yield of hay. The result of that study produced the following regression equation:

$$y = 9.4 + 3.3x_1 + .004x_2$$

Here y denotes the amount of hay in units of 100 pounds per acre, x_1 is the spring rainfall in inches, and x_2 is the accumulated temperature above 42°F in the spring. The data of this study also produced the values $\bar{y} = 28$, $\bar{x}_1 = 4.9$, and $\bar{x}_2 = 594$.

(*a*) If the spring temperature does not change from its mean value, how much increase in hay yield would you expect if the rainfall increases one inch?

(*b*) How many units increase in the spring temperature are required on the average to increase the hay yield by 100 pounds, if the spring rainfall does not change from its mean value?

(*c*) Comment on the relative importance of spring rainfall and temperature on the yield of hay. Are those two variables likely to be correlated?

2 In the table below, data are given on (1) per capita consumption of food (2) per capita disposable income, and (3) price of food (deflated) for the United States

Year	Consumption of food	Disposable income	Price of food	Year	Consumption of food	Disposable income	Price of food
1922	89.0	61.0	83.0	1937	90.4	72.5	84.9
1923	90.9	68.3	84.2	1938	90.6	67.8	80.3
1924	91.5	67.4	83.2	1939	93.8	73.2	79.3
1925	90.9	68.5	87.7	1940	95.5	77.6	79.8
1926	92.1	69.6	89.9	1941	97.5	89.5	83.0
1927	90.9	70.2	88.3				
1928	90.9	71.9	88.4	1948	99.1	100.6	101.3
1929	91.1	75.2	89.5	1949	98.8	100.1	98.2
1930	90.7	68.3	87.4	1950	99.9	106.8	98.4
1931	90.0	64.0	79.1	1951	98.1	106.6	101.4
1932	87.8	53.9	73.3	1952	100.4	107.6	101.0
1933	88.0	53.2	75.2	1953	101.5	110.8	98.6
1934	89.1	58.0	81.1	1954	101.4	110.3	98.1
1935	87.3	63.2	84.7	1955	102.8	115.5	96.9
1936	90.5	70.5	84.5	1956	104.0	118.4	95.9

during the periods 1922–1941 and 1948–1956, where 1947–1949 = 100 and the BLS *Consumers' Price Index* was used for deflating food prices.

The regression equation for food consumption (y) as a function of disposable income (x_1) and the price of food (x_2) is

$$y = 82.94 + .297x_1 - .147x_2$$

(*a*) What percentage increase in food consumption could be expected for an individual whose disposable income increased 20 percent from its earlier value of 70, assuming that the price of food did not change from a value of 80?

(*b*) What percentage decrease in food consumption could be expected for an individual if the price of food increased 20 percent from its earlier value of 80, assuming that his disposable income did not change from a value of 70?

(*c*) What percentage change in food consumption could be expected for an individual if the price of food changed from 80 to 90 and his disposable income changed from 70 to 80?

3 Given the following data, calculate the regression equation.

x_1	2	1	5	8	7	2	1	3	0	9	4	6
x_2	6	5	5	7	3	1	8	2	6	7	7	9
y	13	9	15	16	21	9	15	11	12	30	19	22

4 Below are data on farm prices for turkeys, per capita supplies of turkeys, and per capita supplies of chickens, for the January–August (or "nonholiday") season for the years 1955 to 1965. These data were obtained from a 1966 article by Bluestone and Rojko in *Agricultural Economics Research.*

Year	P (cents/lb)	T (lb)	C (lb)	Year	P (cents/lb)	T (lb)	C (lb)
1955	31.4	1.9	14.1	1960	24.0	2.8	19.2
1956	31.1	2.0	16.0	1961	20.1	3.7	21.2
1957	24.3	2.8	17.0	1962	19.2	3.5	20.7
1958	24.7	2.6	19.0	1963	20.3	3.2	21.3
1959	22.1	2.8	20.4	1964	19.5	3.4	21.8
				1965	18.7	3.4	22.8

Notes: P = weighted farm turkey price per pound in January–August deflated by Consumer Price Index

T = per capita supplies of turkeys available for domestic consumption in January–August

C = per capita supplies of chickens available for domestic consumption in January–August

(*a*) Using the regression function $P = a_0 + a_1T + a_2C$, calculate the regression coefficients for the "demand function" for turkeys during the nonholiday season.

(*b*) If the supplies of chicken are increased by one pound per capita and turkey supplies are held fixed, estimate the amount and direction of change in turkey price that would be expected from the model.

SECTION 3

5 The following data are from a U.S. Department of Agriculture study of the relationship between the annual income of U.S. families and their weekly expenditure for food.

(a) Plot the data with x as income and y as food expenditure.

(b) Sketch a smooth freehand curve that appears to fit the points, ignoring the last entry and assuming the first entry is at $x = 700$.

(c) If you were to fit the least-squares curve to the points in (b), what type of curve would you use?

Annual income	Food expenditure	Annual income	Food expenditure
Under 1,000	11.7	5,000–5,999	33.0
1,000–1,999	16.6	6,000–7,999	36.1
2,000–2,999	22.6	8,000–9,999	39.2
3,000–3,999	27.0	10,000 and over	52.4
4,000–4,999	30.3		

6 Fit a parabola to the points of problem 5, ignoring the last entry in the table of values. Graph it on the point graph in (a) of problem 5.

SECTION 4

7 The regression equation for problem 4 is $P = 50.20 - 4.24T - .752C$. The regression equation of P on T will be found to be $P = 44.54 - 7.31T$.

(a) Calculate the errors of prediction for both regression functions.

(b) Use the results in (a) to calculate the two standard errors of estimate.

(c) Comment about the value of the variable C for predicting P when T is available.

8 For the data of problem 3, (a) calculate the errors of prediction; (b) calculate the standard error of estimate; (c) omit the variable x_2 and use the regression of y on x_1 to calculate the errors of prediction; (d) calculate the standard error of estimate based on (c); (e) compare the answers in (b) and (d) and comment about how valuable x_2 is for prediction. The equations of the regression functions for these data are:

$$y = 5.18 + 1.501x_1 + .877x_2$$

and

$$y = 9.43 + 1.643x_1$$

SECTION 5

9 Forty students were selected at random from a large group of students and their grade-point averages, y, intelligence test scores, x_1, and reading rates,

x_2, recorded. The equation of the least-squares plane based on those data was found to be $y = .016x_1 - .009x_2 - 1.20$. Grade-point averages ranged from 0 to 4, intelligence test scores ranged from 110 to 295, and reading rate scores ranged from 10 to 45.

(*a*) If two students have the same reading rate score but one student scores 20 points more than the other on his intelligence test, how much would his grade-point average be expected to exceed that of the other student?

(*b*) What does the negative coefficient of x_2 imply concerning the use of reading rate scores for predicting grade-point averages?

(*c*) If the regression equation of y on x_1 alone were calculated and turned out to be $y = .015x_1 - 1.50$, what conclusions would you draw concerning the usefulness of x_1 and x_2 for predicting y? If you were given the data on which these equations are based, how would you use them to determine whether your conclusions were justified?

10 The following table gives the world track records, as of 1973 or 1974, for distances from 100 yards to 10 miles. It also gives the speed in yards per second that the runner averaged in breaking the record. Because of the large fraction of the racing time consumed in getting up speed from a dead start in the 100-yard dash, the data for 100 yards do not represent the true speed in running that distance; therefore they will not be used in studying the relation between distance and speed.

Distance	Time	Speed
100 yards	9.0	11.1
220 yards	19.5	11.3
440 yards	44.5	9.9
880 yards	1:44.6	8.4
1 mile	3:51.1	7.6
2 miles	8:13.8	7.1
3 miles	12:47.6	6.9
6 miles	26:47.0	6.6
10 miles	46:37.8	6.3

It is helpful in studying the relation between distance and speed to calculate the logarithm of distance and plot speed against the logarithm of distance. Calculations based on the preceding table of values yield the following relationship.

Speed	11.3	9.9	8.4	7.6	7.1	6.9	6.6	6.3
Log of distance	2.34	2.64	2.94	3.25	3.55	3.72	4.02	4.25

A graph of this relationship suggests that a parabola will not fit the points well but that a third-degree polynomial might do so. A least-squares fitting was carried out and produced the following equation.

$$y = 49.9921 - 29.6962x + 6.9433x^2 - .5583x^3$$

Here x = log of distance and y = speed.

(*a*) On graph paper plot the points that represent the relationship.

(*b*) Assuming that this relationship also holds for the marathon race (26.2 miles), estimate the speed for the marathon race on the basis of the preceding equation.

(*c*) Using your answer in (*b*), estimate the world record for the marathon (time = distance/speed). The actual record is 2 hours, 12 minutes, 11.2 seconds.

(*d*) Explain why it would be unrealistic to expect that a fitted parabola would be satisfactory for solving (*b*) and (*c*).

11 The data of Table XIII in the appendix may be used to construct various multiple regression problems. For example, what is the equation of the linear regression function of variable x_3 on the two variables x_1 and x_2?

Chapter 12 Chi-Square

1 INTRODUCTION

In the preceding chapters, problems of estimation and hypothesis testing were solved by means of the binomial, normal, or Student t distribution. All such problems were concerned with population means or proportions. However, problems related to the normal distribution parameter σ had to be postponed because they could not be solved satisfactorily by means of any of the available distributions. One of the purposes of this chapter is to complete the theory of estimation and hypothesis testing for a normal distribution by presenting methods for treating problems concerning σ.

Problems of the counting type that were solved in the preceding chapters were the kind in which an outcome can be classified as a success or a failure. Such problems gave rise to the binomial distribution and the hypergeometric distribution. There are many counting problems, however, in which an experimental outcome requires more than two categories of classification. Such problems also had to be postponed because the binomial distribution is not capable of solving them. Another purpose of this chapter is to complete that theory by presenting methods for solving multiple classification problems.

A remarkable feature of these two seemingly unrelated classes of problems is that their solutions depend upon the same probability distribution. This distribution, which is called the chi-square distribution, was devised by a famous English statistician Karl Pearson in a publication of 1899 in which he used the Greek letter chi for an index of variation. Since the index is commonly used as a square, it is generally known as chi-square, symbolized by χ^2.

Not only is the chi-square distribution capable of solving the two classes of problems discussed in the preceding paragraphs, but it can also be used to solve certain classes of correlation problems for counting variables. A method for treating some of those problems will also be presented in this chapter. The chi-square distribution is obviously a very versatile distribution if it can handle such diverse problems.

2 CHI-SQUARE

The problem of testing the equality of two proportions, which was solved in Chapter 8, is a special case of the more general problems of this type. The method for solving these more general problems will be considered first and will lead to the introduction of the chi-square distribution. Such problems can be described in the following manner.

There is a finite number, denoted by k, of possible outcomes of an experiment. These possible outcomes are represented by k boxes or *cells*. The experiment is performed n times, and the results are expressed by recording the observed frequencies of outcomes in the corresponding cells. The problem then is to determine whether the frequencies are compatible with those expected from some postulated theory.

Example. As an illustration, if the experiment consists of rolling a single die, there will be 6 cells. The results of performing such an experiment 60 times are recorded in the row of cells labeled o_i in Table 1. If the die is "honest," each face will have the probability $\frac{1}{6}$ of appearing in a single roll. Each face would therefore be expected to appear 10 times, on the average, in an experiment of this kind. These mean frequencies are usually called expected frequencies. They are shown in the row of cells labeled e_i. The problem then is to decide whether the observed frequencies in Table 1 are compatible with the expected frequencies listed there. In terms of the notation of Chapter 8, this is a problem of testing the hypothesis

$$H_0: p_1 = p_2 = \cdots = p_6 = \tfrac{1}{6}$$

Table 1

	1	2	3	4	5	6
o_i	15	7	4	11	6	17
e_i	10	10	10	10	10	10

The general method for testing compatibility is based on a measure of the extent to which the observed and expected frequencies agree. This measure, called *chi-square,* is defined by the formula

$$\chi^2 = \sum_{i=1}^{k} \frac{(o_i - e_i)^2}{e_i} \tag{1}$$

Here o_i and e_i denote the observed and expected frequencies, respectively, for the ith cell, and k denotes the number of cells.

For Table 1 the value of χ^2 is given by

$$\chi^2 = \frac{(15-10)^2}{10} + \frac{(7-10)^2}{10} + \frac{(4-10)^2}{10} + \frac{(11-10)^2}{10} + \frac{(6-10)^2}{10} + \frac{(17-10)^2}{10}$$

$$= 13.6$$

Now it is clear from inspecting formula (1) that the value of χ^2 will be 0 if there is perfect agreement with expectation, whereas its value will be large if the differences from expectation are large. Thus increasingly large values of χ^2 may be thought of as corresponding to increasingly poor experimental agreement. If an honest die were available and if the experiment of rolling the die 60 times were repeated a large number of times and each time the value of χ^2 were computed, a set of χ^2's would be obtained that could be classified into a relative frequency table and histogram. The histogram would tell approximately in what percentage of such experiments various ranges of values of χ^2 could be expected to be obtained. Then one would be able to judge whether the value of $\chi^2 = 13.6$ was unusually large, as compared to the run of χ^2's that are obtained in experiments with an honest die. If the percentage of experiments for which $\chi^2 > 13.6$ was very small, say less than 5 percent, one would judge that the observed frequencies were not compatible with the frequencies expected for an honest die; hence one would conclude that the die is not honest.

3 THE CHI-SQUARE DISTRIBUTION

Just as in the case of other sampling distributions, it is possible to use mathematical methods to arrive at the desired theoretical frequency distribution. Since there is only a limited number of possible values for the cell frequencies in Table 1, there is only a limited number of values of χ^2 possible. Thus the theoretical distribution of χ^2 must be a discrete distribution. Since a discrete distribution with many possible values requires the application of lengthy computations, practical considerations demand a simple continuous approximation to the discrete χ^2 distribution, very much like the normal approximation to the binomial distribution. Such a continuous distribution is available in what is known as the *chi-square distribution.* It is unfortunate that the continuous distribution approximating the discrete chi-square distribution should also be called the chi-square distribution; however, there will be no confusion because the continuous distribution is the only one ever used.

The graph of the continuous chi-square distribution for the die problem is shown in Fig. 1. The experimental value of $\chi^2 = 13.6$ has been located on this graph, together with the value of $\chi^2 = 11.1$, which cuts off the 5 percent right tail of the distribution. Since large values of χ^2 correspond to poor experimental agreement, the values of χ^2 exceeding 11.1 are chosen as the critical region of the test. The experimental value $\chi^2 = 13.6$ falls in the critical region; therefore, the hypothesis that the die is honest is rejected.

Example. As a second illustration of the χ^2 test, consider the following genetics problem. In experiments on the breeding of flowers of a certain species, an experimenter obtained 120 magenta flowers with a green stigma, 48 magenta flowers with a red stigma, 36 red flowers with a green stigma, and 13 red flowers with a red stigma. Mendelian theory predicts that flowers of these types should be obtained in the ratios 9:3:3:1. Are these experimental results compatible with the theory?

In terms of hypothesis testing notation, this is a problem of testing the hypothesis

$$H_0: p_1 = \tfrac{9}{16}, \; p_2 = \tfrac{3}{16}, \; p_3 = \tfrac{3}{16}, \; p_4 = \tfrac{1}{16}$$

For this problem, $n = 217$ and $k = 4$. The expected frequencies for the four cells are obtained by multiplying 217 by each of the H_0 probabilities. The observed frequencies, together with the expected frequencies, correct to the nearest integer, are shown in Table 2.

Table 2

o_i	120	48	36	13
e_i	122	41	41	14

Calculations give

$$\chi^2 = \frac{(120 - 122)^2}{122} + \frac{(48 - 41)^2}{41} + \frac{(36 - 41)^2}{41} + \frac{(13 - 14)^2}{14} = 1.9$$

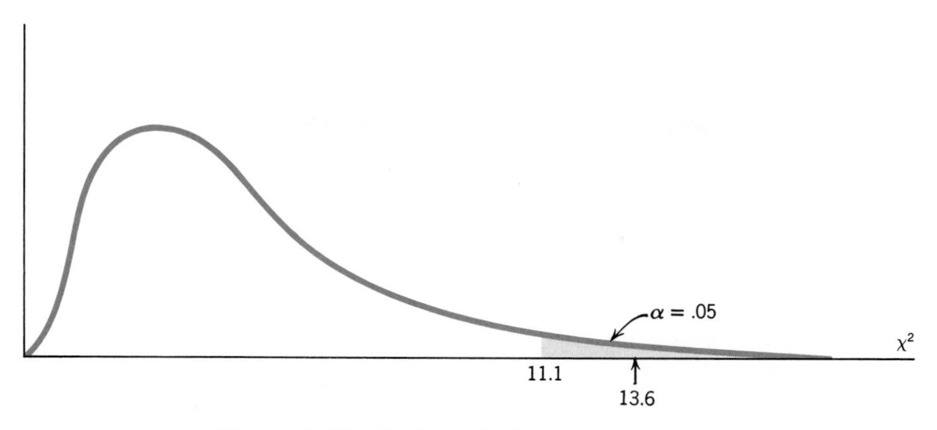

Figure 1 Distribution of χ^2 for die problem.

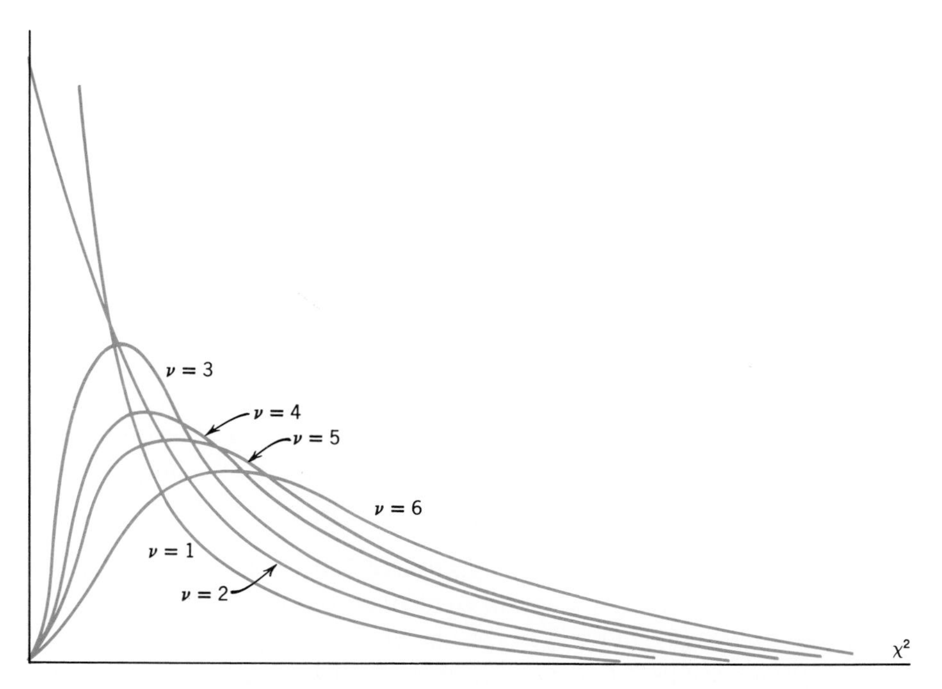

Figure 2 Distribution of χ^2 for various degrees of freedom.

Now the theory of the χ^2 distribution for problems such as these shows that the χ^2 curve in Fig. 1 does not apply to the present problem because the number of cells in this problem is not the same as before. A remarkable feature of the χ^2 distribution is that its form depends only upon the number of cells. Figure 2 gives the graphs of six such χ^2 curves corresponding to the number of cells, ranging from 2 to 7. It is customary to label a χ^2 distribution by means of a parameter $\nu = k - 1$, called the number of degrees of freedom, rather than by the number of cells. The phrase "degrees of freedom" refers to the number of independent cell frequencies. Since the sum of the four observed frequencies in Table 2 must equal 217, the fourth-cell frequency is determined as soon as the first three cell frequencies are specified. Thus there are $\nu = 3$ degrees of freedom for this problem, just as there were $\nu = 5$ degrees of freedom in the earlier problem of the die. The value of χ^2 cutting off the 5 percent right tail of the χ^2 distribution for $\nu = 3$ turns out to be 7.8. Since the value of χ^2 for Table 2, namely 1.9, does not fall in the critical region, the result is not significant. There is no reason on the basis of this test for doubting that Mendelian theory is applicable to the data of Table 2.

The values of χ^2 that determined the 5 percent critical regions in the two preceding illustrations were obtained from Table IX in the appendix. The 5 percent critical value is found in the column headed .05 in the row corresponding to the appropriate number of degrees of freedom, $\nu = k - 1$.

4 LIMITATIONS ON THE CHI-SQUARE TEST

Since a χ^2 curve, such as those in Fig. 2, is only an approximation to the true discrete distribution of χ^2, the χ^2 test should be used only when the approximation is good. Experience and theory indicate that the approximation is usually satisfactory provided that the expected frequencies in all the cells are at least as large as 5. This limitation is similar to that placed on the use of the normal-curve approximation to the binomial distribution, in which np for $p \leq \frac{1}{2}$ was required to exceed 5. If the expected frequency of a cell is not as large as 5, this cell can be combined with one or more other cells until the condition is satisfied.

Example. As an illustration, consider the data of Table 3 on the classification of automobile accidents in a certain community according to the age of the driver for drivers below 25 years of age. The percentages of drivers in these age groups in this community are approximately 10 percent, 20 percent, 20 percent, 25 percent, and 25 percent, respectively. The problem here is to test whether the accident rate among drivers under 25 years of age is independent of age. Since the total number of observations here is $n = 40$, the expected frequencies are obtained by multiplying 40 by these percentages treated as decimals. They have been recorded in Table 3 in the row labeled e_i. Since the expected frequency in the first cell is less than 5, the first two cells are combined to give Table 4.

The value of χ^2 for Table 4 is 5.5. Since $\nu = 3$ here and Table IX yields the 5 percent critical value of 7.8, this result is not significant.

Table 3

	15–16	17–18	19–20	21–22	23–24
o_i	5	12	10	8	5
e_i	4	8	8	10	10

Table 4

	15–18	19–20	21–22	23–24
o_i	17	10	8	5
e_i	12	8	10	10

5 GOODNESS OF FIT

The theoretical distributions that were introduced in Chapter 5 were designed to serve as mathematical models for empirical frequency distributions. For example, the binomial distribution was designed for problems involving repeated trials of an experiment and the normal distribution was assumed to be applicable to many continuous variable problems. Most of the problems of estimation and hypothesis testing that occurred in Chapters 7 and 8 were based on some type of normality assumption. If a model is not justified on strong theoretical grounds, such as the central limit theorem, then it behooves the investigator to check on the reasonableness of his model assumption. This can be done by means of a χ^2 test.

As an illustration, suppose that a highway engineer believes that the number of weekly accidents occurring on a mile stretch of a particular road follows a Poisson distribution. If x denotes the number of accidents per week, the mathematical model is the Poisson probability function

$$f(x) = \frac{e^{-\mu}\mu^x}{x!}, \qquad x = 0, 1, 2, \cdots \tag{2}$$

where μ is the mean number of accidents per week.

The difficulty here is that μ is unknown; therefore it must be estimated from sample data. Suppose that the weekly accident records for the past year for this road yielded the following frequencies:

Number of accidents	0	1	2	3	4	5	6
Frequency	10	12	12	9	5	3	1

The mean of this empirical distribution is 2.0; hence the Poisson probability distribution that should be chosen as a mathematical model here is given by the function

$$f(x) = \frac{e^{-2}2^x}{x!}, \qquad x = 0, 1, 2, \cdot \cdot \cdot$$

Since the total frequency count here is 52, the expected number of observations for each x-value is given by multiplying $f(x)$ by 52. The results of such calculations yield the following pairs of observed and expected frequencies, with the cells beyond 7 producing expected frequencies of less than .1.

x	0	1	2	3	4	5	6	7
o_i	10	12	12	9	5	3	1	0
e_i	7.0	14.1	14.1	9.4	4.7	1.9	1.0	.2

Because the expected frequencies in the tail cells are so small, it is necessary to combine the last four cells in order to realize a cell total of at least 5. This produced the following values:

o_i	10	12	12	9	9
e_i	7.0	14.1	14.1	9.4	7.8

Treating this as an ordinary χ^2 problem of comparing observed and expected frequencies, one finds that $\chi^2 = 2.11$. It is at this stage of the problem that the test of goodness of fit changes. Normally, one would choose $\nu = 4$ here because there are 5 cells and $\nu = n - 1$; however, this problem is slightly different because the sample mean $\bar{x}$ was used in place of the unknown mean μ in (2). As a result, the fit is likely to be better than it would be if the value of μ were known. Fortunately, it can be shown mathematically that the usual χ^2 test is valid provided the degrees of freedom are reduced by 1. More generally, if a mathematical model has more than one unknown parameter that must be estimated from the data, the usual χ^2 test is still valid provided one degree of freedom is subtracted from $\nu = n - 1$ for each such estimated parameter. For example, in fitting a normal curve to an empirical frequency distribution, the number of degrees of freedom would be given by $\nu = n - 3$ because both the mean and the standard deviation would need to be estimated from the data. There is a slight restriction on the type of estimate that may be used; however, it is a minor refinement and the estimates employed in Chapter 7 will usually be satisfactory.

Returning to the problem being discussed, it suffices to look up the 5 percent critical value of χ^2 in Table IX corresponding to $\nu = 3$, which is 7.81. Since $\chi^2 = 2.11$ does not lie in the critical region, there is no reason on the basis of these data for questioning the Poisson assumption.

6 CONTINGENCY TABLES

The preceding sections have been concerned with a single factor of classification and related hypothesis testing problems. In this section the problem is generalized to two variables of classification but the problem is restricted to that of testing whether the two variables are independent. Thus, this section supplements the material of Section 3, Chapter 9, which tests whether there is any correlation between two continuous variables, by presenting a test for correlation between two counting variables.

A two-way table of frequencies corresponding to two factors of classification is commonly called a *contingency table.* Table 5 is an illustration of such a table, where the frequencies corresponding to the indicated classifications for a sample of 400 are recorded.

Table 5

	Marriage adjustment				
Education	Very low	Low	High	Very high	Totals
College	18(27)	29(39)	70(64)	115(102)	232
High school	17(13)	28(19)	30(32)	41(51)	116
Grade school	11(6)	10(9)	11(14)	20(23)	52
Totals	46	67	111	176	400

A contingency table is usually constructed for the purpose of studying the relationship between the two variables of classification. In particular, one may wish to know whether the two variables are at all related. By means of the χ^2 test it is possible to test the hypothesis that the two variables are independent. Thus, in connection with Table 5, the χ^2 test can be used to test the hypothesis that there is no relationship between an individual's educational level and his adjustment to marriage.

This problem differs from the preceding problems in that the probabilities of an observation falling in the various cells are not known. As a result, it is not possible to write down the expected frequencies for the various cells, as was the case in the other problems. This difficulty can be overcome in the following manner.

Consider repeated sampling experiments of this kind in which 400 people are classified in their proper categories. If only those experiments that produce the same marginal totals are considered, then expected frequencies can be obtained. Since the margins are now fixed, the proportion of college graduates in such samples of 400 is always $\frac{232}{400} = .58$. Therefore, if there is no relationship between education and marriage adjustment, 58 percent of the 46 individuals in the very low adjustment category would be expected to be college graduates. Since 58 percent of 46 is 27, to the nearest unit, this is the expected frequency for the first cell in Table 5 on the assumption of independence. The expected frequencies in the remaining three cells of the first row are obtained in a similar manner by taking 58 percent of the column totals. The second- and third-row cell frequencies are obtained by using 29 percent and 13 percent, respectively, corresponding to the proportions $\frac{116}{400}$ and $\frac{52}{400}$. These expected frequencies are recorded in parentheses in Table 5.

The value of χ^2 for Table 5, using the numbers in parentheses as expected frequencies, is 20.7. Although there are twelve cells in this table, one does not choose 11 degrees of freedom in looking up the critical value of χ^2, as was the procedure in most of the earlier problems. The correct value of ν to choose here is quite different and is determined by strictly mathematical arguments; however, it is possible to acquire a feeling for the plausibility of the correct value by the following type of argument. Assume that n is large but fixed in value. Then χ^2 will be expected to decrease in variability if the number of cells is decreased, because there will be fewer possibilities for large differences between observed and expected frequencies. This is observed in Fig. 2 in the shifting of the distribution to the left as ν decreases. Furthermore, if the number of cells is fixed but restrictions are placed on the frequencies that occur in the cells, the variability of χ^2 will also be expected to decrease. Now in repeated experiments of the preceding type the cell frequencies have been rather severely restricted by requiring them to possess the proper row sums and the proper column sums. In the first row of Table 5, for example, the observed frequencies must sum to 232. This essentially says that the frequencies in the first three cells of that row are free to assume any values they please, as long as they do not sum to more than 232, but that then the fourth-cell frequency is completely determined. Thus, if the χ^2 test were to be applied to the first row only, it would employ $\nu = 3$ degrees of freedom. The phrase "degrees of freedom" aptly describes what corresponds to the number of cell frequencies that are free to vary. The same argument would apply to the second row, thus giving a total of six degrees of freedom for the two rows. However, when one looks at the third row, the picture changes because now it is necessary to realize that what applies to rows must also apply to columns; therefore the frequency found in the last cell of the first column is determined when the first two cell frequencies in that column have been specified, because the column sum has been fixed. As a result, the third-row frequencies as well as the fourth-column frequencies are determined when the remaining cell frequencies have been specified. Thus, it seems reasonable that the experiment should behave like one wit [illegible] cells in which the frequencies are free to vary, subject only to the restrictio[illegible]t their sums not be excessive. In view of arguments of this type it should com[illegible]s no surprise to learn that mathematical theory has demonstrated that one should choose $\nu = 6$ here.

From Table IX, the 5 percent critical value of χ^2 with $\nu = 6$ degrees of freedom is 12.6. Since $\chi^2 = 20.7$ here, this result is significant and the hypothesis of independence is rejected. An inspection of Table 5 shows that individuals with some college education appear to adjust themselves to marriage more readily than those with less education.

In the foregoing solution, since only experiments with the same marginal totals are being considered, it is tacitly assumed that any relationship, or lack of it, that exists for restricted sampling experiments will also hold for unrestricted sampling experiments. This seems to be a reasonable assumption here because there appears to be no reason for believing that fixing the marginal totals in this way will influence the relationship.

This problem illustrates very well why Table IX for χ^2 lists the number of degrees of freedom rather than the number of cells to determine which χ^2 curve to use. For a contingency table having r rows and c columns, the number of degrees of freedom is given by the formula

$$\nu = (r - 1)(c - 1)$$

This follows from the earlier arguments that the frequencies in the last row and in the last column are determined by the marginal totals as soon as the other cell frequencies are given. Thus the number of independent cell frequencies is obtained by counting the number of cells after the last row and the last column have been deleted. After the deletion there will be $r - 1$ rows and $c - 1$ columns and therefore $(r - 1)(c - 1)$ cells.

7 HOMOGENEITY

The preceding technique for testing whether two variables of classification are independent can also be used to test the *homogeneity* of sets of classified data. For example, suppose that a business firm wishes to determine whether the quality of its product is the same at all five of its production units. Suppose, further, that it takes a sample of 100 items off the production line at each of its production units and, after testing them for quality, obtains the following data.

Satisfactory items	80	93	89	92	81
Below-quality items	20	7	11	8	19
Totals	100	100	100	100	100

This table differs from the customary χ^2 table because the samples here, which total 500, are not permitted to fall into just any one of the 10 cells available.

Instead, the sample is split into five sets of 100 each and the samples in each set are permitted to fall in only one of two possible cells.

Now, the basic homogeneity problem is to test whether the probability of a production line's turning out a satisfactory item is the same for all five production units. Since this probability is unknown, it must be estimated from the combined data, under the hypothesis assumption that this probability is the same for all units. Calculating expected frequencies on this basis is equivalent to calculating expected frequencies in the usual manner for a contingency table. Furthermore, it can be shown that the usual χ^2 test for contingency tables is applicable to problems of this type, with $\nu = (r - 1)(c - 1)$. The conclusion, therefore, is that one may ignore the fact that the sample has been split into several separate groups and treat a homogeneity test as an ordinary χ^2 contingency table test.

The usual χ^2 calculations produced the following expected frequencies shown in parentheses:

80 (87)	93 (87)	89 (87)	92 (87)	81 (87)
20 (13)	7 (13)	11 (13)	8 (13)	19 (13)

Calculations will show that $\chi^2 = 13.3$ here. The Table IX critical value for $\nu = 4$ is 9.49; hence the hypothesis of homogeneity is rejected. There is too much variability in quality to attribute it to sampling variation.

8 DISTRIBUTION OF THE SAMPLE VARIANCE

As indicated in section 1, the chi-square distribution can also be used to solve problems of estimation and hypothesis testing about the normal distribution parameter σ. For the purpose of explaining how it can be used to solve such problems, consider the following estimation problem.

A market analyst took a sample of 20 markets in a large city in an attempt to determine how much variation there is in meat prices. The 20 prices (cents per pound) that were quoted him for the same cut of meat yielded the sample values $\bar{x} = 192$ and $s = 8$. The problem now is to find a 95 percent confidence interval for the standard deviation of all the market prices.

Suppose a large number of experiments of the same kind was carried out. This means that each time a random sample of 20 markets would be taken and the value of s computed for the 20 quoted prices. The values of $\bar{x}$ are of no interest here, and they are therefore ignored. Suppose, further, that meat prices for this cut are normally distributed and, for the time being, assume that the value of

σ^2 is known for this population. Then each experiment would yield a value of the variable

$$v = \frac{(n-1)s^2}{\sigma^2}$$

If these values of v were classified into a frequency table, one would obtain a good estimate of the limiting, or theoretical, distribution of v. As usual, mathematical methods yield the exact distribution and show that the distribution of v depends only upon the value of n. Surprising as it may seem, the distribution of v turns out to be a χ^2 distribution, with $\nu = n - 1$ degrees of freedom. Since Table IX gives the necessary probabilities for the χ^2 distribution, problems of estimation and hypothesis testing for σ can be solved by methods similar to those used for μ.

For the problem under consideration, since $\nu = n - 1 = 19$; the variable v will possess a χ^2 distribution with 19 degrees of freedom. From Table IX it will be found that $\chi_1^2 = 8.91$ and $\chi_2^2 = 32.85$ are the two values of χ^2 cutting off $2\frac{1}{2}$ percent tail areas of the χ^2 curve for 19 degrees of freedom. Hence the probability is .95 that v will satisfy the inequalities

$$\chi_1^2 < v < \chi_2^2$$

If the value of v is substituted, these inequalities become

$$\chi_1^2 < \frac{(n-1)s^2}{\sigma^2} < \chi_2^2$$

Each of these inequalities may be solved for σ^2 to yield the equivalent inequalities

$$\frac{(n-1)s^2}{\chi_2^2} < \sigma^2 < \frac{(n-1)s^2}{\chi_1^2}$$

When the proper numerical values are inserted, this expression will yield a 95 percent confidence interval for σ^2. For this problem, numerical substitutions yield

$$\frac{19(8)^2}{32.85} < \sigma^2 < \frac{19(8)^2}{8.91}$$

This simplifies to

$$37 < \sigma^2 < 136$$

If the square roots of these numbers are found, the desired 95 percent confidence interval for σ will be obtained, namely

$$6.1 < \sigma < 11.7$$

It is clear from this result that one cannot estimate σ with much precision when the sample is as small as 20. It should also be noted that the sample point estimate $s = 8$ is not in the middle of this interval. This is because the χ^2 curve for small degrees of freedom is heavily skewed to the right.

As a second illustration of the use of the χ^2 distribution for standard deviation problems, consider once more the problem discussed in section 2, Chapter 8, on testing a mean. It was assumed there that the variability of brand B bulbs is the same as that for brand A bulbs. Suppose however that the sample of the brand B bulbs had produced a value of $s = 80$, whereas long experience with brand A bulbs showed a standard deviation of 90 hours. If one assumes that there is no difference between the two brands, then the two brands should possess the same theoretical standard deviations as well as the same means. Consider, therefore, the problem of testing the hypothesis that the population standard deviation of brand B bulbs is equal to that of brand A bulbs. Since the value 90 is based on long experience, it may be treated as the population value for brand A. The hypothesis to be tested then is that the standard deviation of the brand B bulbs is also 90. This is written in the form

$$H_0 : \sigma = 90$$

A two-tailed test is used here because before the sample was taken, there was no reason to believe that the standard deviation of brand B bulbs would be larger, or smaller, than the standard deviation for brand A bulbs.

Since the sample size here is 100, there will be 99 degrees of freedom for this problem. To eliminate the necessity for interpolating in Table IX, assume that the sample size is 101 rather than 100. From Table IX it will be found that the two $2\frac{1}{2}$ percent tail areas of the χ^2 curve for 100 degrees of freedom are cut off by the two χ^2 values given by

$$\chi_1^2 = 74 \qquad \text{and} \qquad \chi_2^2 = 130$$

As a result, the 5 percent two-tailed critical region for the test is chosen as the values of χ^2 lying outside the interval (74,130). Since, for this problem, the value

$$v = \frac{(n-1)s^2}{\sigma^2} = \frac{100(80)^2}{(90)^2} = 79$$

does not fall in the critical region, the hypothesis is accepted.

For degrees of freedom less than 100 but not listed in Table IX, it suffices to interpolate roughly between the two nearest listed values. For degrees of freedom larger than 100, one may treat the variable

$$z = \sqrt{2\chi^2} - \sqrt{2\nu - 1}$$

as an approximately normal variable with zero mean and unit standard deviation. For example, in the preceding problem with $n = 101$ one would obtain

$$\begin{aligned} z &= \sqrt{2(79)} - \sqrt{2(100) - 1} \\ &= \sqrt{158} - \sqrt{199} \\ &= 12.57 - 14.11 \\ &= -1.54 \end{aligned}$$

Since z is a standard normal variable, a two-sided critical region of size .05 consists of the z-values outside the interval from -1.96 to 1.96. As was to be expected, $z = -1.54$ does not lie in the critical region.

Investigations have shown that the preceding methods are not reliable unless one is quite certain that the basic variable possesses a normal distribution. Since it is usually very difficult to determine whether sample values may be treated as properly coming from a normal, or approximately normal, distribution, it is best not to use the preceding methods unless there are theoretical reasons, or past experience, to justify the normality assumption. Other more complicated methods are available that possess the advantage that they are not heavily dependent on the normality assumption.

9 REVIEW ILLUSTRATIONS

Example 1. A merchandising firm was interested in the preferences of housewives in packaging designs. Its marketing department presented four designs to a random sample of 200 housewives with the following results.

Design	*A*	*B*	*C*	*D*
Number preferring	33	42	67	58

Use χ^2 to test the statistical significance of the results.

Under the hypothesis that $p_A = p_B = p_C = p_D$, all these probabilities are equal to $\frac{1}{4}$; hence the expected frequencies are each $200(\frac{1}{4}) = 50$. Then

$$\chi^2 = \frac{(33-50)^2}{50} + \frac{(42-50)^2}{50} + \frac{(67-50)^2}{50} + \frac{(58-50)^2}{50} = 14.1$$

Since the .05 critical point for χ^2 based on 3 degrees of freedom is 7.8, this result is significant; therefore the assumption of equal preferences must be rejected.

Example 2. The following data give the number of absences from work of 318 industrial workers in two consecutive six-month periods. Absenteeism is rated as "low" for less than 3 absences, "medium" if between 3 and 5 inclusive, and "high" if more than 5. These data are from the *South African Council of Scientific and Industrial Research* (1952). Use the χ^2 test to test the hypothesis that "absence proneness" does not exist, which implies that the two variables of classification are independent.

		First 6 months		
		L	M	H
	L	165	35	4
Second 6 months	M	40	24	20
	H	2	13	15

This may be treated as a contingency table; hence calculating row and column totals, and then row proportions, gives the proportions .641, .264, and .095. Taking these proportions of the column totals gives the following expected frequencies.

133	46	25
55	19	10
19	7	4

As a result,

$$\chi^2 = \frac{(165-133)^2}{133} + \frac{(35-46)^2}{46} + \frac{(4-25)^2}{25} + \frac{(40-55)^2}{55}$$
$$+ \frac{(24-19)^2}{19} + \frac{(20-10)^2}{10} + \frac{(2-19)^2}{19} + \frac{(13-7)^2}{7} + \frac{(15-4)^2}{4} \doteq 94$$

Since the .05 critical point for χ^2 based on 4 degrees of freedom is 9.5, this result is significant. There does appear to be absence proneness present in the sense that those employees who have a high absence record in one six-month period tend to have a high record in the next six-month period.

Example 3. The following data are from a book, *Housing Market in a Declining Area* by Grebler. They give the distribution of transfers of ownership by foreclosure and surrender in the years 1900–1949 for 949 parcels of real estate in the lower East Side of New York City. (*a*) Fit a Poisson distribution to the data. (*b*) Graph the observed and theoretical distributions on the same set of axes. (*c*) Use the χ^2 test to determine whether the Poisson model appears to be satisfactory.

Number of transfers	0	1	2	3	4	5	6
Number of parcels	487	300	122	32	5	2	1

(*a*) $\bar{x} = .712$; hence the Poisson function to use is

$$f(x) = \frac{e^{-.712}(.712)^x}{x!}$$

The expected frequencies are given by $949f(x)$. Calculations give, to the nearest whole number,

x	0	1	2	3	4	5	6
$949f(x)$	466	332	118	28	5	1	0

(*b*)

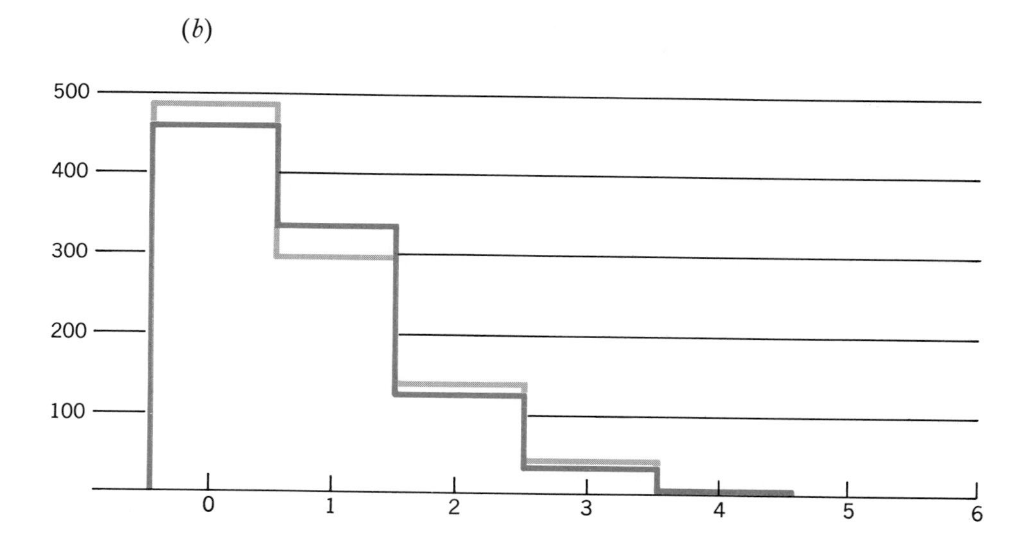

(*c*) $$\chi^2 = \frac{(487 - 466)^2}{466} + \frac{(300 - 332)^2}{332} + \frac{(122 - 118)^2}{118} + \frac{(32 - 28)^2}{28} + \frac{(8 - 6)^2}{6}$$

$$= .95 + 3.08 + .14 + .57 + .67 = 5.41$$

Since there are 5 cells and the mean was estimated from the data, the proper number of degrees of freedom is given by $\nu = 5 - 1 - 1 = 3$. The critical value for $\nu = 3$ is 7.8. Since $5.41 < 7.8$, the hypothesis of compatibility is accepted, which implies that the Poisson is a satisfactory model for these data.

Example 4. A market research staff is concerned about the low percentage of questionnaires that are returned to them after a mailing. Some of the staff felt that a questionnaire that created the impression of personal attention would yield a higher return of answers. For the purpose of testing this belief, they sent out 500 mimeographed copies, 300 offset-printed copies that looked like typewritten copies, and 200 original typed copies. These mailings produced 120 answers from the mimeographed set, 100 answers from the offset-printed set, and 80 answers from the typed set. What can be concluded?

M	P	T	Totals
500	300	200	1000
120	100	80	300

The problem is to test the hypothesis that the return rate is the same for the three types of mailings. This implies that $p_1 = .5, p_2 = .3, p_3 = .2$, and that the expected frequencies are obtained by multiplying these probabilities by 300, the total number of responses. This gives the following table.

o_i	120	100	80
e_i	150	90	60

Hence

$$\chi^2 = \frac{(120 - 150)^2}{150} + \frac{(100 - 90)^2}{90} + \frac{(80 - 60)^2}{60} = 13.8$$

For $\nu = 2$, the critical value of χ^2 is 5.99; therefore the hypothesis of the same return rate for the three mailings is rejected. The typed copies do seem to yield a higher percentage of returns than the mimeographed copies. The offset-printed copies also tend to do so but not to the same extent.

Example 5. An article in the *Journal of Industrial Economics* for November 1972 ranked the 100 largest manufacturing companies in the United Kingdom for the year 1948. It broke this ranking into four groups: the 12 largest, the 13 next largest, the 25 next largest, and the remaining 50. It then determined how many of those companies were still in business (alive) in the year 1968. The results of that study are shown in the following table.

Rank in 1948	Number alive in 1968
1–12	10
13–25	8
26–50	14
51–100	32

The problem is to determine whether it is true that the larger the company, the better is its chance for survival. The chi-square test can be used to assist in such a determination by testing the hypothesis that the survival rate is the same for all four ranking categories. This "homogeneity" assumption permits the calculation of cell probabilities. To do so, it is first necessary to list the number of companies in the various cells. This gives the following table.

Number in cell in 1948	Number alive in 1968
12	10
13	8
25	14
50	32

The probabilities, based on a uniform survival rate, are therefore .12, .13, .25, and .50. The expected frequencies are obtained by multiplying these probabilities by 64, the total number of surviving companies. This gives the following table of observed and expected frequencies.

o_i	10	8	14	32
e_i	8	8	16	32

The value of χ^2 is .75. Since there are 3 degrees of freedom for this problem and the critical value is 7.8, the homogeneity hypothesis is accepted. The actual survivals are surprisingly close to those expected on the assumption that size had nothing to do with the chances of surviving.

Example 6. Certain industrial machines require overhaul when wear on their parts introduces too much variability in the product to pass specifications. If the variability must not exceed $\sigma^2 = 50$ and a sample of 40 gave $s^2 = 60$, is the inspector justified in calling for an overhaul? Solve this problem by testing the hypothesis $H_0{:}\sigma^2 = 50$ against $H_1{:}\sigma^2 > 50$.

Since this is a one-sided hypothesis, the critical region for the test should consist of the right tail of the χ^2 distribution with 39 degrees of freedom. From Table IX, the .05 critical value is approximately 57. But

$$\chi^2 = \frac{(n-1)s^2}{\sigma^2} = \frac{39(60)}{50} = 46.8 < 57$$

Hence, H_0 is accepted and the inspector is not justified. This much variability is not unusual for a sample of this size when $\sigma^2 = 50$.

Further insight is given in this problem by calculating a confidence interval for σ^2. Thus, a 95 percent confidence interval is given by

$$\frac{39(60)}{58} < \sigma^2 < \frac{39(60)}{24}$$

which reduces to $40 < \sigma^2 < 98$. This shows that the value $\sigma^2 = 50$ is well within the interval of possible σ^2 values for a sample of this size.

EXERCISES

SECTION 3

1 The number of automobile accidents per week in a certain community were as follows: 12, 8, 20, 2, 14, 10, 15, 6, 9, 4. Are these frequencies in agreement with the belief that accident conditions were the same over this 10-week period?

2 According to Mendelian inheritance, offspring of a certain crossing should be colored red, black, or white in the ratios 9:3:4. If an experiment gave 72, 34, and 38 offspring in those categories, is the theory substantiated?

3 The number of individuals of a certain race possessing the four blood types should be in the proportions .16, .48, .20, .16. Given the observed frequencies 180, 360, 132, 98 for another race, test to see whether it possesses the same distribution of blood types.

4 A director of real estate sales after several years of experience decided that the best measure of effectiveness of his salesman is the number of sales made, rather than, say, the value of the sales.

Four of his salesmen over the past year had the following records:

	A	*B*	*C*	*D*
Sales	12	16	10	14

Use χ^2 to test whether the salesmen differ significantly.

5 A merchandising firm was interested in knowing which of three coffee flavors housewives prefer. Its marketing research department asked 300 housewives to indicate their preferences and obtained the following results.

Coffee type	*A*	*B*	*C*
Housewives preferring it	73	102	125

Do these data justify the claim that all three flavors are equally acceptable?

6 In a certain crossing of two varieties of peas, genetic theory suggests that half the peas should be wrinkled and half smooth.

(*a*) In a sample of 800 peas that were examined, there were 430 wrinkled ones. Does this sample conform to theory?

(*b*) If the sample had been 400 and the number of wrinkled peas had been 215, what would that have done to the value of χ^2 as compared to its value in (*a*)? Would you have arrived at the same conclusion?

7 The number of commercial aircraft accidents that occurred over a period of time, listed for the various days of the week, are as follows:

Sunday	13
Monday	14
Tuesday	6
Wednesday	10
Thursday	8

Friday	7
Saturday	12

(*a*) Test to see whether the accident rate is uniform over the week.
(*b*) Combine the Saturday, Sunday, and Monday data to see whether the weekend rate is the same as for the rest of the week.

SECTION 4

8 In a city the proportion of car owners who have 0 accidents, 1 accident, and more than 1 accident in a year are, respectively, .75, .21, and .04. An insurance company took a sample of 100 of its insured and found 65, 22, and 13 in those categories. Are these frequencies compatible with the postulated theoretical proportions?

9 In the preceding problem combine the frequencies in the last two cells and test. Comment on any differences in the two tests.

10 A pinball machine has six holes of varying size into which the ball may land. The payoff for the ball landing in a given hole indicates that the probabilities of success for the various holes should be $\frac{6}{21}$, $\frac{5}{21}$, $\frac{4}{21}$, $\frac{3}{21}$, $\frac{2}{21}$, and $\frac{1}{21}$. A contestant plays the game 42 times and obtains the following frequencies: 19, 10, 6, 5, 2, and 0. Test the hypothesis that the probabilities corresponding to the payoffs are correct.

11 Suppose one wishes to test the hypothesis that the payoffs for the more difficult holes are too low as compared to the payoffs for the easier holes in the game of problem 10. One method of doing this is to combine the frequencies of the first two holes and combine those of the last four holes and test the hypothesis $H_0: p = \frac{11}{21}$, where p is the probability of landing in one of the first two holes. Carry out this test.

SECTION 5

12 An instructor who claims that he grades on the "curve" (normal) should expect to give 7 percent A's, 24 percent B's, 38 percent C's, 24 percent D's, and 7 percent F's. On the basis of the following set of grades given out by him in a class of 125 students, can one conclude that he does grade on the curve?

A	25
B	38
C	49
D	7
F	6

13 The number of vacancies that have occurred on the U.S. Supreme Court from 1837 to 1932 is given in the following table, together with the expected number of vacancies based on a Poisson distribution fitted to the data. Is the Poisson distribution a reasonable model here?

Number of vacancies	0	1	2	3
Number of years	59	27	9	1
Expected number	58	29	7	1

14 The following data are for the number of railroad switchmen who had various numbers of accidents on the job over a given period of time and for the expected numbers based on a Poisson distribution fitted to the data. Use a χ^2 test to determine whether the Poisson model is satisfactory here.

Accidents per man	0	1	2	3	4	5	6
Men having this many accidents	121	85	19	1	0	0	1
Expected number of such men	127	74	21	4	1	0	0

15 The following data represent the results of an investigation of the sex distribution of the children of 32 families containing 4 children each. Use the binomial distribution with $n = 4$ and $p = \frac{1}{2}$ to calculate expected frequencies. Then apply the χ^2 test to see whether this binomial distribution model is satisfactory here.

Number of sons	0	1	2	3	4
Number of families	4	9	8	8	3

SECTION 6

16 Test to see whether the two variables of classification in the following contingency table are independent.

20	10	10
10	20	30

17 A certain drug is claimed to be effective in curing colds. In an experiment on 164 people with colds, half of them were given the drug and half of them were given sugar pills. The patients' reactions to the treatment are recorded in the following table. Test the hypothesis that the drug and the sugar pills yield similar reactions.

	Helped	Harmed	No effect
Drug	52	10	20
Sugar	44	12	26

18 A market analyst is concerned whether housewives who are not at home when interviewers call differ in their opinions of a certain product. To check this possibility, interviewers returned to "not-at-home" houses until an interview was

obtained. The results of this study are given in the following table. Test to see whether the "not-at-home" housewives have the same opinion as the "at-home" housewives.

Opinion of product	Number of housewives interviewed	
	First call	Later call
Excellent	62	36
Satisfactory	84	42
Unsatisfactory	24	22

19 In an epidemic of a certain disease 927 children contracted the disease. Of these, 408 received no treatment and 104 of those suffered aftereffects. Of the remainder, who did receive treatment, 166 suffered aftereffects. Test the hypothesis that the treatment was not effective and comment about the conclusion.

20 The following data are taken from a study found in the journal *Biometrika.* Are they consistent with the view that the proportion of intelligent students is the same for athletes and nonathletes?

	Intelligent	Not intelligent
Athletes	581	567
Nonathletes	209	351

21 The following quotation is from the *Los Angeles Times* of January 1973.

"Drivers who die in single-car accidents are younger, more belligerent, less well adjusted, and more likely to be drunk than those involved in multiple-car fatalities, a psychiatric study shows.

"The postmortem study by a five man team in Baltimore, Maryland, involved interviews with the families and associates of 22 persons killed in single-car accidents between August 1968 and June 1969. These case studies were compared with similar investigations into victims of multiple-car accidents."

The relevant data obtained in these studies are the following:

Type of accident	Average age	Condition of driver	
		Drunk	Not drunk
Single car	31	17	5
Multiple car	45	4	7

Do these data justify the conclusion that drunkenness is more common in single car fatalities than in multiple car fatalities?

22 A study was made of the 1948 presidential election to determine whether there was any difference in voting preferences among adults who were at home the first time an interviewer called and those who were visited two or more times before they were available for an interview. The following data give the results of such a study carried out by Philip McCarthy in a 1949 mimeographed Cornell University report entitled, "Sampling Procedure for 1948 Voting Study." Use a χ^2 test to determine whether the sample obtained on a first visit gives a satisfactory sample of voter preferences.

	Interviews			
	First call	Second call	Three or more	Totals
Roosevelt	138	122	95	355
Dewey	124	115	85	324
Did not vote	90	67	75	232
Too young	25	25	30	80
Other	14	9	14	37
Totals	391	338	299	1,028

23 An article in *Science* for January 1971 by Fienberg studied the problem of whether the U.S. Selective Service had used a bias-free method of selecting draft numbers. Its method consisted of selecting capsules from two drums, each drum containing 366 capsules. The number selected from the first drum determined the birth date of the selectee, whereas the second number determined the order call-up for that birth date. A claim was made that this process was not random and favored those who were born early in the year. What conclusions would you draw from these data?

	Call-up Number			
Month	1–122	123–244	245–366	Totals
January	9	12	10	31
February	7	12	10	29
March	5	10	16	31
April	8	8	14	30
May	9	7	15	31
June	11	7	12	30
July	12	7	12	31
August	13	7	11	31
September	10	15	5	30
October	9	15	7	31
November	12	12	6	30
December	17	10	4	31
Totals	122	122	122	366

The call-up numbers were divided into three equal groups. If this double randomization scheme were operationally effective, one would expect about 10 frequencies in each cell. For the purpose of eliminating extensive calculations, combine the first four rows of frequencies, the second four rows of frequencies, and the last four rows of frequencies to obtain a contingency table with three rows and three columns. Then apply a χ^2 test to determine whether the frequencies behave like a set of frequencies for two independent classification variables. Comment.

SECTION 7

24 A supermarket has five stores located in different sections of the city. Management has permitted customers to cash personal checks at all five stores but, on looking over the records for the past year, it appeared that the privilege should

be withdrawn from stores having an excessively large bad-check record. The records were as follows.

	Store				
	A	*B*	*C*	*D*	*E*
Checks cashed	5000	3000	2000	2000	1000
Bad checks	22	25	10	15	8

Use χ^2 to test whether the stores differ significantly with respect to bad checks.

25 The following quotation, concerning heredity and alcoholism, is from an article in the *Los Angeles Times* for March 4, 1973. "Five American and Danish psychiatrists said that a study of adopted children who became alcoholics as adults suggested a tendency that the disease might be biologically inherited. They studied 55 Danish men who had been separated from their biological parents during early infancy, each had one parent who had been diagnosed as alcoholic. These men were compared with 78 adopted men whose biological parents had no known record of alcoholism." After some years, when the men ranged in age from 23 to 45, it was found that 10 of the 55, and 4 of the 78 were alcoholic. Test to see whether these data are compatible with the assumption that alcoholism is not affected by heredity.

26 The following table gives the breakdown of tests made on various cars to see whether they satisfied the emissions standards of California in 1971. These data were reported in the *Los Angeles Times* for November 4, 1971. Apply a χ^2 test to see whether these makes of cars have the same failure rate.

Make of car	Number of cars	Number that failed
Datsun	49	38
Ford	176	110
Chrysler	179	109
GM	162	62
Toyota	37	22
AMC	93	36
VW	64	36

SECTION 8

27 Given that x is normally distributed and given the sample values $n = 15$ and $s = 7$, test the hypothesis that $\sigma = 5$.

28 Test the hypothesis that $\sigma = 8$, given that $s = 10$ for a sample size (*a*) 20, (*b*) 51.

29 Using the data of problem 27, find (*a*) 95 percent, (*b*) 99 percent confidence limits for σ.

30 Find 95 percent confidence limits for σ if $s = 10$ for a sample size (*a*) 20, (*b*) 51.

31 Test the hypothesis that $\sigma = 20$, given that $s = 10$ for a sample of 25.

32 Work problem 30 for a sample of size (*a*) 76, (*b*) 120.

33 The following data represent the amounts of a certain chemical compound obtained in daily analyses of a chemical product. Long-run experience has yielded a standard deviation of .5.
(*a*) Test the hypothesis that $\sigma = .5$ for these analyses.
(*b*) Find 95 percent confidence limits for the σ of these analyses: 12.7, 12.3, 13.2, 12.8, 13.6, 13.1, 12.6, 12.4, 14.1, 13.3, 13.4, 13.1, 12.6, 12.9, 13.0, 12.4, 14.6, 13.8, 13.4, 12.7, 13.5, 12.5.

SECTION 9

34 The tax records in a city show that 70 percent of home owners pay their property taxes on time, that 25 percent are delinquent by at most 6 months, and that 5 percent are delinquent more than 6 months. The tax collector, who is interested in knowing whether delinquencies are increasing, takes a random sample of 100 tax accounts and finds 66, 22, and 12 are in those three categories of delinquency. Are these frequencies compatible with past experience, or is the tax collector justified in his belief?

35 Work problem 19 by the method explained in Chapter 8 for testing the difference of two proportions. It can be shown that the two methods are equivalent for problems such as this.

36 Take a sample of 500 random digits from Table XII and list the observed frequencies in 10 cells.
(*a*) Apply the χ^2 test to see whether the sample is compatible with theory here.
(*b*) Combine the odd-digit cells and the even-digit cells to obtain only two cells and test to see whether odd and even digits possess the same probability of occurrence.

37 Take a sample of 20 random digits and (*a*) calculate the value of s^2, (*b*) test the hypothesis that $\sigma^2 = 8$, (*c*) find a 95 percent confidence interval for σ. Does it include the true value, which is known to be $\sigma = 2.87$?

38 The responses of 96 mothers and their small sons to preferences in children's clothing yielded the following results.

	Mother		
Son	Favor	Dislike	Sometimes
Favor	43	5	34
Dislike	2	1	2
Sometimes	3	2	4

Do mothers understand their son's clothing preferences?

39 As reported in a 1941 article by D. Durand, "Risk Elements in Consumer Installment Financing," *National Bureau of Economic Research,* New York, a study was made to determine whether an individual who possesses life insurance is a better loan risk than one who does not possess such insurance. The following data give the results of this study. The determination as to whether a loan was a good or bad risk was made by certain lending institutions with much experience in that field. Do these data justify the claim that insurance should not

be used as one criterion for determining whether or not to grant a loan to an individual?

	Loan classification	
	Good	Bad
Life insurance	284	249
No life insurance	66	93

40 The Poisson distribution is frequently applied to certain counts taken over successive time intervals. During a period of 60 consecutive minutes, counts were made of the number of taxicabs that arrived at a station during each minute. The results of those observations are given in the first two rows of the following table of values. They were taken from an article by D. G. Kendall in a 1951 volume of the *Journal of the Royal Statistical Society.* The third row gives the expected frequencies based on the assumption that the frequencies follow a postulated Poisson distribution.

Number of taxicab arrivals during a 1-minute interval	0	1	2	3	4	≥ 5
Observed number of 1-minute intervals	18	18	14	7	3	0
Poisson expected frequencies	16	21	13	6	2	2

Use χ^2 to test whether the Poisson model is justified.

41 A mailed survey was carried out by the Veteran's Administration at the close of World War II to determine the educational plans of veterans eligible for the G.I. Bill. The results of the survey, given below, can be found in a 1947 article by J. A. Clausen and R. N. Ford in volume 42 of the *Journal of the American Statistical Association.* Those not responding to the initial mail-out were sent a follow-up appeal, and those still failing to respond were sent another stronger appeal. The results were as follows.

Educational plans	Percentage in each category		
	First mailing	First follow-up	Second follow-up
In school	42	35	29
Planning to enroll	46	44	41
Considering enrolling	10	15	21
Not planning to enroll	2	6	9
Total responding	7,887	3,359	1,607
Total not responding	6,719	3,360	1,752
Totals	14,606	6,719	3,359

Use χ^2 to test whether the percentages may be assumed to be the same for the three sets of responding individuals. Do not make the mistake of treating percentages as frequencies. Comment about the size of χ^2 and the need of a test for such a large sample.

Chapter 13 Analysis of Variance

1 INTRODUCTION

The methods for testing hypotheses that were explained in Chapter 8 involved testing a single parameter, or at most the difference of two parameters. There are, however, numerous situations in industrial experimentation where it is inadvisable and inefficient to consider at most two possibilities. For example, to determine the temperature at which a chemical reaction should take place to maximize the yield of a given compound, it would be advisable to run the experiment at several temperatures. In some experiments of this type it is questionable whether varying the temperature away from a known satisfactory value will have any appreciable effect on the yield. It would be desirable in such situations to have a test to determine whether there has been any significant change in the yield.

Suppose four different temperatures are used in the preceding experiment. Let μ_1, μ_2, μ_3, and μ_4 denote the theoretical mean yields for those temperatures, representing the limiting values of the sample means if the samples were allowed to become increasingly large. Then the hypothesis of no temperature effect could be written as

$$H_0: \mu_1 = \mu_2 = \mu_3 = \mu_4$$

One way out of the difficulty would seem to be to use the methods of Chapter 8 for testing the equality of two means and apply them to the six possible pairs of means that can be formed here. This, however, would be a very inefficient method if one had, say, six means to compare because there would be $\binom{6}{2} = 15$ such possible comparisons. Furthermore, the method derived for testing a single pair is not valid when comparing several means simultaneously. Another disadvantage of a method based upon comparing only pairs of means, even if a correct method were employed, is that experimentalists who are accustomed to comparing only two means at a time may be led into designing poor experiments to accomplish their ultimate objective. For example, the manufacturer of cake mixes who changed only one ingredient at a time and then retained only the better of two mixes each time might well miss out on a much better mix obtained by varying several ingredients by different amounts and considering various combinations of mixes simultaneously. For industrial experiments that are concerned with maximizing the quality characteristics of a product it has been found very inefficient not to consider different combinations of basic factors simultaneously.

In view of the foregoing discussion, it seems clear that a new method is needed to solve some of the problems related to several variables. One of the methods that has been designed to solve problems of this type for continuous variables is known as the *analysis of variance*. It is customary to abbreviate this phrase by writing ANOVA. As the name might indicate, the method consists of analyzing the variance of the sample into useful components. Although the method has been developed to treat a wide variety of problems, only two of the basic applications are considered in this chapter.

2 ONE-WAY ANALYSIS OF VARIANCE

The simplest type of analysis-of-variance model is one in which observations are classified into groups on the basis of a single property. The observational results in each group are recorded in separate columns of a table. There will be as many entries in a column as there are experiments conducted.

Table 1

I	II	III
44	40	54
39	37	50
33	28	40
56	53	55
43	38	45
56	51	66
47	45	49
58	60	65

For the purpose of explaining the analysis-of-variance technique, consider the data of Table 1. These data represent the scores made by 24 typists during an experiment to determine whether there were any differences between three different brands of typewriters. The 24 typists were split into three equal groups by random selection, with each group of 8 then assigned to one of the brands of typewriters.

If there is no advantage to one brand of typewriter over either of the others, and if the population mean scores corresponding to the three brands are denoted by μ_1, μ_2, and μ_3, then the problem reduces to one of testing the hypothesis

$$H_0 : \mu_1 = \mu_2 = \mu_3 \tag{1}$$

When the hypothesis H_0 is true, the classification of the data into three columns is meaningless and the entire set of measurements can be treated as a sample of size 24 from a single population. This assumes that the variation in typing scores is due to the typists and has nothing to do with typewriter differences. Essentially, therefore, if the hypothesis is true, the experiment is assumed to be equivalent to one in which each of the 24 typists used the same typewriter. If σ^2 denotes the population variance, an estimate of σ^2 can be obtained by means of the familiar sample variance based on those 24 measurements. However, there are several other ways of obtaining valid estimates of σ^2. For example, the sample variance of the first column of measurements is a valid unbiased estimate of σ^2 although it is not nearly as good as the estimate based on all the measurements. Similarly, the sample variances of the second and third columns are also valid estimates of σ^2. Furthermore, the mean of the three column estimates of σ^2 is a valid unbiased estimate of σ^2 and nearly as good as the familiar estimate based on combining the three sets of measurements. If s_1^2, s_2^2, and s_2^3 denote the sample variances for the three columns, this last estimate, which will be denoted by V_c, is

$$V_c = \frac{s_1^2 + s_2^2 + s_3^2}{3} \tag{2}$$

The subscript c is used here to indicate that the estimate is based on the column variances.

Another quite different type of estimate of σ^2 can be obtained by using the relationship between the variance of a sample mean and the variance of the population, namely $\sigma_{\bar{x}}^{\ 2} = \sigma^2/n$. It is convenient here to express this relationship in the form

$$\sigma^2 = n\sigma_{\bar{x}}^{\ 2} \tag{3}$$

Suppose several samples of size n each have been taken from some population. If the sample means have been calculated, then the sample variance of those sample means will be a valid estimate of $\sigma_{\bar{x}}^{\ 2}$. In general, the sample variance of a set of measurements is a valid estimate of the population variance of the measurements regardless of whether those measurements happen to be simple measurements, or means of simple measurements, or other functions of simple measurements. From (3) it follows that if an estimate of $\sigma_{\bar{x}}^2$ is available, it may be multiplied by n to yield an estimate of σ^2. In the present problem there are three such sample means that may be used to construct an estimate of $\sigma_{\bar{x}}^2$. They are the three column means, which will be denoted by $\bar{x}_1$, $\bar{x}_2$, and $\bar{x}_3$. Now, the mean of these three column means is equal to the grand mean, $\bar{x}$, of all the measurements; hence an unbiased estimate of $\sigma_{\bar{x}}^2$ is given by $\Sigma_{j=1}^{3}(\bar{x}_j - \bar{x})^2/2$. Since these means are based on samples of size 8 each, it follows that $n = 8$ and hence, from (3), that the desired estimate of σ^2 is given by

$$V_m = 8\frac{\sum_{j=1}^{3}(\bar{x}_j - \bar{x})^2}{2} \tag{4}$$

The subscript m is used here to indicate that the estimate is based on the means of the columns.

Since V_c and V_m are both valid unbiased estimates of σ^2 when H_0 is true, it follows that they should be approximately equal in value, and therefore that their ratio should have a value close to 1. If, however, H_0 is not true and the column means differ considerably, the two estimates V_c and V_m will be seen to differ considerably in value. Because the estimate V_c is based on calculating the variances of each column separately, it will be unaffected by changing the means of the various columns, for the variance of a set of measurements is independent of the value of their mean. It is clear from (4), however, that the estimate V_m will be directly affected and will increase in value as the sample means move apart. Thus it appears that the ratio of V_m to V_c will exceed 1 when H_0 is not true. This ratio will be used as the desired quantity for testing the hypothesis H_0. It is denoted by the letter F; hence

$$F = \frac{V_m}{V_c} \tag{5}$$

In the typewriter experiment, calculations with the data of Table 1 yield the values $\bar{x}_1 = 47$, $\bar{x}_2 = 44$, $\bar{x}_3 = 53$, $s_1^2 = 81.1$, $s_2^2 = 106.3$, and $s_3^2 = 82.3$ As a result, it follows from (2) that $V_c = 89.9$. Additional calculations yield the value $\Sigma_{j=1}^{3}(\bar{x}_j - \bar{x})^2/2 = 21$; hence it follows from (4) that $V_m = 168$. The value of F is therefore given by

$$F = \frac{V_m}{V_c} = \frac{168}{89.9} = 1.87$$

The question now arises: Is this value of F too large when compared to the values of F that might be expected in repeated experiments of this type with identical typewriters, that is under the assumption that H_0 is true? To answer that question it is necessary to determine the sampling distribution of F. This is considered in the next section.

3 THE *F* DISTRIBUTION

The discussion thus far has been mostly of the qualitative type, stating that F can be expected to have a value close to 1 when H_0 is true and a value considerably larger than 1 when H_0 is not true and the population means differ widely. This information is not sufficient for constructing a test based on probability; it is necessary to know what the distribution of F is before such a test can be performed.

Just as in the case of other sampling distributions, it is possible to approximate the distribution of F by carrying out repeated sampling experiments of the type being considered here and constructing the histogram of the resulting F-values; however, the exact sampling distribution of F can be obtained by mathematical methods, provided that the proper assumptions are made. The assumptions needed here are that the 24 cell variables are independent normal variables, all having the same mean μ and variance σ^2.

It turns out that the distribution of F depends only upon how many data were available for the numerator estimate of σ^2 and how many were available for the denominator estimate. Table X in the appendix lists the 5 percent and the 1 percent right-tail critical values of F corresponding to different values of the parameters ν_1 and ν_2, which are called the number of degrees of freedom in the numerator and denominator of F.

The degrees of freedom here are those that one would naturally associate with the sample variances being used. Since the number of degrees of freedom for the usual estimate of σ^2 is given by $\nu = n - 1$, or one less than the number of measurements, the number of degrees of freedom for the numerator of F in this problem is given by $\nu_1 = 2$ because the estimate is based on the three sample means. The number of degrees of freedom for the denominator of F in this problem is $\nu_2 = 21$ because each column variance contributes 7 degrees of freedom and all three column variances are employed.

From Table X it will be found that the 5 percent critical value of F corresponding to $\nu_1 = 2$ and $\nu_2 = 21$ is 3.47. Since $F = 1.87$ for this problem, the hypothesis is accepted. The data are in agreement with the view that typing skill is not affected by which of the three brands of typewriters is used.

Although the approach used to arrive at the F variable for testing the equality of a set of column means seems quite different from that used in Chapter 8 for testing the equality of two column means, it can be shown that the F test, when applied to testing the equality of two column means, is equivalent to the t test for the same problem. Thus, the test based on F is a generalization of the earlier two-column test based on t.

4 ANOVA NOTATION

The foregoing problem is a special case of more general problems of this type in which one has, say, r rows and c columns of data and in which one wishes to test the hypothesis that the column population means are equal. For the purpose of considering such problems, it is convenient to introduce the following notation.

Let x_{ij} denote the ith measurement in the jth column of a table of measurements, of which Table 1 is a special case. Let r denote the number of rows and c the number of columns in this table. This notation is displayed in Table 2.

The mean of the jth column measurements is denoted by the symbol $\bar{x}_{.j}$. The dot is placed in front of the j to indicate that the mean was obtained from summing on the index i (rows). In the next section it is necessary to sum over columns as well; therefore, some notation such as this is needed to keep straight whether one is summing over rows or columns.

Table 2

x_{11}	x_{12}	$\cdots$	x_{1j}	$\cdots$	x_{1c}
x_{21}	x_{22}	$\cdots$	x_{2j}	$\cdots$	x_{2c}
$\vdots$	$\vdots$		$\vdots$		$\vdots$
x_{i1}	x_{i2}	$\cdots$	x_{ij}	$\cdots$	x_{ic}
$\vdots$	$\vdots$		$\vdots$		$\vdots$
x_{r1}	x_{r2}	$\cdots$	x_{rj}	$\cdots$	x_{rc}
$\bar{x}_{.1}$	$\bar{x}_{.2}$	$\cdots$	$\bar{x}_{.j}$	$\cdots$	$\bar{x}_{.c}$

In terms of this new notation, the variance of the column means is given by

$$\frac{\sum_{j=1}^{c} (\bar{x}_{.j} - \bar{x})^2}{c - 1}$$

Since each column mean is based on r measurements, it is necessary to multiple this variance by r to obtain an estimate of the population variance σ^2. This corresponds to formula (3) with n replaced by r. The numerator term in F is therefore given by the formula

$$V_m = r\frac{\sum_{j=1}^{c} (\bar{x}_{.j} - \bar{x})^2}{c - 1} \tag{6}$$

To obtain a formula for V_c, it is necessary to calculate the variance of the measurements in the jth column. This variance, which is denoted by s_j^2, is given by

$$s_j^2 = \frac{\sum_{i=1}^{r} (x_{ij} - \bar{x}_{.j})^2}{r - 1}$$

The desired estimate of σ^2 that is to be used in the denominator of F is the mean of these column variances, which is

$$\frac{s_1^2 + s_2^2 + \cdot \cdot \cdot + s_c^2}{c}$$

In terms of sums of squares, it therefore follows that

$$V_c = \frac{\sum_{i=1}^{r} \sum_{j=1}^{c} (x_{ij} - \bar{x}_{.j})^2}{c(r - 1)} \tag{7}$$

The ratio of (6) and (7) supplies the desired F variable, with the degrees of freedom given by $\nu_1 = c - 1$ and $\nu_2 = c(r - 1)$; hence the desired general formula for F is given by

$$F = \frac{rc(r-1)}{c-1} \frac{\sum_{j=1}^{c} (\bar{x}_{.j} - \bar{x})^2}{\sum_{j=1}^{c} \sum_{i=1}^{r} (x_{ij} - \bar{x}_{.j})^2} \tag{8}$$

It is customary, in carrying out an ANOVA investigation, to display the essential information needed for the test in the form of a table. Since this table form has stayed essentially unchanged over the years, and is therefore well understood by individuals who are interested in ANOVA experiments, it enables a reader of a research article that uses ANOVA methods to understand the outcome of the research very quickly.

To simplify an ANOVA table it is convenient to introduce the following notation for the various sums of squares that occur in the analysis.

$$S_c = \sum_{i=1}^{r} \sum_{j=1}^{c} (\bar{x}_{.j} - \bar{x})^2 = r \sum_{j=1}^{c} (\bar{x}_{.j} - \bar{x})^2$$

$$S_e = \sum_{i=1}^{r} \sum_{j=1}^{c} (x_{ij} - \bar{x}_{.j})^2$$

$$S_T = \sum_{i=1}^{r} \sum_{j=1}^{c} (x_{ij} - \bar{x})^2$$

The first symbol, S_c, denotes the sum of squares of *column means.* The second symbol, S_e, is usually called the sum of squares for *experimental error.* It measures the variation in the data independent of any column mean differences, and hence represents the natural or experimental variation in the x_{ij} when H_0 is true. The third symbol, S_T, denotes the *total* sum of squares. It can be shown by means of some algebra that $S_c + S_e = S_T$. The total sum of squares is not needed here, but it will be needed in the next section. In terms of this notation, the ANOVA table assumes the form shown in Table 3.

Table 3

Source of variation	Sum of squares	Degrees of freedom	Mean square	F
Column means	S_c	$c-1$	$\frac{S_c}{c-1}$	$\frac{S_c}{c-1} \div \frac{S_e}{c(r-1)}$
Experimental error	S_e	$c(r-1)$	$\frac{S_e}{c(r-1)}$	
Totals	S_T	$cr-1$		

The ANOVA table for the typewriter experiment is shown in Table 4.

Table 4

Source of variation	Sum of squares	Degrees of freedom	Mean square	F
Column means	336	2	168	$\frac{168}{89.9} = 1.87$
Experimental error	1888	21	89.9	
Totals	2224	23		

5 TWO-WAY ANALYSIS OF VARIANCE

The foregoing analysis-of-variance problem was relatively simple because there was only one classification variable, namely, the brand of typewriter. In an experiment of this type one could consider many other variables that might influence the scores. For example, one might consider the different amounts of experience of the typists and whether those typists were familiar with the brand of machine assigned to them. Analysis-of-variance methods have been designed to treat any number of classification variables; however, the discussion here is limited to two such variables. The methods for more than two variables are very similar to those for two variables.

The problem that was considered in the preceding section can be modified to yield a problem that would normally be solved by the methods of this section. Assume that instead of having drawn a random sample of 24 typists from those available, a random sample of only 8 typists had been drawn. Then it would have been necessary for each of the 8 typists to become acquainted with all three brands of typewriters. After that, each typist would have been instructed to type the same amount of material with each of the three brands. Thus the three scores in the first row of Table 1 would represent the three scores made by the first typist. In such an experiment one would randomize the order in which the different typewriters were employed by the typists so that none of the brands would have an advantage with respect to practice.

The second variable of classification here is the individual. Since it is well known that there is large variation in the skill of individuals in typing, it would seem desirable to control this feature of the variability of scores so that any differences that might be caused by the different brands could be recognized. Large individual differences among the 24 typists of the earlier experiment might obliterate any moderate differences arising because of the different brands of typewriters.

In the two-variable analysis-of-variance setup one assumes that each of the variables x_{ij} in Table 2 is an independent normal variable with a common variance σ^2. Furthermore, one assumes that the row variables as well as the column variables do not change in repeated experiments. This means, for example, that in repeated typing experiments of the kind being discussed, the same 8 typists would use each of the three brands each time the experiment was run. This differs from the one-variable setup in which fresh sets of 24 typists would be selected each time. For the two-variable situation it is also necessary to make a few more assumptions about the basic variables, but these assumptions are not discussed here.

Just as in the earlier method, one finds two estimates of the common variance σ^2 and then uses the ratio of the two estimates to obtain an F value. The method of finding such estimates is based on taking the natural variance estimate and analyzing it into useful components. This procedure gave rise to the name "analysis of variance." The natural variance estimate for Table 2 is the quantity

$$\frac{\sum_{i=1}^{r}\sum_{j=1}^{c}(x_{ij}-\bar{x})^2}{rc-1}$$

Only the numerator of this estimate is used in the following analysis. Now, it can be shown by simple algebraic manipulations that the following formula holds:

$$\sum_{i=1}^{r}\sum_{j=1}^{c}(x_{ij}-\bar{x})^2 = \sum_{i=1}^{r}\sum_{j=1}^{c}(\bar{x}_{i\cdot}-\bar{x})^2 + \sum_{i=1}^{r}\sum_{j=1}^{c}(\bar{x}_{\cdot j}-\bar{x})^2 + \sum_{i=1}^{r}\sum_{j=1}^{c}(x_{ij}-\bar{x}_{i\cdot}-\bar{x}_{\cdot j}+\bar{x})^2 \tag{9}$$

It can also be shown that each of the three sums of squares on the right side of (9), if divided by the proper constant, is an unbiased estimate of σ^2 when it is assumed that there are no real differences in the row means or the column means. These three estimates are

$$V_r = \frac{\sum_{i=1}^{r}\sum_{j=1}^{c}(\bar{x}_{i\cdot} - \bar{x})^2}{r-1} = c\frac{\sum_{i=1}^{r}(\bar{x}_{i\cdot} - \bar{x})^2}{r-1}$$

$$V_c = \frac{\sum_{i=1}^{r}\sum_{j=1}^{c}(\bar{x}_{\cdot j} - \bar{x})^2}{c-1} = r\frac{\sum_{j=1}^{c}(\bar{x}_{\cdot j} - \bar{x})^2}{c-1}$$

$$V_e = \frac{\sum_{i=1}^{r}\sum_{j=1}^{c}(x_{ij} - \bar{x}_{i\cdot} - \bar{x}_{\cdot j} + \bar{x})^2}{(r-1)(c-1)}$$

The subscripts here on V refer to rows, columns, and experimental error. The expression for V_c is precisely the same as that for V_m given by (6). The expression for V_r is similar, except that it measures row variation rather than column variation; therefore, if there are large differences in typing skill among typists, this quantity will tend to be considerably larger than if there were no individual differences. The expression for V_e essentially measures the variation in the data after the variation caused by column differences and row differences has been eliminated. It serves as an estimate of σ^2 unaffected by brand differences and individual differences.

For the purpose of testing the hypothesis that all the theoretical column means are equal, the estimates to use are V_c and V_e. Thus the test reduces to computing the value of F given by

$$F = \frac{r(r-1)\sum_{j=1}^{c}(\bar{x}_{\cdot j} - \bar{x})^2}{\sum_{i=1}^{r}\sum_{j=1}^{c}(x_{ij} - \bar{x}_{i\cdot} - \bar{x}_{\cdot j} + \bar{x})^2} \tag{10}$$

where $\nu_1 = c - 1$ and $\nu_2 = (r-1)(c-1)$. The values of ν_1 and ν_2 in F are always the denominators needed to make the corresponding sums of squares unbiased estimates of σ^2. The mathematical theory for this problem shows that the F distribution is valid here whether or not there are real differences in the row means.

In the two-variable scheme it is also possible to test the hypothesis that the theoretical row means are equal, which would imply, for example, that there are no differences among the 8 typists with respect to typing skill in the preceding experiment. Here one uses the estimates V_r and V_e, and one forms the F ratio given by

(11)
$$F = \frac{c(c-1)\sum_{i=1}^{r}(\bar{x}_{i.} - \bar{x})^2}{\sum_{i=1}^{r}\sum_{j=1}^{c}(x_{ij} - \bar{x}_{i.} - \bar{x}_{.j} + \bar{x})^2}$$

where $\nu_1 = r - 1$ and $\nu_2 = (r - 1)(c - 1)$.

In view of the fact that the more sources of variation one can control in an experiment the more likely one is to detect differences of experimental interest; the F test based on eliminating variation due to individual differences in typing skill and given by formula (10) should be a more delicate test than the one used in section 4 and given by formula (8). For the purpose of comparing these two tests, consider the application of formula (10) to the data of Table 1.

The numerator sum of squares is the same as in the earlier test; therefore, its numerical value need not be computed. It is usually easier to compute the denominator sum of squares by means of formula (9) than it is to compute it directly from its definition. Thus one computes the left side of (9) as well as the first two sums of squares on the right side and then obtains the desired sum of squares by subtraction. Earlier computations gave

$$\sum_{i=1}^{8}\sum_{j=1}^{3}(\bar{x}_{.j} - \bar{x})^2 = 8(42) = 336$$

The first sum of squares on the right side of (9) was computed in the same manner as the second sum of squares. Computations for the data of Table 1 yielded the values

$$\sum_{i=1}^{8}\sum_{j=1}^{3}(x_{ij} - \bar{x})^2 = 2224$$

and

$$\sum_{j=1}^{3}\sum_{i=1}^{8}(\bar{x}_{i.} - \bar{x})^2 = 1768$$

Consequently, formula (9) yields the value

$$\sum_{i=1}^{8}\sum_{j=1}^{3}(x_{ij} - \bar{x}_{i.} - \bar{x}_{.j} + \bar{x})^2 = 2224 - 1768 - 336 = 120$$

The value of F as given by formula (10) then becomes

$$F = \frac{8 \cdot 7 \cdot 42}{120} = 19.6$$

where $\nu_1 = 2$ and $\nu_2 = 14$. From Table X it will be found that F_0, the 5 percent critical value of F corresponding to $\nu_1 = 2$ and $\nu_2 = 14$, is 3.74. Since $F = 19.6$ is in the critical region, the hypothesis H_0 is rejected.

The conclusion here is contrary to that made for the same data in section 3. Actually, the data for Table 1 were obtained for a group of 8 typists; consequently, only the second method is applicable here. The first method required that the scores be those of 24 randomly selected typists. The purpose of using the same data for both methods was to point out the similarities and the differences of the two methods and to stress the fact that it usually pays to introduce important classification variables in the analysis-of-variance technique if one wishes to obtain a delicate test for testing a set of theoretical means. If there had been no appreciable individual differences in typing skill, nothing would have been gained by designing the experiment to measure and eliminate this source of variation in the test, in which case the method in section 3 would have been preferable. The application of formula (11) shows, however, that $F = 29.5$, $\nu_1 = 7$, $\nu_2 = 14$. Since F_0, the 5 percent critical value of F, is 2.77, this means that there is large variation in the row means, hence large variation in individual typing skill.

For the purpose of constructing a two-way ANOVA table it is necessary to introduce some additional notation and change the definition of S_e:

$$S_r = \sum_{i=1}^{r} \sum_{j=1}^{c} (\bar{x}_{i\cdot} - \bar{x})^2 = c \sum_{i=1}^{r} (\bar{x}_{i\cdot} - \bar{x})^2$$

$$S_e = \sum_{i=1}^{r} \sum_{j=1}^{c} (x_{ij} - \bar{x}_{i\cdot} - \bar{x}_{\cdot j} + \bar{x})^2$$

With this notation the general two-way ANOVA table assumes the form shown in Table 5.

Table 5

Source of variation	Sum of squares	Degrees of freedom	Mean square	F
Column means	S_c	$c - 1$	$\dfrac{S_c}{c-1}$	$\dfrac{S_c}{c-1} \div \dfrac{S_e}{(c-1)(r-1)}$
Row means	S_r	$r - 1$	$\dfrac{S_r}{r-1}$	$\dfrac{S_r}{r-1} \div \dfrac{S_e}{(c-1)(r-1)}$
Error	S_e	$(c-1)(r-1)$	$\dfrac{S_e}{(c-1)(r-1)}$	
Totals	S_T	$rc - 1$		

The ANOVA table for the problem that was just solved is shown in Table 6.

Table 6

Source of variation	Sum of squares	Degrees of freedom	Mean square	F
Column means	336	2	168	$\frac{168}{8.57} = 19.6$
Row means	1768	7	252.6	$\frac{252.6}{8.57} = 29.5$
Error	120	14	8.57	
Totals	2224	23		

6 COMPUTING FORMULAS

If calculating equipment is available for obtaining the sums of squares that are needed to apply the F test, it is usually best to use formulas that require finding only sums and sums of squares of the entries of Table 2. These calculating formulas can be obtained by simple algebraic manipulations. The first two of the following formulas are needed for (8), whereas all the formulas with the exception of (13) are needed for (10) and (11).

$$\sum_{j=1}^{c} (\bar{x}_{\cdot j} - \bar{x})^2 = \frac{1}{r^2} \sum_{j=1}^{c} \left(\sum_{i=1}^{r} x_{ij} \right)^2 - \frac{1}{r^2 c} \left(\sum_{j=1}^{c} \sum_{i=1}^{r} x_{ij} \right)^2 \tag{12}$$

$$\sum_{j=1}^{c} \sum_{i=1}^{r} (x_{ij} - \bar{x}_{\cdot j})^2 = \sum_{j=1}^{c} \sum_{i=1}^{r} x_{ij}^2 - \frac{1}{r} \sum_{j=1}^{c} \left(\sum_{i=1}^{r} x_{ij} \right)^2 \tag{13}$$

$$\sum_{i=1}^{r} (\bar{x}_{i\cdot} - \bar{x})^2 = \frac{1}{c^2} \sum_{i=1}^{r} \left(\sum_{j=1}^{c} x_{ij} \right)^2 - \frac{1}{c^2 r} \left(\sum_{i=1}^{r} \sum_{j=1}^{c} x_{ij} \right)^2 \tag{14}$$

$$\sum_{i=1}^{r} \sum_{j=1}^{c} (x_{ij} - \bar{x})^2 = \sum_{i=1}^{r} \sum_{j=1}^{c} x_{ij}^2 - \frac{1}{rc} \left(\sum_{i=1}^{r} \sum_{j=1}^{c} x_{ij} \right)^2 \tag{15}$$

For the purpose of illustrating the use of these computational formulas, a second set of experimental data will be analyzed by means of the analysis-of-variance technique. The data are displayed in Table 7 and represent the yields of potatoes on four plots of ground, each of which was divided into five subplots.

Table 7

	Fertilizer				
Plot	*A*	*B*	*C*	*D*	*E*
1	310	353	366	299	367
2	284	293	335	264	314
3	307	306	339	311	377
4	267	308	312	266	342

For each plot five different fertilizers were assigned at random to the five subplots. The problem is to test whether the five fertilizers are equally effective with respect to mean yield.

Calculations yield the following sums:

$$\sum_{i=1}^{4}\sum_{j=1}^{5} x_{ij} = 6320, \qquad \sum_{i=1}^{4}\sum_{j=1}^{5} x_{ij}^2 = 2{,}018{,}650$$

Additional calculations yield the values

$$\sum_{j=1}^{5}\left(\sum_{i=1}^{4} x_{ij}\right)^2 = 8{,}039{,}328, \qquad \sum_{i=1}^{4}\left(\sum_{j=1}^{5} x_{ij}\right)^2 = 10{,}017{,}750$$

As a result, formulas (12), (14), and (15) yield

$$\sum_{j=1}^{5} (\bar{x}_{.j} - \bar{x})^2 = \frac{8{,}039{,}328}{16} - \frac{(6320)^2}{80} = 3178$$

$$\sum_{i=1}^{4} (\bar{x}_{i.} - \bar{x})^2 = \frac{10{,}017{,}750}{25} - \frac{(6320)^2}{100} = 1286$$

$$\sum_{i=1}^{4}\sum_{j=1}^{5} (x_{ij} - \bar{x})^2 = 2{,}018{,}650 - \frac{(6320)^2}{20} = 21{,}530$$

The denominator sum of squares in (10) can now be obtained by means of formula (9). Thus

$$\begin{aligned}\sum_{i=1}^{4}\sum_{j=1}^{5} (x_{ij} - \bar{x}_{i.} - \bar{x}_{.j} + \bar{x})^2 &= 21{,}530 - 5(1286) - 4(3178) \\ &= 2388\end{aligned}$$

The value of F in (10) is therefore given by

$$F = \frac{4 \cdot 3 \cdot 3178}{2388} = 16.0$$

where $\nu_1 = 4$ and $\nu_2 = 12$.

From Table X it will be found that the 5 percent critical value of F corresponding to $\nu_1 = 4$ and $\nu_2 = 12$ is 3.26. Since $F = 16.0$ is in the critical region, the hypothesis that the fertilizers are equally effective is rejected. Fertilizers E and C appear to be superior to the other fertilizers. Further experimentation and additional tests might well give more precise information concerning the relative superiority of E and C to the rest and also determine the rating of C relative to E.

The method presented in section 4 is easily generalized to apply to the situation in which the number of measurements in the various columns is not the same. It is also fairly easy to generalize the methods of section 5 so that they apply to the situation in which one has several measurements in each cell of Table 2. A more complicated generalization occurs when one extends these methods to the situation in which there are several variables of classification; however, the methods are quite similar to those just explained.

7 REVIEW ILLUSTRATIONS

Example 1. The following data are from a chemical manufacturing plant and give the yield of a chemical product that resulted from trying four different catalysts in the chemical process. (*a*) Construct an ANOVA table for these data. (*b*) Use formula (8) to test to see whether yields are influenced by the catalysts.

I	II	III	IV
36	35	35	34
33	37	39	31
35	36	37	35
34	35	38	32
32	37	39	34
34	36	38	33

(*a*) The values of $\bar{x}_{.j}$ for $j = 1, 2, 3, 4$ are 34, 36, $37\frac{2}{3}$, and $33\frac{1}{6}$, and the value of $\bar{x}$ is 35.21. These numbers give

$$\sum_{j=1}^{4} (\bar{x}_{.j} - \bar{x})^2 = 12.33$$

Additional calculations give

$$\sum_{j=1}^{4} \sum_{i=1}^{6} (x_{ij} - \bar{x}_{.j})^2 = 36.17$$

The ANOVA table then assumes the form

Source of variation	Sum of squares	Degrees of freedom	Mean square	F
Column means	6(12.33)	3	24.66	$\frac{24.66}{1.81} = 13.6$
Experimental error	36.17	20	1.81	
Totals	110.15	23		

(*b*) Since $\nu_1 = 3$, $\nu_2 = 20$, $F_0 = 3.10$, and $F = 13.6$, the hypothesis of column means being equal is rejected. The yield is clearly affected by the kind of catalyst used.

Example 2. A buyer for a company that uses considerable quantities of fabric in processing was confronted with products made of 4 different types of fibers offered by 4 different manufacturers, each claiming a superior product, whatever the type of fabric being considered. The buyer asked his company's research department to help resolve the problem. The research department obtained specimen products of each fabric type from each manufacturer and made a test run on a quality measure with the following results.

Manufacturer	Fabric type			
	1	2	3	4
A	40	41	46	34
B	28	37	42	22
C	31	40	45	25
D	46	47	52	40

(*a*) Construct a two-way ANOVA table for these data. (*b*) Test the hypothesis that there is no difference in fabric types. (*c*) Test the hypothesis that there is no difference in the quality of the four manufacturers.

(*a*) Calculations give

$$\bar{x}_{1\cdot} = 40.25, \quad \bar{x}_{2\cdot} = 32.25, \quad \bar{x}_{3\cdot} = 35.25, \quad \bar{x}_{4\cdot} = 46.25$$

$$\bar{x}_{\cdot 1} = 36.25, \quad \bar{x}_{\cdot 2} = 41.25, \quad \bar{x}_{\cdot 3} = 46.25, \quad \bar{x}_{\cdot 4} = 30.25$$

and $\bar{x} = 38.50$. These values result in

$$\sum_{i=1}^{4} (\bar{x}_{i\cdot} - \bar{x})^2 = 112.75, \qquad \sum_{j=1}^{4} (\bar{x}_{\cdot j} - \bar{x})^2 = 140.75$$

Additional calculations give

$$\sum_{i=1}^{4} \sum_{j=1}^{4} (x_{ij} - \bar{x})^2 = 1078$$

Hence

$$\sum_{i=1}^{4} \sum_{j=1}^{4} (x_{ij} - \bar{x}_{i\cdot} - \bar{x}_{\cdot j} + \bar{x})^2 = 1078 - 4(112.75) - 4(140.75)$$
$$= 64$$

The ANOVA table then assumes the form

Source of variation	Sum of squares	Degrees of freedom	Mean square	F
Column means	4(140.75)	3	187.67	$\frac{187.67}{7.11} = 26.4$
Row means	4(112.75)	3	150.33	$\frac{150.33}{7.11} = 21.1$
Error	64	9	7.11	
Totals	1078	15		

(*b*) Since $\nu_1 = 3$, $\nu_2 = 9$, $F_0 = 3.86$, and $F = 26.4$, the hypothesis of no fabric differences is rejected.

(*c*) Since $\nu_1 = 3$, $\nu_2 = 9$, $F_0 = 3.86$, and $F = 21.1$, the hypothesis of no manufacturer differences is rejected.

EXERCISES

SECTIONS 3 and 4

1 Four modifications of a drug that is often used to decrease the blood pressure of patients with high blood pressure were given to 10 patients each. The following table summarizes the data that resulted from this experiment. The variable here is the amount that the patient's blood pressure decreased after medication. Determine whether there are significant differences due to the four modifications.

Modification	I	II	III	IV
Sample mean, $\bar{x}_j$	11	9	14	10
Sample variance, s_j^2	25	30	34	28

2 A firm has 3 branches located in different regions but doing rather similar work. An efficiency study was carried out to see if workers of a particular classification differed in effectiveness among the branches. A sample of 50 random time intervals was taken from each branch and certain relevant observations of activities going on during these periods were recorded with the following results.

	Branch		
	A	*B*	*C*
Sample mean, $\bar{x}_j$	49	54	47
Sample variance, s_j^2	270	200	220

Using the *F* test, determine if there is a significant difference in workers' effectiveness among the three branches using an α of .05.

3 The following data give the yields of wheat on some experimental plots of ground corresponding to 4 different sulfur treatments for the control of rust. The treatments consisted of (1) dusting before rains, (2) dusting after rains, (3) dusting once each week, and (4) no dusting. Test to see if there are significant differences in yields due to the dusting methods. Display your results in an ANOVA table.

Plot	Dusting method			
	1	2	3	4
1	5.3	4.4	8.4	7.4
2	3.7	5.1	6.0	4.3
3	14.3	5.4	4.9	3.5
4	6.5	12.1	9.5	3.8

4 In problem 3 use Student's t test to determine whether dusting had any effect upon yield. That is, combine the data for the first three columns and treat the measurements as though they were a sample of size 12 from a normal population. The last-column measurements may be treated as a sample of size 4 from a second normal population.

5 For the purpose of determining the best of 4 proposed schemes for estimating the value of a complicated manufactured product in various stages of fabrication, the control department of a large firm randomly selected 16 men from its staff and split them into 4 teams of 4 each. Each scheme was assigned to 1 of the 4 teams by chance. The following data give the estimates arrived at (in coded values) by these 4 teams in a sequence of 5 such experiments. On the basis of these data, can you conclude that there are no significant differences in these schemes? Display your results in an ANOVA table.

Run	Scheme			
	A	B	C	D
1	55	61	42	169
2	49	113	97	137
3	42	30	81	169
4	21	89	95	85
5	52	63	92	154

6 Two sets of 50 school children were taught to read by 2 different methods. After instruction was completed, a reading test was given and produced the following data: $\bar{x}_1 = 73.4$, $\bar{x}_2 = 70.3$, $s_1 = 8$, $s_2 = 10$. Test the hypothesis that $\mu_1 = \mu_2$ by means of ANOVA.

7 Fill in the missing values in the following incomplete ANOVA table. Find the value of F.

Source	SS	DF	MS	F
Column means		2		
Error		9	10	
Totals	120			

8 Given the algebraic identity

$$x_{ij} - \overline{x} = (x_{ij} - \overline{x}._j) + (\overline{x}._j - \overline{x}), \qquad \begin{array}{l} i = 1, \cdots, r \\ j = 1, \cdots, c \end{array}$$

square both sides and sum over i from 1 to r, and then over j from 1 to c. Show that the cross-product term on the right sums to zero. The resulting relationship will then justify the statement made in the text, just before Table 3, that $S_T = S_c + S_e$.

SECTION 5

9 The following data represent the number of units of production per day turned out by 5 workmen using 4 different types of machines.
   (*a*) Test to see whether the mean productivity is the same for the 4 different machine types.
   (*b*) Test to see whether the 5 men differ with respect to mean productivity. Use formulas (10) and (11) here.

	Machine type			
Workman	1	2	3	4
1	44	38	47	36
2	46	40	52	43
3	34	36	44	32
4	43	38	46	33
5	38	42	49	39

10 Work problem 3 by the two-variable method to see whether eliminating the variation due to plots will affect the test.

11 For the data of Table 7 in the text, test to see whether there are plot differences in yield.

12 The following data give the yields of a chemical in a chemical reaction in which 2 temperatures and 3 concentrations of solvent are used in an attempt to increase the yield. Does temperature affect the yield? Does the concentration of solvent affect the yield? Display your results in an ANOVA table.

	Concentration of solvent		
Temperature	40%	50%	60%
50°	45.1	45.7	44.9
60°	44.8	45.8	44.7

13 The following data give the results of a test made to determine the relative durability of tires made from 3 different types of raw materials. Tires were purchased from 4 different manufacturers, all of whom made tires of the 3 types.
   (*a*) Construct an ANOVA table for these data.
   (*b*) Test to see whether raw materials affect durability.
   (*c*) Test to see whether brands differ with respect to durability.

Brand	Rubber type		
	Natural	Synthetic *A*	Synthetic *B*
I	38	42	43
II	35	38	47
III	35	40	45
IV	40	44	45

14 The following data are from a study made by Griffith, Westman, and Lloyd in the May 1948 issue of *Industrial Quality Control* concerning the calorific value of city gas from day to day over a period of 9 weeks. These are coded values.

Day	Week								
	1	2	3	4	5	6	7	8	9
Mon.	5	1	−4	5	−13	−8	−2	−4	−10
Tues.	3	6	−10	−2	−7	−2	−4	2	2
Wed.	8	4	−14	−3	3	0	5	−11	−12
Thurs.	8	10	−5	−1	4	−2	4	1	−12
Fri.	4	−1	7	−5	5	−3	−7	−3	−6
Sat.	3	−9	3	−8	−6	0	−3	8	−1

(*a*) Is there significant variation from week to week?
(*b*) Is there significant variation from day to day?

SECTION 7

15 Three promotion schemes were tested in a marketing experiment by assigning each scheme to 3 market areas in a city that was divided into 9 market areas. These assignments were made by random selection. The sales results from this experiment, in coded units, are as follows.

Scheme 1	4, 3, 5
Scheme 2	9, 6, 6
Scheme 3	9, 9, 12

Test the hypothesis that there are no significant differences among these 3 schemes.

16 The following data give the scores made by 24 soldiers in an experiment to determine whether shooting accuracy is affected by methods of sighting (*a*) with the right eye open, (*b*) with the left eye open, (*c*) with both eyes open. Test to see whether accuracy is influenced by the type of sighting.

Right	Left	Both
44	40	51
39	37	47
33	28	37
56	53	52
43	38	42
56	51	63
47	45	46
58	60	62

17 An experiment to determine whether the yield of a chemical process can be increased by changing the temperature of the reaction and by changing the amount of a catalyst was carried out with the following results. The entries in the table represent yields under 3 temperatures and 4 catalyst choices.
(*a*) Test to see whether the yield is affected by varying the catalyst.
(*b*) Test to see whether the yield is affected by varying the temperature.

	Catalyst			
Temperature	1	2	3	4
A	53	59	58	50
B	57	65	62	60
C	52	62	54	52

18 If the number of measurements in the *j*th column of Table 2 in the text is denoted by n_j and if the n_j are not all equal, then formula (7) in the text must be replaced by the formula

$$V_c = \frac{\sum_{j=1}^{c} \sum_{i=1}^{n_j} (x_{ij} - \bar{x}_{.j})^2}{\sum_{j=1}^{c} (n_j - 1)}$$

(*a*) Explain how this formula reduces to (7) when the n_j are equal.
(*b*) Show that this formula is a weighted mean of the column variances.

19 For the situation of problem 18, formula (6) in the text must be replaced by the formula

$$V_m = \frac{\sum_{j=1}^{c} n_j(\bar{x}_{.j} - \bar{x})^2}{c - 1}$$

(*a*) Explain the equivalence of this formula to (6) when the n_j *are equal.*
(*b*) Explain why this formula is the natural generalization of (6) for unequal n_j.

20 For the following data on the scores made by pupils taught arithmetic by 3 different methods, use the formulas of problems 18 and 19 to form the appropriate *F* test and test for equality of column means.

Methods		
116	132	108
117	137	96
138	131	131
100	108	130
125	111	111
130	130	126
134	140	
124		
114		

Chapter 14 Survey Designs

1 INTRODUCTION

In the preceding chapters the main objective was to present the basic concepts of statistical methods and show how those concepts can be used to solve certain classes of problems in economics and business. In the illustrations and exercises to which the methods were applied it was assumed that the data available were appropriate for the analysis employed.

In dealing with real-life problems, the data at hand do not always fit the assumptions that are needed for the contemplated theory, and therefore if valid conclusions are to be obtained it is necessary to collect data that are appropriate. It is the purpose of this chapter to study the data collection problem and to present principles that will enable the investigator to obtain data to which statistical techniques can be applied successfully. Moreover, since there are usually a number of valid methods for obtaining satisfactory data, and these may differ considerably with respect to costs in money, time, or resources, it will also be the purpose of this chapter to compare the relative advantages of these various methods.

2 THE STATISTICAL SURVEY

A statistical survey is an investigation of certain quantitative characteristics of a population. Perhaps the largest and best-known statistical survey in the United States is the Population Census, which in 1970 involved the collection of various data from sixty-three million households in the United States. One part of this survey, counting the number of persons residing in each of the states, is required by the U.S. Constitution for determining the number of representatives each state is permitted in Congress. Opinion and attitude polls are examples of statistical surveys, the object of which is to determine the proportion of individuals who hold certain opinions or attitudes concerning political candidates or current issues. In business organizations examples of statistical surveys include investigations of the quality and quantity of services or goods produced, consumer preferences, employee attitudes, inventory determination, customer behavior, advertising effectiveness, TV ratings, and many others.

The survey may be based on either a complete census or a sample, depending upon the size of the population and the relative costs involved. When great accuracy is not necessary in determining some characteristic of the population, a sample will usually be much cheaper than a census, even for fairly small populations. Furthermore, samples often yield more accurate estimates than a census because of the numerous errors that arise in taking a census. If the investigation is to be based on a sample, then the problem arises as to how this sample should be taken.

In most investigations there are usually several alternative ways of choosing the sample that need to be compared for relative efficiency and cost. As an example, suppose a business firm with 200 supervisors and 2000 employees wishes to know what its employees think of a contemplated health benefit program. Since interviewing each of the 2000 members would be too costly and great accuracy is not needed, a sample should suffice. Some of the possible sampling alternatives that come to mind are the following:

1. Prepare a written questionnaire and send it to all employees. Experience indicates that about 30 percent will return usable questionnaires. Thus, a sample of about 600 could be expected.
2. Interview a random sample of 200 employees. Experience indicates that about 95 percent will cooperate; hence a sample of about 190 would be obtained.
3. Post notices throughout the plant presenting the proposed program and inviting employees to submit comments and reactions to the director of personnel or to their union representative.
4. Interview 20 supervisors regarding the opinions of their employees.

It would be rather difficult to determine which of these methods is the best because a relatively cheap method may give biased estimates and therefore it would be necessary to weigh cost against accuracy. If, for example, all that was required was a rough estimate of the percentage of employees who favored the proposed plan, a cheap but biased method might prove satisfactory. In any case, before such a judgment can be made, it is necessary to be able to state what the accuracy is in any proposed sampling method.

When applying sampling principles to practical problems it is helpful to distinguish between the unit on which a measurement is made and the sampling unit that is selected. For example, each of the 2000 employees is a unit of our population of interest, but a sampling unit need not consist of a single employee. For the purpose of taking a sample it is necessary to have a list of all employees. This could be a set of cards in a file or in a computer memory. The list could be ordered alphabetically, or by department, or by some other grouping. This physical set of cards, or names, can be regarded as the *sampling frame* and will have a one-to-one correspondence with the set of population units. However, it is also possible to consider a frame of 200 supervisor names, each of which supervises 10 employees. This frame can be regarded as consisting of 200 frame units of 10 population units each. This is an illustration in which frame units, which are easily sampled, are not identical with population units. It is not obvious here which is the more efficient method of sampling–using a frame based on individuals or a frame based on supervisors. One of the purposes of this chapter is to suggest some principles that may assist one in making such decisions.

As an illustration of the difficulties and possibilities that can arise, suppose the business firm that was concerned about the reception of its new health benefit program contained 20 branches and each branch consisted of 100 employees. Furthermore, assume that each branch was organized into four departments A, B, C, and D, with 10, 20, 30, and 40 employees, respectively, in those departments. Then it is clear that there are several ways of drawing samples from the population of 2000 employees. Since the sampling unit here is the individual employee, it would be necessary to associate one of the numbers from 1 to 2000 with each employee before a sampling scheme could be carried out. If the sampling unit is the supervisor, then there are 200 units rather than 2000 and each sampling unit would contain 10 employees.

In the next few sections, various possibilities for selecting samples will be studied, with a view to discovering which are the most useful techniques to employ in a given situation. All of them will involve random sampling at some stage of the sampling procedure, otherwise, as was pointed out in an earlier chapter, it would not be possible to determine the accuracy of the sample.

3 COMPLETELY RANDOMIZED DESIGN

If the set of population units is also the set of frame units, the selection of a random sample of any desired size is simple. Methods for doing this were given in section 2 of Chapter 6.

As an illustration, suppose the only objective of the business firm that is investigating its health benefit program is to estimate the proportion of its employees who like the program. Then under a completely randomized design it would merely need to use a table of random numbers associated with the numbers from 0000 to 1999 to select a random sample of any desired size. Duplicated numbers would, of course, be discarded so that the sampling will be without replacement. In this sampling scheme all employees are essentially treated as though they were in one large group.

After having obtained a random sample of n population units in this manner, one can proceed with estimation or tests of hypotheses in the manner presented in Chapters 7 and 8. If the size of the population is N and the size of the sample is n, then it follows from formula (9) of Chapter 6 that the standard error of the sample mean, $\bar{x}$, is

$$\sigma_{\bar{x}} = \frac{\sigma}{\sqrt{n}}\sqrt{\frac{N-n}{N-1}}$$

A 95 percent confidence interval for μ, for example, is given by

$$\bar{x} - 1.96\sigma_{\bar{x}} < \mu < \bar{x} + 1.96\sigma_{\bar{x}} \tag{1}$$

It is usually necessary to approximate σ by s, in which case (1) would be only an approximate confidence interval.

In a similar manner, the standard error of a sample proportion, $\hat{p}$, as given by formula (10) of Chapter 6, is

$$\sigma_{\hat{p}} = \sqrt{\frac{pq}{n}}\sqrt{\frac{N-n}{N-1}}$$

If p is approximated by $\hat{p}$ in this formula, an approximate 95 percent confidence interval for the population proportion is given by

$$\hat{p} - 1.96\sigma_{\hat{p}} < p < \hat{p} + 1.96\sigma_{\hat{p}} \tag{2}$$

where p is replaced by $\hat{p}$ in the formula for $\sigma_{\hat{p}}$.

Since problems of finding confidence intervals for means and proportions were illustrated and solved in Chapter 7, no illustrations of these techniques are presented here.

4 STRATIFIED RANDOM SAMPLING

In the preceding problem of determining the proportion of employees favoring a proposed program, if a random sample produced, say, twice as many employees from one branch of the firm as from another, the management might be unhappy and distrustful of the sampling results. They would undoubtedly prefer a sampling scheme in which the various branches of the company are represented equally. Thus, if a sample of 400 were to be taken, this would require that a sample of size 20 be taken from each of the 20 branches. Sampling in which the population is divided into groups, or *strata,* and samples are taken separately from each group is called *stratified sampling.* If the size of the sample to be extracted from each group is proportional to the number of individuals in the group, the sampling is called *proportional sampling.* Thus, proportional sampling is a special kind of stratified sampling and the type that is usually employed in stratified sampling, if it is possible to do so.

Proportional sampling has the advantage not only of appealing to one's sense of fairness, but it can also be shown mathematically that the variance of $\bar{x}$ or $\hat{p}$ is usually smaller for proportional sampling, if selection within strata is random, than for a completely random sample of the same size.

In the case of the multibranch firm there are two natural choices of strata available: branches and departments. As indicated earlier, stratification with respect to branches would require samples of 20 from each of the 20 branches if a sample of 400 were to be taken. If the stratification were made with respect to departments, a sample of 400 would require the firm to choose random samples of sizes 40, 80, 120, and 160 from departments *A, B, C,* and *D*, respectively, provided that it wished to take a proportional sample, because they correspond to the relative sizes of the departments. The selection within departments would then be made randomly over all 20 branches. It seems clear that it would be more difficult to take random samples from such strata than from the strata consisting of branches. A third possibility, of course, is to stratify with respect to both branch and department, in which case a random sample of size 5 might be drawn from each of the 80 such strata.

It is often much easier and cheaper to choose random samples within natural strata than it is to ignore such strata and take a single random sample over the entire population. Thus, in the preceding illustration it should be simpler to let each branch of the firm choose a random sample of a fixed size than to take a random sample from the entire organization.

As indicated earlier, stratified sampling usually leads to estimates that are more accurate than those obtained under simple random sampling, which therefore may give it a double advantage over simple random sampling. The advantage in accuracy is most pronounced when the individuals in a stratum are very homogeneous but the strata differ considerably. For example, if the problem were to estimate the mean wage of the employees in a factory that employs 100 semiskilled workers and 40 highly skilled workers, with the latter group receiving a considerably higher wage, then proportional sampling would yield a much smaller variance for $\bar{x}$ than simple random sampling, provided that all workers in a group receive approximately the same basic wage. Formulas that measure the advantage of stratified sampling over pure random sampling are considered next.

Suppose a population of N units is divided into L strata. Let N_h denote the number of population units in the hth stratum, μ_h the stratum mean, and σ_h^2 its variance. Then $N = \Sigma_{h=1}^{L} N_h$. Let a random sample of size n_h be taken from the hth stratum. This constitutes a stratified-random sample of size $n = \Sigma_{h=1}^{L} n_h$ for the entire population. A weighted estimate of the population mean μ, based on this stratification, is

$$\bar{x} = \sum_{h=1}^{L} \frac{N_h}{N} \bar{x}_h \tag{3}$$

where $\bar{x}_h$ is the sample mean of stratum h. This estimate is unbiased and has the variance given by the formula

$$V_s(\bar{x}) = \sum_{h=1}^{L} \left(\frac{N_h}{N}\right)^2 \frac{\sigma_h^2}{n_h} \tag{4}$$

This formula assumes that the strata are sufficiently large to ignore the finite population correction factor.

In the preceding formulas, n_h can be chosen as desired; however, it can be shown that for a given total sample of size n, the best choice for n_h is given by

$$n_h = n \frac{N_h \sigma_h}{\sum_{h=1}^{L} N_h \sigma_h} \tag{5}$$

This choice is best in the sense that it minimizes the variance in formula (4). This allocation of stratum sample sizes is referred to as *optimal stratification*. It follows from formula (5) that proportional sampling will be optimal if the σ_h are equal.

Unfortunately, the values of the σ_h are seldom available in a practical situation. If this is the case, one should ignore formula (5) and choose the n_h under proportional sampling, that is, choose $n_h = nN_h/N$. This gives the estimate

$$\bar{x} = \frac{\sum_{h=1}^{L} \sum_{j=1}^{n_h} x_{hj}}{n} \tag{6}$$

The variance of this estimate is

$$V_P(\bar{x}) = \sum_{h=1}^{L} \frac{N_h}{N} \frac{\sigma_h^2}{n} \tag{7}$$

Since the σ_h are usually unknown, it is necessary for calculating this variance to replace the σ_h by their sample estimates s_h.

The formula that relates the variance of $\bar{x}$ under proportional stratified sampling to its variance under simple random sampling is given by

$$V_P(\bar{x}) = V_R(\bar{x}) - \sum_{h=1}^{L} (\mu_h - \mu)^2 \frac{\pi_h}{n} \tag{8}$$

In this formula L denotes the number of strata, n denotes the total size of the sample being taken, μ_h is the mean of x in the hth stratum, μ is the mean of x for the entire population, and $\pi_h = N_h/N$ is the proportion of the total population located in the hth stratum. If the various strata possess the same mean, the two variances will be equal; however, if this is not the case then $V_P(\bar{x})$ will be smaller than $V_R(\bar{x})$ by the amount given in the formula. This formula is also valid for discrete variables, such as the binomial, in which case $\bar{x}$ would be replaced by $\hat{p}$ and μ_h by p_h, resulting in the formula,

$$V_P(\hat{p}) = V_R(\hat{p}) - \sum_{h=1}^{L} (p_h - p)^2 \frac{\pi_h}{n}$$

As an illustration, suppose a sample of size 200 is to be taken to estimate the percentage of employees who would like to start their working day an hour earlier. Suppose that 60 percent of the employees are males and 40 percent are females and that 20 percent of the males and 70 percent of the females actually do favor starting earlier. Here there are two strata with $n = 200$, $\pi_1 = .6$, $\pi_2 = .4$, $p_1 = .2$, $p_2 = .7$, $p = \pi_1 p_1 + \pi_2 p_2 = .40$, and

$$V_R(\hat{p}) = \frac{pq}{n} = .0012$$

Calculations give

$$\sum_{h=1}^{2} (p_h - p)^2 \frac{\pi_h}{200} = .0003$$

As a result, the modification of formula (8) for proportions gives

$$V_P(\hat{p}) = .0012 - .0003 = .0009$$

Hence,

$$\frac{V_P(\hat{p})}{V_R(\hat{p})} = \frac{.0009}{.0012} = \frac{3}{4}$$

This shows that the variance under proportional sampling based on formula (6) is only three-fourths as large as under simple random sampling. This large improvement occurred because the two strata differed considerably.

5 CLUSTERED RANDOM SAMPLING

There are situations in which it is much easier and cheaper to select samples in groups, or clusters, rather than selecting them individually. For example, suppose a sample of gidgets is to be checked for their quality. If the gidgets are boxed in sets of 24, it is more convenient to inspect all 24 items in a box, once it has been located and opened, than to inspect just isolated items that have been drawn in a simple random sampling of items. Similarly, in attempting to obtain audience ratings of TV programs over the week, it is less expensive to observe the TV viewing of the same household over a week rather than to choose a fresh household each day for observation. Such sampling is commonly called *cluster sampling.*

When clusters contain a fixed number of elements, for example, 24 gidgets in a box, only a slight modification of simple random sampling theory is needed to determine satisfactory estimates of means and their sampling variability. The theory is considerably more complex when the number of elements in clusters is not constant; therefore that case will not be considered here. Another sophisticated version of cluster sampling, called two-stage cluster sampling, occurs when only a fraction of each cluster is sampled. It also will not be studied here.

Suppose that a population of M units is divided into N clusters of M_0 units each. Let x represent the variable being studied in this population and let μ and σ denote its mean and standard deviation. Furthermore, let x_{ij} denote the observed value of x for element j in the ith cluster.

If a random sample of n clusters is taken from the population of N clusters, an estimate of μ is given by the grand mean of all the elements obtained, that is, by

$$\bar{x} = \frac{\sum_{i=1}^{n} \sum_{j=1}^{M_0} x_{ij}}{nM_0} \tag{9}$$

For the purpose of obtaining cluster sampling formulas, consider each cluster as an element in a population of N clusters. Each cluster will have its own mean to represent it. Let μ_i denote the mean of the ith cluster. This entire set of N cluster means will have a mean and a standard deviation, which will be denoted by μ_c and σ_c. However, since all the clusters have the same number of units, the mean μ of the population of M units must be the same as the mean of the N cluster means; hence

$$\mu_c = \mu$$

If all the cluster means were available, σ_c could be calculated; however, only a sample of n such cluster means is available. Therefore a sample estimate of σ_c must be substituted for σ_c. Such an estimate is given by

$$s_c = \sqrt{\frac{\sum_{i=1}^{n} (\bar{x}_i - \bar{x})^2}{n - 1}}$$

where $\bar{x}_i$ denotes the mean of the ith cluster that was sampled. Each $\bar{x}_i$ is one of the μ_i $(i = 1, \ldots, N)$, but which one it is is unknown.

Since formula (9) can be expressed in the form

$$\bar{x} = \frac{\sum_{i=1}^{n} \bar{x}_i}{n}$$

it follows from properties of variances that

$$V_c(\bar{x}) = \frac{\sum_{i=1}^{n} V(\bar{x}_i)}{n^2} = \frac{\sum_{i=1}^{n} \sigma_c^2}{n^2} = \frac{\sigma_c^2}{n} \tag{10}$$

This formula assumes that N is so large relative to n that the finite population correction factor may be ignored.

A comparison of the accuracy of $\bar{x}$ as an estimate of μ when the estimate is based on a simple random sample or on a cluster sample of the same size cannot be made unless the values of σ and σ_c are available. The variance of $\bar{x}$ for a random sample of n clusters is, by formula (10),

$$V_c(\bar{x}) = \frac{\sigma_c^2}{n} \tag{11}$$

whereas for a simple random sample of nM_0 elements it is given by

$$V(\bar{x}) = \frac{\sigma^2}{nM_0} \tag{12}$$

If the clusters are formed by taking M_0 random units from the population of M units, then each cluster mean is merely the mean of a random sample of size M_0 and

$$\sigma_c^2 = \frac{\sigma^2}{M_0} \tag{13}$$

As a result, it follows from (11), (12), and (13) that $V_c(\bar{x}) = V(\bar{x})$ under these circumstances.

A more common situation, however, is one in which the cluster elements are quite homogeneous, but the cluster means differ considerably. Then the value of σ_c^2 will exceed the value given by formula (13), and cluster sampling will not be as accurate as simple random sampling. Since cluster samples are often so much easier to obtain than random samples, it may be more economical, in spite of this loss of accuracy, to take a large cluster sample rather than a smaller random sample that yields the same accuracy of estimate.

The preceding formulas apply equally well to a sample proportion $\hat{p}$, in which case $\sigma = \sqrt{pq}$.

Example 1. Suppose an inspector is checking on the weight of pound boxes of brown sugar that are shipped in cartons containing 40 boxes each. Suppose, further, that he takes a sample of 20 cartons from a large shipment and finds that the mean weight for all 800 boxes is 15.9 ounces. (*a*) If past experience has shown that the standard deviation of cluster means is approximately .03 ounce, how accurate is $\bar{x}$ as an estimate of μ? (*b*) If experience has shown that the standard deviation of box weights for simple random sampling is approximately .1 ounce, how large a simple random sample would yield the same accuracy of estimate as that given in (*a*)?

(*a*) Here $n = 20$, $M_0 = 40$, $\sigma_c = .03$, $\sigma = .1$, and $\bar{x} = 15.9$. It therefore follows from formula (11) that the standard deviation of $\bar{x}$ for cluster sampling is given by

$$\sigma_c(\bar{x}) = \sqrt{V_c(\bar{x})} = \frac{\sigma_c}{\sqrt{n}}$$

$$= \frac{.03}{\sqrt{20}} = .0067$$

(*b*) If m denotes the size of a simple random sample that will be just as accurate as the preceding cluster sample, then m must satisfy the equation

$$.0067 = \frac{\sigma}{\sqrt{m}} = \frac{.1}{\sqrt{m}}$$

The solution of this equation is $m = 223$; therefore a simple random sample of 223 boxes would do just as well as the 800 boxes obtained in cluster sampling. It is clear, however, that it is a simple matter to weigh 20 cartons of boxes without even opening them, if the weight of an empty carton is known, as contrasted to opening cartons and extracting possibly only one box from each carton for weighing in pure random sampling. The savings in inspection costs would be considerable here. It should be noted that the inspector would be justified in claiming that the boxes in this shipment are underweight.

Example 2. A few years ago the U.S. Bureau of the Census investigated the effects of various cluster sizes on the accuracy of estimating the means of several household characteristics. Among those characteristics were the proportion of unemployed males and the proportion of homes owned. The results of that investigation with respect to those two characteristics are given in the following table, in which the entries give the ratio of the variance of $\hat{p}$ for cluster sampling to the variance of $\hat{p}$ for the same size simple random sample, that is, the ratio $\sigma_c^2 \div \sigma^2/M_0$.

	Cluster size, M_0				
	1	3	9	27	62
Proportion of unemployed	1.00	1.12	1.56	2.17	3.08
Proportion of homes owned	1.00	1.34	2.37	5.30	6.85

(*a*) Suppose a simple random sample of 2700 households is selected and a sample of 100 clusters with 27 households in each cluster is also selected. How would the standard error of the estimated proportion of unemployed compare in cluster sampling with that of simple random sampling? (*b*) How many clusters of size 9 would need to be taken to obtain the same accuracy in estimating the proportion of homeowners as a simple random sample of size 500? (*c*) Suppose that for the same cost, only half as many households can be sampled under simple random sampling as under cluster sampling when the cluster size is 27. Which type of sampling is more accurate for estimating the unemployment proportion? The answers follow.

(*a*) For a simple random sample of size 2700, the standard error of $\hat{p}$ is given by

$$\sigma(\hat{p}) = \frac{\sigma}{\sqrt{2700}} = .019245\sigma \tag{14}$$

For cluster sampling in which $M_0 = 27$ and $n = 100$, formula (11) gives

$$\sigma_c(\hat{p}) = \frac{\sigma_c}{\sqrt{100}} = .10\sigma_c \tag{15}$$

But from the table of values, it follows that

$$\sigma_c^2 = 2.17\frac{\sigma^2}{27} = .08037\sigma^2$$

Hence,

$$\sigma_c = .2835\sigma$$

and therefore, from (15),

$$\sigma_c(\hat{p}) = .02835\sigma$$

Comparing this result with that given in (14) for simple random sampling, it follows that the standard error for this type of cluster sampling is about 1.47 times as large as for a simple random sample of the same size.

(*b*) To determine the value of n, it is necessary to equate the variances under these two types of sampling and solve for n. From formulas (11) and (12), this equation is

$$\frac{\sigma_c^2}{n} = \frac{\sigma^2}{500}$$

But for $M_0 = 9$, the table of values gives

$$\sigma_c^2 = 2.37\frac{\sigma^2}{9} = .2633\sigma^2$$

Therefore, n must satisfy the equation

$$\frac{.2633\sigma^2}{n} = \frac{\sigma^2}{500}$$

The solution of this equation is $n = 132$; therefore 132 clusters of size 9 will suffice to yield the same accuracy of estimate as a simple random sample of size 500.

(*c*) If the number of clusters to be sampled is n and $M_0 = 27$, it is assumed that a simple random sample of size $\frac{1}{2}(27n)$ will cost the same. From formula (11) it follows that

$$\sigma_c^2(\hat{p}) = \frac{\sigma_c^2}{n}$$

For a simple random sample of size $\frac{1}{2}(27n)$, the variance is given by

$$\sigma^2(\hat{p}) = \frac{2\sigma^2}{27n} = .0741\frac{\sigma^2}{n} \tag{16}$$

But from the table of values for $M_0 = 27$,

$$\sigma_c^2 = 2.17\frac{\sigma^2}{27} = .0804\sigma^2$$

Therefore,

$$\sigma_c^2(\hat{p}) = .0804\frac{\sigma^2}{n} \tag{17}$$

A comparison of (16) and (17) shows that simple random sampling is slightly more efficient with the assumed cost ratios.

6 RESPONSE ERRORS

Errors in responses to interviews or questionnaires and errors of counts and measurements in surveys involving direct observation present a problem above and beyond the errors ascribed to sampling variation. Some of those errors are of a random nature and their effects may largely disappear in large samples. However, many of them are in the nature of biases and will not disappear with large samples. Such errors can be isolated and measured by appropriate planning and clever design. Once they have been revealed, it is possible to deal with them and counteract their biasing effect. Sampling in such situations can be a very effective way of solving the problem because it is usually much cheaper to take an adequate sample than it is to pursue a painstaking investigation of possible biases.

7 REVIEW ILLUSTRATIONS

Example 1. Suppose the number of pupils in each of the six grades of an elementary school are 200, 200, 180, 160, 140, and 120, respectively. A proportional sample of total size 200 is to be taken to estimate the mean weight of the schoolchildren. Suppose that the standard deviation of individual weights from the entire school is $\sigma = 18$. Suppose, further, that the sample mean weights turn out to be 50, 60, 70, 80, 90, and 100. If these sample mean weights are substituted for the μ_h, and if the value of $\bar{x}$ calculated by means of formula (3) is substituted for μ, determine the advantage of proportional sampling over simple random sampling here.

Since the sample size is $n = 200$,

$$V(\bar{x}) = \frac{\sigma^2}{200} = 1.62$$

$\pi_h = .2, .2, .18, .16, .14, .12$, and from (3), $\bar{x} = \Sigma \pi_h \bar{x}_h = 72$; therefore, approximately,

$$\sum_{h=1}^{6} \frac{\pi_h}{n}(\mu_h - \mu)^2$$

$$= \frac{1}{200}[.2(22)^2 + .2(12)^2 + .18(2)^2 + .16(8)^2 + .14(18)^2 + .12(28)^2]$$

$$= 1.38$$

$V_P(\bar{x}) = 1.62 - 1.38 = .24$ and $V_P(\bar{x})/V(\bar{x}) = .24/1.62 = .15$. Thus, the proportional variance is about 15 percent of the random sample variance.

Example 2. In "A Statistical Study of Sampling Methods for Tree Nursery Inventories" (Iowa State University, 1941), F. A. Johnson considered the effectiveness of different sized units in estimating the number of seedlings in a tree nursery. One of the alternatives was to use a two-foot-long segment of a planted row of seedlings rather than a one-foot-long segment for counting. By multiplying the mean number of seedlings per segment, $\bar{x}$, in a sample by the total footage of rows, an unbiased estimate of the total number of seedlings can be obtained. The data that resulted in comparing the effectiveness of the one-foot and two-foot segments for white pine seedlings are as follows.

	One-foot segment	Two-foot segment
Number of segments, N	2,604	1,302
Variance among segment means, σ_c^2	2.537	6.746
Length of row counted in 15 minutes	44 feet	62 feet

(a) Calculate the variance of $\bar{x}$ for both the one-foot and the two-foot segments for the same total row length. That is, for n two-foot segments and for $2n$ one-foot segments. (b) Which segment length is more efficient when the counting time is taken into consideration?

(a) Let $\bar{x}_1$ be the sample mean for the one-foot segments and $\bar{x}_2$ for the two-foot segments. Since the comparison is to be for a one-foot segment, it is necessary to compare $\bar{x}_1$ with $\bar{x}_2/2$. From formula (11),

$$V_c(\bar{x}_1) = \frac{\sigma_{c_1}{}^2}{2n} = \frac{2.537}{2n} = \frac{1.2685}{n}$$

Since the variance of a constant times a variable is equal to the square of the constant times the variance of the variable, it follows from this fact and formula (11) that

$$V_c\left(\frac{\bar{x}_2}{2}\right) = \frac{V_c(\bar{x}_2)}{4} = \frac{1}{4}\frac{\sigma_{c_2}{}^2}{n} = \frac{6.746}{4n} = \frac{1.6865}{n}$$

(b) If n two-foot segments can be counted in a given period of time, then only $\frac{44}{62} \cdot 2n$ one-foot segments can be counted in that same time period. Therefore the value of $2n$ in $V_c(\bar{x}_1)$ must be replaced by $\frac{44}{62} \cdot 2n$. This is equivalent to replacing n by $\frac{44}{62}n$. The comparison is then between

$$\frac{1.2685}{\frac{44}{62}n} = \frac{1.7874}{n} \quad \text{and} \quad \frac{1.6865}{n}$$

Because of the ease of counting larger segments, the two-foot segment is now seen to be slightly superior to the one-foot segment with respect to accuracy of estimating $\bar{x}$ and hence with respect to estimating the total number of seedlings.

EXERCISES

SECTION 2

1 One way to obtain a completely random sample of current telephone subscribers (whether listed or not) is to draw random digits from a table of random numbers and dial them. (This sometimes obtains rather surprising results!) Does this supply a random sample of "telephone homes"—that is, households with at least one phone? Discuss.

2 A manufacturer of soft drinks has found four new ways to reduce the calorie content of its drinks but would like to know which is most acceptable, tastewise. It is considering the alternatives of (*i*) a panel of experts to taste-test the products of the four methods, (*ii*) a free "bar" set up in the company's lounge serving the experimental drinks with a short questionnaire form available to be filled out by those wishing to comment, or (*iii*) a sample of "consumers" on which the experimental drinks would be tested. Comment on the adequacy of the several methods for selecting the units on which to carry out an experiment of this sort.

3 The Gas Company in Los Angeles wishes to estimate periodically the number of gas-using appliances by type (e.g., stove, water heater, etc.) among its customers. It has a complete listing of all "hookups," that is a street address, and the number of meters installed at each. The listing looks something like this:

Address	*Number of meters*
1101 "E" St.	1
1103 "E" St.	10
1005 "E" St.	1
1007 "E" St.	2
.	.
.	.
.	.

The company wishes to draw a sample of either addresses or meters and send its field men out to obtain the appropriate information from the occupant or, in the case of apartments, from the manager.

(*a*) Which sampling unit would you suggest be used? Why? (Remember that one or more households may be served gas from the same meter, particularly where "utilities are furnished.")

(*b*) Describe the problem of finding the appropriate respondent under certain circumstances.

SECTION 3

4 The 13 major chain grocers in Los Angeles County have about 700 stores in all. Suppose you have been given the task of determining the nature of practices used in their bakery departments (inventorying, etc.) by means of a survey. A sample of 30 stores is drawn completely at random and without replacement. After the survey of the sample stores was completed an interest developed in stores located in the San Fernando Valley portion of the county and the question arose as to whether the 10 stores in the sample that were also in the San Fernando Valley could be regarded as a random sample of all chain stores in the Valley. Comment.

5 An agency wishes to take a sample of 200 adults in a certain residential section of the city. It proposes to do so by taking a random sample of 200 households obtained from a listing of all households in that district and then selecting at random an adult from each such household. Why, or why not, will this procedure yield random samples?

6 Explain why specifying a certain percentage-sized sample of a population, such as 10 percent of the population, is not adequate to determine the accuracy of a sample estimate of the mean of the population.

SECTION 4

7 Suggest different criteria that might be useful in enabling one to take stratified samples for estimating the mean annual expenditure on meat by families in a large city.

8 How would you take a sample of 25 clerks from the population of clerks working in food markets in a city, using proportional sampling?

9 Give an illustration of a population for which you believe stratified sampling would be considerably cheaper or better than random sampling.

10 At the close of the fiscal year a manufacturing firm normally would close down for 5 to 10 days in order to take inventory required to establish the value of materials and various components and subassemblies in various stages of completion. In order to save costs and still hold the margin of error within tolerable limits, the auditors agreed to the use of a sample that could be taken on a Saturday, when the plant is normally shut down. Without going into details of the sampling plan other than to mention that the 6 departments appeared to be suitable strata, it was decided that within each department all large units were to be listed separately in a sampling scheme and smaller units were to be put into lots (such as bins, boxes, portions of conveyor belts, etc.) and each lot listed in the sampling scheme. A proportional sample of 200 lots was taken, and each lot carefully examined and evaluated. Suppose the population number of lots and mean inventory values were as follows:

Stratum, h:	1	2	3	4	5	6
Number of elements, N_h:	140	150	160	170	180	200
Stratum means, μ_h:	200	200	180	160	150	120

where the overall mean value per lot is $\mu = 165$ with a standard deviation of $\sigma = 35$.

(*a*) Determine the advantage of proportional stratified sampling over simple random sampling.

(*b*) Determine the standard error of the sample mean under proportional sampling.

(*c*) Determine 95 percent confidence limits of the estimated inventory values based on these results.

11 Management wishes to determine the preferences of its employees on a policy question. It wishes to take a sample of 400. Its employees consist of 3000 "white collar" and 6000 "blue collar" workers, and it is believed that 60 percent of the white collar employees and 40 percent of the blue collar employees favor the policy.

(*a*) What is the value of $V(\hat{p})$ if a simple random sample is taken?

(*b*) What is the value of $V(\hat{p})$ if a proportional random sample is taken? Is there much improvement over random sampling?

12 A sample of 500 registered voters is to be taken to estimate the percentage that favors fluoridation of the city water supply. The city consists of four distinct sections: *A*, *B*, *C*, and *D*. The percentages of voters living in those sections are 40, 30, 20, and 10, respectively. Preliminary studies have indicated that the percentages of voters in those sections that favor fluoridation are approximately 70, 50, 20, and 30, respectively.

(*a*) Calculate the value of $V(\hat{p})$ if a simple random sample is taken.

(*b*) Calculate the value of $V(\hat{p})$ if a proportional sample is taken. Here p denotes the proportion of voters in the entire city that favors fluoridation.

(*c*) How much gain in estimation precision resulted from using proportional sampling over simple random sampling?

SECTION 5

13 A sample of eggs is to be obtained to determine their quality. The shipment consists of 100 cases of 36 dozen eggs each. The sample is obtained by selecting 10 cases at random and from each sample case 2 eggs are selected at random.

(*a*) What kind of sample is this?

(*b*) Suppose all 36 eggs in the sample cases were examined; would sampling variance in this case be zero? Why or why not?

14 When samples of households representative of all households in the United States are selected, they are usually obtained in stages. For example, first a sample of counties is obtained; then within these a sample of cities, towns or rural areas, within these a sample of census tracts, within these a sample of blocks (or their equivalent in rural areas), and within these blocks a sample of households. This is usually done in a manner such that each household in the United States has an equal chance of being selected. Is this a random sample of households? Explain your answer.

15 For safety reasons, a taxi company wishes to estimate the proportion of unsafe tires on its cabs. It selects a random sample of 20 cabs from the fleet of 200 cabs and finds the following number of unsafe tires per cab (not including the spare tire) for this sample:

1, 0, 1, 2, 3, 0, 4, 1, 0, 2, 2, 1, 4, 1, 0, 3, 2, 0, 1, 2

(*a*) Estimate the proportion of unsafe tires in the entire population of cab tires.
(*b*) Calculate the standard error of this cluster sampling estimate.
(*c*) Compare the result in (*b*) with the standard deviation of a simple random sample if the estimate in (*a*) is used as the value of p in the binomial formula for $\sigma_{\hat{p}}$.

16 An inspector for the Food and Drug Administration sampled a truckload of canned corn to determine the mean number of corn-borer fragments per can. The truck contained 1000 cases of cans, each case containing 6 cans. Ten cases were selected at random and all 60 cans were inspected. The inspection yielded the following data, which give the number of corn-borer fragments found in each can.

Can	Case 1	2	3	4	5	6	7	8	9	10
1	2	2	0	0	6	6	1	3	4	5
2	5	0	3	3	0	6	3	0	0	2
3	1	2	2	0	2	3	1	0	0	0
4	6	0	4	4	2	0	6	2	0	3
5	5	0	0	2	5	4	4	0	0	0
6	6	1	5	0	6	1	5	0	5	1

(*a*) Estimate the mean number of fragments per can for the entire truck load.
(*b*) Calculate the standard deviation of this estimate.

SECTION 7

17 An estimate is desired of the total amount of wheat in storage in Nebraska elevators. The following data give the average amounts stored in elevators of three different types. The figures are in terms of thousands of bushels.

Stratum	Number of elevators	Average amount per elevator	Standard deviation of amount per elevator
1	146	15.60	20.1
2	610	4.93	8.4
3	192	.60	2.5

(*a*) Suppose that 100 elevators are to be sampled. How should this sample be distributed among the three strata?
(*b*) What standard error should be expected from the sample in (*a*)?

18 A merchandiser wishing to determine the preferences of his customers for 3 types of packaging, A, B, and C, selected 6 market areas in the U.S. and in each studied the preferences of 100 of his customers. He obtained the following numbers of customers who preferred packaging C.

$$50, 45, 38, 65, 20, 52$$

Treat these 6 groups of 100 customers as a random selection of clusters of customers.

(*a*) Estimate p, the proportion of all customers who prefer C.

(*b*) Estimate the standard deviation of $\hat{p}$ in (*a*).

(*c*) Estimate what the standard deviation of $\hat{p}$ would be if the 600 customers had been selected at random from the entire population.

19 The accounting department of a business firm was asked to obtain a reasonably accurate estimate of the current number of uncollectable accounts in its consumer operations. To obtain such an estimate, the department classified its accounts receivable into 4 groups according to the amount outstanding. It then drew a 1 percent sample at random from each group and determined their collectability, with the following results.

Category	1	2	3	4
Sample size	500	300	300	200
Uncollectables	25	75	6	2

(*a*) Estimate the proportion of uncollectable accounts in the population of accounts.

(*b*) Estimate the standard deviation of the estimate obtained in (*a*).

20 As reported in Volume 41 of the *Journal of the American Statistical Association,* a survey was carried out during World War II in the U.S. to determine the number of new truck and bus tires in the hands of dealers. From earlier contacts a considerable amount of information was available concerning these dealers and their inventories. This information enabled the samplers to divide the dealers into strata according to previous inventory sizes. They were also able to use this information to estimate the standard deviations with fair accuracy. The results of this survey were as follows.

Stratum	Number of dealers	Number of dealers sampled	Mean number of tires in sample	Standard deviation of number of tires
1	19,850	3,000	4.1	5.9
2	3,250	600	13.0	9.6
3	1,007	340	25.0	13.2
4	606	230	38.2	17.9
Totals	24,713	4,170		

(*a*) Estimate the total number of tires in the inventories of dealers in the U.S. at the time of the survey.

(*b*) Estimate the standard deviation of the estimate obtained in (*a*).

Chapter 15 Index Numbers

1 INTRODUCTION

There are many concepts in the social sciences that can be measured directly in a satisfactory manner, but many others for which this cannot be done. For example, a bank's prime interest rate on loans, or the percentage of workers in a city on relief, is a fairly well defined concept capable of direct measurement. However, it is not clear how one should measure such things as the intelligence of students, the level of the stock market, or the cost of living.

For things that do not lend themselves to direct measurement it is necessary to introduce some associated quantity that will serve to represent them quantitatively. Thus, psychologists have used certain tests for measuring a student's academic talents and, on the basis of such tests, have introduced a quantity called an intelligence quotient for discussing intelligence in a quantitative manner. Similarly, economists have introduced various quantities for describing the level of the stock market, the most common of which is the Dow-Jones Industrial Average.

These indirect measures, or indicators, are generally known as indexes, or indices, even though they may not carry that name in their title. For example, sociologists have their integration index, psychologists their personality index, and economists their GNP (gross national product). In economics and business the number of such indexes is extremely large. In the field of business change alone one author located and described 449 index series.

This chapter will be concerned with explaining the nature of index numbers, presenting some principles that have been found useful in their construction, and describing a few of the more important such indexes that occur in everyday and business life. It will also point out some of the practical problems that complicate the construction of indexes, particularly those designed to measure such concepts as the level of prices and the cost of living.

2 RELATIVES: SIMPLE AND LINKED

Consider the data of Table 1, which gives the price per box in dollars to growers for oranges produced in Florida during the years 1957 to 1972. Since it is difficult to study the variation in such numbers, it is customary to express each value as a percentage of the value for a particular year called the *reference year* or *reference base.* Hence the reference year is the year on which all calculations are based, and the index for that year is usually set at 100. The numbers shown in column 3 are such percentages, with 1957 as the reference year. They are generally called *relatives,* in this case *value relatives,* with 1957 as the reference year. If neighboring years are to be compared, it is convenient to express each year's value as a percentage of the preceding year's value. Such percentages are called *linked relatives* and are shown for the data of Table 1 in column 4. It should be noted that linked relatives can also be calculated from successive pairs of the *simple* relatives of column 3. These relatives, whether simple or linked, certainly make it much easier to study change than working only with the values themselves.

Table 1 Average seasonal price of Florida oranges, 1957–1972

Year (1)	Price per box (2)	Column 2 as percent of 1957 price (3)	Column 2 as percent of preceding year (4)
1957	1.45	100	—
1958	1.57	108	108
1959	2.65	183	169
1960	1.82	126	69
1961	2.86	197	157
1962	1.93	133	67
1963	2.17	150	112
1964	4.43	306	204
1965	2.57	177	58
1966	1.44	99	56
1967	.81	56	56
1968	1.85	128	228
1969	1.56	108	84
1970	1.15	79	74
1971	1.10	76	96
1972	2.09	144	190

3 INDEXES FOR AGGREGATIONS

The problem of index construction becomes complicated as soon as more than one commodity is involved in the construction. For example, suppose an index for the level of food prices is to be designed based on only four foods and two points in time, and that the study is limited to the microeconomy described in Table 2. Prices are in dollars.

Table 2

Time period	Commodity	Unit of commodity	Quantity	Price	Value
1	Bread	loaf	1,000	.25	250
	Meat	pound	500	.75	375
	Cabbage	bag	500	1.00	500
	Wine	bottle	1,000	1.25	1,250
				3.25	2,375
2	Bread	loaf	650	.50	325
	Meat	pound	400	.75	300
	Cabbage	bag	1,500	.50	750
	Wine	bottle	1,000	1.00	1,000
				2.75	2,375

One simple approach to a price index is to look at the total of the price column and calculate the simple relative of period 2 total prices with respect to period 1 prices. Letting Σ denote summation over the four items, this gives the index value

$$I_1 = \frac{\Sigma p_2}{\Sigma p_1} = \frac{2.75}{3.25} = .846$$

Here p_1 and p_2 represent item prices during the time periods 1 and 2, respectively. A second simple approach is to calculate the simple relative of each period 2 price with respect to its period 1 price and average those relatives. The calculations needed to yield this type of index are shown in Table 3. The corresponding index value is given by

$$I_2 = \frac{1}{4}\sum \frac{p_2}{p_1} = \frac{4.30}{4} = 1.075$$

The first index indicates that prices have dropped, whereas the second indicates that they have risen. However, an inspection of the last column of Table 2 shows that according to the index based on the total value of these commodities, there has been no change in the cost of food for the microeconomy represented by the data of Table 2. It would be difficult to justify the use of I_1 as a measure of the change in the price level. If, for example, the price of wine had been quoted with the gallon as a unit rather than the bottle, the value of I_1 would have been different. The index I_2 is not affected by a change in units; however, both of these indexes ignore the possibility that there may be large changes in the amount of consumption of the various commodities, thereby making the index unrealistic. It does not make much sense, for example, to give equal weight to a set of relative prices if some members of the set correspond to prices of commodities that are never purchased. Thus, it is clear that neither of these indexes is suitable to represent the price level of food.

Table 3

Commodity	p_1	p_2	p_2/p_1
Bread	.25	.50	2.00
Meat	.75	.75	1.00
Cabbage	1.00	.50	.50
Wine	1.25	1.00	.80
			4.30

4 WEIGHTED INDEXES

As was indicated in the preceding section, one of the difficulties in measuring change in price levels is the fact that along with prices, the quantities of each commodity are also changing. A direct measure of average price that ignores the relationship between price and quantity is likely to include some of the unwanted quantity changes. This confounding of prices and quantities can be avoided if, for example, quantities can be held fixed while prices are allowed to vary. Since this cannot be done experimentally it must be done artificially.

In the case of index numbers, this conceptual device of holding one variable constant while allowing another to vary is most useful when constructing an index such as the cost-of-living index. As an illustration, consider the problem of constructing a cost-of-food index for Mr. A, who lives in the microeconomy represented by Table 2. In order to prevent Mr. A's changes in eating habits from confounding the change in prices, he will be required to maintain his period 1 eating pattern through period 2. Suppose Mr. A consumed the following amounts of foods during the first period: 5 loaves of bread, 2 pounds of meat, 2 bags of cabbage, and 5 bottles of wine. His total outlay for food during period 1 is therefore given by

$$V_1 = 5(.25) + 2(.75) + 2(1.00) + 5(1.25) = 11.00$$

If he consumed these same quantities during period 2 his total outlay for that period would be

$$V_2 = 5(.50) + 2(.75) + 2(.50) + 5(1.00) = 10.00$$

Hence Mr. A's cost of food, if his eating habits had not changed, would have decreased from \$11.00 to \$10.00. His cost-of-food index based on these two values would therefore by given by $V_2/V_1 = .91$, showing a 9 percent decrease.

The preceding technique for measuring the change in the cost of food for an individual seems like a reasonable way of solving that problem. Although it is based on fictitious behavior for Mr. A, this behavior is possible and reasonable. Incidentally, this technique for constructing an index is the method commonly used to obtain a cost-of-living index for an entire economy. If, however, one is interested in constructing an index for the price level of an entire economy, then it is not clear that the preceding method is satisfactory.

Price Index. Return now to the basic problem of how to obtain a satisfactory index for the price level. In view of the preceding discussion it seems clear that the index should be a weighted index of prices with the weights being determined by the quantities purchased. Furthermore, if the device of holding such quantities fixed over the two time periods is employed to eliminate the confounding of price and quantity, then an index for the price level should be of the form

$$I_p = \frac{\sum_j q_{ij} p_{2j}}{\sum_j q_{ij} p_{1j}} \tag{1}$$

Here q_{ij} represents the quantity of commodity j purchased during period i, where i is 1 or 2, and p_{ij} denotes the price of that commodity during period i. The summation is over all commodities purchased during those two time periods. It should be noted that the weights q_{ij} are the same for the two time periods, but that they can be chosen to be the quantities purchased during the first time period or during the second time period.

The two special cases of this index are known by the names of their early sponsors. When i is chosen to be 1, which was the choice made in calculating a cost-of-food index for Mr. A, the index is called the *Laspeyres index* and is denoted by the formula

$$I_p(L) = \frac{\sum_j q_{1j} p_{2j}}{\sum_j q_{1j} p_{1j}} \tag{2}$$

When i is chosen to be 2, the index is called the *Paasche index* and is given by the formula

$$I_p(P) = \frac{\sum_j q_{2j} p_{2j}}{\sum_j q_{2j} p_{1j}} \tag{3}$$

Coming back to the cost-of-food problem for Mr. A, the Paasche index would measure the change in the cost of food for Mr. A if he had eaten the same quantity of each food during the first time period as he did during the second time period. This also seems to be a reasonable way of measuring the change in the cost of food for Mr. A.

For the purpose of seeing how these two weighted indexes compare for determining the price level of an economy, their values will be calculated for the microeconomy of Table 2. Calculations yield the values

$$I_p(L) = \frac{2125}{2375} = .895$$

and

$$I_p(P) = \frac{2375}{3213} = .739$$

Neither the Laspeyres nor the Paasche index appears to deal properly with the price-level problem for an entire economy. The constant weights have a very reasonable interpretation when discussing the total cost of food, or living, but not when interest is centered on the price-level problem.

Quantity Index. The price-level index has been of major interest thus far, but the preceding material now suggests a look at quantities also. If one merely interchanges the roles of price and quantity, with the quantities weighted by the corresponding prices and with those weights held constant over the two time periods, then a quantity index comparable to the general price index in (1) is given by the formula

$$I_q = \frac{\sum_j p_{ij} q_{2j}}{\sum_j p_{ij} q_{1j}} \tag{4}$$

As before, i is chosen to be 1 or 2. These two choices give the corresponding Laspeyres and Paasche *quantity indexes.* They yield the formulas

$$I_q(L) = \frac{\sum_j p_{1j} q_{2j}}{\sum_j p_{1j} q_{1j}} \tag{5}$$

and

$$I_q(P) = \frac{\sum_j p_{2j}q_{2j}}{\sum_j p_{2j}q_{1j}}$$

For the microeconomy, calculations give

$$I_q(L) = \frac{3213}{2375} = 1.353$$

and

$$I_q(P) = \frac{2375}{2125} = 1.118$$

These results indicate that there has been an increase in the quantity of food consumed, but that the increase in consumption would be greater if the first-time-period prices prevailed over both periods than if the second-time-period prices prevailed over those periods.

Value Index. In addition to indexes of price and quantity, there are also indexes of value. The value of all the commodities purchased in time period 1 is given by

$$v_1 = \sum_j p_{1j}q_{1j}$$

Similarly, the value of all the commodities purchased in time period 2 is given by

$$v_2 = \sum_j p_{2j}q_{2j}$$

The simple *value index* is then defined by the formula

$$I_v = \frac{v_2}{v_1} = \frac{\sum_j p_{2j}q_{2j}}{\sum_j p_{1j}q_{1j}} \tag{7}$$

For the microeconomy data, calculations give

$$I_v = \frac{2375}{2375} = 1.00 \tag{8}$$

Hence, no change has occurred in the value of those commodities.

Fisher's Index. An index proposed by the economist Irving Fisher has some interesting properties in dealing with the price level of an entire economy. It is defined to be the geometrical mean of the Laspeyres and Paasche index. The respective formulas for prices and quantities are

$$I_p(F) = \sqrt{I_p(L)I_p(P)} \tag{9}$$

and

$$I_q(F) = \sqrt{I_q(L)I_q(P)} \tag{10}$$

Application of the Fisher index to the microeconomy of Table 2 yields the values

$$I_p(F) = \sqrt{(.895)(.739)} = .813$$

and

$$I_q(F) = \sqrt{(1.353)(1.118)} = 1.230$$

From the way they are constructed, these indexes have values lying between those of the corresponding Laspeyres and Paasche indexes.

The Fisher index has some nice properties. For example, since value is the product of price and quantity, it would be desirable to have the value index equal the product of the price and quantity indexes. That is, it would be desirable to have $I_v = I_p I_q$. It can be shown that the Fisher index does possess this property.

It is interesting to compare the various indexes for prices and quantities that have been introduced and discussed in the preceding sections as they apply to the microeconomy of Table 2. Calculations produce the following values.

Type of index	I_p	I_q	$I_p \cdot I_q$
Ratio-of-means	.846	1.183	1.000
Mean-of-relatives	1.075	1.363	1.465
Laspeyres	.895	1.353	1.211
Paasche	.739	1.118	.826
Fisher	.813	1.230	1.000

Although the product of the ratio-of-means indexes for price and quantity is equal to the value index for this problem, which was observed in (8) to be equal to 1.00, this is a coincidence and it is not true in general. The most appealing result here is that associated with the Fisher index, namely that the price level has decreased on the average by 19 percent whereas the physical output has increased by 23 percent, while no change has occurred in the value of the output.

5 ECONOMIC INDEXES IN SERIES

The discussion of indexes thus far has been confined mainly to two periods of time, because then the basic ideas can be explained quite easily. Now consider several time periods and the problems that arise in trying to characterize an economy and life in it by the use of index numbers.

These problems can be illustrated by studying a microeconomy similar to that in section 3, but extending over a longer period of time. Such a microeconomy is presented in Table 4. Two of the basic problems here are (*a*) to characterize the economy by means of its gross national product (GNP) and determine how much of the GNP is due to price inflation or deflation and (*b*) to construct an index for the cost of living in this economy.

The basic data of Table 4 give the prices and quantities of 4 commodities produced and consumed in each of 4 years. These are to be regarded as the complete set of commodities produced, so that no sampling is involved.

Table 4

Time period	Commodity	Unit	Quantity (q)	Price (p)	Value (v)
1	Bread	loaf	1,000	.25	250
	Meat	pound	500	.75	375
	Cabbage	bag	500	1.00	500
	Wine	bottle	1,000	1.25	1,250
					2,375
2	Bread	loaf	900	.25	225
	Meat	pound	600	.70	420
	Cabbage	bag	500	1.00	500
	Wine	bottle	900	1.20	1.080
					2,225
3	Bread	loaf	1,000	.20	200
	Meat	pound	600	.80	480
	Cabbage	bag	600	1.15	690
	Wine	bottle	900	1.20	1,080
					2,450
4	Bread	loaf	900	.30	270
	Meat	pound	700	.75	525
	Cabbage	bag	550	1.30	715
	Wine	bottle	1,000	1.30	1,300
					2,810

(*a*) The GNP consists of the aggregated values of the commodities produced each year. The GNP for each year is obtained by summing the values, v, over the 4 commodities for that year. These values together with a simple-relatives index series with the first year as a base are shown in columns 2 and 3 of Table 5.

Table 5

Year (1)	GNP (current) (2)	GNP (relative) (3)	GNP (year-1 dollars) (4)	Real GNP (5)	Implicit deflator index (6)
1	2,375	100	2,375	100	100
2	2,225	94	2,300	97	97
3	2,450	103	2,425	102	101
4	2,810	118	2,550	107	110

To determine how much of the GNP can be ascribed to real production, the GNP is computed in terms of *constant dollars,* which are prices for the year 1. The results of such computations are shown in column 4 of Table 5. In these computations one uses the year-1 prices and the quantities for the year in question. As an illustration, the second entry in column 4 is given by

$$900(.25) + 600(.75) + 500(1.00) + 900(1.25) = 2300$$

The simple relatives of the column 4 values are shown in column 5. Finally, dividing the current dollar GNP column 2 by the constant dollar GNP column 4 yields the set of indexes in column 6, which are called *implicit deflator* indexes. An index of this type is the official measure of price inflation or deflation in the U.S. economy. These indexes show that there was a slight deflation during the second year and a slight inflation during the third year but a strong inflation during the fourth year.

(*b*) To construct a cost-of-living index for a particular family type, such as the blue collar group, suppose the average blue collar family in year 1 consumed 10 loaves of bread, 3 pounds of meat, and 15 bags of cabbage. How have the blue collar members of the economy fared over the four years? The necessary computations are given in Table 6.

Table 6

Commodity	q_1	p_1	p_2	p_3	p_4	q_1p_1	q_1p_2	q_1p_3	q_1p_4
Bread	10	.25	.25	.20	.30	2.50	2.50	2.00	3.00
Meat	3	.75	.70	.80	.75	2.25	2.10	2.40	2.25
Cabbage	15	1.00	1.00	1.15	1.30	15.00	15.00	17.25	19.50
						19.75	19.60	21.65	24.75

The resulting index values with year 1 as the base are 100, 99, 110, and 125. Comparing these values with the index values 100, 97, 102, and 107, which are those for the real GNP obtained in Table 5, it can be seen that the blue collar families fared worse than the nation as a whole.

6 SOME COMMON ECONOMIC AND BUSINESS INDEXES

In order to see the nature and complexity of indexes and thereby gain some judgment concerning their limitations, five of the more common indexes will be described and discussed.

The Gross National Product (GNP). This is the standard measure of the output of an economy. It appears quarterly in the *Survey of Current Business,* in current and in 1958 dollars.

The Consumer Price Index. This index, abbreviated CPI, is often called the cost-of-living index and is a statistical measure of changes in the prices of goods and services purchased by urban wage earners and clerical workers, who represent about 45 percent of the U.S. population. It measures changes in prices, which are the most important cause of changes in the cost of living, but it does not indicate how much families actually spend to defray their living expenses.

The index includes 400 items, both goods and services, that families buy. The major groups, and their relative importance (in December 1972) are: Food (22.5%), Housing (33.9%), Apparel and Upkeep (10.4%), Transportation (13.1%), Health and Recreation (19.8%), and Miscellaneous (0.4%).

The CPI is used extensively in labor-management contracts to adjust wages to the cost of living. Demands from other population groups for similar arrangements (social security recipients, for example) have brought about the decision to provide a more comprehensive CPI by 1977 to cover about 80% of the population. The CPI is also widely used to measure inflationary trends in the economy.

The most recent weight base is 1960–1961, at which time a sample of 5,497 families or individuals located in 72 urban areas was taken and a record made of the kind, qualities, quantities, and prices of the goods and services they bought.

The Bureau of Labor Statistics uses mail questionnaires to obtain such data as transportation costs, public utility rates, newspaper prices, and prices of certain other items, but most prices are obtained by means of personal visits by Bureau agents. Prices obtained by personal visits come from a representative sample of 18,000 retail stores and service establishments where wage and clerical workers buy goods and services.

The CPI is an index of the Laspeyres type. It is published monthly for major cities and for major classes of items, and as yearly averages for major cities, regions, and the nation.

The Dow-Jones Average of 30 Industrial Stocks (D-J). This index is usually called the Dow-Jones Industrial Average and is the best-known index for measuring common stock prices. It was established in 1897 at which time it was based on 12 stocks, but it was gradually expanded to include 30 stocks. The 30 companies whose stock prices are used in the index are listed in Table 7. Since this index is supposed to be the average value of the 30 stocks comprising it, one would suppose that the index value could be obtained by calculating the arithmetic mean of the prices listed in Table 7; however, because of stock splits and dividends over the years the arithmetic is more complicated than the original arithmetic, when it sufficed to divide by 12. The divisor in September 1968 was 2.011 instead of 30. As a result, calculations based on Table 7 will show that the D-J index value for September 29, 1968, was 936.

Table 7

Closing prices on September 29, 1968, of the 30 stocks in the Dow-Jones Industrial Average

Stock	Price	Stock	Price
1. Allied Chemical	35	16. Int'n'l Nickel	38
2. Alluminum Co. of A.	73	17. Int'n'l Paper	36
3. American Can	49	18. Johns-Mansville	75
4. American Tel. & Tel.	53	19. Owen-Ill. Glass	70
5. American Tobacco	33	20. Procter & Gamble	87
6. Anaconda	49	21. Sears, Roebuck	69
7. Bethelem Steel	31	22. Std Oil of Calif.	66
8. Chrysler	70	23. Std Oil of N.J.	78
9. DuPont	170	24. Swift & Co.	28
10. Eastman Kodak	81	25. Texaco	84
11. General Electric	85	26. Union Carbide	43
12. General Foods	86	27. United Aircraft	63
13. General Motors	83	28. U.S. Steel	43
14. Goodyear	58	29. Westinghouse El.	77
15. Int'n'l Harvester	36	30. Woolworth	32

D-J = 1882/2.011 = 936

Although the 30 stocks contained in the D-J index are not a random or other statistical type of sample of the approximately 1200 stocks that make up the New York Stock Exchange listing, they were judiciously selected to represent somehow the entire listing. The index has survived remarkably well against increasing competition with old and new rivals such as Standard and Poor's 500 index and the New York Stock Exchange's price of an average share based on all issues. This survival undoubtedly results partly from the fact that it is so widely known.

Index of Industrial Production (IIP). This index is a measure of the physical volume of output in those industries that are primarily engaged in manufacturing and mining, together with some utilities such as electric power and gas. These industries account for about three-fourths of the national income production. The index is widely used as a general business barometer.

The IIP is compiled by the Federal Reserve Board and is issued monthly with and without seasonal adjustments. It was started during World War I and became a regular published series in 1927, with major revisions occurring in 1940, 1953, and 1959. Since 1967 both the weight and reference bases have been those of 1967. The IIP is a Laspeyres-type index, although many approximations are made to overcome difficulties arising because of inadequate or nonexistent output data in some industries. The data are not obtained directly by the Federal Reserve Board but come from other governmental agencies and business associations and must be pieced together to form the index. This is a difficult task because the weights are not based on prices but on value added by each manufacturer in order to give each manufacturing process its proper consideration in contributing to output.

The Implicit Deflator Index (IDI). This index is published in the *Survey of Current Business* along with statistics on the GNP and is the official quantification of the average amount of price inflation (or deflation) in the economy. It is obtained by dividing the GNP expressed in constant dollars (1958 dollars, for example) by the corresponding GNP figure expressed in current dollars and multiplying by 100. It is published quarterly and revised as more information becomes available.

7 DEFLATING A TIME SERIES: STANDARDIZED DOLLARS

Although the dollar is the unit used to express the magnitude of financial and many economic and business transactions and operations, it does not remain constant in value through time. We hear the expressions "the dollar does not buy what it used to," "the dollar is depreciated," "the dollar has lost its purchasing power," etc. In comparing the cost of a pound of bread now and 10 years ago we conclude that either the price of bread has gone up or the value of the dollar has gone down. When a number of comparisons of this sort are made and in the preponderance of cases we find that prices are up, we are likely to conclude that the dollar, in terms of what it can buy, has lost some of its value.

A measure of the value of the dollar is simply the inverse of the index of prices. If prices from period 1 to period 2 have doubled, then the value of the dollar has halved. If on the other hand prices have halved, then the dollar has doubled in value. Price indexes are widely applied to *standardize* the dollar, to convert money wages or income into *real* wages or *real* income.

As an illustration, consider once more the microeconomy of Tables 4 and 5. According to the implicit deflator index (IDI), shown in column (6) of Table 5, the price level for the economy as a whole fell 3 percent in year 2, rose to a level of 101 in year 3, and rose to 110 in year 4. In view of this information the value of the dollar can be obtained by calculating the reciprocal of the IDI and multiplying by 100, as follows.

	Year 1	Year 2	Year 3	Year 4
IDI	100	97	101	110
Value of dollar	1.00	1.03	.99	.91

Since the dollar does not maintain the same value throughout the economy, an assessment of its value must take into account the particular sector, geographic region, processing stage, etc., being dealt with. For example the U.S. CPI may be useful for blue collar workers but not necessarily for retired families. In the construction and interpretation of index numbers it is necessary to realize that practical considerations such as these are part of the problem.

8 REVIEW ILLUSTRATIONS

Example 1. Given the following prices and quantities of three commodities purchased by a family for the years 1968 to 1969, use them to calculate the increase, or decrease, in prices from 1968 to 1969 (*a*) if the mean of relatives is employed, (*b*) if the Laspeyres formula is employed, (*c*) if the Paasche formula is employed, (*d*) if the Fisher formula is employed. (*e*) Check to see whether $I_v = I_p I_q$ is satisfied in (*b*) and (*c*). (*f*) Verify numerically that $I_v = I_p I_q$ is approximately satisfied in (*d*).

	1968		1969	
	Price	Quantity	Price	Quantity
Bread	35¢	200 loaves	40¢	180 loaves
Meat	90¢	300 pounds	100¢	250 pounds
Eggs	50¢	100 dozen	40¢	150 dozen

(*a*) $\frac{1}{3}\sum \frac{p_2}{p_1} = 1.02$; hence a 2 percent increase.

(*b*) $\frac{\Sigma q_1 p_2}{\Sigma q_1 p_1} = \frac{42{,}000}{39{,}000} = 1.08$; hence an 8 percent increase.

(c) $\frac{\Sigma q_2 p_2}{\Sigma q_2 p_1} = \frac{38{,}200}{36{,}300} = 1.05$; hence a 5 percent increase.

(d) $\sqrt{(1.08)(1.05)} = 1.065$; hence a $6\frac{1}{2}$ percent increase.

(e) $\frac{\Sigma q_2 p_2}{\Sigma q_1 p_1} = .979$

$\frac{\Sigma q_1 p_2}{\Sigma q_1 p_1} \cdot \frac{\Sigma p_1 q_2}{\Sigma p_1 q_1} = (1.08)(.931) = 1.005 \neq .979$; hence not satisfied in (b).

$\frac{\Sigma q_2 p_2}{\Sigma q_2 p_1} \cdot \frac{\Sigma p_2 q_2}{\Sigma p_2 q_1} = (1.05)(.910) = .956 \neq .979$; hence not satisfied in (c).

(f) $\sqrt{\frac{\Sigma p_1 q_2}{\Sigma p_1 q_1} \cdot \frac{\Sigma p_2 q_2}{\Sigma p_2 q_1}} = \sqrt{(.931)(.910)} = .920$

Since $(.920)(1.065) = .980$ is approximately equal to .979, the test is satisfied to this numerical accuracy.

Example 2. The following table gives the prices and the average weekly quantities of 4 items purchased by children.

	1954		*1964*		*1974*	
Item	Price	Quantity	Price	Quantity	Price	Quantity
Comic books	20¢	2	25¢	4	35¢	5
Candy	1¢	6	2¢	7	3¢	5
Ice cream	7¢	3	10¢	4	12¢	3
Popsicles	5¢	5	7¢	8	10¢	9

(a) Using 1954 as a reference year for prices and 1964 as the base year for quantities, calculate the price index for the three years. (b) What type of index is the one calculated in (a) for 1954, for 1974 (Laspeyres, Paasche, Fisher, etc.)? (c) Calculate the value index for the three years using 1954 as the reference year. (d) Calculate the purchasing power of the dollar for buying the 1964 market basket in each of the 3 years.

(a) Letting 1, 2, and 3 correspond to the years 1954, 1964, and 1974, it is necessary to calculate $I_p(1)$, $I_p(2)$, and $I_p(3)$. Here

$$I_p(1) = \frac{\Sigma q_{2j} p_{1j}}{\Sigma q_{2j} p_{1j}} = 1$$

$$I_p(2) = \frac{\Sigma q_{2j} p_{2j}}{\Sigma q_{2j} p_{1j}} = \frac{4(25) + 7(2) + 4(10) + 8(7)}{4(20) + 7(1) + 4(7) + 8(5)} = 1.35$$

$$I_p(3) = \frac{\Sigma q_{2j} p_{3j}}{\Sigma q_{2j} p_{1j}} = \frac{4(35) + 7(3) + 4(12) + 8(10)}{4(20) + 7(1) + 4(7) + 8(5)} = 1.86$$

(*b*) 1954 is a Paasche index. 1974 is none of these indexes.

(*c*) $v_1 = \Sigma p_{1j} q_{1j} = 20(2) + 1(6) + 7(3) + 5(5) = 92$

$$v_2 = \Sigma p_{2j} q_{2j} = 25(4) + 2(7) + 10(4) + 7(8) = 210$$

$$v_3 = \Sigma p_{3j} q_{3j} = 35(5) + 3(5) + 12(3) + 10(9) = 316$$

Hence the 3 value indexes relative to 1954 are

$$I_v(1) = \frac{v_1}{v_1} = 1, \qquad I_v(2) = \frac{v_2}{v_1} = 2.28, \qquad I_v(3) = \frac{v_3}{v_1} = 3.43$$

(*d*)

q_2	p_1	p_2	p_3	$q_2 p_1$	$q_2 p_1$	$q_2 p_3$
4	20	25	35	80	100	140
7	1	2	3	7	14	21
4	7	10	12	28	40	48
8	5	7	10	40	56	80
				155	210	289

These are the current GNP for the 3 years. In terms of 1954 dollars, the GNP is 155 for all 3 years because the same quantity (q_2) is purchased each year; hence the IDI is obtained by dividing each of these values by 155 and then multiplying by 100 to give 100, 135, 186. The value of the dollar in terms of 1954 dollars is obtained by taking the reciprocals of these IDI numbers and then multiplying by 100. This gives 1.00, .74, and .54.

Example 3. Show that the Fisher index for prices satisfies the relation $I_v = I_p I_q$. From the definition of the Fisher index,

$$I_p(F) I_q(F) = \sqrt{I_p(L) I_p(P)} \sqrt{I_q(L) I_q(P)}$$

$$= \sqrt{\frac{\Sigma q_1 p_2}{\Sigma q_1 p_1} \cdot \frac{\Sigma q_2 p_2}{\Sigma q_2 p_1} \cdot \frac{\Sigma p_1 q_2}{\Sigma p_1 q_1} \cdot \frac{\Sigma p_2 q_2}{\Sigma p_2 q_1}}$$

Canceling terms, this reduces to

$$\sqrt{\frac{\Sigma q_2 p_2}{\Sigma q_1 p_1} \cdot \frac{\Sigma q_2 p_2}{\Sigma q_1 p_1}} = \frac{\Sigma q_2 p_2}{\Sigma q_1 p_1}$$

The last fraction is $\frac{v_2}{v_1}$, which in turn is I_v.

EXERCISES

SECTION 2

1 For the crop year 1965–1966 the production of boxes of oranges by states was as follows:

Florida	100,400,000
California	36,500,000
Texas	1,300,000
Arizona	2,420,000
Total	140,620,000

(*a*) Regarding the state as the occasion, rather than the year, calculate the simple relatives for each state using California as the reference.
(*b*) Is there any justification for linked relatives in this case? Explain.

2 Simple production relatives for sectors A and B of the economy for year i are 140 and 80, respectively. In the reference year, B's production was 20 percent of A's. What is the index of production for the combined sectors A and B in year i?

3 Suppose the price index on rent for the year 1959, with 1950 as the reference year, is 130 for San Francisco and 120 for Philadelphia. What information do these indexes supply concerning relative rental costs in the two cities?

4 Using the data on orange prices given in Table 1, calculate price relatives for the first 5 years with 1959 as the reference year. Plot the results together with those of Table 1, where 1957 was the reference year. How has the reference change affected the price relatives?

5 As reported in the *Engineering News-Records* for June 17, 1971, an investigation was made of the changes in the level of bid prices for roadway excavation on federal highway contracts. The investigation produced the following values, in which each price is listed as a percentage of the preceding year's price. Calculate the price indexes using 1963 as the reference year.

Year	Price index	Year	Price index
1963	100	1967	104
1964	102	1968	104
1965	102	1969	105
1966	111	1970	112

SECTIONS 3 and 4

6 The following data give the prices in dollars per box and the production in millions of boxes of citrus fruit from 1935 to 1938. Calculate price indexes, using 1935 as the reference and base year. Comment about these index values.

	1935		1936		1937		1938	
Oranges	1.37	33	1.04	30	1.55	46	.55	41
Lemons	3.15	8	1.35	8	4.95	9	1.10	11

7 Prices paid (or sold) and quantities consumed (or produced) of three commodities during two time periods are given in the following table.

Commodity	Time period 1		Time period 2	
	p_1	q_1	p_2	q_2
A	10	2	15	1
B	15	3	10	3
C	20	4	15	4

(*a*) Keeping the quantity mix of period 1, what percentage change in prices has occurred between periods?

(*b*) What is the percentage change in prices if the quantity mix of period 2 is used as the base?

(*c*) Same as (*a*) except now find the percentage change in quantities between the two periods, with prices in period 1 as the base.

(*d*) What percentage change in value of consumption (or production) has occurred?

8 Explain why formula (3) is likely to underestimate the increase in prices for the average consumer.

9 Construct a set of fictitious prices and quantities of 2 commodities for 2 years such that formula (2) will show an increase in prices but formula (3) will show a decrease in prices.

10 Do as in problem 9, but now have (2) show a decrease and (3) show an increase. Is a situation like this likely to arise in real life?

11 Using the data of problem 6, calculate the citrus price indexes for the 4 years if 1935 is the reference base and 1936 is the weight base.

12 Show that if the number of items in an economy is reduced to one, the Fisher price, quantity, and value indexes reduce to simple relatives.

13 A report to management stated that over the past 5 years, labor costs per hour had doubled yet the only increases in rates that had been granted were cost-of-living increases of 26 percent. Moreover, it was asserted that labor costs of the final product had actually decreased. Someone raised the question of the possibility that these results were in error. Give reasons why you believe the results could be correct, assuming that there were no computational errors.

SECTION 5

14 Discuss some of the practical difficulties in determining the average price of meat for use in a consumer price index.

15 Test the logic of the Fisher price index for the case where all prices increase by the factor k. Show that $I_p(F) = k$.

16 How applicable is the BLS consumer price index to your own "cost of living"? Discuss.

17 The two most common indexes used to measure price levels in the U.S. are the CPI (consumer price index) and the IDI (implicit deflator index). The following

data give the values of these two indexes for the years 1960–1973. For the purpose of comparing them, (*a*) convert the IPI series to a 1967 base, (*b*) plot the two series with this common base on the same graph. Although they move together, are they consistent over the entire period? Comment.

Year	CPI 1967 base	IDI 1958 base
1960	88.7	103.3
1961	89.6	104.6
1962	90.6	105.8
1963	91.7	107.2
1964	92.7	108.8
1965	94.5	110.9
1966	97.2	113.9
1967	100.0	117.6
1968	104.2	122.3
1969	109.8	128.2
1970	116.3	135.2
1971	121.3	141.4
1972	125.3	146.1
1973	133.1	154.3

SECTIONS 6 and 7

18 The following data give the GNP (gross national product) in current dollars per capita for the U.S. for the indicated years. They also give the value of the IDI (implicit deflator index) for the various years.

Year	1950	1955	1960	1965	1970	1972
GNP	1,877	2,408	2,788	3,528	4,769	5,531
IDI (1958 base)	80	91	103	111	135	146

(*a*) Calculate the value of the dollar in terms of the IDI with 1970 as the reference base.

(*b*) Calculate the per capita GNP in 1970 dollars.

19 The following data give the per capita GNP (deflated) and the CPI for the period 1960–72. The indexes indicate that the cost of living increased faster than per capita consumption. Explain what these two indexes seem to be describing.

Year	GNP (1958 dollars)	GNP (1967 = 100)	CPI (1967 = 100)
1967	3,400	100	100
1968	3,520	104	104
1969	3,580	105	110
1970	3,530	104	116
1971	3,610	106	121
1972	3,800	112	125

20 The consumer price indexes for the U.S. for the years 1956–1958, if the reference period is 1947–1949, are as follows.

Year	1956	1957	1958
CPI	116	120	124

(*a*) Calculate the CPI if 1956 is used as the reference year.

(*b*) Calculate the purchasing power of the dollar during this period if 1956 is used as the reference year.

21 If each of the 30 stocks used in the Dow-Jones Industrial Average increases by $1 in price, how many points will the D-J increase? (Use the divisor for September, 1968.)

22 Below are presented two indexes of physical production of the United States economy for comparison. The IQI (Implicit Quantity Index of the GNP) was derived. The IIP is the well-known Index of Industrial Production. The GNP (Gross National Product) as a relative index using current prices is also presented. All use 1940 as the reference year.

Although both indexes presumably measure physical production, they do not move together. In fact, in some cases, year-to-year changes may be in opposite directions. (*a*) Graph the IQI and IIP series. (*b*) Write a short statement of your views on the matter.

	GNP				GNP		
Year	Relative	IQI	IIP	Year	Relative	IQI	IIP
1940	100	100	100	1955	398	193	220
1941	125	116	128	1956	419	196	228
1942	158	131	158	1957	441	199	229
1943	192	148	189	1958	447	197	213
1944	211	159	186	1959	484	210	241
1945	212	156	161	1960	504	215	248
1946	209	138	136	1961	520	219	250
1947	232	137	150	1962	560	233	270
1948	258	142	156	1963	590	242	283
1949	257	143	147	1964	632	256	301
1950	286	156	171				
1951	329	169	185				
1952	346	174	192				
1953	366	182	208				
1954	366	179	194				

23 During World War II, a policy for the settlement of wages in a number of industries was based on the BLS consumer price index. This had the effect of limiting wage increases to changes in the level of the index. The unions charged that this index did not properly measure the cost-of-living changes because: (1) quality of goods and services were deteriorating; (2) actual prices were not accurately reported because of official "price ceilings," thus "black market" prices, etc., are not likely to be reported; (3) many goods and services in the index were just not available in some price categories—particularly the cheaper ones; (4) there

was a decrease in the prevalence of price discounting and special sales (such as groceries) because of lesser competition. Comment on the validity of those criticisms in view of the structure of the index.

24 The following item appeared in *Newsweek* magazine, September 30, 1968. Critically evaluate.

THE BLS: UNDER FIRE

The Labor Department's Bureau of Labor Statistics (BLS) is coming under increasing fire in Washington for its method of calculating the consumer price index.

Most recent example: a study by the Department of Health, Education and Welfare that sharply disputes a BLS finding that drug prices have declined by 11% over the past ten years.

Actually, the HEW study says, drug prices have nearly doubled during that period.

According to HEW, the BLS fails to take into account "the changes in consumer expenditures which constantly occur when new and more costly products are introduced—and replace less costly products."

Critics of the BLS on Capitol Hill say the bureau is similarly off the mark when it tries to measure price changes in automobiles and other consumer products.

SECTION 8

25 Which of the following are measurable characteristics and which should be regarded as an index of some type?
(*a*) The unemployment rate in the U.S.
(*b*) The temperature read from a thermometer.
(*c*) Your score on a midterm test.
(*d*) Your score on a driving test.
(*e*) The U.S. Gross National Product.

26 An executive in Kansas City is offered a position in Cleveland and wishes to compare the cost of living in the two cities. The latest Consumer Price Index for Kansas City is 115.3 and it is 108.1 for Cleveland. He therefore concludes that living costs are somewhat lower in Cleveland. Is he justified in this conclusion?

27 If money wages go up 50 percent and the Consumer Price Index goes up 25 percent, (*a*) what is the percent increase in real wages? (*b*) What percent change has occurred in the value of the dollar as far as the wage earner is concerned?

28 The following data give the PDI (personal disposable income) per person in the U.S. for the indicated years. Values of the CPI (consumer price index) are also given for those years.

Year	1950	1955	1960	1965	1970	1972
PDI	1,364	1,666	1,937	2,436	3,376	3,816
CPI (1967 base)	72	80	89	94	116	125

(*a*) Calculate the value of the dollar in terms of the CPI with 1950 as the reference base.
(*b*) Calculate the PDI in 1950 dollars.
(*c*) Calculate the PDI as an index in 1950 dollars and with 1950 as the reference base.

Chapter 16 Time Series

1 INTRODUCTION

It was pointed out in the chapter on correlation that samples taken over time often do not behave like random samples and that therefore the standard statistical techniques cannot be applied to them. In particular, the serial correlation coefficient was introduced as a device for helping to check on the randomness of such samples. This lack of randomness is typical of certain sets of economic data, such as the price of stocks, the cost of living, or the consumption of tobacco. It is the purpose of this chapter to consider methods for treating data of this type.

A set of observations taken over a period of time is called a *time series.* Economists in particular have studied such series extensively because so many of the interesting problems of economics involve them. They have also been studied in the physical sciences in connection with periodic phenomena of various kinds. A few illustrations of time-series data are given in Tables 1, 2, and 3.

Table 1 Yields of 30-year corporate bonds from 1930 to 1972

1930	1940	1950	1960	1970
4.40	2.70	2.58	4.55	7.60
4.10	2.65	2.67	4.22	7.12
4.70	2.65	3.00	4.42	7.20
4.15	2.65	3.15	4.16	
3.99	2.60	3.00	4.33	
3.50	2.55	3.04	4.35	
3.20	2.43	3.09	4.75	
3.08	2.50	3.68	4.95	
3.00	2.80	3.61	5.93	
2.75	2.74	4.10	6.54	

Table 2 Private domestic investment in billions of dollars based on 1958 dollars (1930–1970)

1930	1940	1950	1960	1970
27.4	33.0	69.3	72.4	102.2
16.8	41.6	70.0	69.0	
4.7	21.4	60.5	79.4	
5.3	12.7	61.2	82.5	
9.4	14.0	59.4	87.8	
18.0	19.6	75.4	99.2	
24.0	52.3	74.3	109.3	
29.9	51.5	68.8	101.2	
17.0	60.4	60.9	105.2	
24.7	48.0	73.6	109.6	

Table 3 Annual precipitation in inches in Los Angeles from 1908 to 1973

14	24	5	18	10	17	23	17	23	8
17	9	11	20	15	6	8	9	19	19
9	8	13	19	11	19	15	14	18	18
27	12	20	31	7	23	17	13	16	4
8	11	7	14	25	4	14	12	14	13
17	6	10	6	15	12	8	27	13	24
8	26	17	9	7	17				

The graphs of these three time series, in which neighboring points have been joined by straight-line segments, are given in Figs. 1, 2, and 3. The first two are typical economic time series, whereas the third should be of interest to anyone concerned about the weather and its predictability. The series given by Tables 2 and 3 will be used to illustrate some of the methods that are applied in time-series analysis. They were chosen because of their length and because they differ considerably in some of their properties.

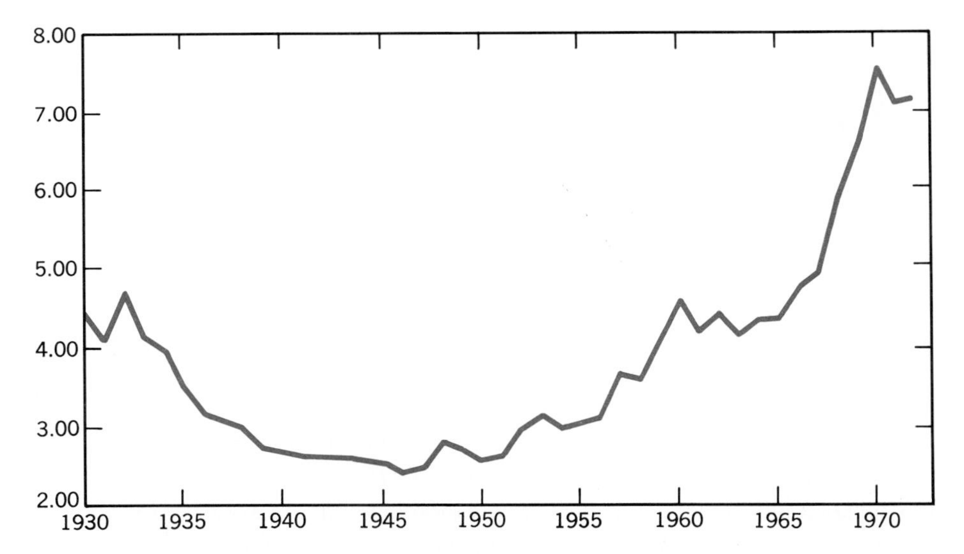

Figure 1 Yields on 30-year corporate bonds (1930–1972).

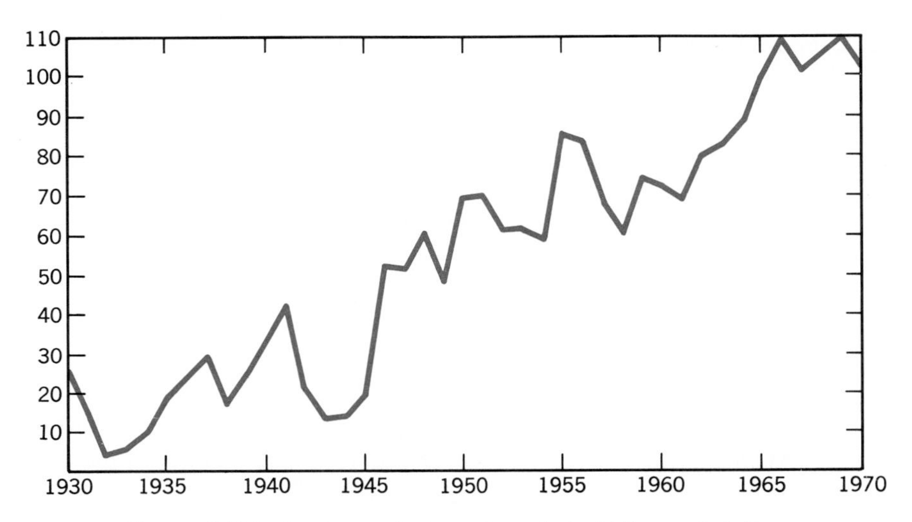

Figure 2 Private domestic investments in billions of dollars.

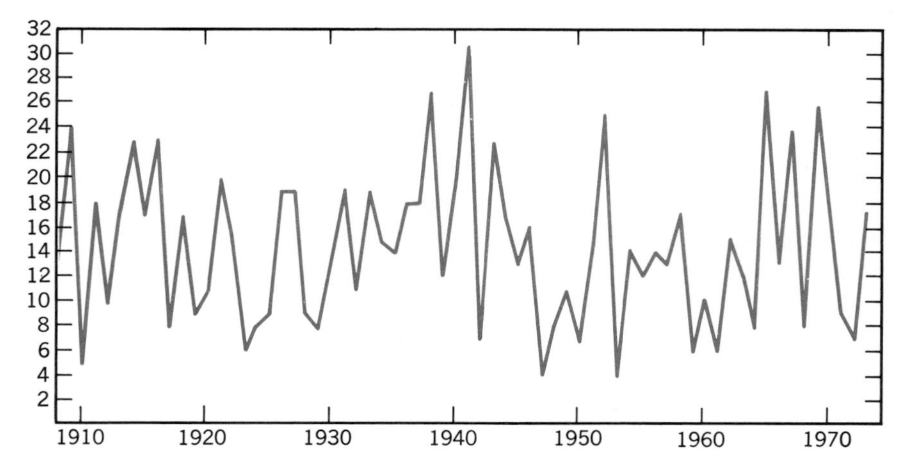

Figure 3 Precipitation in inches in Los Angeles from 1908 to 1973.

In studying time series, the first question that needs to be answered is whether the series really depends on time. A number of statistical tests, such as the one in Chapter 9 based on serial correlation, can be used to answer this question.

If it is quite certain that the series is time-dependent, the next step is to estimate the nature of this dependence on time. Since only the larger movements of the series over time are of major interest, this means fitting a fairly smooth curve to the graph of the series, similar to the fitting of a regression line to a set of points, as explained in Chapters 10 and 11. Regression methods are available for obtaining such an estimate; however, another method based on moving averages is also commonly employed to carry out this estimation.

After the major portion of the time dependence of the series has been estimated, the final problem is to determine whether the resulting relationship can be used to predict the future course of the series with any reliability. The ability to predict the future of economic time series is, of course, the dream and desire of every economist. Many statistical techniques have been proposed for making such predictions; some of these are discussed briefly in a later section.

The preceding three phases in the study of time series are considered in their natural order in the following sections.

2 TIME DEPENDENCE IN A SERIES

It is often clear by merely inspecting their graphs that many familiar time series are changing in a fairly regular manner with time. For example, if the series consisted of the weekly height measurements of a growing plant, it is obvious that the measurements would be time-dependent because they would be increasing with time. The graph of the purchasing power of the dollar for the last fifty years would also be expected to show a strong dependence on time because, as any old-timer will tell you, the value of the dollar has been shrinking over that period of time. However, if the series consisted of the yields of common stocks, the production of coal, or the percentage of the laboring force employed, all taken annually during the last fifty years, then it is not so clear that there would be a definite dependence on time.

The difficulty with inspecting the graph of a time series, and thereby trying to determine whether the series behaves like a random sample from some stable population or whether it depends on time, is that the variation of a random series will often deceive one into believing that the series is dependent on time. Most individuals looking at a time series tend to see regular patterns of movement in the series, whether such regularity is present or not. For example, if one lets his imagination have a free hand, he will undoubtedly see some fairly regular movements in the series shown graphically in Fig. 4. This series was obtained by taking random samples from a normal population with zero mean and unit standard deviation, and therefore there should be no time dependence in it. The data that yielded Fig. 4 are given in Table 4. For series that are not obviously time-dependent, it is necessary to apply some statistical test for randomness before one is justified in proceeding further in the analysis of the series. In selecting a randomization test, it is desirable to consider what alternatives to randomization should be postulated; therefore, such alternatives are considered next.

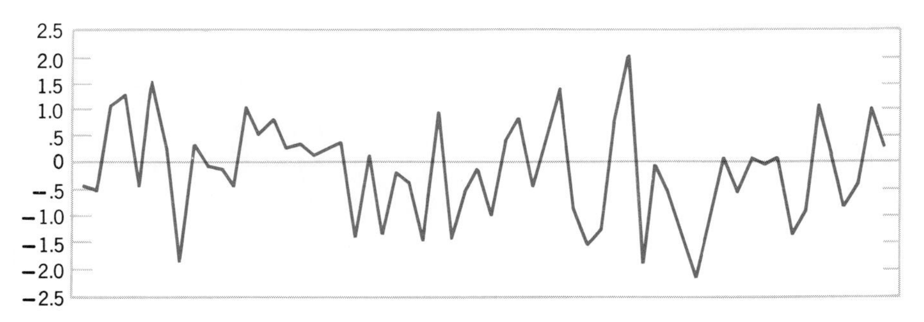

Figure 4 The graph of random samples from a standard normal population.

For many years students of time-series analysis have been concerned principally in determining whether cycles exist in various economic time series. The more ardent advocates of business-cycle analysis attempt to decompose a time series into a long-time trend, cycles, and random effects. The seasonal variation in a time series, which would be found, for example, in monthly department store sales, is not treated as a business cycle and is often removed from the series before studying the series. To qualify as a business cycle, the cycle should normally exceed a year in length.

Table 4 Random samples taken from a standard normal population

−.4	−.1	−1.4	−.9	2.1	.0
−.5	−.4	.2	.5	−1.9	.1
1.1	1.1	−1.3	.8	.0	−1.3
1.3	.6	−.2	−.4	−.6	−.8
−.4	.8	−.3	.5	−1.4	1.1
1.7	.3	−1.4	1.5	−2.2	.2
.3	.4	1.0	−.8	−1.0	−.8
−1.9	.2	−1.4	−1.5	.1	−.4
.4	.3	−.5	−1.2	−.5	1.1
.0	.4	−.1	.8	.1	.3

In view of the preceding remarks, the natural alternative to randomness in an economic time series is a dependence on time that may involve a long-time trend, together with some reasonably long cycles. This is illustrated in the sketch shown in Fig. 5, which was constructed by taking a two-wave cycle and superimposing it upon a straight-line trend and then adding some random deviations.

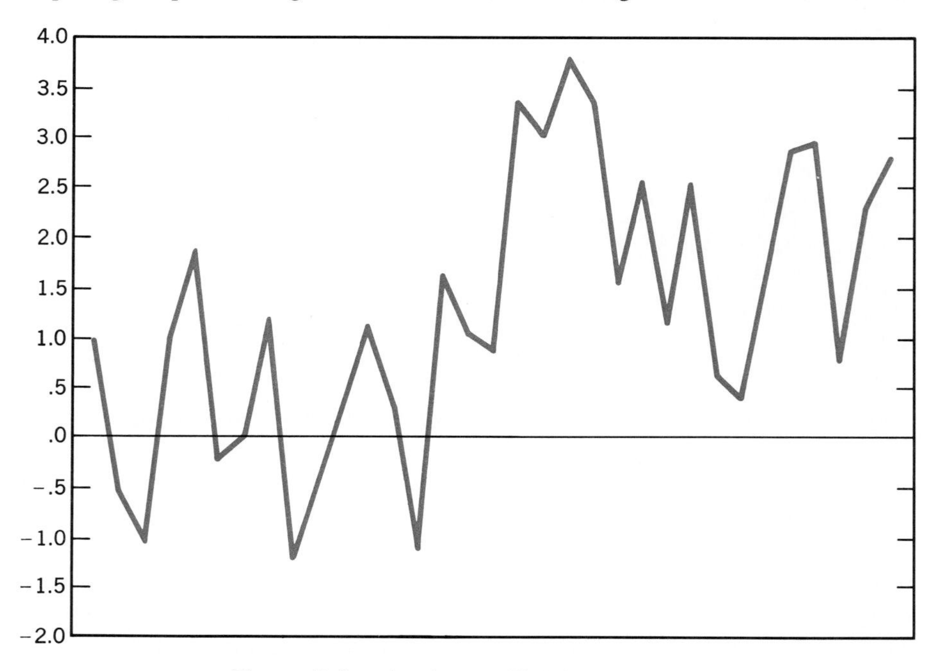

Figure 5 Graph of an artificial time series.

The data corresponding to this sketch are given in Table 5. In this table one reads down columns to obtain consecutive values. It would be very naïve, of course, to expect an actual economic time series to possess a strict cycle; all that can be hoped for is that there will be some fair degree of approximation to one or more cycles. It would be more realistic to speak of the type of cycle considered in economic time series as an undulation or oscillation rather than as a cycle, because the latter term implies regular recurrence and such regularity is seldom found in economic data.

Now that some more or less intuitive notions as to what constitutes reasonable alternatives to randomness in economic time series are available, it is time to consider appropriate tests for randomness. This is done in the next section.

Table 5 Artificial time series consisting of a linear trend, a cycle, and random deviations

1.0	.0	.3	3.1	2.6	.8
−.5	1.2	−1.1	3.8	.7	2.3
−1.0	−1.2	1.7	3.4	.4	2.8
1.0	−.4	1.1	1.6	1.7	
1.9	.4	.9	2.6	2.9	
−.2	1.2	3.4	1.2	3.0	

3 TESTING FOR RANDOMNESS

The problem of testing to see whether a set of observations taken over time behaves like a random sample from some stable population was considered briefly in section 4 of Chapter 9. There it was handled by means of the serial correlation coefficient with lag 1.

If a time series possesses the property that its elements are gradually increasing, or decreasing, in value but otherwise behaves like a random sequence, then the correlation between successive pairs of elements is not likely to reveal the long-time trend in the data. A trend of this type can, however, be revealed by regression methods by showing that the slope of a fitted least-squares line is not zero. This type of a lack of randomness will be considered later.

If a series possesses cyclical movements, the serial correlation with lag 1 may be very effective in revealing this fact because consecutive pairs of observations will tend to be larger than average when observations are taken near the crest of a cyclical wave and to be smaller than average when taken from the trough of such a wave. As a result neighboring pairs will be positively correlated except in the vicinity of the general trend of the series. It is assumed here that observations are taken with sufficient regularity to insure that there will be several observations per wave of a cycle. Thus, the serial correlation coefficient with lag 1 can be effective in rejecting randomness when the lack of randomness is due to cyclical-type movements, without the necessity of knowing the nature of those movements. This is quite different from using the serial correlation coefficient to ferret out a particular length cyclical movement by means of a lag that corresponds to the cycle length.

One of the advantages of using serial correlation methods to check on randomness is that it possesses a certain amount of flexibility. It is possible to calculate the serial correlation for various lags other than lag 1 and thereby sometimes discover an inherent lack of randomness that was not revealed by the lag 1 calculation.

Since the technique for applying the serial correlation coefficient to test for randomness was discussed and illustrated in section 4 of Chapter 9, it will not be repeated here. Incidentally, the illustration used there was based upon the data of Table 3.

There are, of course, numerous other tests available to check on the randomness of a time series. Each such test will be found to be effective in ferreting out some types of nonrandomness but ineffective against other types. If there is reason to believe that a series is nonrandom in a certain way, then a test that it is particularly suited to discover that type of nonrandomness should be employed. These matters will not be pursued further here because the objective of this section is merely to point out that the problem exists and to present one simple method for treating it.

4 SMOOTHING BY MOVING AVERAGES

If a correlation test, or any similar test, has verified that the time series being studied does depend on time, then the next phase of the investigation is to estimate this time dependence. Now most economic time series are quite erratic in appearance, so that it is difficult to determine by inspection any underlying regularity that may exist. Since short-term variation is often of little interest and, except for seasonal variation, is not likely to possess enough regularity to be meaningful anyway, one attempts to estimate the longer movements only. The problem, therefore, is to eliminate the erratic and short-term fluctuations so that the remaining long-range dependence on time can be recognized.

One of the favorite methods for eliminating erratic and short-term movements in a time series is that of *moving averages.* As its name implies, successive averages are computed while moving along in the series. An operation that eliminates erratic and short-term movements is called a *smoothing* operation because it tends to make the graph of the time series appear smooth. Thus moving averages are basically smoothing devices. The number of terms used in a moving average will determine the degree of smoothness that results. Ordinarily, the more terms one uses in the average, the smoother the outcome.

Let the terms of the time series be denoted by $x_1, x_2, x_3, \ldots, x_n$. Then, for example, the three-term moving average series is constructed from it by taking successive averages of three consecutive terms and placing the average opposite the middle term of the three terms being averaged. The first few terms in such a moving average series, together with their location with respect to the original series, are

$$\begin{array}{cc} & x_1 \\ \dfrac{x_1 + x_2 + x_3}{3} & x_2 \\ \dfrac{x_2 + x_3 + x_4}{3} & x_3 \\ \dfrac{x_3 + x_4 + x_5}{3} & x_4 \\ \cdot & \cdot \\ \cdot & \cdot \\ \cdot & \cdot \end{array}$$

A twelve-term moving average is often used on economic time series that consist of monthly data because a moving average of this length is especially effective in smoothing out seasonal variation as well as erratic variation. The terms of a moving average of this type should be located halfway between the sixth and seventh terms used in the average; however, for the sake of convenience, the average is usually placed opposite the seventh term. A more precise location method is to take the mean of two consecutive twelve-term moving averages and place it opposite the seventh term of the thirteen terms involved in the two moving averages. Although this device gives the correct location of such terms, it requires additional lengthy computations, and, since it seldom alters the moving average series appreciably, it is usually omitted. A similar convention of placing the average opposite the first term beyond the middle is often used for other moving averages based on an even number of terms. Shorter moving averages, such as a five-term moving average, are usually long enough to eliminate erratic variation and produce satisfactory smoothness and are to be preferred when working with annual data or with nonseasonal data.

To illustrate the smoothing effects of moving averages, a three-term and also a six-term moving average were computed for the data of Table 2. The computations needed for moving averages are most easily carried out by first finding the proper sums by addition and subtraction. For example, after the first sum, $x_1 + x_2 + x_3$, is obtained for a three-term moving average, the next sum, which is $x_2 + x_3 + x_4$, is obtained by adding x_4 and subtracting x_1 from the preceding sum. This technique of adding the next term of the time series and subtracting the first element of the moving average sum to obtain the next sum for the moving average is especially helpful when computing, say, a twelve-term moving average.

The computations for the first five terms of the three-term moving average of Table 2 would proceed as follows.

x	Sum of three terms	Average
27.4	·	·
16.8	48.9	16.3
4.7	26.8	8.9
5.3	19.4	6.5
9.4	32.7	10.9
18.0	51.4	17.1
24.0	·	·
·	·	·
·	·	·
·	·	·

The results of these computations, together with similar computations for the rest of the series of Table 2, are given in Table 6 and are displayed graphically in Fig. 6. The three-term moving average appears to have eliminated much of the erratic variation in the series; however, there is still a considerable amount of short-term variation remaining. Similar calculations for a six-term moving average yield the graph displayed in Fig. 7. This smoothing operation has certainly eliminated most of the short-term variation that still remained after the three-term moving average was applied.

Table 6 Three-term moving average applied to Table 2

16.3	32.0	66.6	73.6
8.9	25.2	63.9	77.0
6.5	16.0	60.4	83.2
10.9	15.4	65.3	89.8
17.1	28.6	69.7	98.8
24.0	41.1	72.8	103.2
23.6	54.7	68.0	105.2
23.9	53.3	67.8	105.3
24.9	59.2	69.0	105.7
33.1	62.4	71.7	

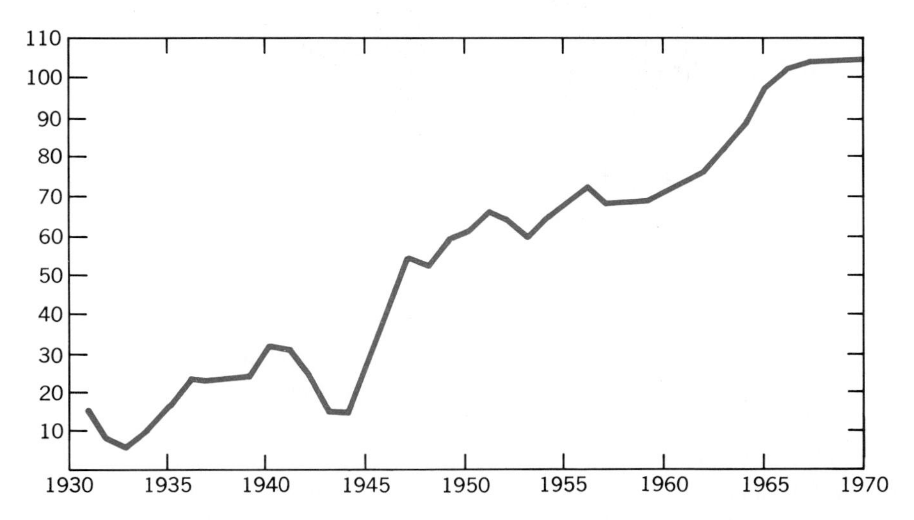

Figure 6 The graph of a three-term moving average associated with Figure 2.

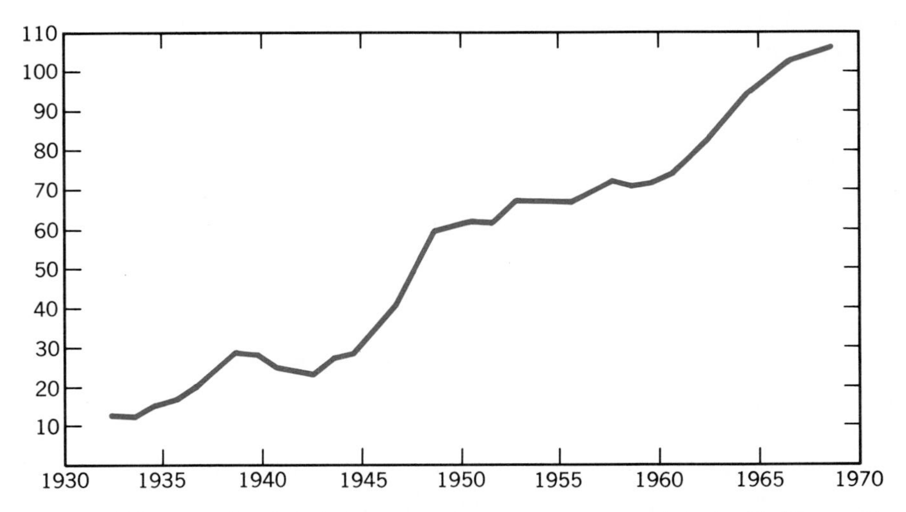

Figure 7 The graph of a six-term moving average associated with Figure 2.

5 WEAKNESSES OF MOVING AVERAGES

Although Figs. 6 and 7 would seem to indicate that moving averages are very effective in smoothing a time series and thus enable one to estimate the underlying long-range time dependence of a time series, there are several properties of moving averages that must be understood before one can hope to use them intelligently.

Moving averages tend to deflate, or depress, the magnitude of oscillations in a series. The larger the number of terms employed in the average, the greater the deflation. A comparison of Figs. 2, 6, and 7 will show that the high oscillations of the original series are damped somewhat by the three-term moving average and are thoroughly deflated by the six-term moving average. Thus, the greater the degree of smoothing, the less realistic the resulting movements.

Not only do moving averages deflate the magnitudes of movements, but they also anticipate rapid changes before the changes occur. For example, the large rise in investments for the year 1950 showed its effect on the three-term moving average in the year 1949. Thus one receives the impression from the moving average that investments rose one year in advance of their actual rise, when in fact they fell in 1949. This effect is due to the fact that the three-term moving average for 1949 uses the data for 1948 and 1950 and thus is heavily influenced by the large 1950 value. Of course, if moving averages have a tendency to anticipate rapid changes, they must also have a tendency to prolong such changes.

Although the two preceding properties of moving averages are not at all desirable, they can be taken into account when interpreting such averages. A third property that can cause considerable difficulty is the tendency of moving averages to produce cyclical movements in data, even though such movements do not exist in the original data. In other words, if one starts with a set of random data and applies a moving average to it, one may generate a new series that appears to have fairly regular cyclical movements in it. The practical implication of this property of moving averages is that the business analyst who uses moving averages to smooth his data, while in the process of trying to discover business cycles, is likely to come up with some nonexistent cycles.

As an illustration of this tendency of moving averages to produce what appear to be cyclical movements in data, a five-term moving average was applied to the data of Table 4. Since these data are random samples from a standard normal population, there is no time dependence in the data, and therefore there are no underlying cyclical movements present. An inspection of Fig. 4 certainly does not reveal any systematic movements in these data, although some cycle enthusiasts with clairvoyant powers might discern some regularity. The results of applying a five-term moving average to the data of Table 4 are shown in Fig. 8. This graph seems to show some rather pronounced oscillations. A comparison of this graph with that of Fig. 6 would seem to indicate that there is not a great deal of difference between the degree of regularity of movement in the two series; yet the series giving rise to Fig. 8 is a random series, whereas that giving rise to Fig. 6 is an actual economic time series.

Some of the minor faults of moving averages can be remedied to a considerable extent by introducing weighted moving averages. Instead of calculating the

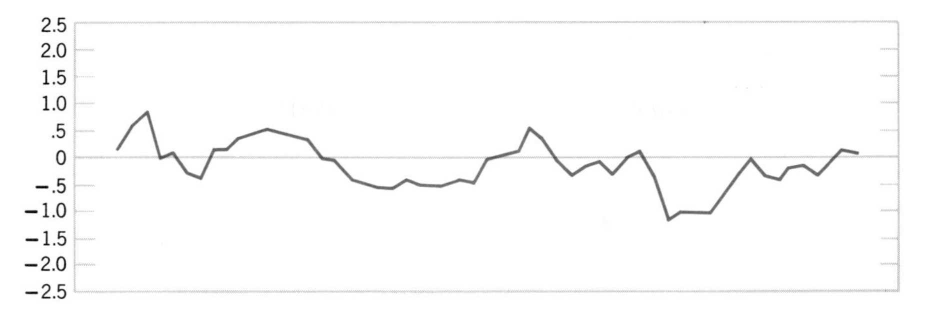

Figure 8 The graph of a five-term moving average associated with Figure 4.

mean of, say, five consecutive terms for a five-term moving average, one might attach weights such as 1, 2, 4, 2, 1 to the respective terms on the grounds that the middle term should have the greatest influence on the average that is to replace it. These weights are quite arbitrary, and other sets of weights can be chosen. Weighted moving averages do not deflate a time series nearly so much as a simple moving average does. Furthermore, they do not stretch out the influence of a sudden change so much as a simple moving average of the same number of terms. They are, however, subject to the same weakness as simple moving averages in that they tend to introduce spurious cyclical movements.

In view of the earlier discussion about the cycle-producing effects of moving averages, it follows that one needs to show extreme caution in claiming cyclical movements in a time series merely because they appear to exist in the moving average series. It is necessary to check the original series very carefully for any movement that is suggested when a moving average is applied to the series. There are methods available for checking on postulated cyclical movements, one of which is discussed very briefly in section 9. Thus it is possible to circumvent the cycle-producing effects of moving averages.

The estimation of the basic movements of a time series by means of moving averages is quite subjective because the estimate depends considerably on the number of terms used in the moving average. For example, the estimate given by Fig. 6 differs considerably from that given by Fig. 7, yet the same series is being estimated in each case. An estimate such as that given by Fig. 7 should be looked upon as an estimate of an underlying structure that is not defined explicitly and whose economic interpretation is often obscure. Such an estimate differs from the customary estimates of statistics that have been studied in the preceding chapters because one cannot state explicitly what is being estimated and how accurate the estimate is. Nevertheless, moving averages are very useful in helping to eliminate the uninteresting features of a time series and thus assist in the study of more basic features of the series.

6 ESTIMATING TRENDS

The method of moving averages that has been described in the two preceding sections is effective in eliminating the erratic and short-run components in a series and thereby revealing any long-run trend in the series. However, when theory or experience indicates that the trend should possess a simple structure, it is usually preferable to determine the trend directly and treat the erratic components as random deviations. For example, in some series it is to be expected that the terms will increase by a constant amount each year, in which case the series will possess a linear trend.

The standard method for determining a linear trend is the method of least squares that was used for linear regression. This is accomplished by replacing the variable x by the variable t in formulas (1) and (2) of Chapter 9. As an illustration, consider the data of Table 7 on orange production in Florida for the years 1930–1971. The graph of this time series is given in Fig. 9. If the years 1962–1964 are ignored (a disastrous freeze occurred in 1962 and seriously affected the crop that year and, because of tree damage, the following two years), it appears that production has been following a linear trend. Excluding the data for 1962–1964, calculations based on formula (2) of Chapter 9 gave the equation

$$y = 3.1875(t - 1930) + 2.924$$

This trend line is the line that is shown in Fig. 9.

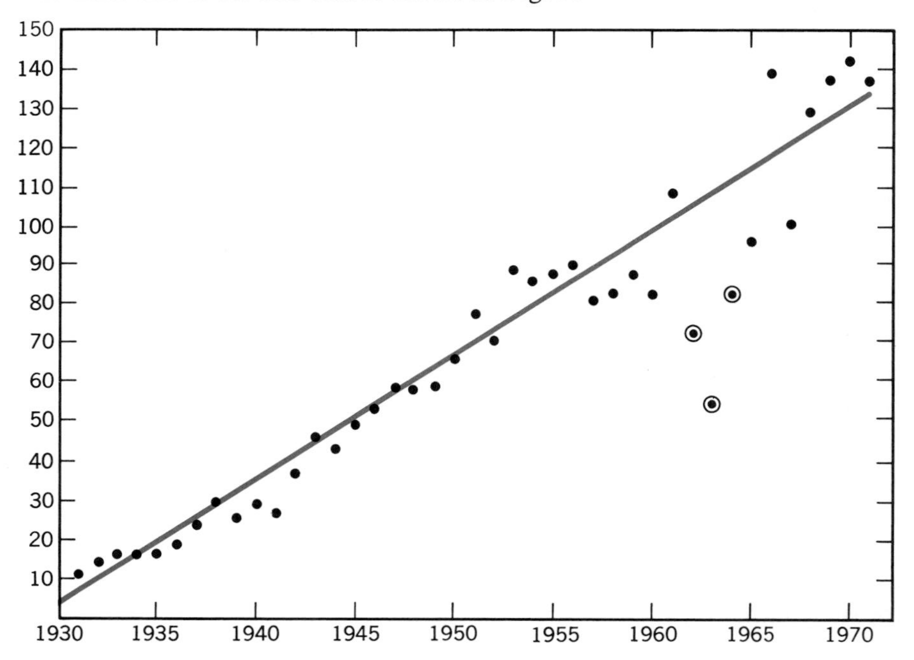

Figure 9 Production of Florida oranges from 1930 to 1971.

Table 7 Orange production in Florida from 1930 to 1971

1930	1940	1950	1960	1970
16.8	28.6	66.2	82.7	142.3
12.2	27.2	76.9	108.8	137.0
14.5	37.2	70.5	72.5	
15.9	46.2	89.1	54.9	
15.6	42.8	85.9	82.4	
15.9	49.8	88.2	95.9	
19.1	53.7	90.3	139.5	
23.9	58.4	81.0	100.5	
29.9	58.3	83.0	129.7	
25.6	58.5	87.6	137.7	

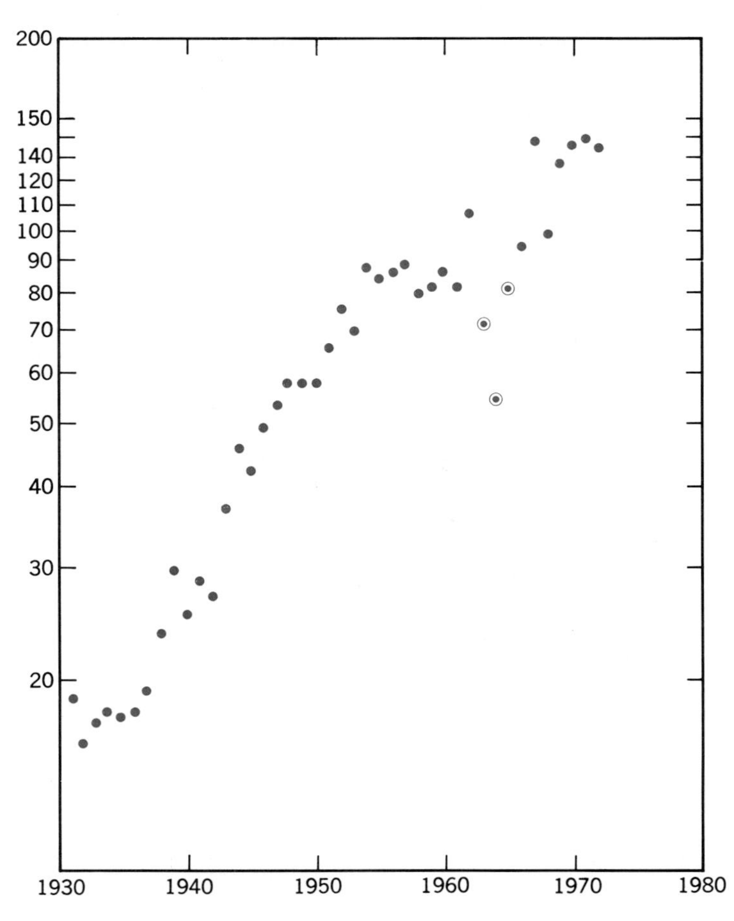

Figure 10 Production of Florida oranges plotted on semi-logarithmic paper.

In some series the terms are expected to increase by a constant percentage each year. This possibility can be discovered by graphing the series on semilogarithmic paper. If a trend is linear on this type of paper, it is of the form $y = ab^t$ on ordinary paper, where b denotes the growth factor. To check the possibility that orange production may be increasing at a constant percentage, the data of Table 7 were plotted on a semilogarithmic scale in Fig. 10. This graph does not appear to be linear over the entire period of time. There appears to have been a constant but high growth rate from 1931 to about 1950, after which the growth dropped to a lower level. The simple linear regression shown in Fig. 9, representing a constant amount of increase per year, appears to be a more accurate and reasonable explanation of the production trend.

Another type of series that occurs frequently is one in which there is a rapid initial growth followed by a gradually decreasing rate of growth. Trends of this type can sometimes be described satisfactorily by means of a parabola or by means of some other simple type curve. An example of this kind of series is that of the sales of U.S. automobiles during the period 1946–1972. A graph of this series is shown in Fig. 11. A fourth-degree polynomial would probably give a better fit than a parabola here; however, it would then incorporate some of the oscillations present and therefore would not represent only a long-run trend.

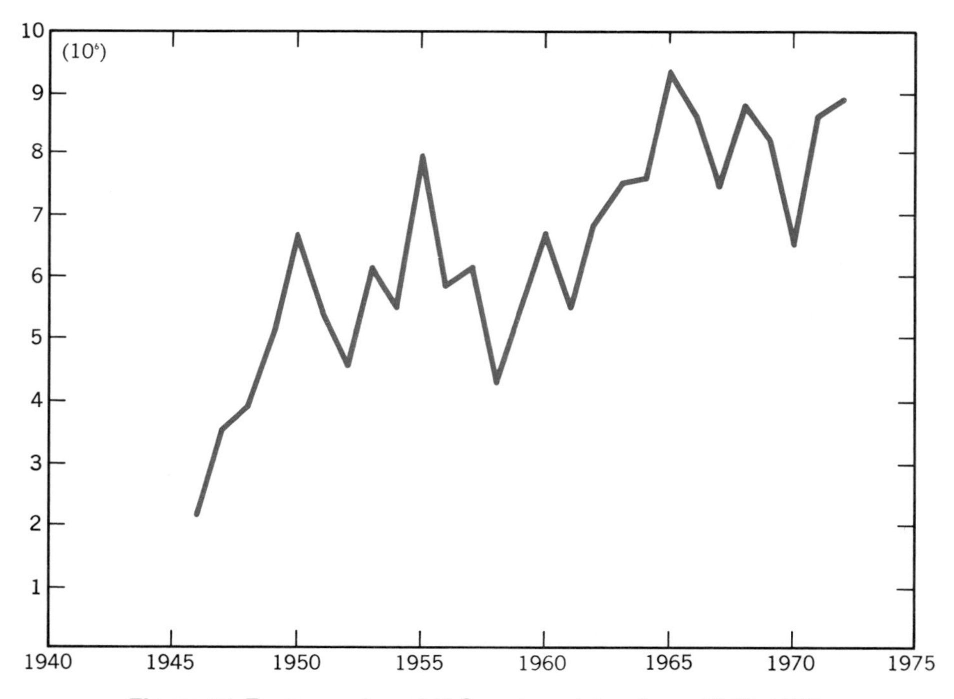

Figure 11 Factory sales of U.S. automobiles from 1946–1972.

7 ADJUSTMENT FOR SEASONAL

In the preceding discussions of time series, only annual data were considered. If the series consists of monthly data and if one wishes to estimate only the more important movements in the series, it is desirable first to eliminate any seasonal variation that may be present. A second reason for eliminating seasonal variation is that one can study short-term trends more easily when pronounced seasonal fluctuations are missing. For example, suppose one wished to know whether department store sales were increasing during 1954, as compared to earlier years, and that only the data of Table 8 were available. It is difficult to inspect Table 8 and determine whether sales during 1954 were rising or falling because of the heavy seasonal pattern in sales; it would be much easier to do so if the seasonal variation were not present.

One of the simplest methods for eliminating seasonal variation is to apply a twelve-term moving average to the series. However, if one does not wish to smooth the data so much, this method is not completely satisfactory. There are methods that are not subject to the smoothing criticism and that, at the same time, yield more reliable results. One of the most popular of these methods is discussed next.

Consider the problem of adjusting the data of department store sales (index values) shown in Table 8 for seasonal variation. This means replacing each monthly sales index by a value that will represent what the sales are that month if allowance is made for seasonal variation. The mechanics of the proposed method is explained first, after which the reasoning behind the method is given.

Table 8 x_{ij}: Index values for department store sales

	J	F	M	A	M	J	J	A	S	O	N	D
1951	96	90	98	99	103	99	84	93	112	112	134	184
1952	83	83	92	103	108	105	84	98	112	120	134	196
1953	85	88	103	115	108	108	89	98	112	115	136	192
1954	83	86	89	110	106	106	88	98	112	118	137	200

Let x_{ij} denote the index value for the year i and the month j. Then one first applies a twelve-term moving average to the data, centering each value opposite the seventh term in the average. As explained in section 4, a more precise location method is available, but it hardly justifies the extra computational labor involved. More data than those shown in Table 8 are needed to find moving averages for the first half-year of 1951 and the last-half year of 1954. Such additional data obtained from other sources were used to compute the moving averages for the data of Table 8. The results of those computations are given in Table 9. These values represent approximate average yearly sales for the middle of the year located at the indicated month. They are denoted by the symbol $\bar{x}_{ij}$.

Table 9 $\bar{x}_{ij}$: 12-Month moving averages

	J	F	M	A	M	J	J	A	S	O	N	D
1951	109	108	108	107	108	109	109	108	107	106	107	107
1952	108	108	108	108	109	109	110	110	110	111	112	112
1953	113	113	113	113	113	113	112	112	112	111	110	110
1954	110	110	110	110	110	110	111	111	111	111	111	111

Each entry in Table 8 is now divided by its corresponding entry in Table 9 and multiplied by 100 to yield the set of crude seasonal-index numbers shown in Table 10. These values are denoted by r_{ij}, where $r_{ij} = 100x_{ij}/\bar{x}_{ij}$.

Table 10 r_{ij}: Percentage of x_{ij} to $\bar{x}_{ij}$, and $\tilde{x}_{.j}$

	J	F	M	A	M	J	J	A	S	O	N	D
1951	88	83	91	93	95	91	77	86	105	106	125	172
1952	77	77	85	95	99	96	76	89	102	108	120	175
1953	75	78	91	102	96	96	79	88	100	104	124	175
1954	75	78	81	100	96	96	79	88	101	106	123	180
$\tilde{x}_{.j}$	79	79	87	98	96	95	78	88	102	106	123	175

The next step in the adjustment consists in calculating a mean seasonal index for each month. This is accomplished by calculating the mean of each column in Table 10. These means are shown in the last row of Table 10 and are denoted by the symbol $\tilde{x}_{.j}$, where $\tilde{x}_{.j} = \Sigma_{i=1}^{r} r_{ij}/r$ and r represents the number of rows (years). If these means do not total 1200 it is necessary to adjust them until they do. In this illustration the means sum to 1206; hence one should multiply each mean by $\frac{1200}{1206}$. Since 1206 differs from 1200 by only $\frac{1}{2}$ percent, this refinement has been ignored here because the other computations have been carried only to the nearest 1 percent. The resulting means are called *seasonal index numbers* because they represent the average monthly rate of sales. If the index numbers were divided by 12, they would yield an average percentage of annual sales corresponding to each month.

The final step in the seasonal adjustment is to divide each entry in Table 8 by the correct monthly seasonal index number from Table 10, treating the latter as a decimal fraction. For example, all the entries in the January column of Table 8 would be divided by .79 because 79 is the seasonal index number for January in Table 10. The resulting values, which are denoted by x'_{ij} and are shown in Table 11, represent sales for the corresponding dates when allowance is made for seasonal variation.

Table 11 $x_{ij}' = x_{ij}/\tilde{x}_{\cdot j}$: Seasonally adjusted department store sales:

	J	F	M	A	M	J	J	A	S	O	N	D
1951	122	114	113	101	107	104	108	106	110	106	109	105
1952	105	105	106	105	112	111	108	111	110	113	109	112
1953	108	111	118	117	112	114	114	111	110	108	111	110
1954	105	109	102	112	110	112	113	111	110	111	111	114

Data that have been adjusted for seasonal variation enable one to determine whether a series is increasing or decreasing during the period of a year without being deceived by apparent increases or decreases caused by natural seasonal variation. Thus one is not able to compare the sales for November and December of 1953 and determine whether department store sales are really picking up, or dropping, because the annual Christmas buying splurge masks any underlying increase or decrease. The adjusted values of Table 11, however, show that there might have been a very slight decrease then, when account is taken of the natural seasonal variation in the data.

The reasoning behind the foregoing adjustment methods runs somewhat as follows. Suppose the value of any entry in Table 8 is assumed to be expressible as the product of three quantities: (1) a number that represents the underlying basic series value, (2) a number that represents the seasonal-index value, and (3) a number that represents the erratic, or random, part of the series, all for the date in question. If these components are denoted by B, S, and E, respectively, then the value of the entry is assumed to be given by $B \cdot S \cdot E$.

The application of a twelve-term moving average to a series of this kind will usually eliminate most of the seasonal component (S) and also most of the erratic component (E), and leave the underlying basic component (B) only slightly changed. Thus the entries in Table 9 represent estimates of B values. Since the first step in the preceding adjustment is to divide the original series values by their twelve-term moving average values, the resulting series should consist essentially of $S \cdot E$ values because one is essentially performing the division $B \cdot S \cdot E/B$. Thus the entries in Table 10 are estimates of $S \cdot E$ values. Since the seasonal (S) values for a given column of this table are constant, by taking the mean of the column, most of the erratic component (E) will be eliminated, thereby leaving only the desired S value. This follows from the assumption that the erratic element E should possess a mean of 1 because it is a factor of series terms and that therefore the mean of a sample of such quantities should be close to 1. Thus the entries in the last row of Table 10 represent estimates of S values. Finally, if the original series values ($B \cdot S \cdot E$) are divided by their S values, the resulting series should represent $B \cdot E$ values, that is, original series values with the seasonal component eliminated.

The assumption that a time series can be decomposed into the product of three such factors is sometimes replaced by the assumption that the series can be expressed as the sum of three such components. In the preceding adjustment method one would then subtract quantities rather than divide by them.

8 PREDICTION (FORECASTING)

There are two major reasons for attempting to estimate the principal movements of an economic time series. First, it is hoped that the estimate will reveal trends and oscillations that can be accounted for on the basis of reasonable economic factors. Economists would like to be able to explain the larger oscillations of economic time series on the basis of economic theory. It would be rather frustrating to them to find that such series were not reacting to the impact of supposedly important economic factors. Thus the analysis of economic time series is a kind of laboratory work for economists who are interested in explaining the dynamics of economic systems.

The second major interest in the estimation of economic time-series movements is the hope that the estimate can be used for predicting the future course of the series. Predicting the stock market with some degree of reliability would, of course, satisfy most individuals, but few time-series analysts are optimistic enough to believe that they can do much with such a sensitive and variable series.

An estimate that has been obtained by means of a moving average is useful for studying the effects of economic factors in the past, but it is not capable of predicting the future unless some method for extending the smoothed series is given. Various methods have been suggested for analyzing the smoothed series further so that it can be extended into the future. For many years the commonest method of performing such an analysis was that of decomposing the series into various parts such that the sum, or the product, of the parts would yield the original series. This is similar to what was done in section 7 in explaining the logic behind the seasonal adjustment procedure. These parts, or components, have usually consisted of a trend, cycles, and erratic elements. For monthly data, seasonal variation is usually first eliminated in a manner such as that described in section 7.

If a series can be decomposed into the sum (or product) of such components, then it is easy to extend the series into the future by merely extending the trend and the cycles and adding (or multiplying) the extended components.

The possibility of decomposing a series in this manner is discussed first for artificial series and then for economic series. Methods for treating economic time series have evolved from those that were created for handling periodic data in the physical sciences, which in turn were patterned after techniques for analyzing artificial series.

9 DETERMINING CYCLES

If a time series has been built up from strictly mathematical components, as in the artificial series shown in Fig. 5, then methods exist for decomposing the series into its basic parts. The major difficulty is, of course, to discover the lengths of the underlying cycles. One simple method for attempting to do this is based on serial correlation.

Suppose a time series consists of a strict cycle such as that displayed in Fig. 12. Then if one chooses a lag of 8, which is the number of time units between the high points on two consecutive waves and also the distance between pairs of points that occupy the same relative position on two neighboring waves, the values x_t and x_{t+8}, which represent the values of x_t and y_t in the serial correlation formula, would be identical. For example, the value of x_7 is the distance to the trough on the first wave, whereas x_{15} is the distance to the trough on the second wave; therefore, these two values are identical. But when corresponding x- and y-values are equal, the value of the correlation coefficient is 1. This is easily seen by plotting the scatter diagram and observing that all the points will lie on the line whose equation is $y = x$. If a lag of 16 had been chosen, the same conclusion would have been obtained; however, there would have been considerably fewer pairs of values available for calculating the correlation coefficient with this lag. Thus it appears that a value near 1 of the serial correlation coefficient with lag k may indicate the existence of a cycle with a distance of approximately k time units or some simple fraction of k, such as $k/2$ or $k/3$, between crests of neighboring waves. Other possibilities could also give rise to a value near 1, but they have been ignored here.

In order to be able to use serial correlation to discover cyclical components in a series, it is necessary to calculate the value of the serial correlation coefficient for all lags from $k = 1$ up to the largest value of k that is considered reasonable and possible. The graph of these serial correlation coefficient values as a function of the lag is called the *correlogram*. From it one can observe the lags that seem to

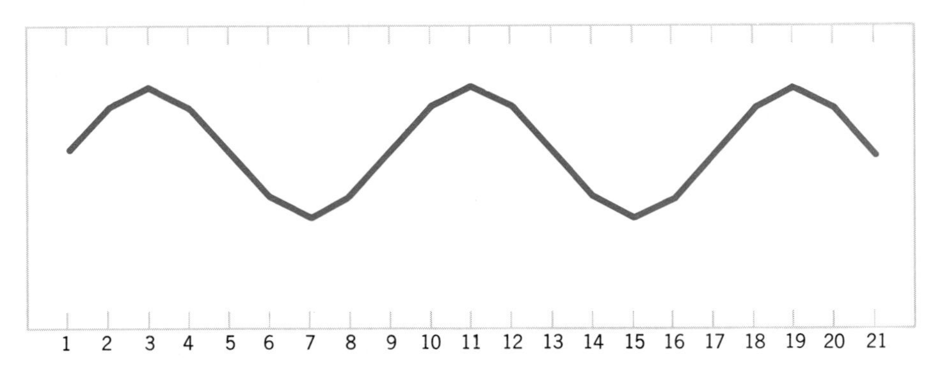

Figure 12 A cyclical series of period 8.

correspond to cycles in the series. If there are several different cycles hidden in the time series, the correlogram might be expected to show a value close to 1 for those lags that correspond to the cycles, and it should have small values for lags that are not close to cycle lags. Serial correlation methods have proved very useful in the physical sciences for discovering cycles and for analyzing the nature of certain types of time series.

After the lengths of the cycles have been discovered in an artificial time series that consists of, say, a trend and the sum of several cyclical components plus some random deviations, it is not difficult to decompose the series. One such method, for example, is based on applying the proper moving averages to the series and isolating one cyclical component at a time.

This discussion has been concerned with a technique that is used in the decomposition of a mathematical time series. The interesting question, of course, is whether methods such as this one, which have been designed for ideal series, are also capable of decomposing economic time series. The answer, unfortunately, is that these methods are successful only if the series is quite regular in behavior. The serial correlation technique, for example, will discover a cyclical movement only if it is fairly regular. Similarly, moving averages will be effective in the decomposition of a series only if the components are quite stable.

In most economic time series the distance between crests of waves is usually highly variable, so that even the larger movements would not respond very well to the serial correlation and moving averages techniques of decomposition. Furthermore, the erratic component is often very large compared to much of the oscillation, with the result that even fairly regular movements are often masked by such irregularities. In view of the lack of stability in most economic time series, it is not surprising to learn that the decomposition methods designed for physical phenomena have not proved very effective when applied to economic data. Economic time series are much more difficult to treat than those encountered in the physical sciences.

To illustrate the difficulty that arises when methods which have been designed to discover regular cycles are applied to data in which there exists a varying length of cyclical movement, a time series was constructed by alternating a cycle of length 4 with one of length 8. This series is shown graphically in Fig. 13. The nonzero ordinates were chosen to be ± 1.

The calculation of the correlogram is easily carried out for this series and yields the following results, where r is the correlation coefficient.

lag	1	2	3	4	5	6	7	8	9	10
r	.46	−.29	−.49	−.37	−.07	.20	.07	−.23	−.24	.24

From these results, which are also shown in Fig. 14, it is clear that r is not capable of discovering the alternating cycles present in this series. The only values of r

that are statistically significant here are the values that correspond to lags 1 and 3, and these serve merely to demonstrate the lack of randomness in the series rather than to indicate the presence of cyclical movements.

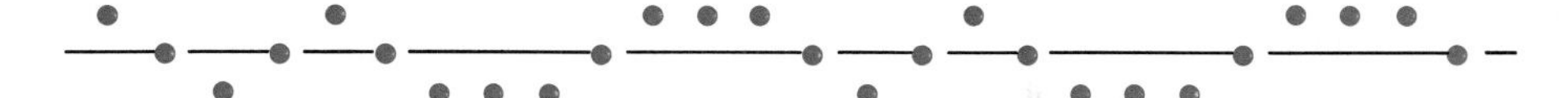

Figure 13 A series with varying length cycles.

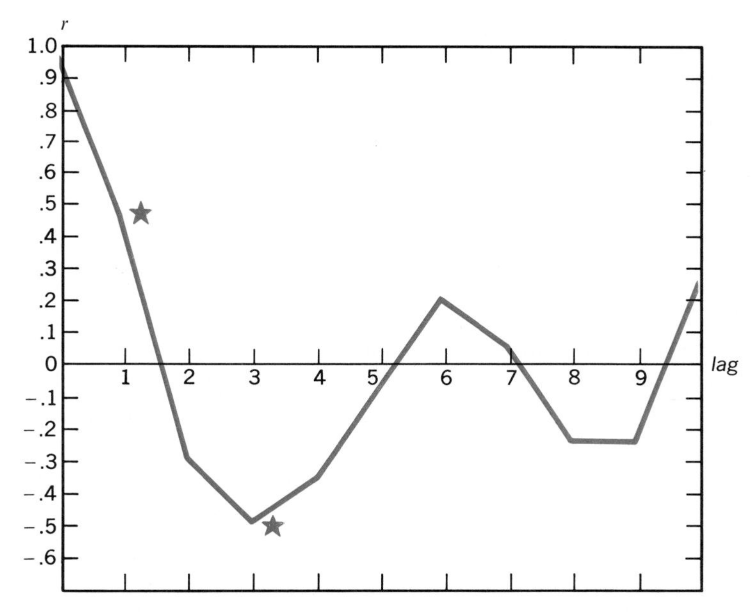

Figure 14 Correlogram for the data of Figure 13.

10 DECOMPOSING ECONOMIC TIME SERIES

Because of the failure of the physical sciences' decomposition methods to yield reliable forecasts for economic time series, economists have devised their own decomposition techniques. The most popular of these assumes that a time series can be expressed as a product of four components, which will be denoted by the letters T, C, S, and E, and which are called trend, cyclical, seasonal, and erratic components. Another popular model assumes that these four components are added rather than multiplied to produce the series. Since a product can be expressed as a sum by means of logarithms, it will suffice to discuss the product model only. Thus, assume that the value of any term in a time series can be obtained from the corresponding term in the product $T \cdot C \cdot S \cdot E$. This model differs from the one introduced in section 7 only in that the quantity B there has been expressed as the product of a trend component T and a cyclical component C. The latter component is conceived of as being made up of various business cycles.

The first step in the decomposition of this product is to determine the trend factor T. If the word trend is understood to mean the overall gradual upward or downward tendency of the series, then T is treated as a linear or quadratic function of the time t and it can be found by fitting the corresponding curve to the points of the graph of the time series. This was considered in section 6. The second step in the decomposition is to divide the time series values by their corresponding fitted trend values. This will yield the values of $C \cdot S \cdot E$. One can now proceed to determine the seasonal component S and the cyclical component C by means of the techniques used in section 7, because C merely plays the role of B there. Since the erratic component E can be determined by division, all four components are now determined.

After the decomposition has been performed, the problem is how to use it to predict future values of the series. The predicted value of T is obtained by substituting the desired value of t in the equation of the fitted trend line. The corresponding value of S is also readily obtained from the table of S values that results from determining the seasonal component. Since the erratic component is assumed to be unpredictable, it is ignored. The remaining cyclical component C is also unpredictable unless further assumptions are made concerning its nature. If it is assumed to be made up of several fairly regular cycles, then one can try to determine the nature of those cycles and project them forward to the desired point in time to obtain a predicted C value. The product of the three predicted component values would then yield the desired forecast of the series.

Although the preceding analysis sounds promising, the difficulty lies in forecasting the value of C because, for most economic time series, this component cannot be expressed in terms of regular cycles. This is precisely why the decomposition methods of the physical sciences failed on economic time series. Unless C can be expressed satisfactorily in terms of a few fairly well defined business cycles, there is not much that one can do to predict a time series by these methods other than to predict on the basis of its trend and seasonal component. Unfortunately, the problem of how to discover business cycles, if indeed they exist, is a very difficult one and one about which there is very little agreement among economists. It follows, therefore, that the decomposition procedure should not be treated as a well-established technique for predicting a time series. Any statistical device that sheds light on the structure of a time series should prove useful to someone who understands all the economic factors operating to generate the series, and therefore should enable him to forecast with more assurance than before. The decomposition technique is such a device but it certainly is not a tool for the amateur forecaster.

Another disturbing feature of the decomposition procedure for prediction, besides that of lacking reliability, is the difficulty in interpreting the components that have been isolated. In a dynamic economic system it is difficult to pick out economic factors that contribute only to the gradual growth, or decline, of the series, other factors that contribute only to cyclical movements, and still other factors that have only a purely random effect on the series. Thus many inventions and new processes in industry have been introduced at random times, but they certainly have contributed to the steady growth, or decline, of certain industries. It is very difficult to decompose dynamic factors in an economic time series in this artificial mathematical manner.

11 STOCHASTIC MODELS; AUTOREGRESSION

In recent years economists have become interested in new methods for studying time series; these methods are based on investigating the relationship between successive terms of the series. In this new approach the terms of a series are treated as a set of random variables that possess certain probability properties. For mathematical convenience a time series is assumed to extend to infinity in both past and future time; therefore it is conveniently expressed in the form

$$\ldots, x_{-2}, x_{-1}, x_0, x_1, x_2, \ldots$$

The subscript on any x denotes the number of time units in the future, if it is positive, and in the past, if it is negative. A set of random variables such as this is called a *stochastic process.*

In terms of the preceding notation, various assumptions concerning these random variables can be introduced that will correspond to how the terms of an economic time series might be expected to be related to each other. One of the simplest models of this type assumes that the relationship between consecutive terms is given by the formula

$$x_t = \beta x_{t-1} + e_t \tag{1}$$

The quantity β is a constant and the quantities $\ldots, e_{-2}, e_{-1}, e_0, e_1, e_2, \ldots$ are a set of random variables that are usually assumed to be independent and to possess a common distribution. In words, this model states that the value of any term in the series depends only upon the value of the term immediately preceding it in time and upon an erratic, that is, random, additive component. One would hardly expect that very many time series can be explained by such a simple model because this model assumes that the entire past history of the series, except for the term one time unit back, is of no value in predicting the future course of the series. This weakness can easily be overcome, however, by assuming a more sophisticated model, such as

$$x_t = \beta_1 x_{t-1} + \beta_2 x_{t-2} + \cdot \cdot \cdot + \beta_k x_{t-k} + e_t \tag{2}$$

Here the value of any term depends upon the values of the k terms immediately preceding it, where k may be chosen sufficiently large to include enough past history to satisfy one's desires.

The preceding stochastic-process model has been applied with much success to some of the problems of engineering communication theory. Although economic time series do not possess properties as attractive as those of the series of communication theory, there is enough similarity to give some hope that these newer methods may yield better forecasts than those obtained by the decomposition technique.

One obvious weakness of the stochastic-process approach, which also applies to the decomposition method, is that the prediction of a future value of a time series is based exclusively upon the values of that same series in the past. One would expect that a knowledge of the behavior of other related series should contribute to a better forecast. This certainly would be true, for example, if one variable had a tendency to follow the movements of another variable one time unit later, because then the latest value of the leading variable could be used to forecast the value of the lagging variable. Variables such as wholesale and retail prices are an illustration of such possibilities. Considerations such as the foregoing would lead one to generalize model (2) to include additional sets of pertinent variables on the right side of the equation. This would give rise to a grand stochastic linear-regression model, which presumably should do better at forecasting than the earlier simpler models. The calculations needed to determine a regression function of this magnitude would be rather formidable. However, now that high-speed computing facilities are available, the opportunities for trying out such other approaches to the forecasting problem have increased remarkably. A prominent economic statistician was overheard waxing enthusiastic about his success with an elaborate linear-regression model for prediction. His enthusiasm, however, may possibly have been based on his personal success in the stock market rather than upon a scientific appraisal of his methods. The best that one can hope for in a method of predicting economic time series is to outdo those who do not possess such methods, because the economic factors upon which any predictive scheme is based are not likely to remain static very long. The problem of prediction is a difficult but fascinating one, and new approaches to a solution will always find a sympathetic audience among those who have not succeeded with the aid of current methods. Problems are plentiful and await the attack of imaginative minds.

Example. As an illustration of how an autoregression model is applied, consider the application of the model given in (1) to the data on the prices of American railroad stocks that are given in Table 12. Graphing the value of x_t against the value of x_{t-1} produces Fig. 15. Except for the year 1931, the relationship appears to be linear and therefore the model appears to be a reasonable one. Calculations based on the usual least-squares techniques of linear regression yield the equation

$$x_t = 10.8 + .828x_{t-1}$$

To forecast the stock price for a given year t, it is merely necessary to multiply the price of the stock for the preceding year, $t - 1$, by .828 and add the constant value 10.8.

For the purpose of observing how well this forecasting function performs, it was used to forecast the stock price for each of the years from 1913 through 1936. These forecasts are shown in Fig. 16 by means of the dots. It should be noted that his forecasting function is not capable of

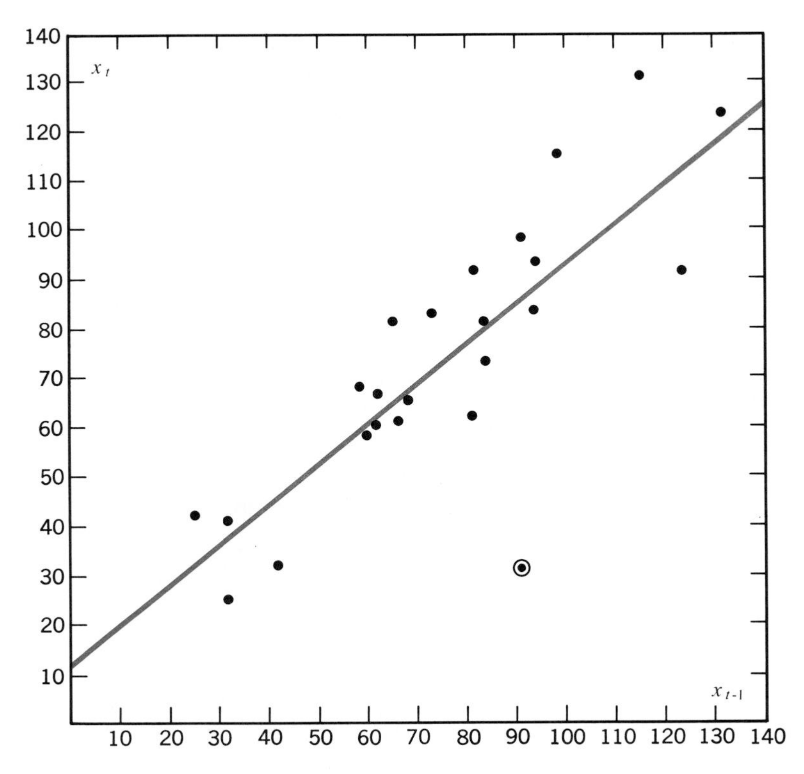

Figure 15 The relation between x_{t-1} and x_t for American railroad stocks (1912–1936).

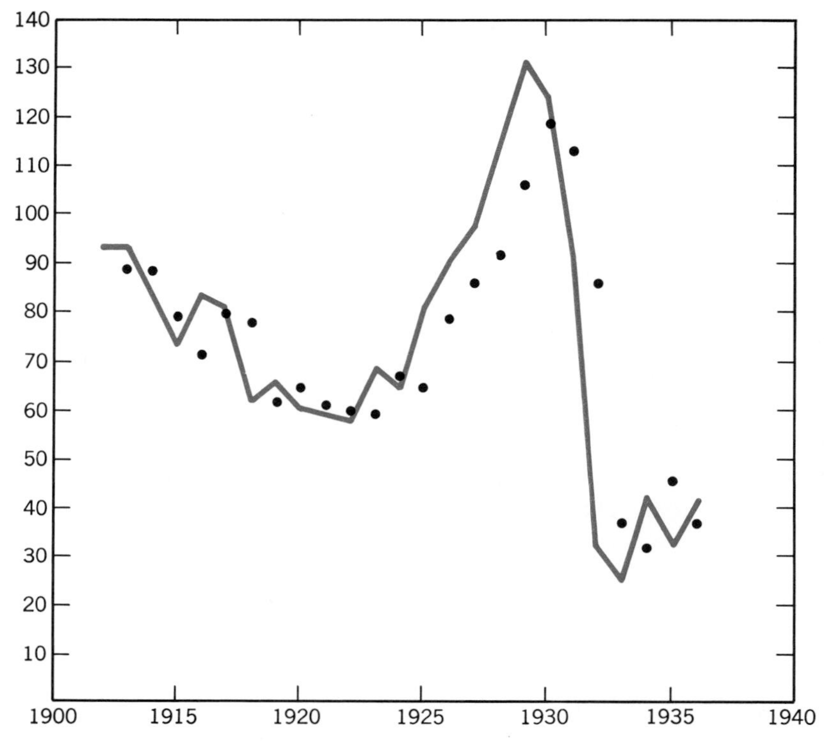

Figure 16 American railroad stock prices from 1912–1936 and their forecasts.

anticipating sharp changes; however, that would be expecting too much of a forecasting formula, since such changes are so irregular.

It would be interesting to see whether a forecast that is based on using a linear function of x_{t-1} and x_{t-2} would yield better forecasts than the formula employed here. It would also be interesting to see whether forecasts

Table 12 Prices of American railroad stocks from 1912 to 1936

1910	1920	1930
·	61	124
·	60	91
93	58	32
93	68	25
83	65	42
73	81	32
83	91	41
81	98	
62	115	
66	131	

based on linear trends involving the preceding data would do as well as this stochastic model.

12 TWO-VARIABLE TIME SERIES

Studies are frequently made of the relationship between pairs of economic variables. If these variables are dependent on time and the variables are denoted by x_t and y_t, the problem is one of attempting to measure the correlation between those variables or of attempting to predict one in terms of the other. As was pointed out in the chapter on correlation, the value of the correlation coefficient may depend heavily upon time in such situations and does not represent a true basic underlying relationship between the variables. Similarly, if one attempts to predict y_t from a knowledge of x_t by means of a linear regression function of the form

$$y_t = \alpha + \beta x_t$$

the relationship between these variables is an artificial one because of their dependence on time. In such situations it is often useful to treat t as a third variable and use the multiple regression techniques of Chapter 11 to study the relationship. Thus, one might use the regression model

$$y_t = a + bt + cx_t$$

Such techniques can also be applied to studying the relationship between lagged variables. For example, they could be used to study the relationship between wholesale prices at time t and retail prices at time $t + k$. This device of attempting to eliminate the affect of time on the relationship enables one to study the desired underlying relationship between the two variables, independent of time.

The time series of economics are a fascinating but difficult field to study. They are so highly irregular that it is difficult to construct a mathematical model that will yield realistic results. The preceding methods are only a beginning in that direction.

REVIEW ILLUSTRATIONS

Example 1. The following data give the production (P) of Florida grapefruit for the period 1940–1971 in units of millions of boxes.

Year	P	Year	P	Year	P	Year	P
1940	25	1950	33	1960	32	1970	43
1941	19	1951	36	1961	35	1971	47
1942	27	1952	32	1962	30		
1943	31	1953	42	1963	26		
1944	22	1954	35	1964	32		
1945	32	1955	38	1965	35		
1946	29	1956	37	1966	44		
1947	33	1957	31	1967	33		
1948	30	1958	35	1968	40		
1949	24	1959	30	1969	37		

(*a*) Plot the data as a time series. Comment on the nature of a trend that might prove satisfactory. (*b*) Calculate a 5-year moving average and graph it. (*c*) Do there appear to be any cycles present? (*d*) Calculate the serial correlation coefficient with lag 1 year and test it for significance. Comment.

(*a*)

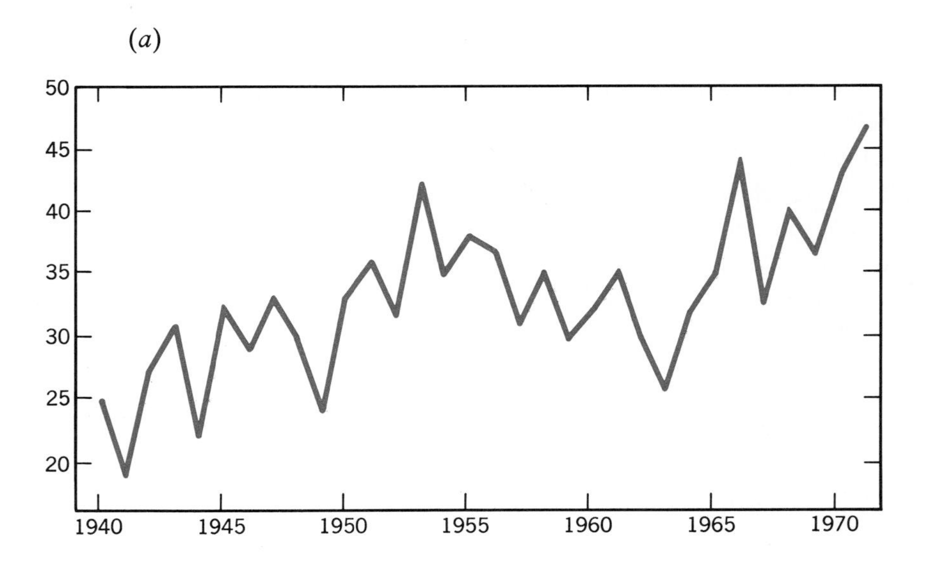

A linear trend might prove satisfactory, however a third-degree polynomial would probably give a better fit.

(*b*) The 5-year moving average values are: 24.8, 26.2, 28.2, 29.4, 29.2, 29.6, 29.8, 31.2, 31.0, 33.4, 35.6, 36.6, 36.8, 36.6, 35.2, 34.2, 33.0, 32.6, 32.4, 30.6, 31.0, 31.6, 33.4, 34.0, 36.8, 37.8, 39.4, 40.0.

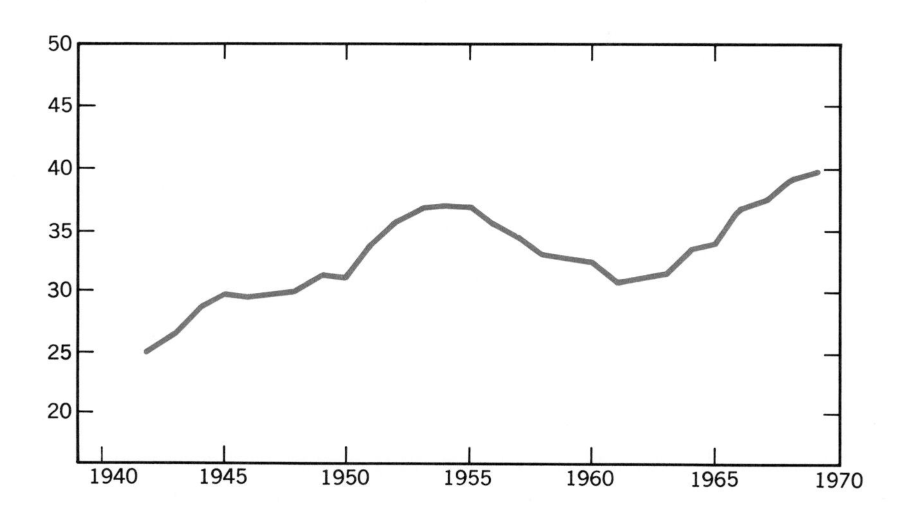

(*c*) No, except possibly a long cycle of about 16 years. The latter, however, is not compatible with the start of the series, that is, from about 1940 to 1944.

(*d*) Here $r = .50$. From Table VIII in the appendix with $N = 32 + 2 = 34$, the value of r_0 for a two-tailed test is .339. Hence

$$\pm r_0 - \frac{1}{n-1} = \pm .339 - \frac{1}{31} = \pm .339 - .032$$

The critical region is therefore the values of r that lie outside the interval $(-.371, .307)$. Since $r = .50$ lies in the critical region, the hypothesis of randomness is rejected. This positive correlation could be caused by the gradual rise in production over the years, because at the start small production values for one year are followed by relatively small values the following year, whereas near the end of the series larger productions occur with the result that a large production is likely to be followed by a relatively large production the following year. If the trend were eliminated from this series, it might behave like a random series, or it might reveal a two-year cycle.

Example 2. The following data from *Statistical Abstracts* give the ratio of total Republican votes to total Democratic votes for candidates for the House of Representatives from 1920 to 1972.

Year	R/D	Year	R/D	Year	R/D
1920	1.65	1940	.89	1960	.82
1922	1.16	1942	1.10	1962	.90
1924	1.38	1944	.93	1964	.74
1926	1.41	1946	1.21	1966	.95
1928	1.33	1948	.88	1968	.96
1930	1.18	1950	1.00	1970	.84
1932	.76	1952	1.00	1972	.90
1934	.78	1954	.90		
1936	.71	1956	.95		
1938	.95	1958	.77		

(*a*) Plot the data. Do they appear to form a random sequence? (*b*) Test for randomness. (*c*) Looking at the graph in (*a*), what would you have predicted for the 1974 election? (*d*) As a forecasting device, calculate the autoregression equation based on a two-year lag. (*e*) Use your result in (*d*) to predict the 1974 value. Compare with your result in (*c*).

(*a*)

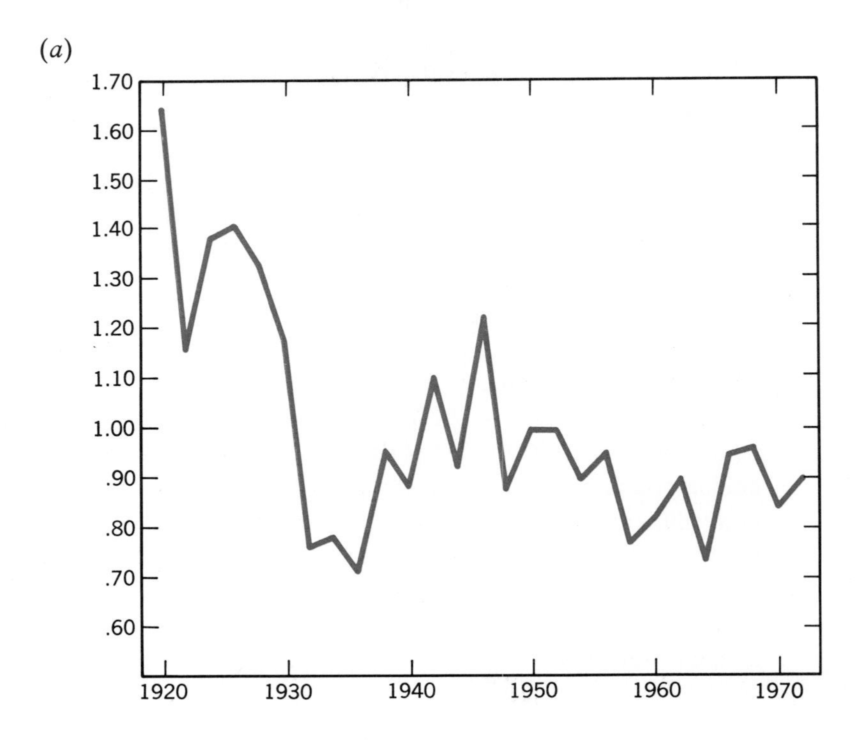

The data do not appear to form a random sequence, unless the first 10 or 20 years are eliminated.

(*b*) Here $r = .60$. From Table VIII with $N = 27 + 2 = 29$, the value of r_0 for a two-tailed test is .367; hence

$$\pm r_0 - \frac{1}{n-1} = \pm .367 - \frac{1}{26} = \pm .367 - .038$$

The critical region is therefore the values of r that lie outside the interval $(-.405, .329)$. Since $r = .60$ lies in the critical region, the hypothesis of randomness is rejected.

(*c*) Between .90 and .95.

(*d*) $x_t = .473 + .501x_{t-2}$

(*e*) $x_{1974} = .473 + .501x_{1972} = .473 + .501(.90) = .924$

EXERCISES

SECTION 3

1 Apply the serial correlation coefficient test for randomness to the time series given in Table 4 in the text. Comment on your result.

2 The following data give the monthly call-money rates for five consecutive years. Each column corresponds to a year. Test for randomness by means of the serial correlation coefficient with lag 1.

4.6	5.8	2.3	2.2	8.6
2.4	2.9	1.8	2.2	4.6
3.9	6.0	1.8	3.2	4.9
5.1	4.2	1.4	3.2	9.5
5.6	2.4	1.6	2.4	4.2
2.8	3.1	1.1	2.5	3.2
3.5	2.5	1.0	2.3	3.0
3.8	2.0	.9	2.1	4.4
10.8	2.3	1.5	3.6	9.4
7.6	2.7	2.0	5.3	5.2
4.9	5.2	2.8	7.7	7.5
6.8	5.5	3.1	16.0	14.0

3 The climate in the Los Angeles area during the Pleistocene period some 15,000 years ago has been studied by examining plants taken from the tar pits at Hancock Park. Basing rainfall estimates on the sizes of growth rings of a tree that according to radiocarbon dating has been dead about 14,400 years, the

data below are for a 40-year period. They are from the *Los Angeles Times* of March 22, 1965. Test the series for randomness.

Year (B.C.)	Rain (inches)	Year (B.C.)	Rain (inches)	Year (B.C.)	Rain (inches)	Year (B.C.)	Rain (inches)
12,935	40	12,925	60	12,915	15	12,905	20
12,934	20	12,924	20	12,914	15	12,904	20
12,933	60	12,923	25	12,913	10	12,903	40
12,932	40	12,922	25	12,912	20	12,902	50
12,931	40	12,921	25	12,911	20	12,901	50
12,930	50	12,920	20	12,910	30	12,900	30
12,929	60	12,919	40	12,909	40	12,899	20
12,928	70	12,918	15	12,908	40	12,898	40
12,927	70	12,917	15	12,907	20	12,897	15
12,926	75	12,916	10	12,906	40	12,896	20

SECTION 4

4 The following data give the prices of American railroad stocks from 1880 to 1936. Read down columns. Apply a 3-term moving average to the data and graph both series. Do any cyclical movements appear to be present?

45	47	52	103	61	124
56	42	64	94	60	91
53	49	78	93	58	32
52	48	84	93	68	25
47	37	68	83	65	42
38	36	86	73	81	32
47	37	100	83	91	41
50	36	98	81	98	
47	42	74	62	115	
45	51	96	66	131	

5 Apply a 3-term moving average to the data of Table 5 in the text and graph the result.

6 Draw 50 one-digit random numbers from Table XII in the appendix and apply a 5-term moving average to them. Graph the result and observe whether any cyclical movements have been generated.

7 Apply a 5-term moving average to the data of Table 5 in this chapter and graph the result. Compare this result with that of problem 5.

8 What moving average would you apply to the data of Table 5 if you wished to eliminate the 2-cycle wave as well as the random component and thus estimate the underlying trend?

9 Apply a 3-year moving average to the data of Table 3 in this chapter. Do there appear to be any cycles in precipitation over that time period?

10 Apply a 5-year moving average to the data of Table 3 in this chapter. Does there appear to be any trend in the amount of precipitation over that time period? Any possible long cycles?

SECTION 6

11 Fit a straight line by the method of Chapters 10 and 11 to the data of Table 5 in the text to obtain an estimate of the underlying trend.

12 The following data give the mortgage loans paid off during the year as a percentage of the gross mortgage portfolio from 1929 to 1958.
(*a*) Graph the data.
(*b*) Does a linear trend appear to be present? How would you test for a linear trend? Read the data a row at a time.

16.8, 16.6, 19.2, 19.8, 21.1, 27.9, 28.3, 23.0, 21.3, 18.3,
20.4, 21.4, 21.0, 22.1, 25.0, 25.9, 26.2, 28.6, 26.0, 22.4,
21.1, 25.5, 22.9, 22.4, 20.9, 20.1, 21.5, 18.4, 16.1, 15.8

13 The following data give the worldwide average yields of major grains (excluding rice) for the period 1960–1973. The yields are in quintals per hectare (metric units in agriculture).

Year	Yield	Year	Yield	Year	Yield
1960	13.9	1965	14.7	1970	17.3
1961	13.4	1966	16.2	1971	18.8
1962	14.3	1967	16.2	1972	18.5
1963	13.9	1968	16.7	1973	19.4
1964	14.5	1969	16.9		

(*a*) Find the equation of the trend line, assuming that the trend is linear.
(*b*) Plot the data and graph the line obtained in (*a*). Does a linear trend appear to be a satisfactory type of trend?

SECTION 7

14 Use the method explained in the text to obtain seasonal index numbers for the data of problem 2. In performing the computations, obtain entries for the 3 middle years only.

15 The following data represent the monthly amounts of magazine advertising in ten-thousands of lines for 4 consecutive years. Use the method explained in the text to adjust these data for seasonal variation. In performing the computations, obtain entries for the 2 middle years only. Each column corresponds to one year.

254	245	268	250
301	285	316	302
326	321	360	342
350	368	408	388
358	343	387	364
301	306	355	335
242	258	286	245
223	216	243	206
276	281	316	260
341	350	376	302
357	349	383	304
318	317	362	282

16 The following data give the number of persons killed in automobile accidents in the U.S. in selected years from 1960 to 1972. They also give the percentages of deaths that occurred during each of 4 six-hour periods of the day.

Year	Number	12P.M.–6A.M.	6A.M.–12M.	12M.–6P.M.	6P.M.–12P.M.
1960	38,000	22.3	15.6	27.8	34.3
1965	49,000	22.2	14.8	28.7	34.3
1970	55,000	21.6	15.6	28.4	34.4
1971	54,000	21.5	15.7	28.5	34.3
1972	56,000	21.4	15.1	28.1	35.4

Calculate the period-of-day indexes for automobile accident deaths in the U.S. by treating these percentages as r_{ij} in the example given in section 7.

SECTION 8

17 The trend of sales for a new department store was found to be given by the formula

$$y = 120{,}000 + 1{,}000t$$

where t denotes the number of months after December 1973, which means that $t = 0$ corresponds to December 1973, and where y is sales in dollars. Some of the seasonal indexes for the company's sales, based on January sales, are the following:

January—100, February—80, March—90, April—120

Ignoring possible shifts in economic conditions, use this trend to predict sales for (*a*) February 1975, (*b*) April 1976.

18 The following data are the U.S. R&D obligations to universities and colleges in millions of dollars for the period 1963–1973.

Year	1963	1964	1965	1966	1967	1968	1969	1970	1971	1972	1973
Amount	900	1077	1197	1351	1455	1490	1526	1464	1496	1757	1953

(*a*) Plot the data.
(*b*) Predict the 1974 value based on a straight-line trend fitted to the data by means of judgment and a ruler.

19 The following data are numbers of passenger cars (in millions) sold by U.S. factories per year from 1950 to 1967.

Year	Number	Year	Number	Year	Number	Year	Number
1950	6.7	1955	7.9	1960	6.7	1965	9.3
1951	5.3	1956	5.8	1961	5.5	1966	8.6
1952	4.3	1957	6.1	1962	6.9	1967	7.4
1953	6.1	1958	4.3	1963	7.6		
1954	5.6	1959	5.6	1964	7.8		

(*a*) Plot the data by year. Does the series appear to be random apart from a trend?
(*b*) Fit a straight line trend to the data through 1966 only.
(*c*) Use your result in (*b*) to predict production in 1967 and compare with the actual production.

20 Look at the graph of the annual rainfall in Los Angeles over a 66-year period as given by Fig. 3. A test indicates that the series is random. What sort of prediction function would you suggest for the next year's rainfall? Why?

21 Refer to the grapefruit data of problem 12, Chapter 10. Plot the data and comment on the problem of predicting next year's production.

SECTION 9

22 Calculate the serial correlation coefficient with lag 12 for the data of Table 8 in the text and comment about its value.

23 Calculate the serial correlation coefficient with lag 12 for the data of problem 15. Comment on your result.

24 For the series of problem 12, does a fairly regular cycle appear to be present? If you believe it does, apply a moving average with the number of terms equal to half the distance between waves and observe whether it kills off the assumed cycle.

25 In problem 24 calculate the serial correlation for a lag equal to the assumed cycle length and comment on your result.

26 Calculate the serial correlation coefficient with lag 16 for the data of Table 5 in this chapter. Does there appear to be a cycle of this length present? If not, look at Fig. 5 and guess what the underlying cycle is. Then calculate the serial correlation for that lag to see whether your guess was a good one. Since this time series was constructed with a two-wave cycle, the serial correlation should be able to pick it out if all lags from 1 to, say, 20 were considered. How could you improve on the chances of the serial correlation technique picking up a cycle when it exists in a situation like this?

SECTION 11

27 Suppose that observations have been taken over the past 20 years on some characteristic x. To determine if the phenomenon is time dependent, the data are plotted and are judged not to possess any trend. However, the autocorrelation coefficient with lag 1 year is computed and yields the value $r = -.52$. Discuss these results in terms of what type of mechanism might be producing the observations.

28 Suppose that the terms of a time series yield the least-squares equation $x_{i+1} = ax_i$, where a is some positive constant.

(*a*) Plot a set of points that you believe would produce this type of equation.

(*b*) What type of curve would you use if you were to fit a curve to your points?

SECTION 12

29 The following data are GNP (gross national product), in billions of current dollars, and revenue-passenger miles flown on U.S. domestic airline routes, in billions, by year from 1945 to 1973.

Year	GNP	P-Miles	Year	GNP	P-miles
1945	212	3.4	1960	504	30.6
1946	209	5.9	1961	520	31.1
1947	231	6.1	1962	560	33.6
1948	258	6.0	1963	591	38.5
1949	257	6.8	1964	632	44.1
1950	285	8.0	1965	685	51.9
1951	328	10.6	1966	750	60.6
1952	346	12.5	1967	794	75.5
1953	365	14.8	1968	864	87.5
1954	365	16.8	1969	921	102.7
1955	398	19.8	1970	977	104.1
1956	419	22.4	1971	1055	106.4
1957	441	25.3	1972	1158	118.1
1958	447	25.3	1973	1295	126.3
1959	484	29.1			

(*a*) Using a scale of (GNP − 200)/10 in place of GNP, plot the two series by year on the same graph. Note the similarities of the two trends.

(*b*) If a scatter diagram of P-miles and GNP were plotted, would you expect the relationship between the two to be approximately linear, based on the appearance of the plot in (*a*)? Why or why not?

30 The following data give the values of the SPI (securities price index) for 500 general stocks and the values of the IDI (implicit deflator index) for the years from 1940 to 1972.

(*a*) Plot the two series on the same chart.

(*b*) Does it appear that the two series are related?

(*c*) If linear trends were fitted to each series by using a ruler and good judgment, would the deviations appear to be related?

Year	SPI (1941–43 base)	IDI (1958 base)	Year	SPI (1941–43 base)	IDI (1958 base)
1940	11.0	43.9	1960	55.9	103.3
1941	9.8	47.2	1961	66.3	104.6
1942	8.7	53.0	1962	62.4	105.8
1943	11.5	56.8	1963	69.9	107.2
1944	12.5	58.2	1964	81.4	108.8
1945	15.2	59.7	1965	88.2	110.9
1946	17.1	66.7	1966	85.3	113.9
1947	15.2	74.6	1967	91.9	117.6
1948	15.5	79.6	1968	98.7	122.3
1949	15.2	79.1	1969	97.8	128.2
1950	18.4	80.2	1970	83.2	135.2
1951	22.2	85.6	1971	98.3	141.4
1952	24.5	87.5	1972	109.2	146.1
1953	24.7	88.3			
1954	29.7	89.6			
1955	40.5	90.9			
1956	42.6	94.0			
1957	44.4	97.5			
1958	46.2	100.0			
1959	57.4	101.6			

SECTION 13

31 The following data give the number of games that were played, in order to determine a winner, in each of baseball's World Series from 1905 to 1974.

Year	Games	Year	Games	Year	Games	Year	Games	Year	Games
1905	5	1920	—	1935	6	1950	4	1965	7
1906	6	1921	—	1936	6	1951	6	1966	4
1907	4	1922	4	1937	5	1952	7	1967	7
1908	5	1923	6	1938	4	1953	6	1968	7
1909	7	1924	7	1939	4	1954	4	1969	5
1910	5	1925	7	1940	7	1955	7	1970	5
1911	6	1926	7	1941	5	1956	7	1971	7
1912	7	1927	4	1942	5	1957	7	1972	7
1913	5	1928	4	1943	5	1958	7	1973	7
1914	4	1929	5	1944	6	1959	6	1974	5
1915	5	1930	6	1945	7	1960	7		
1916	5	1931	7	1946	7	1961	5		
1917	6	1932	4	1947	7	1962	7		
1918	6	1933	5	1948	6	1963	4		
1919	—	1934	7	1949	5	1964	7		

(*a*) Plot the data as a time series.
(*b*) Smooth by calculating a 5-year moving average, and graph your results.
(*c*) Compare the first half of the series with the second half, say, 1905–1940 with 1941–1974. Does the smoothing in (*b*) suggest the possibility that the averages over the two periods are not the same? Comment.

32 The following data give the number of workers in transportation and public utilities in ten-thousands of employees from 1929 to 1958. Use the data (*a*) to test for randomness by means of serial correlation with lag 1, (*b*) to calculate a 3-term moving average, (*c*) to fit a straight-line trend, and (*d*) to construct a graph and comment on the trend of the data. Read the data a row at a time.

391, 368, 324, 280, 266, 274, 277, 296, 311, 284,
291, 301, 325, 343, 362, 380, 387, 402, 412, 414,
395, 398, 417, 418, 422, 401, 406, 416, 415, 390

33 For the data of Table 3, on precipitation in Los Angeles, calculate the serial correlation with lag 11. Test the hypothesis that there is no significant correlation for this lag. This lag corresponds to a time period that some individuals claim is a natural weather cycle, because of high sunspot activity that occurs about this often.

34 Arrange the Los Angeles rainfall data of Table 3 into 6 rows and 11 columns. The data are to be read a row at a time; hence the last element in the first of the 6 rows will be 17. Calculate the means of the 11 columns and graph them. Do there appear to be any obvious differences among these means? If the sunspot theory (11-year lag) were correct, what would you expect to observe in a graph such as this?

35 Fit a straight-line trend to the data of Example 1 (Florida grapefruit) by means of a ruler and good judgment and calculate the deviations from this trend line. Then calculate the serial correlation coefficient of these deviations with lag 1 year and test it for significance. Comment.

Chapter 17 Statistical Decision

1 INTRODUCTION

In the preceding chapters the emphasis has been largely on statistical inference, with very little attention having been given to interpretation and decision making. This is because the action to be taken after a statistical inference has been made usually depends on outside economic considerations that are not available to the statistician. The businessman who instigates a sample survey, for example, must use his business judgment in deciding how best to use the results of that survey.

There are certain types of problems, however, where the statistician can indicate a procedure for making an economically sound decision, provided the rationale on which it is based is accepted. The simplest such problem is one in which there are only two possible states or events to consider and two possible actions available. For example, a truck gardener may be concerned about the possibility of a heavy freeze ruining his lettuce crop and must decide whether or not to protect his crop against the possibility of a freeze. The two possible states or events here are those of a freeze occurring and no freeze occurring, and the two possible actions are taking protection or not doing so. A more general version of this type of problem occurs when they are, say, $k \geq 2$ possible events and $l \geq 2$ possible actions. For example, a business firm may classify a new product into one of three categories: a big success, a moderate success, or a failure. Its choice of actions may consist of advertising on television, advertising on radio, advertising in newspapers, or advertising by handbills. Since the methods that are about to be proposed do not depend upon the values of k and l, provided that they both exceed 1, it will suffice to study the simplest version of the problem in which $k = 2$ and $l = 2$.

2 THE PAYOFF TABLE

For the purpose of explaining the reasoning behind a proposed solution for problems of the preceding type, consider first the following particular problem, which is similar to the one discussed in section 4 of Chapter 4. Suppose the promoter of a sporting event cannot decide whether or not to take out insurance against the possibility of rain. Assume that he will net \$20,000 provided that it does not rain but only \$2,000 if it does, and that he is contemplating a \$20,000 insurance policy costing \$5,000. This information is conveniently displayed in the following two-by-two table, called a *payoff table,* which lists his net profit under the various possibilities (Table 1).

Table 1

	Action	
Weather	Insure	Don't insure
Clear	\$15,000	\$20,000
Rain	\$17,000	\$ 2,000

The basis for preferring one of these actions, or decisions, over the other will be that of expected profit. Thus, the decision to take out insurance will be made if, and only if, the expected profit in doing so is larger than the expected profit when no insurance is purchased. For the purpose of calculating these two expected profits, let p denote the probability that the weather will be clear on the day of the sporting event. Then $1 - p$ is the probability that it will rain on that day. Recalling the definition of expectation given in Chapter 4, the calculation of the expected profits under the two possible actions, which are denoted by E_I and E_N, can be carried out as shown in Table 2.

Table 2

Weather	Probability	Insure		Don't insure	
		Profit	Expected profit	Profit	Expected profit
Clear	p	15,000	$15{,}000p$	20,000	$20{,}000p$
Rain	$1 - p$	17,000	$17{,}000(1 - p)$	2,000	$2{,}000(1 - p)$
Totals	1		E_I		E_N

Hence,

$$E_I = 15{,}000p + 17{,}000(1 - p) \tag{1}$$

and

$$E_N = 20{,}000p + 2{,}000(1 - p) \tag{2}$$

Assume that the value of p is not known. Then these two possible actions will be equally good provided that p satisfies the equation obtained by setting $E_I = E_N$. Thus, p must satisfy the equation

$$15{,}000p + 17{,}000(1 - p) = 20{,}000p + 2{,}000(1 - p)$$

The solution of this equation is easily seen to be $p = .75$. The value of p that produces equality between the two expected values of a problem such as this is called the *break-even point.*

It is clear intuitively that if $p > .75$ the promoter should not take out insurance because the better the weather the less the need for insurance and the two possible actions are equally good when $p = .75$. This fact is readily verified algebraically, however, by solving the inequality $E_N > E_I$, namely,

$$20{,}000p + 2{,}000(1 - p) > 15{,}000p + 17{,}000(1 - p)$$

Since inequalities are manipulated in the same manner as equalities, except when multiplying by a negative number, this inequality reduces to

$$20p + 2(1 - p) > 15p + 17(1 - p)$$

or

$$5p > 15(1 - p)$$

or

$$p > \frac{15}{20}$$

In view of the preceding calculations, the sports promoter should consult the weather records for the period of time when his event is to be held to see whether the proportion of clear (no rain) days exceeds .75. If it does and he is willing to use expected profit for making decisions, then he should not buy insurance. This assumes that the event is to take place so far in the future that no reliable weather forecast is available and that the insurance rates are based on this fact also. Insurance rates are usually based on past experience, but they could be based on a forecast if the event were to occur within a sufficiently short time interval for which a valid forecast is feasible.

Suppose weather records for that period of time revealed that 80 percent of the days were clear. The correct decision in that case would be to not buy insurance. The question then arises as to how much more profit the promoter can expect to obtain than if he had bought insurance. Calculations using $p = .80$ and formulas (1) and (2) give the values

$$E_I = 15{,}400 \quad \text{and} \quad E_N = 16{,}400$$

Hence, the promoter can expect his profits to be $1,000 higher if he chooses the correct decision. For a promoter who feels that he cannot afford to risk making only $2,000 if it rains, the philosophy of basing decisions on expected profit may not be altogether appealing here. An alternative rationale for such a promoter will be considered in section 5.

3 OPPORTUNITY LOSS; COST OF UNCERTAINTY

In this section a slightly different approach to a solution of the foregoing problem will be taken. Now, corresponding to each row of a payoff table there is a largest payoff that represents the maximum amount to be gained if the correct decision for that event is made. Thus, in Table 1 the maximum value for the first row is 20,000 and that for the second row is 17,000. The difference between a possible maximum amount and the payoff when the incorrect decision is made is a measure of the loss due to making the wrong decision for that event. This difference is called an *opportunity loss* because one had the opportunity of a higher payoff but failed to achieve it. These opportunity losses corresponding to the payoff table given in Table 1 are shown in Table 3. The entries in the first row resulted from subtracting the entries in the first row of Table 1 from 20,000, whereas those in the second row were obtained by subtracting the second-row entries of Table 1 from 17,000.

Table 3 Opportunity losses

Weather	Insure	Don't insure
Clear	5,000	0
Rain	0	15,000

At first glance it would appear from Table 3 that refraining from taking out insurance would be a serious mistake because of the large opportunity loss, namely \$15,000, that is possible compared to that when insurance is purchased; however, since the basis for action is expected value, it is necessary to calculate the expected values of these opportunity losses before making a decision. As before, let p denote the probability of clear weather. Then the two expectations, denoted by E_{OI} and E_{ON}, are obtained by carrying out calculations similar to those of Table 2, as shown in Table 4.

Table 4

Weather	Probability	Insure		Don't insure	
		Loss	Expected loss	Loss	Expected loss
Clear	p	5,000	$5{,}000p$	0	0
Rain	$1 - p$	0	0	15,000	$15{,}000(1 - p)$
Totals			E_{OI}		E_{ON}

Hence,

$$E_{OI} = 5{,}000p \tag{3}$$

and

$$E_{ON} = 15{,}000(1 - p) \tag{4}$$

If a decision is to be made on the basis of the magnitude of the expected value of opportunity loss, one can proceed just as in the preceding section; however, now it is necessary to minimize expected opportunity loss rather than to maximize expected gain. As before, the value of p that makes the two possible actions equally desirable is obtained by equating E_{OI} and E_{ON} and solving for p. The solution is the same as before: $p = .75$. Now, it is readily shown by algebraic manipulations that the break-even point is always the same in these two approaches and furthermore that $E_{ON} < E_{OI}$ if, and only if, $E_N > E_I$. Thus, it makes no difference whether one uses the principle of maximizing expected profit or minimizing expected opportunity loss for making a decision. The advantage of using opportunity loss is that it serves as a means of measuring the loss that results from not knowing what the weather is going to be. If it is assumed, as was done in section 2, that records yielded the value $p = .8$, in which case no insurance would be purchased, the expected opportunity loss is obtained by replacing p by .8 in formula (4). As a result,

$$E_{ON} = 3{,}000$$

This value is called the *cost of uncertainty*. It represents the expected loss to the promoter because he does not possess the occult power of perfect prediction of the weather. If he could make perfect predictions each time, which implies that he would insure when it is going to rain and he wouldn't insure when it is going to be clear, then his opportunity loss in each such situation would be 0 and therefore his expected opportunity loss would be 0.

4 PERFECT INFORMATION

A slightly different approach to that explained in the preceding section for measuring the cost of uncertainty is an approach that uses expected profits rather than opportunity losses. It requires the calculation of what is called the *expected profit with perfect information.*

Suppose that a weather forecaster could predict the weather perfectly whenever he was hired to do so. If the promoter bought the forecaster's services, he would always make the correct decision. That is, he would buy insurance when the forecast was for rain and he wouldn't when the forecast was for clear weather. Since it is being assumed here that clear weather will occur 80 percent of the time and rain 20 percent of the time, and therefore that the forecaster will predict those events 80 percent of the time and 20 percent of the time, respectively, the expected profit to the promoter who bought the forecaster's services, based on Table 1, would be

$$\begin{aligned} E &= 20{,}000(.8) + 17{,}000(.2) \\ &= 19{,}400 \end{aligned}$$

This is the promoter's expected profit with perfect information. The promoter's expected profit without such information is his expected profit when using the better of the two possible decisions of Table 1. That was calculated in section 2 and found to be $E_N = 16{,}400$. The difference between this expected profit and the expected profit with perfect information, which is \$3,000 here, is called the *expected value of perfect information.* It is the amount that the promoter should be willing to pay the forecaster for producing perfect forecasts. It is to be noticed that the value obtained here is the same as the value obtained in the preceding section in calculating the cost of uncertainty. It can be shown that this equality holds in general. That is, the cost of uncertainty calculated on the basis of opportunity loss is always equal to the expected value of perfect information based on expected profit. Since it is easier to work directly with profits rather than convert them into opportunity losses, the expected value of perfect information technique will be used hereafter whenever the cost of uncertainty is to be obtained.

Example. To illustrate the method of calculating the expected value of perfect information on a more complicated problem, consider the setup in Table 5.

This table represents the profits that can be expected to be realized by a well-known author under three different contractual arrangements that have been proposed to him by his publishing company. The first of them (A) is based on a flat royalty percentage; the second (B) on a lower royalty percentage for the first ten thousand copies sold and a higher royalty percentage thereafter; and the third (C) on a substantial initial payment and a relatively low royalty percentage for all copies sold. The degree of success is determined by the number of copies sold; however, to simplify the problem it is assumed that the number sold will be one of three numbers that represent high, medium, and low success, respectively.

Table 5

Degree of success	Probability of success	Contract A	B	C
High	.3	60,000	73,000	56,000
Medium	.5	33,000	32,000	31,000
Low	.2	15,000	13,000	20,000

The three success probabilities are those estimated by the author on the basis of the success of his other books and his confidence in the success of his latest book.

The expected profits for the three types of agreement are given by

$$E_A = 60{,}000(.3) + 33{,}000(.5) + 15{,}000(.2) = 37{,}500$$

$$E_B = 73{,}000(.3) + 32{,}000(.5) + 13{,}000(.2) = 40{,}500$$

$$E_C = 56{,}000(.3) + 31{,}000(.5) + 20{,}000(.2) = 36{,}300$$

Hence the author should accept the second type of agreement. Although the third type of agreement would undoubtedly find favor with many authors because it gives the author a fixed amount regardless of how well the book sells, it is the least desirable of the three for this problem.

To calculate the expected value of perfect information, it is first necessary to multiply the largest entry in each row by its corresponding probability and then to sum the resulting products. This gives

$$E = 73{,}000(.3) + 33{,}000(.5) + 20{,}000(.2) = 42{,}400$$

The expected value of perfect information is therefore given by

$$42{,}400 - 40{,}500 = 1{,}900$$

This represents the amount of money that the author could expect to gain on the average if, instead of always signing up with the company to a type

B contract, he knew in advance how well the book would do and could sign a type of contract that would maximize his profits for that book. As was pointed out earlier, this value is the same as the cost of uncertainty for this problem.

5 UTILITY

The principle of expected profit as a basis for making business decisions is very useful in many business situations, but it may be quite inappropriate for personal decisions and for those business decisions that involve the risk of heavy losses. Even if maximizing expected profit is a desirable objective in many business ventures, it can have its limitations when a single important decision is to be made. For example, a strategy in which there is a large probability of making a large profit but also a moderate probability of going bankrupt may not be nearly as desirable as one for which the expected profit is much smaller but for which the probability of bankruptcy is very small. For situations such as these it is necessary to replace money as the criterion for making decisions by some function that will represent, at various money levels, the value of money to the individual or business firm making the decision. Such a function is called a *utility function.*

For the purpose of explaining the concept of utility as it pertains to simple business decision making, consider once more the problem of the sports promoter that was discussed in section 2. Suppose the promoter is presented with the following gambling situation. He will be permitted to take one chance at a lottery in which the probability is π that he will win \$20,000 and is $1 - \pi$ that he will win \$2,000. Corresponding to various possible values of π, he is requested to decide what cash settlement he would accept in lieu of the privilege of taking a chance at the lottery.

If the promoter looks upon the expected value of the lottery as a reasonable measure of the value to him of a chance at the lottery; then he will accept the expected value as the cash settlement. If he loves to gamble, he will undoubtedly require a cash settlement somewhat higher than the expected value of the lottery. If, however, he is the cautious type, he will likely be willing to settle for a sure amount somewhat below expectation. A cautious type of promoter who desperately needs several thousand dollars to meet his obligations would undoubtedly require a cash settlement considerably higher than the expected value for small values of π and lower than the expected value for large values of π.

The cash amount that the promoter will accept, corresponding to the various possible values of π, when plotted against π yields a curve that represents the promoter's evaluation of money at the time of the proposition. Such a curve could be constructed by requiring the promoter to produce values corresponding to, say, $\pi = .1, .2, \ldots, .9$ and then drawing a smooth curve through the resulting plotted points. Since $E = 20{,}000\pi + 2{,}000(1 - \pi)$, a promoter who looks upon decision making from a long-range point of view, or who is happy with expected value as representing his value of money, would submit the following values, which are obtained by using the formula for E.

A

π	0	.1	.2	.3	.4	.5	.6	.7	.8	.9	1.0
Cash	2,000	3,800	5,600	7,400	9,200	11,000	12,800	14,600	16,400	18,200	20,000

If, however, the promoter is the cautious type who prefers a cash settlement somewhat below what he could expect to win by taking his chances, he might submit the following values. It will be noted that these values, except for the first and last, are smaller than the preceding corresponding expected values.

B

π	0	.1	.2	.3	.4	.5	.6	.7	.8	.9	1.0
Cash	2,000	2,800	4,000	5,400	6,400	8,000	9,500	11,500	14,000	17,000	20,000

In using information of the preceding type to construct a curve that represents the promoter's evaluation of money in a gambling situation, it is convenient to represent values of π along the y-axis and values of cash along the x-axis rather than in the normal reverse order. Several curves of this kind are shown in Fig. 1. The two smooth curves that correspond to the preceding two sets of data are labeled A and B, respectively, in Fig. 1. The curve labeled C represents an individual who loves to gamble for high stakes rather than to settle for an amount he should expect to win on the average, whereas curve D represents a promoter who

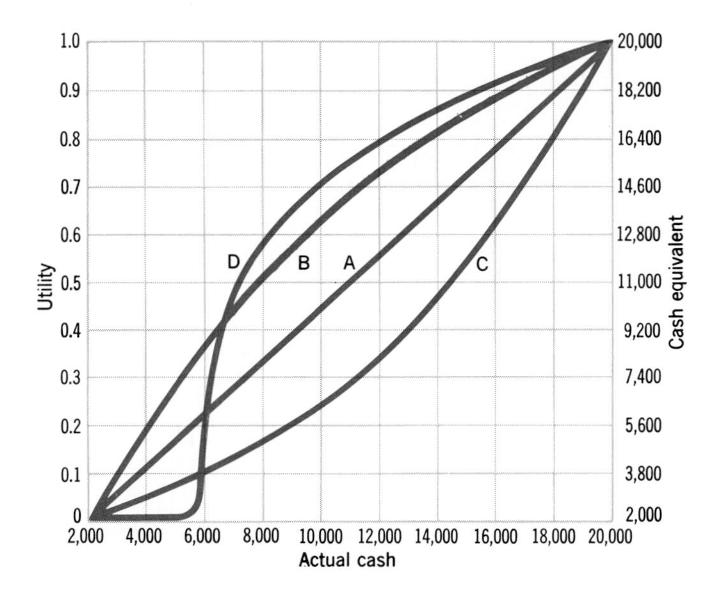

Figure 1 Utility curves.

is normally very conservative but who feels compelled to gamble unless he can be assured of obtaining at least \$5,000.

A curve of the preceding types is called a *utility curve* and the function that corresponds to it is called the *utility function for money* of the promoter. The symbol $u(M)$ will be used to represent this function, where M denotes money in dollars. Corresponding to any fixed amount of cash M, the function $u(M)$ gives the value of π such that the promoter would just as soon take a chance at the lottery at this π value as to receive a cash settlement of M dollars. It is in this sense that $u(M)$ measures the value of money to a promoter who is faced with a decision situation where the outcome depends on chance. In this formulation $u(M)$ takes on values between 0 and 1 only. If desired one can convert utility values over to corresponding dollar values by scaling the y-axis from 2,000 to 20,000, rather than from 0 to 1. The utility dollar scaling is shown in Fig. 1 on the right vertical line. In making comparisons based on utility values, it is irrelevant which scaling is used.

Now that the utility of money is being substituted for money itself as a more meaningful unit of value, it is necessary to revise the basis for making decisions. Instead of comparing the expected values of two or more competing possible actions in a profit payoff table, it is necessary to compare the expected values in the corresponding utility payoff table. Thus, all that needs to be done in the new formulation is to replace the usual payoff table values by their corresponding utility values and proceed as usual. The utility values can be obtained from the utility function $u(M)$ if it is given, or by reading them off the graph of the utility curve if only it is available.

Example. As an illustration, assume that the sports promoter's utility curve is the one labeled B in Fig. 1. From that graph, or more accurately from the B values that were used to graph the curve, it will be seen that the utility values corresponding to the Table 1 entries are given by $u(2{,}000) = 0$, $u(20{,}000) = 1$, $u(15{,}000) = .83$, and $u(17{,}000) = .90$. As a result, Table 1 must be replaced by the following table of values.

	Action	
Weather	Insure	Don't insure
Clear	.83	1
Rain	.90	0

The calculation of expected utility values, which are denoted by $\tilde{E}_I$ and $\tilde{E}_N$, are similar to the calculations in Table 2 for expected profit values and are shown in Table 6.

Table 6

Weather	Probability	Insure		Don't insure	
		Utility	Expected utility	Utility	Expected utility
Clear	p	.83	$.83p$	1	p
Rain	$1 - p$	.90	$.90(1 - p)$	0	0
Totals	1		$\tilde{E}_I$		$\tilde{E}_N$

Hence,

$$\tilde{E}_I = .83p + .90(1 - p)$$

and

$$\tilde{E}_N = p$$

As a result, the break-even point is the solution of the equation

$$.83p + .90(1 - p) = p$$

The solution is given by $p = .84$. The cautious promoter whose utility function is given by curve B will therefore take out insurance if weather records show that the probability of clear weather is below .84, whereas in the earlier solution he would have done so if this probability were below .75. If, for example, the records showed that $p = .83$, he would insure and his expected profit in dollars as given by formula (1) would be $15,340. If he had gambled on the weather, his expected profit as given by formula (2) would have been $16,940. Thus he is willing to take a $1,600 decrease in expected profit when $p = .83$ in return for decreasing his chances of earning only $2,000 from a probability of .17 to a zero probability.

If utility dollar values had been preferred to utility values, it would have sufficed to read off the right-hand scale of Fig. 1 and obtain the following payoff table values.

17,000	20,000
18,200	2,000

The calculations would then proceed as usual and the break-even point would be given as the solution of the equation

$$17{,}000p + 18{,}200(1 - p) = 20{,}000p + 2{,}000(1 - p)$$

The solution is, of course, $p = .84$. Any differences occurring in these two solutions would be due to differences in graph-reading accuracy or computational rounding errors, because the methods are theoretically equivalent.

The preceding technique of employing the expected value of utility as a basis for making decisions should enhance the usefulness of expected value as a decision tool for business problems because it enables the business man to incorporate into the calculations the value to him of different amounts of money. One difficulty with the technique is that the decision maker is usually hard pressed to state with any precision what cash settlement he would take in lieu of a given chance opportunity. Furthermore, any utility function that he may feel is appropriate in one business situation may not satisfy him at the next venture.

It often happens that an individual's utility function must be determined by means of gambles that involve possible payoffs other than the two extreme values listed in a payoff table. Thus, the promoter might state that he will accept \$9,000 cash in lieu of the respective probabilities .4 and .6 of winning \$6,000 and \$16,000. For such situations it suffices to calculate the expected value of the gamble, namely, $.4(6{,}000) + .6(16{,}000) = 12{,}000$, and treat it as the y-coordinate of a point on the utility curve for which the x-coordinate is 9,000. This follows from the fact that the expected value $E = 20{,}000\pi + 2{,}000(1 - \pi)$ assumes the form $E = 1 \cdot \pi + 0 \cdot (1 - \pi) = \pi$ when utility values are used in place of money values; consequently the y-coordinate of a point really represents the expected value of the gamble in the utility scaling, and hence also in the equivalent money scaling.

6 PROBABILITIES FOR BUSINESS EVENTS

In the discussion thus far it has been assumed that there is a probability p associated with the occurrence of the event in question. This normally implies that one can conceive of a sequence of similar situations and that p represents the long-run proportion of occurrences of the event. However, as pointed out in Chapter 3, when dealing with business events one may be forced to use an estimate of p that is based on very little experience. For example, if a business firm is undecided whether to market a new product and needs to have an estimate of the probability that the venture will be a success for the purpose of calculating expected gains or losses, then there is very little past experience available to help in obtaining a good estimate because experience with one new product is not necessarily applicable to another new product.

Some business ventures are so new or possess so little past experience that one is forced to use an estimate of p based on one's personal belief in what the chances are that the event will occur. A value of p that represents an individual's degree of belief in the occurrence of an event is called *personal probability*. Another name for it is *subjective probability*. The latter name arises from an attempt to discriminate between probabilities whose values depend on the person or subject who assigns those values and objective probabilities, whose values are determined only by the object and the experiment in which it is involved. Thus, the probability $p = \frac{1}{6}$ that a six spot will show in rolling a die is an objective probability because it is not based on an individual's personal judgment, whereas the probability $p = \frac{1}{2}$ that a new business venture will be successful is a subjective probability because it depends on the personal judgment of the individual who assigned it. A personal probability of $\frac{1}{4}$, for example, should correspond to the individual's willingness to wager $\frac{1}{4}$ dollar in order to win $\frac{3}{4}$ dollar if the event does occur. There is a degree of subjectivity to all probabilities obtained from real-life experience; therefore the application of probability to real problems is bound to involve subjective probability. The principal difference in problems such as those related to games of chance and those concerned with business activities is the degree of experience and information available for making an assessment of pertinent probabilities.

The interpretation of probability as long-run relative frequency for experiments that are repeatable, at least conceptually, implies that values arrived at from business experience are estimates of underlying theoretical values. There are many business situations, however, where it is unrealistic to assume that the situation is just one of similar situations to be met in the future. Or it may be one for which there is no experience available to assist in assigning probabilities. The assigned value of a probability then represents the individual businessman's betting odds that the event will occur. This measure is strictly personal and for that individual represents a legitimate probability for him in the sense that he is willing to take bets based on that probability. Such a probability need not be interpreted in terms of theoretical relative frequency.

The conclusions to be drawn from calculations based on highly subjective probabilities are the responsibility of the individual who submitted the probabilities. If his probability assessment was poor he will suffer the consequences, but that is his concern and not that of the statistician who designs decision-making models based on the assumption of correct probability values. Although decision theory treats an assigned probability as exact, numerous assumptions, approximations, and subjective judgments must be made in applying the theory to actual business problems.

7 EVALUATING INFORMATION

In view of the preceding discussion, the question naturally arises as to how one can utilize information that comes from experience or judgment to improve one's decision-making accuracy. For the problem of section 2 there is not much that can be done to improve the estimate of p obtained by consulting the records, unless one waits until a short time before the event and obtains a reliable weather forecast. If the insurance rates do not change then, such information might be valuable. For problems that involve probabilities that can be estimated from samples, the decision-making process can be improved by utilizing information from such samples.

This section will consider the evaluation of nonsample information and the next section will study the evaluation of information obtained from samples.

In the problem of section 2, the probability of clear weather was based on past weather records for that time of year and yielded the value $p = .8$. Suppose a meteorologist who runs a weather service proposes to provide the promoter with a weather forecast one week before the event is to take place for a fee of $200. The meteorologist's records reveal that he is correct in such forecasts 75 percent of the time. The problem therefore is to determine whether the cost of the forecast is justified by increased profits. There are now three possible actions, rather than two, because the promoter may buy the meteorologist's advice and insure only when the prediction is for rain. The other two possibilities are as before, to insure or to not insure, without purchasing a forecast. Instead of having two possible weather situations to contend with, there are now four weather-prediction situations that need to be considered. These are displayed in the payoff table shown in Table 7.

Table 7

Event	Probability	Action		
		Insure	Don't insure	Follow forecast
Clear, clear predicted	(.8)(.75) = .60	15,000	20,000	20,000
Clear, rain predicted	(.8)(.25) = .20	15,000	20,000	15,000
Rain, clear predicted	(.2)(.25) = .05	17,000	2,000	2,000
Rain, rain predicted	(.2)(.75) = .15	17,000	2,000	17,000

The calculation of the expected profit under these three possible actions is carried out in the usual manner and gives rise to the following three expected values.

$$E_I = 15{,}000(.60) + 15{,}000(.20) + 17{,}000(.05) + 17{,}000(.15) = 15{,}400$$

$$E_N = 20{,}000(.60) + 20{,}000(.20) + 2{,}000(.05) + 2{,}000(.15) = 16{,}400$$

$$E_P = 20{,}000(.60) + 15{,}000(.20) + 2{,}000(.05) + 17{,}000(.15) = 17{,}650$$

The values of E_I and E_N are, of course, the same as the corresponding values obtained in section 2 for $p = .8$ because the predictions of the meteorologist were not used for those two possible actions. Since the best action before was not to insure, the gain in expected profit by utilizing the services of the meterologist is given by

$$17{,}650 - 16{,}400 = 1{,}250$$

From these calculations it follows that the advice of the meteorologist is worth \$1,250 to the promoter in terms of expected profit, and therefore he should be happy to pay only \$200 for that advice.

The expected profit with perfect information, that is, with the advice of a forecaster who is 100 percent correct in his forecasts, was calculated in section 4 to be \$19,400. Hence, the expected value of perfect information in this setup is given by

$$19{,}400 - 17{,}650 = 1{,}750$$

Hence, there is still a considerable amount of profit lost because of the inability of the forecaster to make perfect predictions.

In these calculations it has been assumed that the insurance company is willing to sell insurance one week before the event at the specified rates, and therefore does not buy the meterologist's services to determine new rates.

The preceding procedure is applicable to many business situations in which the problem is to determine whether expected profits can be increased by buying information and whether the gain in such expected profits exceeds the cost of obtaining the information.

8 UTILIZING NEW INFORMATION; SAMPLING

The preceding section considered the problem of how to utilize and evaluate information that is obtained from sources other than sampling. In this and the following sections, the problem will be to determine how to improve a decision-making process by means of sampling.

For the purpose of explaining how one proceeds to do this, consider the following artificial problem. After that problem has been solved, the same methods will be applied to solve a realistic business problem.

Example 1. A box contains 3 coins, 2 of which are regular coins, whereas the third has a head on both sides. A single coin is to be drawn from the box. The problem is to decide whether the drawn coin is a regular coin or the two-headed coin, if the payoff table for the two possibilities is that shown in Table 8.

Table 8

Coin type	Action: Decide regular	Action: Decide two-headed
Regular	9	−6
Two-headed	−12	15

The calculations needed to determine the expected profits under the two possible decisions are carried out in the usual manner and are shown in Table 9.

Table 9

Coin type	Prior probabilities	Decide regular		Decide two-headed	
		Profit	Expected profit	Profit	Expected profit
Regular	$\frac{2}{3}$	9	6	−6	−4
Two-headed	$\frac{1}{3}$	−12	−4	15	5
Totals	1		2		1

On the basis of these calculations, it is best to decide that the drawn coin is a regular coin.

Now consider how one proceeds when additional information is obtained from a sampling experiment. In particular, suppose the drawn coin is tossed 3 times and 3 heads are obtained. This additional information can be incorporated into the analysis by calculating a new set of probabilities to replace the probabilities in the second column of Table 9. The initial probabilities, namely, $\frac{2}{3}$ and $\frac{1}{3}$, are called *prior probabilities* for this problem because they are the probabilities that were available before the sampling experiment was performed. After the results of the sampling experiment are in, it is possible to use Bayes' formula from Chapter 3 to obtain improved estimates for those probabilities. Such probabilities are called *posterior probabilities* because they are probabilities that apply after the experiment has been performed; the sample results are used to calculate these new probabilities. If the two events in the first column of Table 9 are denoted by R and T, the prior probabilities are given by

$$P\{R\} = \frac{2}{3} \quad \text{and} \quad P\{T\} = \frac{1}{3}$$

Let S denote the result of the sampling experiment, that is, the result of getting 3 heads in the 3 tosses of the coin. Then the posterior probabilities are given by the symbols

$$P\{R \mid S\} \quad \text{and} \quad P\{T \mid S\}$$

These posterior probabilities can be calculated by using Bayes' formula in the form

$$P\{R \mid S\} = \frac{P\{R\}P\{S \mid R\}}{P\{R\}P\{S \mid R\} + P\{T\}P\{S \mid T\}}$$

Although the value of $P\{T \mid S\}$ can be obtained by subtracting $P\{R \mid S\}$ from 1, it is advisable to calculate $P\{T \mid S\}$ directly from the corresponding Bayes' formula to guard against possible errors of calculation.

The conditional probabilities $P\{S \mid R\}$ and $P\{S \mid T\}$ that are needed here are based on 3 successes in 3 independent trials and are given by

$$P\{S \mid R\} = \frac{1}{8} \qquad \text{and} \qquad P\{S \mid T\} = 1$$

Consequently,

$$P\{R \mid S\} = \frac{\frac{2}{3} \cdot \frac{1}{8}}{\frac{2}{3} \cdot \frac{1}{8} + \frac{1}{3} \cdot 1} = \frac{1}{5}$$

and

$$P\{T \mid S\} = \frac{\frac{1}{3} \cdot 1}{\frac{2}{3} \cdot \frac{1}{8} + \frac{1}{3} \cdot 1} = \frac{4}{5}$$

The analysis now proceeds as before with the prior probabilities of the second column of Table 9 replaced by the posterior probabilities just calculated. These calculations are shown in Table 10.

Table 10

Coin type	Posterior probabilities	Decide regular		Decide two-headed	
		Profit	Expected profit	Profit	Expected profit
Regular	$\frac{1}{5}$	9	1.8	−6	−1.2
Two-headed	$\frac{4}{5}$	−12	−9.6	15	12
Totals	1		−7.8		10.8

It is now necessary to reverse the earlier decision in favor of a regular coin to the decision that the drawn coin is the two-headed coin. The difference in expectation is quite pronounced here as compared to the earlier difference, before the sampling experiment was concluded.

The calculation of the expected value of perfect information is easily carried out here for Tables 9 and 10. Since the decision in Table 9 was to decide in favor of a regular coin, the expected value of perfect information before sampling is given by

$$9\left(\frac{2}{3}\right) + 15\left(\frac{1}{3}\right) - 2 = 9$$

Since the decision in Table 10 was to decide in favor of the two-headed coin, the expected value of perfect information after sampling is given by

$$9\left(\frac{1}{5}\right) + 15\left(\frac{4}{5}\right) - 10.8 = 3$$

Thus, the sampling experiment has decreased the expected value of perfect information from 9 to 3. In a practical situation this reduction would need to be weighed against the cost of conducting the sampling experiment.

Example 2. The methods employed to solve the preceding artificial problem will now be used to solve a realistic problem. A manufacturer of a fairly complex product, such as an automobile, is confronted with the problem of how much inspection and testing should be done at the factory before shipping the product to the dealer. He may give the product a superficial inspection and agree to pay the dealer for all costs arising from customers returning the product for repairs because of faulty parts or workmanship, or he may subject the product to thorough inspection and testing at the factory to eliminate most of the dealer's rebates.

Assume that the product is an automobile and suppose that it costs $24 to inspect and test each car at the factory for faulty assembly and parts and an additional $20, on the average per faulty car, to put it in order. Suppose further that it costs, on the average, $80 per car for the dealer to correct faults found in the car by the customer when the factory does not carry out the preceding careful inspection-testing program. It will also be assumed that the factory inspectors and the customers detect the same proportion of faulty cars.

Suppose now that the manufacturer has kept weekly records of the percentage of defective cars that have come through the production line, as shown in Table 11. Here the quantity $P\{p_i\}$ is the proportion of weeks when the fraction of faulty cars was p_i, $i = 1, \ldots, 5$. For example for $i = 5$, in 10 percent of those weekly periods it was found that 60 percent of the cars had one or more defects. All such fractions and proportions have been expressed in decimal form to the nearest tenth. It should be noted that at least 20 percent of the cars had one or more defects during each of these weeks.

Table 11

i	1	2	3	4	5
p_i	.2	.3	.4	.5	.6
$P\{p_i\}$	.2	.2	.3	.2	.1

If it is assumed that the quality of current production is similar to that based on past experience, then the values of Table 11 may be used to calculate expected losses under the two possible actions: factory inspection (F) or customer complaint (C). In doing so, however, it is understood that no consideration is being given to such important factors as manufacturing space, schedule requirements, or consumer-dealer relations. These are intangibles that arise in any decision-making procedure but that are too complicated to be considered here.

The calculation of expected losses consists in calculating such losses for each value of p_i and summing those losses. Assuming that N cars are produced during a week's production run and a fraction p_i of them contains one or more defects, the loss to the manufacturer for inspection is $24N$ dollars and for repairing the Np_i defective cars is $20Np_i$; hence the total loss for that week is

$$L_F = 24N + 20Np_i$$

If no inspection is made, these defective cars will cost the manufacturer $80Np_i$ dollars in dealer reimbursements; hence

$$L_C = 80Np_i$$

The calculation of these expected losses for one week is shown in Table 12.

Table 12

		Factory inspection		Customer complaint	
p_i	$P\{p_i\}$	Loss	Expected loss	Loss	Expected loss
.2	.2	$28N$	$5.6N$	$16N$	3.2N
.3	.2	$30N$	$6.0N$	$24N$	$4.8N$
.4	.3	$32N$	$9.6N$	$32N$	$9.6N$
.5	.2	$34N$	$6.8N$	$40N$	$8.0N$
.6	.1	$36N$	$3.6N$	$48N$	$4.8N$
Totals	1.0		$31.6N$		$30.4N$

On the basis of these calculations, the appropriate decision here is to allow the dealer to correct the faulty cars and to reimburse him for his costs.

Example 3. If there is reason to believe that the quality of current production differs somewhat from that of past production runs, then the preceding procedure can be improved by taking a sample from current production and incorporating it into the decision-making process by means of Bayes' formula, in the same manner as in the earlier artificial coin problem. To illustrate how this is done, suppose that a random sample of 50 cars is taken from current production and that after careful factory inspection and testing, 25 of them are judged to be faulty in some respect. These 50 cars may be treated as 50 trials of an experiment for which the probability of success (here finding a car to be faulty) is p and for which 25 successes occurred. As a result, the binomial distribution may be applied to this sampling problem to assist in the calculation of posterior probabilities. The probabilities $P\{p_i\}$ found in the second column of Table 12 are the prior probabilities for this problem. If S denotes the result of getting 25 successes in 50 trials of the sampling experiment, then the desired posterior probabilities that are needed to replace the prior probabilities are the probabilities $P\{p_i \mid S\}, i = 1, \ldots, 5$. The appropriate version of Bayes' formula for this problem is therefore

$$(5) \qquad P\{p_i \mid S\} = \frac{P\{p_i\}P\{S \mid p_i\}}{\sum_{j=1}^{5} P\{p_j\}P\{S \mid p_j\}}, \qquad i = 1, \ldots, 5$$

The summation index j is used in place of i in the denominator to avoid confusion with the index i of the numerator. Since the binomial distribution is applicable to this sampling experiment, the conditional probabilities needed for this formula are given by the formula

$$(6) \qquad P\{S \mid p_i\} = \frac{50!}{25!25!} p_i^{25}(1 - p_i)^{25}, \qquad i = 1, \ldots, 5$$

This formula, when substituted in (5) for each value of p_i in Table 12, will produce the desired posterior probabilities that are needed to replace the prior probabilities $P\{p_i\}$ of Table 12. As an illustration, consider the calculation of $P\{p_3 \mid S\}$. From (5) and (6) it follows that

$$P\{p_3 \mid S\} = \frac{(.3)\frac{50!}{25!25!}(.4)^{25}(.6)^{25}}{\sum_{j=1}^{5} P\{p_j\}\frac{50!}{25!25!}(p_j)^{25}(1 - p_j)^{25}}$$

$$= \frac{(.3)(.4)^{25}(.6)^{25}}{\sum_{j=1}^{5} P\{p_j\}p_j^{25}(1 - p_j)^{25}}$$

$$= \frac{(.3)(.24)^{25}}{\sum_{j=1}^{5} P\{p_j\}[p_j(1 - p_j)]^{25}} = .312$$

Quantities such as these are readily calculated with electronic hand calculators. The results of such calculations are shown in the second column of Table 13. This table also shows the calculations needed to obtain the revised weekly expected losses based on the posterior probabilities.

Table 13

p_i	$P\{p_i \mid S\}$	Factory Inspection		Customer complaint	
		Loss	Expected loss	Loss	Expected loss
.2	.000	28N	.00N	16N	.00N
.3	.007	30N	.21N	24N	.17N
.4	.312	32N	9.98N	32N	9.98N
.5	.577	34N	19.62N	40N	23.08N
.6	.104	36N	3.74N	48N	4.99N
Totals	1.000		33.55N		38.22N

It will be observed that the information supplied by the sample has reversed the earlier recommended decision. Because of the high proportion of faulty cars in the sample, it now appears to be more economical to carry out a careful inspection at the factory, whereas before the sample had been taken, the reverse was true. It appears that quality of production deteriorated from that previously experienced.

9 BAYESIAN ESTIMATION FOR CONTINUOUS VARIABLES

The preceding section showed how decision making could be improved by imcorporating information concerning a parameter p into the decision-making process. There were two kinds of information employed there: that obtained from knowing the distribution of p and that obtained from a sample estimate of p. Since the more information one can apply to a problem the more reliable one would expect the solution to be, the decision based on using both kinds of information ought to be more reliable than a decision arrived at using just one of them.

Most of the statistical methods presented in the preceding chapters were based on information supplied by the sample only. Such methods are usually referred to as *classical methods.* Statistical methods that require knowledge of the probability distribution of a parameter such as p, are usually called *Bayesian methods* because of their connection with Bayes' formula. The methods employed in the preceding section were Bayesian methods because they employed the distribution of p to arrive at a solution.

In this and the next section, the preceding Bayesian techniques, which were concerned with discrete variables, will be extended to include continuous-variable problems. Since such methods are based on Bayesian estimates, the problem of estimation will be treated first. Toward this objective consider the following problem in quality control.

Example. The manufacturer of a metal product regularly purchases, from the same firm, steel rods used in his manufacturing process. A new shipment of rods arrives. The manufacturer wishes to estimate the mean of a quality characteristic, such as strength or hardness, for the shipment by means of a sample of size n taken from the shipment. Suppose experience has shown that the quality characteristic, denoted by x, is a normal variable with mean μ and standard deviation σ and that the shipment mean μ may be treated as a normal variable, corresponding to successive shipments. Let μ_0 denote the grand mean (that is, the mean of the various shipment means) and assume that experience with shipments has yielded a standard deviation σ_0 for such shipment means. This latter normal distribution of the parameter μ with mean μ_0 and standard deviation σ_0 is the prior distribution for the problem.

The classical estimate of μ is, of course, given by $\bar{x}$. The accuracy of this estimate can be expressed by means of the following familiar 95 percent confidence interval for μ.

$$\bar{x} - 1.96\frac{\sigma}{\sqrt{n}} < \mu < \bar{x} + 1.96\frac{\sigma}{\sqrt{n}} \tag{7}$$

The Bayesian approach uses the normal distribution of $\bar{x}$ and the prior normal distribution of μ to obtain the posterior distribution of μ for a given $\bar{x}$. It can be shown by calculus techniques that this posterior distribution is also normal with its mean μ_1 and standard deviation σ_1 given by the following formulas:

$$\mu_1 = \frac{\bar{x} + \delta\mu_0}{1 + \delta} \tag{8}$$

and

$$\sigma_1^2 = \frac{\sigma_{\bar{x}}^{\ 2}}{1 + \delta} \tag{9}$$

Here $\delta = \sigma_{\bar{x}}^{\ 2}/\sigma_0^2$, which is the ratio of the variance of $\bar{x}$ to the variance of μ in the prior normal distribution. The Bayesian estimate of μ is the mean of this posterior distribution, namely μ_1. Since μ, for a fixed $\bar{x}$, possesses a normal distribution with mean μ_1 and standard deviation σ_1, the probability is .95 that μ satisfies the inequality

$$\mu_1 - 1.96\sigma_1 < \mu < \mu_1 + 1.96\sigma_1 \tag{10}$$

For the purpose of comparing these two estimates in a given situation, suppose experience or other information has produced the values $\sigma = 50$, $\mu_0 = 190$, $\sigma_0 = 5$, and that a sample of size $n = 25$ yielded the value $\bar{x} = 180$. From computations based on formula (8), it then follows that the classical and the Bayesian estimates are, respectively,

$$\bar{x} \quad \text{and} \quad \mu_1 = \frac{1}{5}\bar{x} + \frac{4}{5}(190)$$

This shows that the Bayesian estimate is a weighted average of the sample mean $\bar{x}$ and the prior mean μ_0 with the prior mean weighted four times as heavily as the sample mean. Since $\bar{x} = 180$ here, the numerical values of these two estimates are 180 and 188, respectively.

The relative accuracy of these two estimates may be determined by comparing the lengths of the two intervals given in (7) and (10). Calculations with these data yield the classical interval

$$\bar{x} - 1.96(10) < \mu < \bar{x} + 1.96(10)$$

and the Bayesian interval

$$\mu_1 - 1.96\sqrt{20} < \mu < \mu_1 + 1.96\sqrt{20}$$

The ratio of the lengths of these two .95 probability intervals is

$$\frac{\sqrt{20}}{10} = .45$$

Thus the Bayesian interval is less than one-half as long as the classical interval.

From formula (9) it follows that σ_1 is always smaller than $\sigma_{\bar{x}}$; consequently the Bayesian interval is always shorter than the classical interval. As the sample size n increases, δ approaches 0 because $\sigma_{\bar{x}} = \sigma/\sqrt{n}$ approaches 0. Thus the Bayesian estimate given by (8) approaches the classical estimate $\bar{x}$, which means that the information supplied by the sample is overpowering the information supplied by the prior distribution. The relative advantage of the Bayesian estimate over the classical estimate for various combinations of parameter values and sample sizes is readily determined by calculating the ratio of the lengths of the .95 probability intervals given by (10) and (7). From (9) it follows that this ratio is equal to $1/\sqrt{1 + \delta}$.

The relationship between the classical estimate $\bar{x}$ and the Bayesian estimate μ_1 and their relative accuracies can be shown geometrically for a given numerical situation. As an illustration, suppose that the value of μ for the new shipment is $\mu = 183$, and assume as before that $\mu_0 = 190$, $\sigma_0 = 5$, $\sigma = 50$, $n = 25$, and $\bar{x} = 180$. Then the graphs of the normal prior distribution of μ, the normal distribution of $\bar{x}$, and the resulting normal posterior distribution of μ may be displayed as shown in Fig. 2.

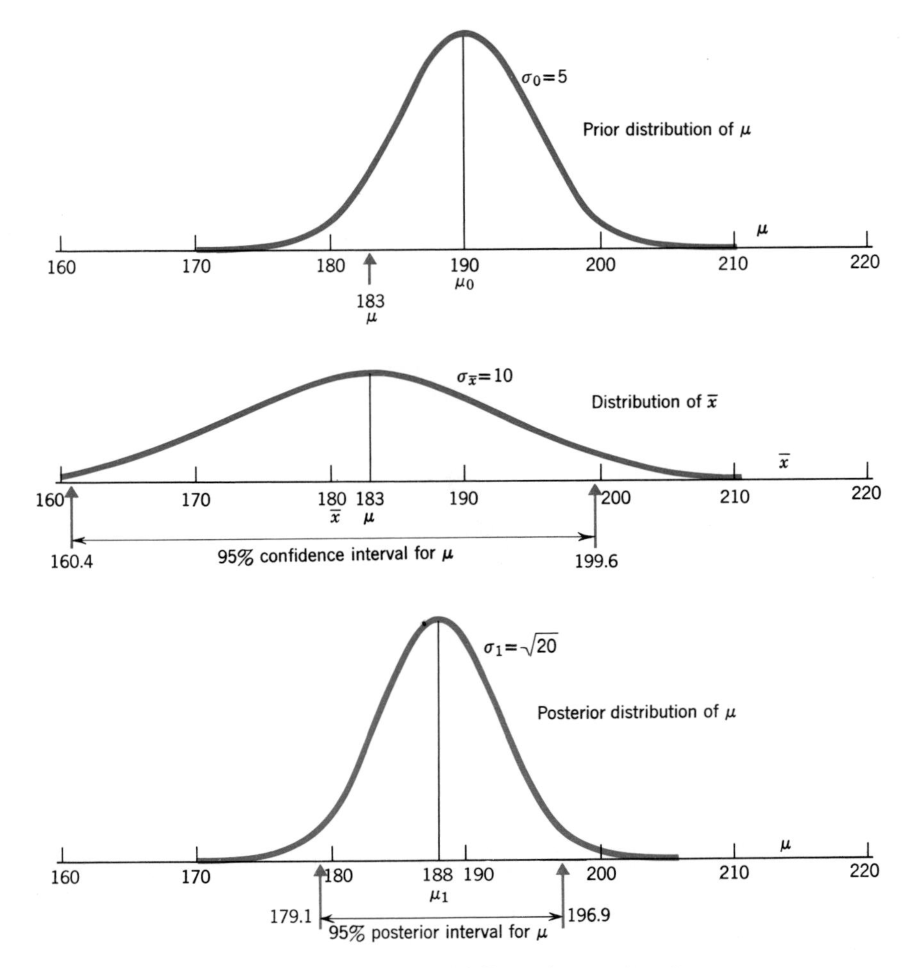

Figure 2 Classical and Bayesian estimation.

10 BAYESIAN DECISION FOR CONTINUOUS VARIABLES

In the preceding section a comparison was made between classical and Bayesian methods of estimation. In some problems a good estimate of a location parameter together with a measure of its accuracy will suffice to solve the problem. This is often true of testing problems where one of two actions is to be made on the basis of the value of a location parameter. In more complex problems where there are several possible actions or varying degrees of loss or gain depending on the value of the location parameter, an estimation interval will not suffice to solve the problem. For such problems a more sophisticated approach based on measuring economic gains or losses is needed. The methods introduced in section 8 are of this type; however, they were applied only to discrete-variable problems. In this section the preceding material on Bayesian estimation will be used to extend the techniques of section 8 to continuous-variable problems. As was pointed out before, such methods are Bayesian methods because they employ the distribution of a parameter.

Example 1. As a first illustration of Bayesian methods for solving continuous-variable problems, consider the problem of deciding whether the manufacturer discussed in section 9 should accept or reject a shipment of steel rods on the basis of having tested a sample of 25 of them, assuming that he has the information concerning the parameters that was given there. Thus, he knows from past experience that a shipment mean μ behaves like a normal variable with mean $\mu_0 = 190$ and $\sigma_0 = 5$. Suppose that the sample of 25 yielded the values $\bar{x} = 180$ and $s = 50$, and that the shipment contains 500 rods.

Because of the central limit theorem, $\bar{x}$ may be treated as a normal variable with mean μ and standard deviation $\sigma_{\bar{x}} = \sigma/\sqrt{25} = \sigma/5$. The value of σ is not known here; therefore it will be replaced by its sample estimate, $s = 50$. Since $n = 25$ is not a very large sample, this estimate is probably of only fair accuracy. Employing this replacement, it will therefore be assumed that $\bar{x}$ possesses a normal distribution with mean μ and standard deviation $\sigma_{\bar{x}} = 10$.

In view of the preceding information and assumptions, the posterior distribution of μ will be normal with mean μ_1 and variance σ_1^2 given by formulas (8) and (9).

Now suppose that if the mean μ of a shipment is less than 185, a high percentage of the manufactured product will be defective and large losses will result. Suppose further that if $185 \leq \mu \leq 195$, a fairly small percentage of the product will be defective and a modest profit will be obtained, but that if $\mu > 195$ there will be very few defectives and profits will be near their maximum. Let the following payoff table represent the profits to be made on the product manufactured with a shipment of 500 rods. If a shipment is rejected, a small loss will occur due to scheduling difficulties and related costs.

	Action	
μ	Accept	Reject
$\mu < 185$	−10,000	−1,000
$185 \leq \mu \leq 195$	5,000	−1,000
$\mu > 195$	10,000	−1,000

Since Bayesian methods that include information from a sample employ posterior probabilities, those probabilities will be calculated directly without first performing an analysis based on prior probabilities as was done in section 8. Using the information that $\mu_0 = 190$, $\sigma_0 = 5$, $\bar{x} = 180$, and $\sigma_{\bar{x}} = 10$, it follows from formulas (8) and (9) that

$$\mu_1 = 188 \qquad \text{and} \qquad \sigma_1 = 4.47$$

Hence, using the normality of μ and the given value of $\bar{x}$, with these values for the mean and the standard deviation of μ, it follows that

$$\begin{aligned} P\{\mu < 185\} &= P\left\{\frac{\mu - \mu_1}{\sigma_1} < \frac{185 - \mu_1}{\sigma_1}\right\} \\ &= P\left\{z < \frac{185 - 188}{4.47}\right\} \\ &= P\{z < -.67\} = .251 \\ P\{\mu > 195\} &= P\left\{\frac{\mu - \mu_1}{\sigma_1} > \frac{195 - \mu_1}{\sigma_1}\right\} \\ &= P\left\{z > \frac{195 - 188}{4.47}\right\} \\ &= P\{z > 1.57\} = .058 \end{aligned}$$

$$P\{185 \le \mu \le 195\} = 1 - (.251 + .058) = .691$$

The calculation of the expected profits under the two possible actions, based on posterior probabilities, is shown in Table 14.

Table 14

		Accept		Reject	
μ	Posterior probability	Profit	Expected profit	Profit	Expected profit
$\mu < 185$	.251	−10,000	− 2,510	−1,000	−251
$185 \le \mu \le 195$	.691	5,000	3,455	−1,000	−691
$\mu > 195$	.058	10,000	580	−1,000	−58
Totals	1.000		1,525		−1,000

On the basis of these calculations and assumptions, the manufacturer should accept the shipment.

Linear Profit Functions. Decision problems for continuous variables are quite easy to solve if the profit function for the various possible actions is a linear function of the mean μ. The technique for solving such problems will be considered next.

Example 2. As an illustration of a linear profit function problem, consider a refinement of the preceding problem. It was assumed there that the profit depended only on the interval in which μ lay, rather than on the actual value of μ. For a more realistic model, assume that the break-even point between accepting or rejecting a shipment is given by $\mu = 187$. Further, assume that the profit function under acceptance is the following linear function of μ:

$$\text{profit} = 700(\mu - 187) - 1000 \tag{11}$$

It is assumed, as before, that if the shipment is rejected, there will be a \$1000 loss attributed to scheduling difficulties and related costs; hence the profit function under rejection is the special linear function

$$\text{profit} = -1000 \tag{12}$$

When the two profit functions are linear in μ, it can be shown by calculus methods that the expected profit is obtained by replacing μ by its expected value $E[\mu]$. This is true whether one is interested in calculating $E[\mu]$ on the basis of prior probabilities or on the basis of posterior probabilities.

Assume, as in the earlier problem, that μ possesses a normal distribution with mean $\mu_0 = 190$. Then, using formula (11), the expected profit is obtained by replacing μ by μ_0 to give

$$E_0[\text{profit}] = 700(190 - 187) - 1{,}000 = 1{,}100$$

Since this is larger than the expected profit under rejection, which is $-$\$1,000, the shipment should be accepted, provided that no sample is taken.

If, however, a sample is taken, the expected profit must be computed using posterior probabilities. But as indicated before, this merely requires one to replace μ by μ_1. Using the sample results obtained before, it follows that

$$E_1[\text{profit}] = 700(188 - 187) - 1{,}000 = -300$$

Since the profit under rejection is still $-$\$1,000, it follows that the shipment should be accepted. It should be noted, however, that the expected profit is considerably less than it was estimated to be before the sample was taken. The value of μ_1 that was used here is the value obtained earlier under the assumption that both x and μ possess normal distributions; hence these same assumptions are needed here.

Example 3. As a second illustration of how quickly continuous-variable problems can be solved when the profit functions under both alternatives are linear and when both the variable x and its mean μ possess normal distributions, consider once more the second problem that was solved in section 8. That problem is a discrete-variable problem with its prior distribution given by Table 11 and its sample variable S possessing a binomial distribution; however, the problem can be converted to a continuous-variable problem in the following manner.

Since $n = 50$ is a large sample, the normal approximation to the binomial distribution justifies treating the variable $S/50$ as a normal variable with the mean $\mu = p$ and standard deviation $\sigma = \sqrt{p(1-p)/50}$. Further, the empirical distribution of p given in Table 11 has the appearance of a sample distribution that might have been obtained from sampling a normal population. An estimate of such a population distribution is obtained by choosing μ_0 and σ_0 as the mean and the standard deviation of the empirical distribution given in Table 11. Calculations on this basis gave $\mu_0 = .38$ and $\sigma_0 = .125$.

Since the profit functions for this problem, as given in section 8, are

$$L_F = 24N + 20Np$$

and

$$L_C = 80Np$$

it follows that they satisfy the linearity assumption.

Assuming, therefore, that both $S/50$ and p may be treated as normal variables, it follows that the expected profits before sampling can be obtained by replacing p by $\mu_0 = .38$; hence

$$L_F = 24N + 20N(.38) = 31.6N$$

and

$$L_C = 80N(.38) = 30.4N$$

It will be observed that these are the same values as those obtained from the calculations of Table 12.

With these same assumptions, it follows that the expected profits after sampling can be obtained by replacing p by μ_1. The value of μ_1 can be found by applying formula (8). In that formula $\bar{x} = S/50$, that is, it is the mean of the sample outcomes, where 1 represents success and 0 failure. Since the sampling experiment gave 25 successes in 50 trials, $\bar{x} = \frac{1}{2}$. Further,

$$\delta = \frac{\sigma_{\bar{x}}^2}{\sigma_0^2} = \frac{p(1-p)/50}{(.125)^2}$$

Since p is unknown, it must be replaced by its distribution mean, which is .38 here, or by its sample estimate; which is .50. Assuming that the prior distribution is not under suspicion, the value .38 will be used, even though the value .50 may be more reliable. This gives $\delta = .302$; hence formula (8) gives

$$\mu_1 = \frac{.5 + (.38)(.302)}{1 + .302} = .472$$

The expected profits after sampling are therefore given by

$$L_F = 24N + 20N(.472) = 33.4N$$

and

$$L_C = 80N(.472) = 37.8N$$

These values should be compared with the values obtained in section 8, namely $L_F = 33.6N$ and $L_C = 38.2N$. There is very little difference in these results. The differences are due to using the fitted normal distribution in place of the empirical distribution of Table 11.

If the sample estimate of p, namely .50, had been used to calculate the value of δ, and hence of μ_1, it would have been found that the values of L_F and L_C differ very little from those based on using the estimate .38.

The preceding problems illustrate some of the difficulties that can arise when attempting to apply a mathematical model to a realistic situation. Estimates, approximations, and assumptions are required at the various stages of the analysis. This is true of all scientific methods when they are applied to the real world, but they are especially true for problems of the business world. Practitioners must be prepared to make such compromises; however, in doing so, they must realize that the decisions obtained by using a mathematical model and its consequences are highly subjective and that the usefulness of the results are no better than the quality of the judgment employed.

11 REMARKS

The preceding examples are merely two illustrations of the kinds of decision-making problems that can be solved if one is willing to accept the principle of maximizing expected gain or, equivalently, minimizing expected opportunity loss as a basis for making decisions. Here gain or loss may be in either monetary or utility units. The techniques that were used to solve these problems can also be applied to solving problems in which there are more than two alternative decisions at issue.

The foregoing discussion of the elements of statistical decision making should help to indicate the kinds of problems that can be dealt with and the power of the method in general. It is particularly useful in those cases where the outcomes of the decision-event table can be quantified in some reasonable manner. The possibilities for doing this seem to be particularly rich in the area of business and economics where money is a usual measure of effectiveness and where expected profit may be a reasonable basis for ranking alternatives. The problem of obtaining valid data is frequently a difficult one, and there is much controversy about how far one can go in using subjective probabilities to arrive at valid decisions for action. However, since some decision will be taken no matter how subjective the available information may be, it may be better to base that decision on explicit decision-making techniques, such as those presented in this chapter, than to rely on unspecified intuitive judgments.

An analysis of the techniques used in the solution of the problems of this chapter will show that a large share of the basic statistical tools developed in the earlier chapters were needed for their solution. As problems become more complex, additional sections of the earlier theory enter into the solution. An investigator who hopes to solve the various kinds of business problems that arise will need all of the techniques developed in this book and many more besides.

12 REVIEW ILLUSTRATIONS

Example 1. A company specializing in the creation and development of new ideas in toys has come up with a novel toy, the disposition of which it is considering. The company can either sell the rights to the toy or it can manufacture and sell the toy itself. The management of the company believes that it can sell the rights for \$100,000 with a bonus of an additional \$100,000 if the toy is a hit. If, however, the company manufactures the toy itself, it believes that it can earn \$500,000 if the toy is a hit but that it will lose \$300,000 if it is not. For simplicity a toy will be considered a hit if it captures at least a certain percentage of the market and not a hit if it captures less than that percentage of the market. This oversimplification of the problem is made for computational convenience. If the company is interested in maximizing profits, what should it do if (*a*) management is willing to wager 7 to 10 that the toy will be a hit, (*b*) management is uncertain about the potential of the toy and compromises on even odds that it will be a hit? (*c*) What is the break-even probability p of the toy being a hit? (*d*) A market survey was made, and it was found that the probability of obtaining the survey outcome is .4 if the toy is a hit, whereas this probability is .3 if the toy is not a hit. What is the appropriate decision using this new information?

(*a*) The payoff table is given by

Event	Action	
	Sell rights	Manufacture
A hit	\$200,000	\$500,000
Not a hit	\$100,000	−\$300,000

The expected payoffs based on management's personal prior probability of $p = .7$ are

$$E_S = 200{,}000(.7) + 100{,}000(.3) = 170{,}000$$

and

$$E_M = 500{,}000(.7) - 300{,}000(.3) = 260{,}000$$

Hence the company should retain its rights, and manufacture and sell the toy itself.

(*b*) With $p = .5$ these expectations become

$$E_S = 200{,}000(.5) + 100{,}000(.5) = 150{,}000$$

and

$$E_M = 500{,}000(.5) - 300{,}000(.5) = 100{,}000$$

Now it is better to sell the rights.

(*c*) The break-even probability p is obtained by solving the equation

$$200{,}000p + 100{,}000(1 - p) = 500{,}000p - 300{,}000(1 - p)$$

This equation is equivalent to the equation

$$2p + (1 - p) = 5p - 3(1 - p)$$

whose solution is $p = \frac{4}{7}$.

(*d*) Let S denote the survey outcome; then the new information gives the conditional probabilities $P\{S \mid H\} = .4$ and $P\{S \mid F\} = .3$, where H denotes a hit and F denotes a failure (not a hit). The posterior probabilities of a hit and a failure are obtained by means of Bayes' formula; hence, using the management's prior probability of a hit, namely $P\{H\} = .7$, the posterior probabilities become

$$P\{H \mid S\} = \frac{P\{H\}P\{S \mid H\}}{P\{H\}P\{S \mid H\} + P\{F\}P\{S \mid F\}} = \frac{(.7)(.4)}{(.7)(.4) + (.3)(.3)} = \frac{28}{37}$$

and

$$P\{F \mid S\} = \frac{P\{F\}P\{S \mid F\}}{P\{H\}P\{S \mid H\} + P\{F\}P\{S \mid F\}} = \frac{(.3)(.3)}{(.7)(.4) + (.3)(.3)} = \frac{9}{37}$$

Using these posterior probabilities in place of the prior probabilities gives

$$E_S = 200{,}000 \cdot \frac{28}{37} + 100{,}000 \cdot \frac{9}{37} = 175{,}676$$

and

$$E_M = 500{,}000 \cdot \frac{28}{37} - 300{,}000 \cdot \frac{9}{37} = 305{,}405$$

The decision in favor of manufacturing is now more pronounced than it was in (*a*).

All of these decisions revolve on the use of personal probabilities, and the contrast between (*a*) and (*b*) is intended to point out this fact. The conditional probabilities in part (*d*) do not add to 1 because a market survey result may be compatible with a given value of p without having a large probability of being obtained with that value of p.

Example 2. Experimentics, Inc. is in the business of making laboratory tests on the effectiveness of experimental pharmaceuticals. One of its clients, Eureka Associates, recently sent them a pharmaceutical, X-100, which scored a mean of 150 and a standard deviation of 15 on a basic testing procedure that always uses 16 test animals.

This result is cause for excitement because rarely have such high values of the mean been achieved on previous formulations submitted to this test. Experimentics looked at their records and found that the grand mean of the means of all previous tests relevant to this test was 120 and that the standard deviation of the earlier means was 10. Assuming that the record data are suitable for establishing a prior distribution of the true mean for this test and that it is normal: (*a*) compute the mean μ_1 of the posterior distribution of μ, which quantity is a measure of the true effectiveness of X-100; (*b*) compute the standard deviation σ_1 of the posterior distribution; (*c*) compute the probability that $\mu > 135$, given that $\bar{x} = 150$. (*d*) Using the experimental results only, what is a 95 percent confidence interval for μ? (*e*) Using the posterior distribution find a corresponding interval for μ and compare its length with that of the interval found in (*d*). The solutions follow.

(*a*) $\mu_0 = 120$, $\sigma_0 = 10$, $\bar{x} = 150$, $s = 15$, $n = 16$; hence $\sigma_{\bar{x}} = \sigma/\sqrt{n} = \sigma/4$. Approximating σ with s, $\sigma_{\bar{x}} \doteq \frac{15}{4}$, hence $\delta = \sigma_{\bar{x}}^2/\sigma_0^2 = (\frac{15}{4})^2/100 = .14$, and

$$\mu_1 = \frac{\bar{x} + \delta\mu_0}{1 + \delta} \doteq \frac{150 + (.14)(120)}{1.14} \doteq 146$$

(*b*) $\sigma_1^2 = \sigma_{\bar{x}}^2/(1 + \delta) \doteq (\frac{15}{4})^2/1.14 = 12.34$

$$(c)\ P\{\mu > 135\} = P\left\{\frac{\mu - \mu_1}{\sigma_1} > \frac{135 - \mu_1}{\sigma_1}\right\} = P\left\{z > \frac{135 - 146}{\sqrt{12.34}}\right\}$$

$$= P\{z > -3.1\} \doteq 1$$

(*d*) Using large-sample methods, $150 - 1.96(\frac{15}{4}) < \mu < 150 + 1.96(\frac{15}{4})$, or $142.65 < \mu < 157.35$.

(*e*) Use $\mu_1 - 1.96\sigma_1 < \mu < \mu_1 + 1.96\sigma_1$, or $146 - 1.96\sqrt{12.34} < \mu < 146 + 1.96\sqrt{12.34}$, or $139.11 < \mu < 152.89$. Lengths are 14.7 and 13.8, respectively.

Example 3 An insurance company is attempting to determine the additional costs that would arise if it included additional types of illness in its coverage. A sample of 100 of its policy holders was taken and the additional coverage was offered to them without increasing their premiums. The prior distribution of the average cost per policy holder was estimated by management to be normal with a mean μ_0 of \$60 per year and a standard deviation σ_0 of \$10.

The result of the sampling experiment, based on two years of experience, was $\bar{x} = \$80$ and $s = \$30$.

Suppose management decided that the cost of offering the new coverage would be \$900,000 to set it up and promote it, and they are quite certain 100,000 policies can be sold at \$90 per policy. (*a*) Calculate the profit functions for the two possible actions: offering the insurance, and not offering it. (*b*) If a decision is made on the basis of the prior information, what decision should be made? (*c*) Calculate the mean, μ_1, and standard deviation, σ_1, of the posterior distribution. (*d*) What should the decision be if it is based on the posterior distribution? Comment on the difference in the expected profit when it is based on the posterior distribution instead of the prior distribution. The solutions follow.

(*a*) If the insurance is offered the profit is given by

$$100{,}000(90) - 900{,}000 - 100{,}000\mu = 8{,}100{,}000 - 100{,}000\mu$$

where μ is the average cost per policy holder. The profit if the insurance is not offered is, of course, zero. Thus, the function to be maximized is a linear function of μ.

(*b*) Using the formula in (*a*), the expected profit can be obtained by replacing μ by μ_0 in the profit function; hence

$$E_0[\text{profit}] = 8{,}100{,}000 - 100{,}000(60) = 2{,}100{,}000$$

The company should therefore offer the new coverage.

(*c*) Using formula (8),

$$\mu_1 = \frac{\bar{x} + \delta\mu_0}{1 + \delta} = \frac{80 + \delta 60}{1 + \delta}$$

But $\delta = \sigma_{\bar{x}}^2/\sigma_0^2 = \sigma^2/n\sigma_0^2 \doteq s^2/n\sigma_0^2 = (30)^2/100(10)^2 = .09$. Hence

$$\mu_1 = \frac{80 + .09(60)}{1 + .09} = 78.3$$

From formula (9),

$$\sigma_1^2 = \frac{\sigma_{\bar{x}}^2}{1 + \delta} = \frac{\sigma^2}{n(1 + \delta)} \doteq \frac{s^2}{n(1 + \delta)} = \frac{(30)^2}{100(1.09)} = 8.26$$

Hence

$$\sigma_1 = 2.87$$

(*d*) Using formula (11),

$$E_1[\text{profit}] = 8{,}100{,}000 - 100{,}000(78.3) = 270{,}000$$

The same decision should be made as before, however it is less decisive than in (*b*). The prior distribution is quite unrealistic here. It would lead to unjustified optimism on the part of management.

EXERCISES

SECTION 2

1 Suppose you have $10,000 to invest in either bonds or common stocks and you would like to obtain the highest expected value of this investment at the end of 5 years. Your securities counselor tells you that bonds will increase in value about 25 percent over the next 5 years, but that the appreciation in value and dividends of stocks will depend on economic conditions. He says that if there is deflation, they will decrease about 10 percent. If there is considerable inflation, he would estimate that growth would be about 100 percent, and if the economy is rather normal, a growth of about 15 percent. He is also willing to conjecture that the probabilities of deflation, normalcy, and inflation are .1, .7, and .2, respectively. Assuming this adivce is reasonable, should you buy bonds or stocks?

2 A farmer must decide when to plant a crop that is quite sensitive to damage by frost. If he plants on April 15 and there is no frost, he will be able to sell at a premium and make a profit of $15,000. If there is a frost he will lose $3,000 for labor and seed. By waiting two weeks, the later crop will yield a profit of $3,000

if no frost interferes. The probability of frost on or after April 15 is $\frac{8}{10}$; whereas it is $\frac{1}{10}$ on or after April 29. The farmer wishes to maximize his expected profit. What should he do?

3 Refer to problem 2. At what probability of frost on or after April 15 will the farmer be indifferent between planting early or late?

4 An oil company is contemplating leasing a piece of land in a region where oil was recently discovered. The lease will cost $50,000 and the cost of drilling a well is $100,000. If oil is discovered the company can expect to earn $400,000 on the well, not counting leasing and drilling costs.
(*a*) Calculate the payoff table.
(*b*) What is the smallest probability of finding oil that will justify the company's proceeding with the lease and drilling?

5 Each of two stock brokers has proposed a small portfolio of 4 stocks and bonds for your consideration. In order to compare their suggestions you have asked them to give you their best estimates of the value of their portfolios after 2 years under 3 possible economic conditions. An economist gives you his estimates of the probabilities that each of the 3 conditions will occur. Assembled, your data are as follows:

Economic condition	Estimated probabilities	Estimated value of securities							
		Broker *A*				Broker *B*			
		1	2	3	4	1	2	3	4
Decline	.2	30	40	70	110	90	100	108	110
No change	.5	102	103	103	110	100	120	108	110
Expansion	.3	160	230	120	110	110	140	108	110
	1.0								

If you wish to maximize the expected value of your portfolio, which broker has the better selection?

6 A street vendor can buy an item for 60 cents and sell it for $1.50. The probability of any specified demand number is given by the following table.

x	0	1	2	3	4	5 or more
$P\{x\}$	.05	.15	.30	.25	.15	.10

(*a*) Construct a payoff table.
(*b*) Determine the optimal number that he should buy.

7 In buying a particular piece of equipment whose suitability to your needs was particularly attractive, you are aware that spare parts may be a problem since the product is not widely known and has few distributions. In assessing the spare-parts problem you are told that only one particular part is likely to present a problem of replacement since it can only be obtained from the manufacturer, but that during the lifetime of the equipment it will practically never need more than 3 replacements. If replacements are bought at the time the equipment is bought, this part costs $100 per unit. If, however, it is ordered later from the manufacturer it will cost $300 per unit including handling charges. The problem facing the buyer is: Should he order spare parts now or wait until a need

develops? Construct a payoff (in this case, negative) table for the event-action matrix appropriate for this problem. Assume unused spare parts are a complete loss.

8 The manufacturer's records for problem 7 show that repacement requirements of the spare part under consideration are as follows:

Number of replacements required:	0	1	2	3
Relative frequency:	.4	.3	.2	.1

Set up a standard decision analysis table and calculate the expected payoffs for each possible order size. What is the best decision?

9 A flower vendor prepares a "specialty of the house," which is a particularly beautiful bouquet freshly prepared during the night and offered to his patrons the next day. Unsold specialties are destroyed at the end of the day because of their perishability. The artist-vendor figures his labor and material costs for each bouquet to be $10; the selling price is $20.

How many bouquets should be prepared each day if the probabilities of demand for 0, 1, 2, 3, or 4 bouquets are .1, .2, .5, .1, and .1, respectively?

SECTION 3

10 (*a*) Compile the conditional opportunity loss table for the following data. The entries in the table are conditional profits.

Events	Event probabilities	Alternative Actions 1	2	3
A	.2	+500	−3000	+1500
B	.5	+300	+5000	−1000
C	.3	+800	−4500	+2100

(*b*) Use the results of the analysis in part (*a*) to recommend the best action.

11 A businessman has decided that he is going to invest in one or the other of two ventures, *A* and *B*. The two ventures are completely independent of one another. The businessman estimates that the probability of success for venture *A* is .35. If *A* succeeds the total profit is estimated at $50,000. If *A* fails, the total profit is estimated at −$12,000. The probability of success for venture *B* is judged to be .50. If *B* succeeds, the total profit estimate is $20,000. If *B* fails, the total profit estimate is −$15,000.

(*a*) Carry out an opportunity loss analysis of the businessman's decision problem as stated above. Indicate the decision he should make and why. Show all calculations.

(*b*) What is the cost of uncertainty in the problem?

12 Compute the opportunity loss table for problem 5 (note that stocks within portfolios can be aggregated to simplify computations). What is the cost of uncertainty in this problem?

13 For problem 9 calculate the vendor's cost of uncertainty.

14 Calculate an opportunity loss table for the data of problem 4. Use it to determine whether the company should lease and drill if $p = .4$ of discovering oil on that property.

15 A wealthy farmer owns a piece of land on which a new school may be built. If his land is chosen as the school site, it will be worth \$20,000, otherwise it will be worth only \$8,000. He has the opportunity to exchange this piece of land plus \$2,000 for the only other site on which the school could be built. The second site will also be worth \$20,000 or \$3,000, depending on whether or not that site is selected for the school. The farmer believes that there is 1 chance in 4 that his present piece of land will be selected for the school.
(*a*) Carry out an expected monetary value analysis and recommend what the farmer should do.
(*b*) Carry out an opportunity loss analysis.

SECTION 4

16 Calculate the expected value of perfect information for problem 4 if the probability of discovering oil is estimated to be .2.

17 Suppose there are 2 chances in 5 that oil will be present if the company decides to drill in problem 4. What is the expected value of perfect information under those conditions?

18 Calculate the expected value of perfect information for the spare parts data of problem 8.

19 A manufacturing company needs a lathe for various jobs. There are 4 models being considered. The profit potential of each depends on the number of jobs on which the lathe can be used. Management came up with the following payoff table for these 4 models on the basis of assuming that the number of jobs would be 0, 10, 30, or 50.

Jobs	*A*	*B*	*C*	*D*
0	−3	−7	−25	−22
10	3	1	6	6
30	5	−4	5	−2
50	3	6	13	12

(*a*) If management estimates the probabilities for these four job possibilities to be $P\{0\} = .05$, $P\{10\} = .45$, $P\{30\} = .40$, and $P\{50\} = .10$, which lathe should be chosen?
(*b*) What is the value of perfect information for this problem?

SECTION 5

20 Suppose that the farmer in problem 2 has the same utility curve for money in the range −3,000 to 15,000 as curve *B* in the range 2,000 to 20,000 in Fig. 1.
(*a*) Construct the farmer's payoff table in utility.

(*b*) What decision should the farmer make now?

(*Hint:* The new scale has a range of 18,000, or 2,000 for each of the 9 intervals in Fig. 1. Hence read −3,000 (for 2,000), −1,000 (for 4,000), etc.)

21 Refer to problem 20. At what probability of frost on or after April 15 will the farmer be indifferent between planting early or later, assuming he has a utility curve similar to curve *B* in Fig 1?

22 Suppose the farmer in problem 2 is the gambling type and has the same utility curve as *C* in Fig. 1 except the money scale ranges from −7,000 to 20,000.

(*a*) Construct the farmer's payoff table in utility.

(*b*) What decision should this farmer make regarding the date of planting?

23 Mr. F, after due deliberation, decided that he would be willing to pay the amounts shown below for the gambles indicated. Assuming a utility scale ranging from 0 to 1 for certainty payoffs of −10,000 to +10,000, compute the utility index for the gambles presented and sketch the appropriate utility scale in a graph over the whole range.

Gamble $\pi(x_1) + (1 - \pi)x_2$	Willing to pay
1(−10,000) + 0(+10,000)	−10,000
0(−10,000) + 1(+10,000)	+10,000
.5(−10,000) + .5(+10,000)	+3,000
.5(−5,000) + .5(−9,000)	−5,000
.4(−4,000) + .6(+6,000)	+5,000
.5(+2,000) + .5(−8,000)	0
.2(−6,000) + .8(+9,000)	+8,000

24 Businessmen *A*, *B*, *C*, and *D* were given the following facts on two business ventures, V_1 and V_2.

	Venture V_1		Venture V_2	
	Success	Failure	Success	Failure
Payoff:	+50,000	−20,000	+5,000	−5,000
Probability, $P\{E\}$	.25	.75	.55	.45

Suppose the businessmen possess the utilities for money within the range −50,000 to 50,000 for the relevant payoffs as shown below.

Businessman	−20,000	−5000	+5000	+50,000
A	.30	.45	.55	1.00
B	.55	.70	.77	1.00
C	.13	.25	.35	1.00
D	.20	.50	.69	1.00

(*a*) Compute the expected utilities for each venture for each businessman.

(*b*) What actions should each businessman take?

(*c*) What are the expected payoffs (in dollars) for each businessman's preferred action in (*b*)?

SECTION 7

25 A geologist is hired to study the particular site that the oil company of problem 4 is contemplating leasing. Suppose that the geologist has a 70 percent chance of being correct when he decides that oil is present and it is, but that 40 percent of the time he believes oil to be present when it is not present. The tests required by the geologist cost $20,000.

(*a*) If the oil company believes that the probability of striking oil is .3, which decision should it make: turn down the lease, drill without employing the geologist, or hire the geologist and follow his advice?

(*b*) How much is the geologist's advice worth?

26 A firm that sells products that are somewhat unpredictable in consumer acceptance uses 3 of its buyers, A, B, and C, to assist it in deciding what items to stock, or not to stock, each season. A record was kept of each buyer's decision on each item offered and the payoffs that resulted, or would have resulted if the alternative decision had been made. This record, together with the firm's experience $P\{E\}$ in producing acceptable items, is as follows:

		Item accepted by buyer $P\{\text{acceptable} \mid E\}$				Item rejected by buyer $P\{\text{acceptable} \mid E\}$			
Event	$P\{E\}$	A	B	C	Payoff	A	B	C	Payoff
Acceptable to consumer	.3	.8	.9	.7	1,000	.2	.1	.3	−5,000
Unacceptable to consumer	.7	.1	.2	.01	−5,000	.9	.8	.99	1,000

(*a*) Which of the buyers is the most valuable decision maker? How much is he worth per item presented?

(*b*) Suppose that it was not possible to determine the payoffs for rejected items, in which case these payoffs would be zero. Which buyer is now the most valuable? What is his value?

SECTION 8

27 A geologist assigns a prior probability of .2 that a particular piece of land will contain oil in producible quantities. This value is based on his past experience with prospecting for oil. A seismic experiment is then performed to assist in a determination of whether oil exists there. Assume, for simplicity, that this experiment gives only a yes or no answer and that the probability of its correctly predicting the presence of oil is .7, whereas its probability of predicting oil when none exists is .25. If the seismic experiment is performed and predicts oil, what posterior probability should the geologist assign to the existence of oil?

28 You are told that a coin is either a fair coin, or is two-headed, or is two-tailed. Since there is no other information given, you assign equal probabilities to these

3 possibilities. Suppose that the coin is tossed n times and each toss produces a head.

(*a*) What posterior probabilities will you assign to the 3 possibilities?

(*b*) How large must n be before you will state that the chances of the coin being a fair coin are less than one in a hundred?

29 Flip Jackson has a coin identification game that he likes to play. He states: "I have in my hand a coin that is either regular or two-headed. For \$5 you can call it either way, whereupon I will pay you \$10 if you are correct; otherwise you lose your \$5. As an extra inducement I will allow you to take one demonstration flip for the price of \$1. Would you like to play?"

Suppose your prior probabilities are .5 and .5 and you buy the demonstration, with the result that a head shows. What is your choice of coin and what is its expected payoff? Was the demonstration worth \$1?

30 A business firm has developed a new product that it is contemplating marketing. The cost of marketing it is estimated at \$500,000. If this product is preferred to its competitor and is marketed, the company can expect to realize sales of \$2,000,000. If it is worse than its competitor and is marketed, the company can expect sales of only \$300,000. If they do not market the product, they can sell the idea, including patents and the process, to their competitor for \$900,000 if it is a superior product and for \$200,000 if it is inferior.

(*a*) If the probability is .6 that the new product is superior to the competitor's product, what decision should the firm make?

(*b*) Suppose a market survey is made and the result, denoted by S, has a probability of .04 of being realized if the product is superior and a probability of .10 of being realized if it is inferior. With this new evidence, what decision should the firm make?

31 Alex has three friends, A, B, and C, whom he would like to compare in regard to their personal probabilities in a little experiment he conducts. "See here, my friends," he says, "I have a coin here that I shall toss and we shall observe the results. Then I wish to know how much each of you is willing to bet against my dollar that a tail will occur on the next toss. But before we go further, I wish each of you to record your opinion on possible events and their probabilities." In response to this the three friends record the following results:

Event	Probability assigned by		
	A	B	C
The coin is true	1/3	.95	1
The coin is two-headed	1/3	.05	0
The coin is two-tailed	1/3	0	0
	1	1.00	1

(*a*) Each of these probability distributions is plausible. Would you have a different one? If so, record it and explain why you prefer yours, or any one of the above over the others.

(*b*) Alex tosses the coin and obtains 4 heads in a row. What are the probabilities that each would now bet against a tail occurring on the next toss?

SECTIONS 9 and 10

32 Suppose the prior probability distribution of the mean water content of a raw material used in food processing is based on a historical record covering what appears to be the same general underlying condition of supply. You calculate the standard deviation of this distribution and compare it with the average standard deviation obtained from sampling within lots purchased during the same period, and find that $\sigma_0 < \sigma$. Is this possible? Explain.

33 Everglades died, leaving an orange grove of 100,000 trees that will be ready for harvest in about 6 months. Everglades' widow would like to settle the estate immediately since the cash is needed to settle other matters. She offered to sell the fruit on the trees for $400,000. Sittrus, a professional orange buyer, took a look at the grove and decided it would pick-out about 3.9 boxes per tree and with the generally accepted on-tree price at the time of $1 per box, the crop was worth about $390,000. Sittrus is aware that his estimates in the past had a standard deviation of 0.8 boxes per tree and it may be that he is underestimating the crop at this time. At a cost of $25 per tree, Sittrus has an alternative way of estimating the yield per tree with a standard deviation of 2 boxes per tree. He decided to use this method in this case and applied it to a random sample of 100 trees, obtaining a mean of 4.1 boxes per tree. Assuming the judgment and special methods to have estimates that are distributed normally, what should Sittrus do, ignoring costs?

34 A prior normal distribution has a mean $\mu_0 = 50$ and standard deviation $\sigma_0 = 10$. If the basic normal population has a standard deviation $\sigma = 30$, how large a random sample is required to obtain a posterior distribution with a standard deviation $\sigma_1 = 5$?

35 A land development corporation is considering a development that will take 3 years to complete. Revenue from the development will come from sales of developed units. Management believes that 5,000 sales can be made and it is willing to wager 2 to 1 that the true sales will be somewhere between 4,000 and 6,000. It is also willing to have its complete prior distribution on sales be represented by a normal distribution with these properties. The corporation's accountants have worked out a profit function that is given approximately by the formula

$$\text{profit} = 3{,}000N - 14{,}000{,}000$$

Here N denotes the number of sales.

(*a*) On the basis of the preceding information concerning the prior distribution, should the corporation proceed with the venture? What would its expected profit be?

(*b*) A market research group was consulted and, after an intensive study of 100 possible customers, estimated that the total market is approximately 4,500 with a standard deviation of 200 for that total. Now what should the corporation do? What would its expected profit be?

SECTION 12

36 A company that manufactures novelty items has kept records of past sales and, on the basis of those records, arrived at the following probabilities for the corresponding number of sales units.

Units sold	Probability
2,000	.1
2,500	.3
3,000	.4
3,500	.1
4,000	.1

The cost in cents of manufacturing a new item, as a function of the number manufactured, is as follows.

Units produced	Cost per unit
2,000	60
2,500	50
3,000	45
3,500	42
4,000	40

These items will be sold for one dollar each. If more items are produced than are sold, the excess, up to a maximum of 1,000 items, may be sold for 10 cents each to a wholesaler. Any excess items beyond that number are worthless. Items must be purchased in blocks of 500 units.

(*a*) Construct a payoff table for the profit function.

(*b*) Determine the number of units that should be manufactured.

37 You are operating an aerial-photographic service. You own an airplane with camera equipment that is bought and paid for, but you hire a pilot and expert photographer to do certain specialized photography of an area from overflights. On sunny (that is, cloudless) days you are able to take pictures without constraint, on cloudy days not at all, and on partly cloudy days with partial success. Suppose you can gross $800 on sunny days, $300 on partly cloudy days, and nothing at all on cloudy days. The pay rates for the crew depend on which of two commitments you make with them: the first requires 2 days notice and $200, whether they are required to work or not (which depends on the weather); the second requires 2 days notice and $300 if they do work and $100 penalty if they don't work. The possible events, then, are the 3 kinds of weather that can occur, and the possible actions are the two contractual arrangements. Construct the appropriate payoff table.

38 Suppose the aerial photographic contractor of problem 37 (you!) has kept some records on the weather—particularly the type of day that follows the weather forecast by 2 days—with the following relative frequencies:

If the forecast is	The relative frequency of actual weather is: Sunny	P-Cloudy	Cloudy	Total
Sunny	.80	.15	.05	1.00
Partly Cloudy	.25	.50	.25	1.00
Cloudy	.10	.30	.60	1.00

Suppose the contractor decides to hire the pilot and photographer only if the forecast is sunny. Which of the two possible labor arrangements of problem 37 should he adopt?

39 Suppose the contractor of problem 38 is considering flying on forecasted partly cloudy days. Would this be wise? Which is the best labor arrangement in this case?

Appendix Tables

The following tables are modifications or condensations of tables found in other books or journals. We wish to take this opportunity to express our appreciation to the authors and publishers from whom these tables were obtained.

Tables I, IV, and X are modifications of similar tables found in the author's book, *Introduction to Mathematical Statistics,* 4th edition, John Wiley & Sons (1971). Table X originally came from *Statistical Methods* by G. Snedecor, Collegiate Press, Iowa State College.

Tables V and IX are from the book *Statistics, An Introduction* by D. A. S. Fraser and published by John Wiley & Sons (1958). These tables originally came from the journal articles "Tables of percentage points of the incomplete beta function" and "Tables of the percentage points of the χ^2 distribution" by C. M. Thompson in *Biometrika,* vol. 32 (1941).

USE OF THE SQUARE ROOT TABLE

To obtain the square root of a number, mark off the digits in blocks of 2, starting at the decimal point. If the number is larger than 1 and the block farthest to the left contains only one digit, use the column $\sqrt{N}$; if it contains two digits use $\sqrt{10N}$. If the number is smaller than 1 and the first nonzero digit to the right of the decimal point is preceded by a zero in its block, use the column $\sqrt{N}$, otherwise use $\sqrt{10N}$. Each block of digits in the number corresponds to a single digit in the answer. Examples:

$$\sqrt{241.67} = \sqrt{2|41.|67}; \quad \text{hence use } \sqrt{N} \text{ column for } \sqrt{242.} \doteq 15.6$$

$$\sqrt{.0024167} = \sqrt{.00|24|16|7}; \quad \text{hence use } \sqrt{10N} \text{ column for } \sqrt{.00242} \doteq .0492$$

Table 1 Squares and square roots

N	N^2	$\sqrt{N}$	$\sqrt{10N}$
1.00	1.0000	1.00000	3.16228
1.01	1.0201	1.00499	3.17805
1.02	1.0404	1.00995	3.19374
1.03	1.0609	1.01489	3.20936
1.04	1.0816	1.01980	3.22490
1.05	1.1025	1.02470	3.24037
1.06	1.1236	1.02956	3.25576
1.07	1.1449	1.03441	3.27109
1.08	1.1664	1.03923	3.28634
1.09	1.1881	1.04403	3.30151
1.10	1.2100	1.04881	3.31662
1.11	1.2321	1.05357	3.33167
1.12	1.2544	1.05830	3.34664
1.13	1.2769	1.06301	3.36155
1.14	1.2996	1.06771	3.37639
1.15	1.3225	1.07238	3.39116
1.16	1.3456	1.07703	3.40588
1.17	1.3689	1.08167	3.42053
1.18	1.3924	1.08628	3.43511
1.19	1.4161	1.09087	3.44964
1.20	1.4400	1.09545	3.46410
1.21	1.4641	1.10000	3.47851
1.22	1.4884	1.10454	3.49285
1.23	1.5129	1.10905	3.50714
1.24	1.5376	1.11355	3.52136
1.25	1.5625	1.11803	3.53553
1.26	1.5876	1.12250	3.54965
1.27	1.6129	1.12694	3.56371
1.28	1.6384	1.13137	3.57771
1.29	1.6641	1.13578	3.59166
1.30	1.6900	1.14018	3.60555
1.31	1.7161	1.14455	3.61939
1.32	1.7424	1.14891	3.63318
1.33	1.7689	1.15326	3.64692
1.34	1.7956	1.15758	3.66060
1.35	1.8225	1.16190	3.67423
1.36	1.8496	1.16619	3.68782
1.37	1.8769	1.17047	3.70135
1.38	1.9044	1.17473	3.71484
1.39	1.9321	1.17898	3.72827
1.40	1.9600	1.18322	3.74166
1.41	1.9881	1.18743	3.75500
1.42	2.0164	1.19164	3.76829
1.43	2.0449	1.19583	3.78153
1.44	2.0736	1.20000	3.79473
1.45	2.1025	1.20416	3.80789
1.46	2.1316	1.20830	3.82099
1.47	2.1609	1.21244	3.83406
1.48	2.1904	1.21655	3.84708
1.49	2.2201	1.22066	3.86005
1.50	2.2500	1.22474	3.87298

N	N^2	$\sqrt{N}$	$\sqrt{10N}$
1.50	2.2500	1.22474	3.87298
1.51	2.2801	1.22882	3.88587
1.52	2.3104	1.23288	3.89872
1.53	2.3409	1.23693	3.91152
1.54	2.3716	1.24097	3.92428
1.55	2.4025	1.24499	3.93700
1.56	2.4336	1.24900	3.94968
1.57	2.4649	1.25300	3.96232
1.58	2.4964	1.25698	3.97492
1.59	2.5281	1.26095	3.98748
1.60	2.5600	1.26491	4.00000
1.61	2.5921	1.26886	4.01248
1.62	2.6244	1.27279	4.02492
1.63	2.6569	1.27671	4.03733
1.64	2.6896	1.28062	4.04969
1.65	2.7225	1.28452	4.06202
1.66	2.7556	1.28841	4.07431
1.67	2.7889	1.29228	4.08656
1.68	2.8224	1.29615	4.09878
1.69	2.8561	1.30000	4.11096
1.70	2.8900	1.30384	4.12311
1.71	2.9241	1.30767	4.13521
1.72	2.9584	1.31149	4.14729
1.73	2.9929	1.31529	4.15933
1.74	3.0276	1.31909	4.17133
1.75	3.0625	1.32288	4.18330
1.76	3.0976	1.32665	4.19524
1.77	3.1329	1.33041	4.20714
1.78	3.1684	1.33417	4.21900
1.79	3.2041	1.33791	4.23084
1.80	3.2400	1.34164	4.24264
1.81	3.2761	1.34536	4.25441
1.82	3.3124	1.34907	4.26615
1.83	3.3489	1.35277	4.27785
1.84	3.3856	1.35647	4.28952
1.85	3.4225	1.36015	4.30116
1.86	3.4596	1.36382	4.31277
1.87	3.4969	1.36748	4.32435
1.88	3,5344	1.37113	4.33590
1.89	3.5721	1.37477	4.34741
1.90	3.6100	1.37840	4.35890
1.91	3.6481	1.38203	4.37035
1.92	3.6864	1.38564	4.38178
1.93	3.7249	1.38924	4.39318
1.94	3.7636	1.39284	4.40454
1.95	3.8025	1.39642	4.41588
1.96	3.8416	1.40000	4.42719
1.97	3.8809	1.40357	4.43847
1.98	3.9204	1.40712	4.44972
1.99	3.9601	1.41067	4.46094
2.00	4.0000	1.41421	4.47214

Squares and square roots (*continued*)

N	N^2	$\sqrt{N}$	$\sqrt{10N}$
2.00	4.0000	1.41421	4.47214
2.01	4.0401	1.41774	4.48330
2.02	4.0804	1.42127	4.49444
2.03	4.1209	1.42478	4.50555
2.04	4.1616	1.42829	4.51664
2.05	4.2025	1.43178	4.52769
2.06	4.2436	1.43527	4.53872
2.07	4.2849	1.43875	4.54973
2.08	4.3264	1.44222	4.56070
2.09	4.3681	1.44568	4.57165
2.10	4.4100	1.44914	4.58258
2.11	4.4521	1.45258	4.59347
2.12	4.4944	1.45602	4.60435
2.13	4.5369	1.45945	4.61519
2.14	4.5796	1.46287	4.62601
2.15	4.6225	1.46629	4.63681
2.16	4.6656	1.46969	4.64758
2.17	4.7089	1.47309	4.65833
2.18	4.7524	1.47648	4.66905
2.19	4.7961	1.47986	4.67974
2.20	4.8400	1.48324	4.69042
2.21	4.8841	1.48661	4.70106
2.22	5.9284	1.48997	4.71169
2.23	4.9729	1.49332	4.72229
2.24	5.0176	1.49666	4.73286
2.25	5.0625	1.50000	4.74342
2.26	5.1076	1.50333	4.75395
2.27	5.1529	1.50665	4.76445
2.28	5.1984	1.50997	4.77493
2.29	5.2441	1.51327	4.78539
2.30	5.2900	1.51658	4.79583
2.31	5.3361	1.51987	4.80625
2.32	5.3824	1.52315	4.81664
2.33	5.4289	1.52643	4.82701
2.34	5.4756	1.52971	4.83735
2.35	5.5225	1.53297	4.84768
2.36	5.5696	1.53623	4.85798
2.37	5.6169	1.53948	4.86826
2.38	5.6644	1.54272	4.87852
2.39	5.7121	1.54596	4.88876
2.40	5.7600	1.54919	4.89898
2.41	5.8081	1.55252	4.90918
2.42	5.8564	1.55563	4.91935
2.43	5.9049	1.55885	4.92950
2.44	5.9536	1.56205	4.93964
2.45	6.0025	1.56525	4.94975
2.46	6.0516	1.56844	4.95984
2.47	6.1009	1.57162	4.96991
2.48	6.1054	1.57480	4.97996
2.49	6.2001	1.57797	4.98999
2.50	6.2500	1.58114	5.00000

N	N^2	$\sqrt{N}$	$\sqrt{10N}$
2.50	6.2500	1.58114	5.00000
2.51	6.3001	1.58430	5.00999
2.52	6.3504	1.58745	5.01996
2.53	6.4009	1.59060	5.02991
2.54	6.4516	1.59374	5.03984
2.55	6.5025	1.59687	5.04975
2.56	6.5536	1.60000	5.05964
2.57	6.6049	1.60312	5.06952
2.58	6.6564	1.60624	5.07937
2.59	6.7081	1.60935	5.08920
2.60	6.7600	1.61245	5.09902
2.61	6.8121	1.61555	5.10882
2.62	6.8644	1.61864	5.11859
2.63	6.9169	1.62173	5.12835
2.64	6.9696	1.62481	5.13809
2.65	7.0225	1.62788	5.14782
2.66	7.0756	1.63095	5.15752
2.67	7.1289	1.63401	5.16720
2.68	7.1824	1.63707	5.17687
2.69	7.2361	1.64012	5.18652
2.70	7.2900	1.64317	5.19615
2.71	7.3441	1.64621	5.20577
2.72	7.3984	1.64924	5.21536
2.73	7.4529	1.65227	5.22494
2.74	7.5076	1.65529	5.23450
2.75	7.5625	1.65831	5.24404
2.76	7.6176	1.66132	5.25357
2.77	7.6729	1.66433	5.26308
2.78	7.7284	1.66733	5.27257
2.79	7.7841	1.67033	5.28205
2.80	7.8400	1.67332	5.29150
2.81	7.8961	1.67631	5.30094
2.82	7.9524	1.67929	5.31037
2.83	8.0089	1.68226	5.31977
2.84	8.0656	1.68523	5.32917
2.85	8.1225	1.68819	5.33854
2.86	8.1796	1.69115	5.34790
2.87	8.2369	1.69411	5.35724
2.88	8.2944	1.69706	5.36656
2.89	8.3521	1.70000	5.37587
2.90	8.4100	1.70294	5.38516
2.91	8.4681	1.70587	5.39444
2.92	8.5264	1.70880	5.40370
2.93	8.5849	1.71172	5.41295
2.94	8.6436	1.71464	5.42218
2.95	8.7025	1.71756	5.43139
2.96	8.7616	1.72047	5.44059
2.97	8.8209	1.72337	5.44977
2.98	8.8804	1.72627	5.45894
2.99	8.9401	1.72916	5.46809
3.00	9.0000	1.73205	5.47723

Squares and square roots (*continued*)

N	N^2	$\sqrt{N}$	$\sqrt{10N}$
3.00	9.0000	1.73205	5.47723
3.01	9.0601	1.73494	5.48635
3.02	9.1204	1.73781	5.49545
3.03	9.1809	1.74069	5.50454
3.04	9.2416	1.74356	5.51362
3.05	9.3025	1.74642	5.52268
3.06	9.3636	1.74929	5.53173
3.07	9.4249	1.75214	5.54076
3.08	9.4864	1.75499	5.54977
3.09	9.5481	1.75784	5.55878
3.10	9.6100	1.76068	5.56776
3.11	9.6721	1.76352	5.57674
3.12	9.7344	1.76635	5.58570
3.13	9.7969	1.76918	5.59464
3.14	9.8596	1.77200	5.60357
3.15	9.9225	1.77482	5.61249
3.16	9.9856	1.77764	5.62139
3.17	10.0489	1.78045	5.63028
3.18	10.1124	1.78326	5.63915
3.19	10.1761	1.78606	5.64801
3.20	10.2400	1.78885	5.65685
3.21	10.3041	1.79165	5.66569
3.22	10.3684	1.79444	5.67450
3.23	10.4329	1.79722	5.68331
3.24	10.4976	1.80000	5.69210
3.25	10.5625	1.80278	5.70088
3.26	10.6276	1.80555	5.70964
3.27	10.6929	1.80831	5.71839
3.28	10.7584	1.81108	5.72713
3.29	10.8241	1.81384	5.73585
3.30	10.8900	1.81659	5.74456
3.31	10.9561	1.81934	5.75326
3.32	10.0224	1.82209	5.76194
3.33	11.0889	1.82483	5.77062
3.34	11.1556	1,82757	5.77927
3.35	11.2225	1.83030	5.78792
3.36	11.2896	1.83303	5.79655
3.37	11.3569	1.83576	5.80517
3.38	11.4244	1.83848	5.81378
3.39	11.4921	1.84120	5.82237
3.40	11.5600	1.84391	5.83095
3.41	11.6281	1.84662	5.83952
3.42	11.6964	1.84932	5.84808
3.43	11.7649	1.85203	5.85662
3.44	11.8336	1.85472	5.86515
3.45	11.9025	1.85742	5.87367
3.46	11.9716	1.86011	5.88218
3.47	12.0409	1.86279	5.89067
3.48	12.1104	1.86548	5.89915
3.49	12.1801	1.86815	5.90762
3.50	12.2500	1.87083	5.91608

N	N^2	$\sqrt{N}$	$\sqrt{10N}$
3.50	12.2500	1.87083	5.91608
3.51	12.3201	1.87350	5.92453
3.52	12.3904	1.87617	5.93296
3.53	12.4609	1.87883	5.94138
3.54	12.5316	1.88149	5.94979
3.55	12.6025	1.88414	5.95819
3.56	12.6736	1.88680	5.96657
3.57	12.7449	1.88944	5.97495
3.58	12.8164	1.89209	5.98331
3.59	12.8881	1.89473	5.99166
3.60	12.9600	1.89737	6.00000
3.61	13.0321	1.90000	6.00833
3.62	13.1044	1.90263	6.01664
3.63	13.1769	1.90526	6.02495
3.64	13.2496	1.90788	6.03324
3.65	13.3225	1.91050	6.04152
3.66	13.3956	1.91311	6.04949
3.67	13.4689	1.91572	6.05805
3.68	13.5424	1.91833	6.06630
3.69	13.6161	1.92094	6.07454
3.70	13.6900	1.92354	6.08276
3.71	13.7641	1.92614	6.09098
3.72	13.8384	1.92873	6.09918
3.73	13.9129	1.93132	6.10737
3.74	13.9876	1.93391	6.11555
3.75	14.0625	1.93649	6.12372
3.76	14.1376	1.93907	6.13188
3.77	14.2129	1.94165	6.14003
3.78	14.2884	1.94422	6.14817
3.79	14.3641	1.94679	6.15630
3.80	14.4400	1.94936	6.16441
3.81	14.5161	1.95192	6.17252
3.82	14.5924	1.95448	6.18061
3.83	14.6689	1.95704	6.18870
3.84	14.7456	1.95959	6.19677
3.85	14.8225	1.96214	6.20484
3.86	14.8996	1.96469	6.21289
3.87	14.9769	1.96723	6.22093
3.88	15.0544	1.96977	6.22896
3.89	15.1321	1.97231	6.23699
3.90	15.2100	1.97484	6.24500
3.91	15.2881	1.97737	6.25300
3.92	15.3664	1.97990	6.26099
3.93	15.4449	1.98242	6.26897
3.94	15.5236	1.98494	6.27694
3.95	15.6025	1.98746	6.28490
3.96	15.6816	1.98997	6.29285
3.97	15.7609	1.99249	6.30079
3.98	15.8404	1.99499	6.30872
3.99	15.9201	1.99750	6.31644
4.00	16.0000	2.00000	6.32456

Squares and square roots (*continued*)

N	N^2	$\sqrt{N}$	$\sqrt{10N}$
4.00	16.0000	2.00000	6.32456
4.01	16.0801	2.00250	6.33246
4.02	16.1604	2.00499	6.34035
4.03	16.2409	2.00749	6.34823
4.04	16.3216	2.00998	6.35610
4.05	16.4025	2.01246	6.36396
4.06	16.4836	2.01494	6.37181
4.07	16.5649	2.01742	6.37966
4.08	16.6464	2.01990	6.38749
4.09	16.7281	2.02237	6.39531
4.10	16.8100	2.02485	6.40312
4.11	16.8921	2.02731	6.41093
4.12	16.9744	2.02978	6.41872
4.13	17.0569	2.03224	6.42651
4.14	17.1396	2.03470	6.43428
4.15	17.2225	2.03715	6.44205
4.16	17.3056	2.03961	6.44981
4.17	17.3889	2.04206	6.45755
4.18	17.4724	2.04450	6.46529
4.19	17.5561	2.04695	6.47302
4.20	17.6400	2.04939	6.48074
4.21	17.7241	2.05183	6.48845
4.22	17.8084	2.05426	6.49615
4.23	17.8929	2.05670	6.50384
4.24	17.9776	2.05913	6.51153
4.25	18.0625	2.06155	6.51920
4.26	18.1476	2.06398	6.52687
4.27	18.2329	2.06640	6.53452
4.28	18.3184	2.06882	6.54217
4.29	18.4041	2.07123	6.54981
4.30	18.4900	2.07364	6.55744
4.31	18.5761	2.07605	6.66506
4.32	18.6624	2.07846	6.57267
4.33	18.7489	2.08087	6.58027
4.34	18.8356	2.08327	6.58787
4.35	18.9225	2.08567	6.59545
4.36	19.0096	2.08806	6.60303
4.37	19.0969	2.09045	6.61060
4.38	19.1844	2.09284	6.61816
4.39	19.2721	2.09523	6.62571
4.40	19.3600	2.09762	6.63325
4.41	19.4481	2.10000	6.64078
4.42	19.5364	2.10238	6.64831
4.43	19.6249	2.10476	6.65582
4.44	19.7136	2.10713	6.66333
4.45	19.8025	2.10950	6.67083
4.46	19.8916	2.11187	6.67832
4.47	19.9809	2.11424	6.68581
4.48	20.0704	2.11660	6.69328
4.49	20.1601	2.11896	6.70075
4.50	20.2500	2.12132	6.70820

N	N^2	$\sqrt{N}$	$\sqrt{10N}$
4.50	20.2500	2.12132	6.70820
4.51	20.3401	2.12368	6.71565
4.52	20.4304	2.12603	6.72309
4.53	20.5209	2.12838	6.73053
4.54	20.6116	2.13073	6.73795
4.55	20.7025	2.13307	6.74537
4.56	20.7936	2.13542	6.75278
4.57	20.8849	2.13776	6.76018
4.58	20.9764	2.14009	6.76757
4.59	21.0681	2.14243	6.77495
4.60	21.1600	2.14476	6.78233
4.61	21.2521	2.14709	6.78970
4.62	21.3444	2.14942	6.79706
4.63	21.4369	2.15174	6.80441
4.64	21.5296	2.15407	6.81175
4.65	21.6225	2.15639	6.81909
4.66	21.7156	2.15870	6.82642
4.67	21.8089	2.16102	6.83374
4.68	21.9024	2.16333	6.84105
4.69	21.9961	2.16564	6.84836
4.70	22.0900	2.16795	6.85565
4.71	22.1841	2.17025	6.86294
4.72	22.2784	2.17256	6.87023
4.73	22.3729	2.17486	6.87750
4.74	22.4676	2.17715	6.88477
4.75	22.5625	2.17945	6.89202
4.76	22.6576	2.18174	6.89928
4.77	22.7529	2.18403	6.90652
4.78	22.8484	2.18632	6.91375
4.79	22.9441	2.18861	6.92098
4.80	23.0400	2.19089	6.92820
4.81	23.1361	2.19317	6.93542
4.82	23.2324	2.19545	6.94262
4.83	23.3289	2.19773	6.94982
4.84	23.4256	2.20000	6.95701
4.85	23.5225	2.20227	6.96419
4.86	23.6196	2.20454	6.97137
4.87	23.7169	2.20681	6.97854
4.88	23.8144	2.20907	6.98570
4.89	23.9121	2.21133	6.99285
4.90	24.0100	2.21359	7.00000
4.91	24.1081	2.21585	7.00714
4.92	24.2064	2.21811	7.01427
4.93	24.3049	2.22036	7.02140
4.94	24.4036	2.22261	7.02851
4.95	24.5025	2.22486	7.03562
4.96	24.6016	2.22711	7.04273
4.97	24.7009	2.22935	7.04982
4.98	24.8004	2.23159	7.05691
4.99	24.9001	2.23383	7.06399
5.00	25.0000	2.23607	7.07107

Squares and square roots (*continued*)

N	N^2	$\sqrt{N}$	$\sqrt{10N}$
5.00	25.0000	2.23607	7.07107
5.01	25.1001	2.23830	7.07814
5.02	25.2004	2.24054	7.08520
5.03	25.3009	2.24277	7.09225
5.04	25.4016	2.24499	7.09930
5.05	25.5025	2.24722	7.10634
5.06	25.6036	2.24944	7.11337
5.07	25.7049	2.25167	7.12039
5.08	25.8064	2.25389	7.12741
5.09	25.9081	2.25610	7.13442
5.10	26.0100	2.25832	7.14143
5.11	26.1121	2.26053	7.14843
5.12	26.2144	2.26274	7.15542
5.13	26.3169	2.26495	7.16240
5.14	26.4196	2.26716	7.16938
5.15	26.5225	2.26936	7.17635
5.16	26.6256	2.27156	7.18331
5.17	26.7289	2.27376	7.19027
5.18	26.8324	2.27596	7.19722
5.19	26.9361	2.27816	7.20417
5.20	27.0400	2.28035	7.21110
5.21	27.1441	2.28254	7.21803
5.22	27.2484	2.28473	7.22496
5.23	27.3529	2.28692	7.23187
5.24	27.4576	2.28910	7.23838
5.25	27.5625	2.29129	7.24569
5.26	27.6676	2.29347	7.25259
5.27	27.7729	2.29565	7.25948
5.28	27.8784	2.29783	7.26636
5.29	27.9841	2.30000	7.27324
5.30	28.0900	2.30217	7.28011
5.31	28.1961	2.30434	7.28697
5.32	28.3024	2.30651	7.29383
5.33	28.4089	2.30868	7.30068
5.34	28.5156	2.31084	7.30753
5.35	28.6225	2.31301	7.31437
5.36	28.7296	2.31517	7.32120
5.37	28.8369	2.31733	7.32803
5.38	28.9444	2.31948	7.33485
5.39	29.0521	2.32164	7.34166
5.40	29.1600	2.32379	7.34847
5.41	29.2681	2.32594	7.35527
5.42	29.3764	2.32809	7.36206
5.43	29.4849	2.33024	7.36885
5.44	29.5936	2.33238	7.37564
5.45	29.7025	2.33452	7.38241
5.46	29.8116	2.33666	7.38918
5.47	29.9209	2.33880	7.39594
5.48	30.0304	2.34094	7.40270
5.49	30.1401	2.34307	7.40945
5.50	30.2500	2.34521	7.41620

N	N^2	$\sqrt{N}$	$\sqrt{10N}$
5.50	30.2500	2.34521	7.41620
5.51	30.3601	2.34734	7.42294
5.52	30.4704	2.34947	7.42967
5.53	30.5809	2.35160	7.43640
5.54	30.6916	2.35372	7.44312
5.55	30.8025	2.35584	7.44983
5.56	30.9136	2.35797	7.45654
5.57	31.0249	2.36008	7.46324
5.58	31.1364	2.36220	7.46994
5.59	31.2481	2.36432	7.47663
5.60	31.3600	2.36643	7.48331
5.61	31.4721	2.36854	7.48999
5.62	31.5844	2.37065	7.49667
5.63	31.6969	2.37276	7.50333
5.64	31.8096	2.37487	7.50999
5.65	31.9225	2.37697	7.51665
5.66	32.0356	2.37908	7.52330
5.67	32.1489	2.38118	7.52994
5.68	32.2624	2.38328	7.53658
5.68	32.3761	2.38537	7.54321
5.70	32.4900	2.38747	7.54983
5.71	32.6041	2.38956	7.55645
5.72	32.7184	2.39165	7.56307
5.73	32.8329	2.39374	7.56968
5.74	32.9476	2.39583	7.57628
5.75	33.0625	2.39792	7.58288
5.76	33.1776	2.40000	7.58947
5.77	33.2929	2.40208	7.59605
5.78	33.4084	2.40416	7.60263
5.79	33.5241	2.40624	7.60920
5.80	33.6400	2.40832	7.61577
5.81	33.7561	2.41039	7.62234
5.82	33.8724	2.41247	7.62889
5.83	33.9889	2.41454	7.63544
5.84	34.1056	2.41661	7.64199
5.85	34.2225	2.41868	7.64853
5.86	34.3396	2.42074	7.65506
5.87	34.4569	2.42281	7.66159
5.88	34.5744	2.42487	7.66812
5.89	34.6921	2.42693	7.67463
5.90	34.8100	2.42899	7.68115
5.91	34.9281	2.43105	7.68765
5.92	35.0464	2.43311	7.69415
5.93	35.1649	2.43516	7.70065
5.94	35.2836	2.43721	7.70714
5.95	35.4025	2.43926	7.71362
5.96	35.5216	2.44131	7.72010
5.97	35.6409	2.44336	7.72658
5.98	35.7604	2.44540	7.73305
5.99	35.8801	2.44745	7.73951
6.00	36.0000	2.44949	7.74597

Squares and square roots (*continued*)

N	N^2	$\sqrt{N}$	$\sqrt{10N}$
6.00	36.0000	2.44949	7.74597
6.01	36.1201	2.45153	7.75242
6.02	36.2404	2.45357	7.75887
6.03	36.3609	2.45561	7.76531
6.04	36.4816	2.45764	7.77174
6.05	36.6025	2.45967	7.77817
6.06	36.7236	2.46171	7.78460
6.07	36.8449	2.46374	7.79102
6.08	36.9664	2.46577	7.79744
6.09	37.0881	2.46779	7.80385
6.10	37.2100	2.46982	7.81025
6.11	37.3321	2.47184	7.81665
6.12	37.4544	2.47386	7.82304
6.13	37.5769	2.47588	7.82943
6.14	37.6996	2.47790	7.83582
6.15	37.8225	2.47992	7.84219
6.16	37.9456	2.48193	7.84857
6.17	38.0689	2.48395	7.85493
6.18	38.1924	2.48596	7.86130
6.19	38.3161	2.48797	7.86766
6.20	38.4400	2.48998	7.87401
6.21	38.5641	2.49199	7.88036
6.22	38.6884	2.49399	7.88670
6.23	38.8129	2.49600	7.89303
6.24	38.9376	2.49800	7.89937
6.25	39.0625	2.50000	7.90569
6.26	39.1876	2.50200	7.91202
6.27	39.3129	2.50400	7.91833
6.28	39.4384	2.50599	7.92465
6.29	39.5641	2.50799	7.93095
6.30	39.6900	2.50998	7.93725
6.31	39.8161	2.51197	7.94355
6.32	39.9424	2.51396	7.94984
6.33	40.0689	2.51595	7.95613
6.34	40.1956	2.51794	7.96241
6.35	40.3225	2.51992	7.96869
6.36	40.4496	2.52190	7.97496
6.37	40.5769	2.52389	7.98123
6.38	40.7044	2.52587	7.98749
6.39	40.8321	2.52784	7.99375
6.40	40.9600	2.52982	8.00000
6.41	41.0881	2.53180	8.00625
6.42	41.2164	2.53377	8.01249
6.43	41.3449	2.53574	8.01873
6.44	41.4736	2.53772	8.02496
6.45	41.6025	2.53969	8.03119
6.46	41.7316	2.54165	8.03741
6.47	41.8609	2.54362	8.04363
6.48	41.9904	2.54558	8.04984
6.49	42.1201	2.54755	8.05605
6.50	42.2500	2.54951	8.06226

N	N^2	$\sqrt{N}$	$\sqrt{10N}$
6.50	42.2500	2.54951	8.06226
6.51	42.3801	2.55147	8.06846
6.52	42.5104	2.55343	8.07465
6.53	42.6409	2.55539	8.08084
6.54	42.7716	2.55734	8.08703
6.55	42.9025	2.55930	8.09321
6.56	43.0336	2.56125	8.09938
6.57	43.1649	2.56320	8.10555
6.58	43.2964	2.56515	8.11172
6.59	43.4281	2.56710	8.11788
6.60	43.5600	2.56905	8.12404
6.61	43.6921	2.57099	8.13019
6.62	43.8244	2.57294	8.13634
6.63	43.9569	2.57488	8.14248
6.64	44.0896	2.57682	8.14862
6.65	44.2225	2.57876	8.15475
6.66	44.3556	2.58070	8.16088
6.67	44.4889	2.58263	8.16701
6.68	44.6224	2.58457	8.17313
6.69	44.7561	2.58650	8.17924
6.70	44.8900	2.58844	8.18535
6.71	45.0241	2.59037	8.19146
6.72	45.1584	2.59230	8.19756
6.73	45.2929	2.59422	8.20366
6.74	45.4276	2.59615	8.20975
6.75	45.5625	2.59808	8.21584
6.76	45.6976	2.60000	8.22192
6.77	45.8329	2.60192	8.22800
6.78	45.9684	2.60384	8.23408
6.79	46.1041	2.60576	8.24015
6.80	46.2400	2.60768	8.24621
6.81	46.3761	2.60960	8.25227
6.82	46.5124	2.61151	8.25833
6.83	46.6489	2.61343	8.26438
6.84	46.7856	2.61534	8.27043
6.85	46.9225	2.61725	8.27647
6.86	47.0596	2.61916	8.28251
6.87	47.1969	2.62107	8.28855
6.88	47.3344	2.62298	8.29458
6.89	47.4721	2.62488	8.30060
6.90	47.6100	2.62679	8.30662
6.91	47.7481	2.62869	8.31264
6.92	47.8864	2.63059	8.31865
6.93	48.0249	2.63249	8.32466
6.94	48.1636	2.63439	8.33067
6.95	48.3025	2.63629	8.33667
6.96	48.4416	2.63818	8.34266
6.97	48.5809	2.64008	8.34865
6.98	48.7204	2.64197	8.35464
6.99	48.8601	2.64386	8.36062
7.00	49.0000	2.64575	8.36660

Squares and square roots (*continued*)

N	N^2	$\sqrt{N}$	$\sqrt{10N}$
7.00	49.0000	2.64575	8.36660
7.01	49.1401	2.64764	8.37257
7.02	49.2804	2.64953	8.37854
7.03	49.4209	2.65141	8.38451
7.04	49.5616	2.65330	8.39047
7.05	49.7025	2.65518	8.39643
7.06	49.8436	2.65707	8.40238
7.07	49.9849	2.65895	8.40833
7.08	50.1264	2.66083	8.41427
7.09	50.2681	2.66271	8.42021
7.10	50.4100	2.66458	8.42615
7.11	50.5521	2.66646	8.43208
7.12	50.6944	2.66833	8.43801
7.13	50.8369	2.67021	8.44393
7.14	50.9796	2.67208	8.44985
7.15	51.1225	2.67395	8.45577
7.16	51.2656	2.67582	8.46168
7.17	51.4089	2.67769	8.46759
7.18	51.5524	2.67955	8.47349
7.19	51.6961	2.68142	8.47939
7.20	51.8400	2.68328	8.48528
7.21	51.9841	2.68514	8.49117
7.22	52.1284	2.68701	8.49706
7.23	52.2729	2.68887	8.50294
7.24	52.4176	2.69072	8.50882
7.25	52.5625	2.69258	8.51469
7.26	52.7076	2.69444	8.52056
7.27	52.8529	2.69629	8.52643
7.28	52.9984	2.69815	8.53229
7.29	53.1441	2.70000	8.53815
7.30	53.2900	2.70185	8.54400
7.31	53.4361	2.70370	8.54985
7.32	53.5824	2.70555	8.55570
7.33	53.7289	2.70740	8.56154
7.34	53.8756	2.70924	8.56738
7.35	54.0225	2.71109	8.57321
7.36	54.1696	2.71293	8.57904
7.37	54.3169	2.71477	8.58487
7.38	54.4644	2.71662	8.59069
7.39	54.6121	2.71846	8.59651
7.40	54.7600	2.72029	8.60233
7.41	54.9081	2.72213	8.60814
7.42	55.0564	2.72397	8.61394
7.43	55.2049	2.72580	8.61974
7.44	55.3536	2.72764	8.62554
7.45	55.5025	2.72947	8.63134
7.46	55.6516	2.73130	8.63713
7.47	55.8009	2.73313	8.64292
7.48	55.9504	2.73496	8.64870
7.49	56.1001	2.73679	8.65448
7.50	56.2500	2.73861	8.66025

N	N^2	$\sqrt{N}$	$\sqrt{10N}$
7.50	56.2500	2.73861	8.66025
7.51	56.4001	2.74044	8.66603
7.52	56.5504	2.74226	8.67179
7.53	56.7009	2.74408	8.67756
7.54	56.8516	2.74591	8.68332
7.55	57.0025	2.74773	8.68907
7.56	57.1536	2.74955	8.69483
7.57	57.3049	2.75136	8.70057
7.58	57.4564	2.75318	8.70632
7.59	57.6081	2.75500	8.71206
7.60	57.7600	2.75681	8.71780
7.61	57.9121	2.75862	8.72353
7.62	58.0644	2.76043	8.72926
7.63	58.2169	2.76225	8.73499
7.64	58.3696	2.76405	8.74071
7.65	58.5225	2.76586	8.74643
7.66	58.6756	2.76767	8.75214
7.67	58.8289	2.76948	8.75785
7.68	58.9824	2.77128	8.76356
7.69	59.1361	2.77308	8.76926
7.70	59.2900	2.77489	8.77496
7.71	59.4441	2.77669	8.78066
7.72	59.5984	2.77849	8.78635
7.73	59.7529	2.78029	8.79204
7.74	59.9076	2.78209	8.79773
7.75	60.0625	2.78388	8.80341
7.76	60.2176	2.78568	8.80909
7.77	60.3729	2.78747	8.81476
7.78	60.5284	2.78927	8.82043
7.79	60.6841	2.79106	8.82610
7.80	60.8400	2.79285	8.83176
7.81	60.9961	2.79464	8.83742
7.82	61.1524	2.79643	8.84308
7.83	61.3089	2.79821	8.84873
7.84	61.4656	2.80000	8.85438
7.85	61.6225	2.80179	8.86002
7.86	61.7796	2.80357	8.86566
7.87	61.9369	2.80535	8.87130
7.88	62.0944	2.80713	8.87694
7.89	62.2521	2.80891	8.88257
7.90	62.4100	2.81069	8.88819
7.91	62.5681	2.81247	8.89382
7.92	62.7264	2.81425	8.89944
7.93	62.8849	2.81603	8.90505
7.94	63.0436	2.81780	8.91067
7.95	63.2025	2.81957	8.91628
7.96	63.3616	2.82135	8.92188
7.97	63.5209	2.82312	8.92749
7.98	63.6804	2.82489	8.93308
7.99	63.8401	2.82666	8.93868
8.00	64.0000	2.82843	8.94427

Squares and square roots (*continued*)

N	N^2	$\sqrt{N}$	$\sqrt{10N}$
8.00	64.0000	2.82843	8.94427
8.01	64.1601	2.83019	8.94986
8.02	64.3204	2.83196	8.95545
8.03	64.4809	2.83373	8.96103
8.04	64.6416	2.83549	8.96660
8.05	64.8025	2.83725	8.97218
8.06	64.9636	2.83901	8.97775
8.07	65.1249	2.84077	8.98332
8.08	65.2864	2.84253	8.98888
8.09	65.4481	2.84429	8.99444
8.10	65.6100	2.84605	9.00000
8.11	65.7721	2.84781	9.00555
8.12	65.9344	2.84956	9.01110
8.13	66.0969	2.85132	9.01665
8.14	66.2596	2.85307	9.02219
8.15	66.4225	2.85482	9.02774
8.16	66.5856	2.85657	9.03327
8.17	66.7489	2.85832	9.03881
8.18	66.9124	2.86007	9.04434
8.19	67.0761	2.86182	9.04986
8.20	67.2400	2.86356	9.05539
8.21	67.4041	2.86531	9.06091
8.22	67.5684	2.86705	9.06642
8.23	67.7329	2.86880	9.07193
8.24	67.8976	2.87054	9.07744
8.25	68.0625	2.87228	9.08295
8.26	68.2276	2.87402	9.08845
8.27	68.3929	2.87576	9.09395
8.28	68.5584	2.87750	9.09945
8.29	68.7241	2.87924	9.10494
8.30	68.8900	2.88097	9.11045
8.31	69.0561	2.88271	9.11592
8.32	69.2224	2.88444	9.12140
8.33	69.3889	2.88617	9.12688
8.34	69.5556	2.88791	9.13236
8.35	69.7225	2.88964	9.13783
8.36	69.8896	2.89137	9.14330
8.37	70.0569	2.89310	9.14877
8.38	70.2244	2.89482	9.15423
8.39	70.3921	2.89655	9.15969
8.40	70.5600	2.89828	9.16515
8.41	70.7281	2.90000	9.17061
8.42	70.8964	2.90172	9.17606
8.43	71.0649	2.90345	9.18150
8.44	71.2336	2.90517	9.18695
8.45	71.4025	2.90689	9.19239
8.46	71.5716	2.90861	9.19783
8.47	71.7409	2.91033	9.20326
8.48	71.9104	2.91204	9.20869
8.49	72.0801	2.91376	9.21412
8.50	72.2500	2.91548	9.21954

N	N^2	$\sqrt{N}$	$\sqrt{10N}$
8.50	72.2500	2.91548	9.21954
8.51	72.4201	2.91719	9.22497
8.52	72.5904	2.91890	9.23038
8.53	72.7609	2.92062	9.23580
8.54	72.9316	2.92233	8.24121
8.55	73.1025	2.92404	9.24662
8.56	73.2736	2.92575	9.25203
8.57	73.4449	2.92746	9.25743
8.58	73.6164	2.92916	9.26283
8.59	73.7881	2.93087	9.26823
8.60	73.9600	2.93258	9.27362
8.61	74.1321	2.93428	9.27901
8.62	74.3044	2.93598	9.28440
8.63	74.4769	2.93769	9.28978
8.64	74.6496	2.93939	9.29516
8.65	74.8225	2.94109	9.30054
8.66	74.9956	2.94279	9.30591
8.67	75.1689	2.94449	9.31128
8.68	75.3424	2.94618	9.31665
8.69	75.5161	2.94788	9.32202
8.70	75.6900	2.94958	9.32738
8.71	75.8641	2.95127	9.33274
8.72	76.0384	2.95296	9.33809
8.73	76.2129	2.95466	9.34345
8.74	76.3876	2.95635	9.34880
8.75	76.5625	2.95804	9.35414
8.76	76.7376	2.95973	9.35949
8.77	76.9129	2.96142	9.36483
8.78	77.0884	2.96311	9.37017
8.79	77.2641	2.96479	9.37550
8.80	77.4400	2.96648	9.38083
8.81	77.6161	2.96816	9.38616
8.82	77.7924	2.96985	9.39149
8.83	77.9689	2.97153	9.39681
8.84	78.1456	2.97321	9.40213
8.85	78.3225	2.97489	9.40744
8.86	78.4996	2.97658	9.41276
8.87	78.6769	2.97825	9.41807
8.88	78.8544	2.97993	9.42338
8.89	79.0321	2.98161	9.42868
8.90	79.2100	2.98329	9.43398
8.91	79.3881	2.98486	9.43928
8.92	79.5664	2.98664	9.44458
8.93	79.7449	2.98831	9.44987
8.94	79.9236	2.98998	9.45516
8.95	80.1025	2.99166	9.46044
8.96	80.2816	2.99333	9.46573
8.97	80.4609	2.99500	9.47101
8.98	80.6404	2.99666	9.47629
8.99	80.8201	2.99833	9.48156
9.00	81.0000	3.99999	9.48683

Squares and square roots (*continued*)

N	N^2	$\sqrt{N}$	$\sqrt{10N}$
9.00	81.0000	3.00000	9.48683
9.01	81.1801	3.00167	9.49210
9.02	81.3604	3.00333	9.49737
9.03	81.5409	3.00500	9.50263
9.04	81.7216	3.00666	9.50789
9.05	81.9025	3.00832	9.51315
9.06	82.0836	3.00998	9.51840
9.07	82.2649	3.01164	9.52365
9.08	82.4464	3.01330	9.52890
9.09	82.6281	3.01496	9.53415
9.10	82.8100	3.01662	9.53939
9.11	82.9921	3.01828	9.54463
9.12	83.1744	3.01993	9.54987
9.13	83.3569	3.02159	9.55510
9.14	83.5396	3.02324	9.56033
9.15	83.7225	3.02490	9.56556
9.16	83.9056	3.02655	9.57079
9.17	84.0889	3.02820	9.57601
9.18	84.2724	3.02985	9.58123
9.19	84.4561	3.03150	9.58645
9.20	84.6400	3.03315	9.59166
9.21	84.8241	3.03480	9.59687
9.22	85.0084	3.03645	9.60208
9.23	85.1929	3.03809	9.60729
9.24	85.3776	3.03974	9.61249
9.25	85.5625	3.04138	9.61769
9.26	85.7476	3.04302	9.62289
9.27	85.9329	3.04467	9.62808
9.28	86.1184	3.04631	9.63328
9.29	86.3041	3.04795	9.63846
9.30	86.4900	3.04959	9.64365
9.31	86.6761	3.05123	9.64883
9.32	86.8624	3.05287	9.65401
9.33	87.0489	3.05450	9.65919
9.34	87.2356	3.05614	9.66437
9.35	87.4225	3.05778	9.66954
9.36	87.6096	3.05941	9.67471
9.37	87.7969	3.06105	9.67988
9.38	87.9844	3.06268	9.68504
9.39	88.1721	3.06431	9.69020
9.40	88.3600	3.06594	9.69536
9.41	88.5481	3.06757	9.70052
9.42	88.7364	3.06920	9.70567
9.43	88.9249	3.07083	9.71082
9.44	89.1136	3.07246	9.71597
9.45	89.3025	3.07409	9.72111
9.46	89.4916	3.07571	9.72625
9.47	89.6809	3.07734	9.73139
9.48	89.8704	3.07896	9.73653
9.49	90.0601	3.08058	9.74166
9.50	90.2500	3.08221	9.74679

N	N^2	$\sqrt{N}$	$\sqrt{10N}$
9.50	90.2500	3.08221	9.74679
9.51	90.4401	3.08383	9.75192
9.52	90.6304	3.08545	9.75705
9.53	90.8209	3.08707	9.76217
9.54	91.0116	3.08869	9.76729
9.55	91.2025	3.09031	9.77241
9.56	91.3936	3.09192	9.77753
9.57	91.5849	3.09354	9.78264
9.58	91.7764	3.09516	9.78775
9.59	91.9681	3.09677	9.79285
9.60	92.1600	3.09839	9.79796
9.61	92.3521	3.10000	9.80306
9.62	92.5444	3.10161	9.80816
9.63	92.7369	3.10322	9.81326
9.64	92.9296	3.10483	9.81835
9.65	93.1225	3.10644	9.82344
9.66	93.3156	3.10805	9.82853
9.67	93.5089	3.10966	9.83362
9.68	93.7024	3.11127	9.83870
9.69	93.8961	3.11288	9.84378
9.70	94.0900	3.11448	9.84886
9.71	94.2841	3.11609	9.85393
9.72	94.4784	3.11769	9.85901
9.73	94.6729	3.11929	9.86408
9.74	94.8676	3.12090	9.86914
9.75	95.0625	3.12250	9.87421
9.76	95.2576	3.12410	9.87927
9.77	95.4529	3.12570	9.88433
9.78	95.6484	3.12730	9.88939
9.79	95.8441	3.12890	9.89444
9.80	96.0400	3.12050	9.89949
9.81	96.2361	3.13209	9.90454
9.82	96.4324	3.13369	9.90959
9.83	96.6289	3.13528	9.91464
9.84	96.8256	3.13688	9.91968
9.85	97.0225	3.13847	9.92472
9.86	97.2196	3.14006	9.92975
9.87	97.4169	3.14166	9.93479
9.88	97.6144	3.14325	9.93982
9.89	97.8121	3.14484	9.94485
9.90	98.0100	3.14643	9.94987
9.91	98.2081	3.14802	9.95490
9.92	98.4064	3.14960	9.95992
9.93	98.6049	3.15119	9.96494
9.94	98.8036	3.15278	9.96995
9.95	99.0025	3.15436	9.97497
9.96	99.2016	3.15595	9.97998
9.97	99.4009	3.15753	9.98499
9.98	99.6004	3.15911	9.98999
9.99	99.8001	3.16070	9.99500
10.0	100.000	3.16228	10.0000

Table II Binomial coefficients: $\frac{n!}{x!(n-x)!} = \binom{n}{x}$

n	x = 2	3	4	5	6	7	8	9	10
2	1								
3	3	1							
4	6	4	1						
5	10	10	5	1					
6	15	20	15	6	1				
7	21	35	35	21	7	1			
8	28	56	70	56	28	8	1		
9	36	84	126	126	84	36	9	1	
10	45	120	210	252	210	120	45	10	1
11	55	165	330	462	462	330	165	55	11
12	66	220	495	792	924	792	495	220	66
13	78	286	715	1,287	1,716	1,716	1,287	715	286
14	91	364	1,001	2,002	3,003	3,432	3,003	2,002	1,001
15	105	455	1,365	3,003	5,005	6,435	6,435	5,005	3,003
16	120	560	1,820	4,368	8,008	11,440	12,870	11,440	8,008
17	136	680	2,380	6,188	12,376	19,448	24,310	24,310	19,448
18	153	816	3,060	8,568	18,564	31,824	43,758	48,620	43,758
19	171	969	3,876	11,628	27,132	50,388	75,582	92,378	92,378
20	190	1,140	4,845	15,504	38,760	77,520	125,970	167,960	184,756

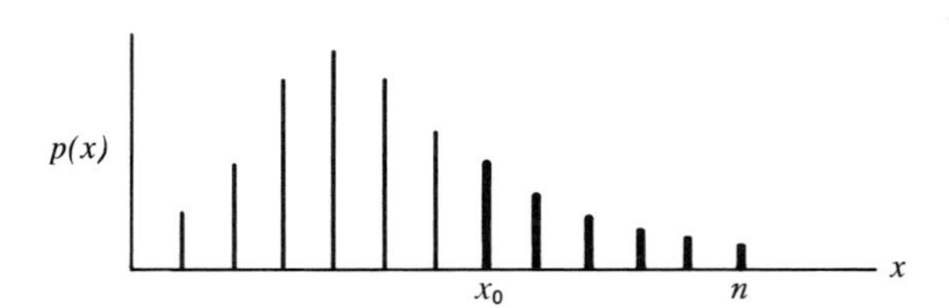

Table III Cumulative binomial probabilities in right-hand tail

n	x_0	π = .05	.10	.15	.20	.25	.30	.35	.40	.45	.50
2	1	.0975	.1900	.2775	.3600	.4375	.5100	.5775	.6400	.6975	.7500
	2	.0025	.0100	.0225	.0400	.0625	.0900	.1225	.1600	.2025	.2500
3	1	.1426	.2710	.3859	.4880	.5781	.6570	.7254	.7840	.8336	.8750
	2	.0072	.0280	.0608	.1040	.1562	.2160	.2818	.3520	.4252	.5000
	3	.0001	.0010	.0034	.0080	.0156	.0270	.0429	.0640	.0911	.1250
4	1	.1855	.3439	.4780	.5904	.6836	.7599	.8215	.8704	.9085	.9375
	2	.0140	.0523	.1095	.1808	.2617	.3483	.4370	.5248	.6090	.6875
	3	.0004	.0037	.0120	.0272	.0508	.0837	.1265	.1792	.2415	.3125
	4	.0000	.0001	.0005	.0016	.0039	.0081	.0150	.0256	.0410	.0625

Table III Cumulative binomial probabilities in right-hand tail (*continued*)

n	x_0	π = .05	.10	.15	.20	.25	.30	.35	.40	.45	.50
5	1	.2262	.4095	.5563	.6723	.7627	.8319	.8840	.9222	.9497	.9688
	2	.0226	.0815	.1648	.2627	.3672	.4718	.5716	.6630	.7438	.8125
	3	.0012	.0086	.0266	.0579	.1035	.1631	.2352	.3174	.4069	.5000
	4	.0000	.0005	.0022	.0067	.0156	.0308	.0540	.0870	.1312	.1875
	5	.0000	.0000	.0001	.0003	.0010	.0024	.0053	.0102	.0185	.0312
6	1	.2649	.4686	.6229	.7379	.8220	.8824	.9246	.9533	.9723	.9844
	2	.0328	.1143	.2235	.3447	.4661	.5798	.6809	.7667	.8364	.8906
	3	.0022	.0158	.0473	.0989	.1694	.2557	.3529	.4557	.5585	.6562
	4	.0001	.0013	.0059	.0170	.0376	.0705	.1174	.1792	.2553	.3438
	5	.0000	.0001	.0004	.0016	.0046	.0109	.0223	.0410	.0692	.1094
	6	.0000	.0000	.0000	.0001	.0002	.0007	.0018	.0041	.0083	.0156
7	1	.3017	.5217	.6794	.7903	.8665	.9176	.9510	.9720	.9848	.9922
	2	.0444	.1497	.2834	.4233	.5551	.6706	.7662	.8414	.8976	.9375
	3	.0038	.0257	.0738	.1480	.2436	.3529	.4677	.5801	.6836	.7734
	4	.0002	.0027	.0121	.0333	.0706	.1260	.1998	.2898	.3917	.5000
	5	.0000	.0002	.0012	.0047	.0129	.0288	.0556	.0963	.1529	.2266
	6	.0000	.0000	.0001	.0004	.0013	.0038	.0090	.0188	.0357	.0625
	7	.0000	.0000	.0000	.0000	.0001	.0002	.0006	.0016	.0037	.0078
8	1	.3366	.5695	.7275	.8322	.8999	.9424	.9681	.9832	.9916	.9961
	2	.0572	.1869	.3428	.4967	.6329	.7447	.8309	.8936	.9368	.9648
	3	.0058	.0381	.1052	.2031	.3215	.4482	.5722	.6846	.7799	.8555
	4	.0004	.0050	.0214	.0563	.1138	.1941	.2936	.4059	.5230	.6367
	5	.0000	.0004	.0029	.0104	.0273	.0580	.1061	.1737	.2604	.3633
	6	.0000	.0000	.0002	.0012	.0042	.0113	.0253	.0498	.0885	.1445
	7	.0000	.0000	.0000	.0001	.0004	.0013	.0036	.0085	.0181	.0352
	8	.0000	.0000	.0000	.0000	.0000	.0001	.0002	.0007	.0017	.0039
9	1	.3698	.6126	.7684	.8658	.9249	.9596	.9793	.9899	.9954	.9980
	2	.0712	.2252	.4005	.5638	.6997	.8040	.8789	.9295	.9615	.9805
	3	.0084	.0530	.1409	.2618	.3993	.5372	.6627	.7682	.8505	.9102
	4	.0006	.0083	.0339	.0856	.1657	.2703	.3911	.5174	.6386	.7461
	5	.0000	.0009	.0056	.0196	.0489	.0988	.1717	.2666	.3786	.5000
	6	.0000	.0001	.0006	.0031	.0100	.0253	.0536	.0994	.1658	.2539
	7	.0000	.0000	.0000	.0003	.0013	.0043	.0112	.0250	.0498	.0898
	8	.0000	.0000	.0000	.0000	.0001	.0004	.0014	.0038	.0091	.0195
	9	.0000	.0000	.0000	.0000	.0000	.0000	.0001	.0003	.0008	.0020
10	1	.4013	.6513	.8031	.8926	.9437	.9718	.9865	.9940	.9975	.9990
	2	.0861	.2639	.4557	.6242	.7560	.8507	.9140	.9536	.9767	.9893
	3	.0115	.0702	.1798	.3222	.4744	.6172	.7384	.8327	.9004	.9453
	4	.0010	.0128	.0500	.1209	.2241	.3504	.4862	.6177	.7340	.8281
	5	.0001	.0016	.0099	.0328	.0781	.1503	.2485	.3669	.4956	.6230
	6	.0000	.0001	.0014	.0064	.0197	.0473	.0949	.1662	.2616	.3770
	7	.0000	.0000	.0001	.0009	.0035	.0106	.0260	.0548	.1020	.1719
	8	.0000	.0000	.0000	.0001	.0004	.0016	.0048	.0123	.0274	.0547
	9	.0000	.0000	.0000	.0000	.0000	.0001	.0005	.0017	.0045	.0107
	10	.0000	.0000	.0000	.0000	.0000	.0000	.0000	.0001	.0003	.0010

Table IV Areas of a standard normal distribution

An entry in the table is the proportion under the entire curve that is between $z = 0$ and a positive value of z. Areas for negative values of z are obtained by symmetry.

z	.00	.01	.02	.03	.04	.05	.06	.07	.08	.09
0.0	.0000	.0040	.0080	.0120	.0160	.0199	.0239	.0279	.0319	.0359
0.1	.0398	.0438	.0478	.0517	.0557	.0596	.0636	.0675	.0714	.0753
0.2	.0793	.0832	.0871	.0910	.0948	.0987	.1026	.1064	.1103	.1141
0.3	.1179	.1217	.1255	.1293	.1331	.1368	.1406	.1443	.1480	.1517
0.4	.1554	.1591	.1628	.1664	.1700	.1736	.1772	.1808	.1844	.1879
0.5	.1915	.1950	.1985	.2019	.2054	.2088	.2123	.2157	.2190	.2224
0.6	.2257	.2291	.2324	.2357	.2389	.2422	.2454	.2486	.2517	.2549
0.7	.2580	.2611	.2642	.2673	.2703	.2734	.2764	.2794	.2823	.2852
0.8	.2881	.2910	.2939	.2967	.2995	.3023	.3051	.3078	.3106	.3133
0.9	.3159	.3186	.3212	.3238	.3264	.3289	.3315	.3340	.3365	.3389
1.0	.3413	.3438	.3461	.3485	.3508	.3531	.3554	.3577	.3599	.3621
1.1	.3643	.3665	.3686	.3708	.3729	.3749	.3770	.3790	.3810	.3830
1.2	.3849	.3869	.3888	.3907	.3925	.3944	.3962	.3980	.3997	.4015
1.3	.4032	.4049	.4066	.4082	.4099	.4115	.4131	.4147	.4162	.4177
1.4	.4192	.4207	.4222	.4236	.4251	.4265	.4279	.4292	.4306	.4319
1.5	.4332	.4345	.4357	.4370	.4382	.4394	.4406	.4418	.4429	.4441
1.6	.4452	.4463	.4474	.4484	.4495	.4505	.4515	.4525	.4535	.4545
1.7	.4554	.4564	.4573	.4582	.4591	.4599	.4608	.4616	.4625	.4633
1.8	.4641	.4649	.4656	.4664	.4671	.4678	.4686	.4693	.4699	.4706
1.9	.4713	.4719	.4726	.4732	.4738	.4744	.4750	.4756	.4761	.4767
2.0	.4772	.4778	.4783	.4788	.4793	.4798	.4803	.4808	.4812	.4817
2.1	.4821	.4826	.4830	.4834	.4838	.4842	.4846	.4850	.4854	.4857
2.2	.4861	.4864	.4868	.4871	.4875	.4878	.4881	.4884	.4887	.4890
2.3	.4893	.4896	.4898	.4901	.4904	.4906	.4909	.4911	.4913	.4916
2.4	.4918	.4920	.4922	.4925	.4927	.4929	.4931	.4932	.4934	.4936
2.5	.4938	.4940	.4941	.4943	.4945	.4946	.4948	.4949	.4951	.4952
2.6	.4953	.4955	.4956	.4957	.4959	.4960	.4961	.4962	.4963	.4964
2.7	.4965	.4966	.4967	.4968	.4969	.4970	.4971	.4972	.4973	.4974
2.8	.4974	.4975	.4976	.4977	.4977	.4978	.4979	.4979	.4980	.4981
2.9	.4981	.4982	.4982	.4983	.4984	.4984	.4985	.4985	.4986	.4986
3.0	.4987	.4987	.4987	.4988	.4988	.4989	.4989	.4989	.4990	.4990
3.1	.4990	.4991	.4991	.4991	.4992	.4992	.4992	.4992	.4993	.4993
3.2	.4993	.4993	.4994	.4994	.4994	.4994	.4994	.4995	.4995	.4995
3.3	.4995	.4995	.4995	.4996	.4996	.4996	.4996	.4996	.4996	.4997

Table V Student's t distribution

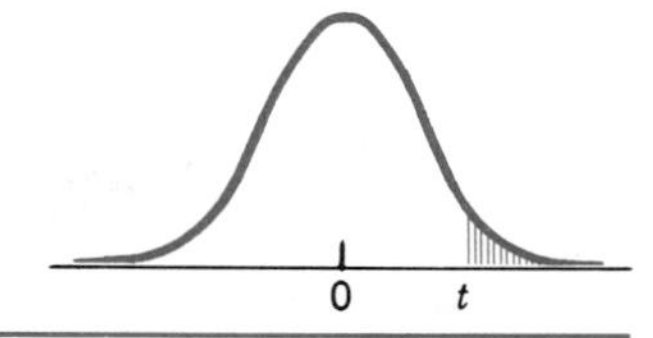

The first column lists the number of degrees of freedom (ν). The headings of the other columns give probabilities (P) that t exceeds the entry value. Use symmetry for negative t-values.

	P				
ν	.10	.05	.025	.01	.005
1	3.078	6.314	12.706	31.821	63.657
2	1.886	2.920	4.303	6.965	9.925
3	1.638	2.353	3.182	4.541	5.841
4	1.533	2.132	2.776	3.747	4.604
5	1.476	2.015	2.571	3.365	4.032
6	1.440	1.943	2.447	3.143	3.707
7	1.415	1.895	2.365	2.998	3.499
8	1.397	1.860	2.306	2.896	3.355
9	1.383	1.833	2.262	2.821	3.250
10	1.372	1.812	2.228	2.764	3.169
11	1.363	1.796	2.201	2.718	3.106
12	1.356	1.782	2.179	2.681	3.055
13	1.350	1.771	2.160	2.650	3.012
14	1.345	1.761	2.145	2.624	2.977
15	1.341	1.753	2.131	2.602	2.947
16	1.337	1.746	2.120	2.583	2.921
17	1.333	1.740	2.110	2.567	2.898
18	1.330	1.734	2.101	2.552	2.878
19	1.328	1.729	2.093	2.539	2.861
20	1.325	1.725	2.086	2.528	2.845
21	1.323	1.721	2.080	2.518	2.831
22	1.321	1.717	2.074	2.508	2.819
23	1.319	1.714	2.069	2.500	2.807
24	1.318	1.711	2.064	2.492	2.797
25	1.316	1.708	2.060	2.485	2.787
26	1.315	1.706	2.056	2.479	2.779
27	1.314	1.703	2.052	2.473	2.771
28	1.313	1.701	2.048	2.467	2.763
29	1.311	1.699	2.045	2.462	2.756
30	1.310	1.697	2.042	2.457	2.750
40	1.303	1.684	2.021	2.423	2.704
60	1.296	1.671	2.000	2.390	2.660
120	1.289	1.658	1.980	2.358	2.617
∞	1.282	1.645	1.960	2.326	2.576

Table VI Poisson distribution function:

$$F(k) = \sum_{x=0}^{k} \frac{\mu^x}{x!} e^{-\mu}$$

	μ									
k	.50	1.0	2.0	3.0	4.0	5.0	6.0	7.0	8.0	9.0
0	.607	.368	.135	.050	.018	.007	.002	.001	.000	.000
1	.910	.736	.406	.199	.092	.040	.017	.007	.003	.001
2	.986	.920	.677	.423	.238	.125	.062	.030	.014	.006
3	.998	.981	.857	.647	.433	.265	.151	.082	.042	.021
4	1.000	.996	.947	.815	.629	.440	.285	.173	.100	.055
5	1.000	.999	.983	.961	.785	.616	.446	.301	.191	.116
6	1.000	1.000	.995	.966	.889	.762	.606	.450	.313	.207
7	1.000	1.000	.999	.988	.949	.867	.744	.599	.453	.324
8	1.000	1.000	1.000	.996	.979	.932	.847	.729	.593	.456
9	1.000	1.000	1.000	.999	.992	.968	.916	.830	.717	.587
10	1.000	1.000	1.000	1.000	.997	.986	.957	.901	.816	.706
11	1.000	1.000	1.000	1.000	.999	.995	.980	.947	.888	.803
12	1.000	1.000	1.000	1.000	1.000	.998	.991	.973	.936	.876
13	1.000	1.000	1.000	1.000	1.000	.999	.996	.987	.966	.926
14	1.000	1.000	1.000	1.000	1.000	1.000	.999	.994	.983	.959
15	1.000	1.000	1.000	1.000	1.000	1.000	.999	.998	.992	.978
16	1.000	1.000	1.000	1.000	1.000	1.000	1.000	.999	.996	.989
17	1.000	1.000	1.000	1.000	1.000	1.000	1.000	1.000	.998	.995
18	1.000	1.000	1.000	1.000	1.000	1.000	1.000	1.000	.999	.998
19	1.000	1.000	1.000	1.000	1.000	1.000	1.000	1.000	1.000	.999
20	1.000	1.000	1.000	1.000	1.000	1.000	1.000	1.000	1.000	1.000

Table VI Poisson Distribution function (*continued*)

k	μ 10.0	11.0	12.0	13.0	14.0	15.0
2	.003	.001	.001	.000	.000	.000
3	.010	.005	.002	.001	.000	.000
4	.029	.015	.008	.004	.002	.001
5	.067	.038	.020	.011	.006	.003
6	.130	.079	.046	.026	.014	.008
7	.220	.143	.090	.054	.032	.018
8	.333	.232	.155	.100	.062	.037
9	.458	.341	.242	.166	.109	.070
10	.583	.460	.347	.252	.176	.118
11	.697	.579	.462	.353	.260	.185
12	.792	.689	.576	.463	.358	.268
13	.864	.781	.682	.573	.464	.363
14	.917	.854	.772	.675	.570	.466
15	.951	.907	.844	.764	.669	.568
16	.973	.944	.899	.835	.756	.664
17	.986	.968	.937	.890	.827	.749
18	.993	.982	.963	.930	.883	.819
19	.997	.991	.979	.957	.923	.875
20	.998	.995	.988	.975	.952	.917
21	.999	.998	.994	.986	.971	.947
22	1.000	.999	.997	.992	.983	.967
23	1.000	1.000	.999	.996	.991	.981
24	1.000	1.000	.999	.998	.995	.989
25	1.000	1.000	1.000	.999	.997	.994
26	1.000	1.000	1.000	1.000	.999	.997
27	1.000	1.000	1.000	1.000	.999	.998
28	1.000	1.000	1.000	1.000	1.000	.999
29	1.000	1.000	1.000	1.000	1.000	1.000

Table VII Confidence limits (95%) for p

Number observed x	Size of sample 10		15		20		30		50		100	
0	0	31	0	22	0	17	0	12	0	07	0	4
1	0	45	0	32	0	25	0	17	0	11	0	5
2	3	56	2	40	1	31	1	22	0	14	0	7
3	7	65	4	48	3	38	2	27	1	17	1	8
4	12	74	8	55	6	44	4	31	2	19	1	10
5	19	81	12	62	9	49	6	35	3	22	2	11
6	26	88	16	68	12	54	8	39	5	24	2	12
7	35	93	21	73	15	59	10	43	6	27	3	14
8	44	97	27	79	19	64	12	46	7	29	4	15
9	55	100	32	84	23	68	15	50	9	31	4	16
10	69	100	38	88	27	73	17	53	10	34	5	18
11			45	92	32	77	20	56	12	36	5	19
12			52	96	36	81	23	60	13	38	6	20
13			60	98	41	85	25	63	15	41	7	21
14			68	100	46	88	28	66	16	43	8	22
15			78	100	51	91	31	69	18	44	9	24
16					56	94	34	72	20	46	9	25
17					62	97	37	75	21	48	10	26
18					69	99	40	77	23	50	11	27
19					75	100	44	80	25	53	12	28
20					83	100	47	83	27	55	13	29
21							50	85	28	57	14	30
22							54	88	30	59	14	31
23							57	90	32	61	15	32
24							61	92	34	63	16	33
25							65	94	36	64	17	35
26							69	96	37	66	18	36
27							73	98	39	68	19	37
28							78	99	41	70	19	38
29							83	100	43	72	20	39
30							88	100	45	73	21	40
31									47	75	22	41
32									50	77	23	42
33									52	79	24	43
34									54	80	25	44
35									56	82	26	45
36									57	84	27	46
37									59	85	28	47
38									62	87	28	48
39									64	88	29	49
40									66	90	30	50
41									69	91	31	51
42									71	93	32	52
43									73	94	33	53
44									76	95	34	54
45									78	97	35	55
46									81	98	36	56
47									83	99	37	57
48									86	100	38	58
49									89	100	39	59
50									93	100	40	60

If $x > 50$, use $100 - x$ in place of x and subtract each confidence limit from 100.

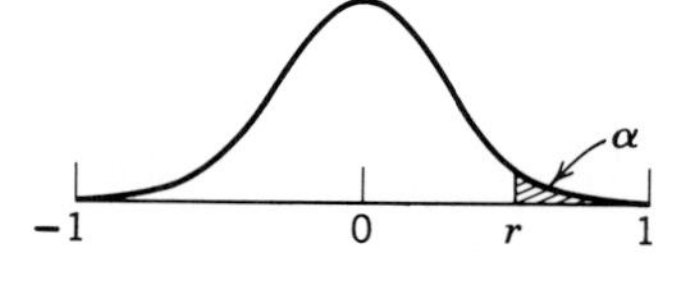

Table VIII Critical values of r for testing $\rho = 0$

For a two-tailed test, α is twice the value listed at the heading of a column of critical r-values, hence for $\alpha = .05$ choose the .025 column.

	α		
n	.05	.025	.005
5	.805	.878	.959
6	.729	.811	.917
7	.669	.754	.875
8	.621	.707	.834
9	.582	.666	.798
10	.549	.632	.765
11	.521	.602	.735
12	.497	.576	.708
13	.476	.553	.684
14	.457	.532	.661
15	.441	.514	.641
16	.426	.497	.623
17	.412	.482	.606
18	.400	.468	.590
19	.389	.456	.575
20	.378	.444	.561
25	.337	.396	.505
30	.306	.361	.463
35	.283	.334	.430
40	.264	.312	.402
50	.235	.279	.361
60	.214	.254	.330
80	.185	.220	.286
100	.165	.196	.256

Table IX The χ^2 distribution

The first column lists the number of degrees of freedom (ν). The headings of the other columns give probabilities (P) that χ^2 exceeds the entry value. For $\nu > 100$, treat $\sqrt{2\nu^2} - \sqrt{2\nu - 1}$ as a standard normal variable.

ν	P = 0.995	0.975	0.050	0.025	0.010	0.005
1	$0.0^4 3927$	$0.0^3 9821$	3.84146	5.02389	6.63490	7.87944
2	0.010025	0.050636	5.99147	7.37776	9.21034	10.5966
3	0.071721	0.215795	7.81473	9.34840	11.3449	12.8381
4	0.206990	0.484419	9.48773	11.1433	13.2767	14.8602
5	0.411740	0.831211	11.0705	12.8325	15.0863	16.7496
6	0.675727	1.237347	12.5916	14.4494	16.8119	18.5476
7	0.989265	1.68987	14.0671	16.0128	18.4753	20.2777
8	1.344419	2.17973	15.5073	17.5346	20.0902	21.9550
9	1.734926	2.70039	16.9190	19.0228	21.6660	23.5893
10	2.15585	3.24697	18.3070	20.4831	23.2093	25.1882
11	2.60321	3.81575	19.6751	21.9200	24.7250	26.7569
12	3.07382	4.40379	21.0261	23.3367	26.2170	28.2995
13	3.56503	5.00874	22.3621	24.7356	27.6883	29.8194
14	4.07468	5.62872	23.6848	26.1190	29.1413	31.3193
15	4.60094	6.26214	24.9958	27.4884	30.5779	32.8013
16	5.14224	6.90766	26.2962	28.8454	31.9999	34.2672
17	5.69724	7.56418	27.5871	30.1910	33.4087	35.7185
18	6.26481	8.23075	28.8693	31.5264	34.8053	37.1564
19	6.84398	8.90655	30.1435	32.8523	36.1908	38.5822
20	7.43386	9.59083	31.4104	34.1696	37.5662	39.9968
21	8.03366	10.28293	32.6705	35.4789	38.9321	41.4010
22	8.64272	10.9823	33.9244	36.7807	40.2894	42.7956
23	9.26042	11.6885	35.1725	30.0757	41.6384	44.1813
24	9.88623	12.4001	36.4151	39.3641	42.9798	45.5585
25	10.5197	13.1197	37.6525	40.6465	44.3141	46.9278
26	11.1603	13.8439	38.8852	41.9232	45.6417	48.2899
27	11.8076	14.5733	40.1133	43.1944	46.9630	49.6449
28	12.4613	15.3079	41.3372	44.4607	48.2782	50.9933
29	13.1211	16.0471	42.5569	45.7222	49.5879	52.3356
30	13.7867	16.7908	43.7729	46.9792	50.8922	53.6720
40	20.7065	24.4331	55.7585	59.3417	63.6907	66.7659
50	27.9907	32.3574	67.5048	71.4202	76.1539	79.4900
60	35.5346	40.4817	79.0819	83.2976	88.3794	91.9517
70	43.2752	48.7576	90.5312	95.0231	100.425	104.215
80	51.1720	57.1532	101.879	106.629	112.329	116.321
90	59.1963	65.6466	113.145	118.136	124.116	128.299
100	67.3276	74.2219	124.342	129.561	135.807	140.169

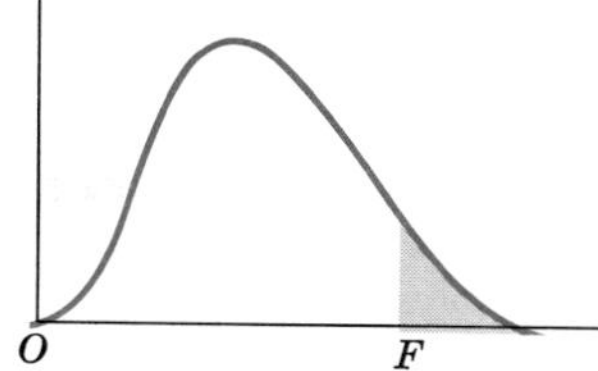

Table X *F* Distribution
5% (roman type) and 1% (boldface type) points for the distribution of *F*

Degrees of freedom for denominator (ν_2)	Degrees of freedom for numerator (ν_1)											
	1	2	3	4	5	6	7	8	9	10	11	12
1	161	200	216	225	230	234	237	239	241	242	243	244
	4052	**4999**	**5403**	**5625**	**5764**	**5859**	**5928**	**5981**	**6022**	**6056**	**6082**	**6106**
2	18.51	19.00	19.16	19.25	19.30	19.33	19.36	19.37	19.38	19.39	19.40	19.41
	98.49	**99.01**	**99.17**	**99.25**	**99.30**	**99.33**	**99.34**	**99.36**	**99.38**	**99.40**	**99.41**	**99.42**
3	10.13	9.55	9.28	9.12	9.01	8.94	8.88	8.84	8.81	8.78	8.76	8.74
	34.12	**30.81**	**29.46**	**28.71**	**28.24**	**27.91**	**27.67**	**27.49**	**27.34**	**27.23**	**27.13**	**27.05**
4	7.71	6.94	6.59	6.39	6.26	6.16	6.09	6.04	6.00	5.96	5.93	5.91
	21.20	**18.00**	**16.69**	**15.98**	**15.52**	**15.21**	**14.98**	**14.80**	**14.66**	**14.54**	**14.45**	**14.37**
5	6.61	5.79	5.41	5.19	5.05	4.95	4.88	4.82	4.78	4.74	4.70	4.68
	16.26	**13.27**	**12.06**	**11.39**	**10.97**	**10.67**	**10.45**	**10.27**	**10.15**	**10.05**	**9.96**	**9.89**
6	5.99	5.14	4.76	4.53	4.39	4.28	4.21	4.15	4.10	4.06	4.03	4.00
	13.74	**10.92**	**9.78**	**9.15**	**8.75**	**8.47**	**8.26**	**8.10**	**7.98**	**7.87**	**7.79**	**7.72**
7	5.59	4.74	4.35	4.12	3.97	3.87	3.79	3.73	3.68	3.63	3.60	3.57
	12.25	**9.55**	**8.45**	**7.85**	**7.46**	**7.19**	**7.00**	**6.84**	**7.71**	**6.62**	**6.54**	**6.47**
8	5.32	4.46	4.07	3.84	3.69	3.58	3.50	3.44	3.39	3.34	3.31	3.28
	11.26	**8.65**	**7.59**	**7.01**	**6.63**	**6.37**	**6.19**	**6.03**	**5.91**	**5.82**	**5.74**	**5.67**
9	5.12	4.26	3.86	3.63	3.48	3.37	3.29	3.23	3.18	3.13	3.10	3.07
	10.56	**8.02**	**6.99**	**6.42**	**6.06**	**5.80**	**5.62**	**5.47**	**5.35**	**5.26**	**5.18**	**5.11**
10	4.96	4.10	3.71	3.48	3.33	3.22	3.14	3.07	3.02	2.97	2.94	2.91
	10.04	**7.56**	**6.55**	**5.99**	**5.64**	**5.39**	**5.21**	**5.06**	**4.95**	**4.85**	**4.78**	**4.71**
11	4.84	3.98	3.59	3.36	3.20	3.09	3.01	2.95	2.90	2.96	2.82	2.79
	9.65	**7.20**	**6.22**	**5.67**	**5.32**	**5.07**	**4.88**	**4.74**	**4.63**	**4.54**	**4.46**	**4.40**
12	4.75	3.88	3.49	3.26	3.11	3.00	2.92	2.85	2.80	2.76	2.72	2.69
	9.33	**6.93**	**5.95**	**5.41**	**5.06**	**4.82**	**4.65**	**4.50**	**4.39**	**4.30**	**4.22**	**4.16**
13	4.67	3.80	3.41	3.18	3.02	2.92	2.84	2.77	2.72	2.67	2.63	2.60
	9.07	**6.70**	**5.74**	**5.20**	**4.86**	**4.62**	**4.44**	**4.30**	**4.19**	**4.10**	**4.02**	**3.96**
14	4.60	3.74	3.34	3.11	2.96	2.85	2.77	2.70	2.65	2.60	2.56	2.53
	8.86	**6.51**	**5.56**	**5.03**	**4.69**	**4.46**	**4.28**	**4.14**	**4.03**	**3.94**	**3.86**	**3.80**
15	4.54	3.68	3.29	3.06	2.90	2.79	2.70	2.64	2.59	2.55	2.51	2.48
	8.68	**6.36**	**5.42**	**4.89**	**4.56**	**4.32**	**4.14**	**4.00**	**3.89**	**3.80**	**3.73**	**3.67**
16	4.49	3.63	3.24	3.01	2.85	2.74	2.66	2.59	2.54	2.49	2.45	2.42
	8.53	**6.23**	**5.29**	**4.77**	**4.44**	**4.20**	**4.03**	**3.89**	**3.78**	**3.69**	**3.61**	**3.55**
17	4.45	3.59	3.20	2.96	2.81	2.70	2.62	2.55	2.50	2.45	2.41	2.38
	8.40	**6.11**	**5.18**	**4.67**	**4.34**	**4.10**	**3.93**	**3.79**	**3.68**	**3.59**	**3.52**	**3.45**
18	4.41	3.55	3.16	2.93	2.77	2.66	2.58	2.51	2.46	2.41	2.37	2.34
	8.28	**6.01**	**5.09**	**4.58**	**4.25**	**4.01**	**3.85**	**3.71**	**3.60**	**3.51**	**3.44**	**3.37**
19	4.38	3.52	3.13	2.90	2.74	2.63	2.55	2.48	2.43	2.38	2.34	2.31
	8.18	**5.93**	**5.01**	**4.50**	**4.17**	**3.94**	**3.77**	**3.63**	**3.52**	**3.43**	**3.36**	**3.30**
20	4.35	3.49	3.10	2.87	2.71	2.60	2.52	2.45	2.40	2.35	2.31	2.28
	8.10	**5.85**	**4.94**	**4.43**	**4.10**	**3.87**	**3.71**	**3.56**	**3.45**	**3.37**	**3.30**	**3.23**

Table X *F* Distribution (*continued*)
5% (roman type) and 1% (boldface type) points for the distribution of *F*

Degrees of freedom for denominator (ν_2)	Degrees of freedom for numerator (ν_1)											
	14	16	20	24	30	40	50	75	100	200	500	∞
1	245	246	248	249	250	251	252	253	253	254	254	254
	6142	**6169**	**6208**	**6234**	**6258**	**6286**	**6302**	**6323**	**6334**	**6352**	**6361**	**6366**
2	19.42	19.43	19.44	19.45	19.46	19.47	19.47	19.48	19.49	19.49	19.50	19.50
	99.43	**99.44**	**99.45**	**99.46**	**99.47**	**99.48**	**99.48**	**99.49**	**99.49**	**99.49**	**99.50**	**99.50**
3	8.71	8.69	8.66	8.64	8.62	8.60	8.58	8.57	8.56	8.54	8.54	8.53
	26.92	**26.83**	**26.69**	**26.60**	**26.50**	**26.41**	**26.30**	**26.27**	**26.23**	**26.18**	**26.14**	**26.12**
4	5.87	5.84	5.80	5.77	5.74	5.71	5.70	5.68	5.66	5.65	5.64	5.63
	14.24	**14.15**	**14.02**	**13.93**	**13.83**	**13.74**	**13.69**	**13.61**	**13.57**	**13.52**	**13.48**	**13.46**
5	4.64	4.60	4.56	4.53	4.50	4.46	4.44	4.42	4.40	4.38	4.37	4.36
	9.77	**9.68**	**9.55**	**9.47**	**9.38**	**9.29**	**9.24**	**9.17**	**9.13**	**9.07**	**9.04**	**9.02**
6	3.96	3.92	3.87	3.84	3.81	3.77	3.75	3.72	3.71	3.69	3.68	3.67
	7.60	**7.52**	**7.39**	**7.31**	**7.23**	**7.14**	**7.09**	**7.02**	**6.99**	**6.94**	**6.90**	**6.88**
7	3.52	3.49	3.44	3.41	3.38	3.34	3.32	3.29	3.28	3.25	3.24	3.23
	6.35	**6.27**	**6.15**	**6.07**	**5.98**	**5.90**	**5.85**	**5.78**	**5.75**	**5.70**	**5.67**	**5.65**
8	3.23	3.20	3.15	3.12	3.08	3.05	3.03	3.00	2.98	2.96	2.94	2.93
	5.56	**5.48**	**5.36**	**5.28**	**5.20**	**5.11**	**5.06**	**5.00**	**4.96**	**4.91**	**4.88**	**4.86**
9	3.02	2.98	2.93	2.90	2.86	2.82	2.80	2.77	2.76	2.73	2.72	2.71
	5.00	**4.92**	**4.80**	**4.73**	**4.64**	**4.56**	**4.51**	**4.45**	**4.41**	**4.36**	**4.33**	**4.31**
10	2.86	2.82	2.77	2.74	2.70	2.67	2.64	2.61	2.59	2.56	2.55	2.54
	4.60	**4.52**	**4.41**	**4.33**	**4.25**	**4.17**	**4.12**	**4.05**	**4.01**	**3.96**	**3.93**	**3.91**
11	2.74	2.70	2.65	2.61	2.57	2.53	2.50	2.47	2.45	2.42	2.41	2.40
	4.29	**4.21**	**4.10**	**4.02**	**3.94**	**3.86**	**3.80**	**3.74**	**3.70**	**3.66**	**3.62**	**3.60**
12	2.64	2.60	2.54	2.50	2.46	2.42	2.40	2.36	2.35	2.32	2.31	2.30
	4.05	**3.98**	**3.86**	**3.78**	**3.70**	**3.61**	**3.56**	**3.49**	**3.46**	**3.41**	**3.38**	**3.36**
13	2.55	2.51	2.46	2.42	2.38	2.34	2.32	2.28	2.26	2.24	2.22	2.21
	3.85	**3.78**	**3.67**	**3.59**	**3.51**	**3.42**	**3.37**	**3.30**	**3.27**	**3.21**	**3.18**	**3.16**
14	2.48	2.44	2.39	2.35	2.31	2.27	2.24	2.21	2.19	2.16	2.14	2.13
	3.70	**3.62**	**3.51**	**3.43**	**3.34**	**3.26**	**3.21**	**3.14**	**3.11**	**3.06**	**3.02**	**3.00**
15	2.43	2.39	2.33	2.29	2.25	2.21	2.18	2.15	2.12	2.10	2.08	2.07
	3.56	**3.48**	**3.36**	**3.29**	**3.20**	**3.12**	**3.07**	**3.00**	**2.97**	**2.92**	**2.89**	**2.87**
16	2.37	2.33	2.28	2.24	2.20	2.16	2.13	2.09	2.07	2.04	2.02	2.01
	3.45	**3.37**	**3.25**	**3.18**	**3.10**	**3.01**	**2.96**	**2.89**	**2.86**	**2.80**	**2.77**	**2.75**
17	2.33	2.29	2.23	2.19	2.15	2.11	2.08	2.04	2.02	1.99	1.97	1.96
	3.35	**3.27**	**3.16**	**3.08**	**3.00**	**2.92**	**2.86**	**2.79**	**2.76**	**2.70**	**2.67**	**2.65**
18	2.29	2.25	2.19	2.15	2.11	2.07	2.04	2.00	1.98	1.95	1.93	1.92
	3.27	**3.19**	**3.07**	**3.00**	**2.91**	**2.83**	**2.78**	**2.71**	**2.68**	**2.62**	**2.59**	**2.57**
19	2.26	2.21	2.15	2.11	2.07	2.02	2.00	1.96	1.94	1.91	1.90	1.88
	3.19	**3.12**	**3.00**	**2.92**	**2.84**	**2.76**	**2.70**	**2.63**	**2.60**	**2.54**	**2.51**	**2.49**
20	2.23	2.18	2.12	2.08	2.04	1.99	1.96	1.92	1.90	1.87	1.85	1.84
	3.13	**3.05**	**2.94**	**2.86**	**2.77**	**2.69**	**2.63**	**2.56**	**2.53**	**2.47**	**2.44**	**2.42**

Table X *F* distribution (*continued*)
5% (roman type) and 1% (boldface type) points for the distribution of *F*

Degrees of freedom for denominator (ν_2)	Degrees of freedom for numerator (ν_1)											
	1	2	3	4	5	6	7	8	9	10	11	12
21	4.32	3.47	3.07	2.84	2.68	2.57	2.49	2.42	2.37	2.32	2.28	2.25
	8.02	**5.78**	**4.87**	**4.37**	**4.04**	**3.81**	**3.65**	**3.51**	**3.40**	**3.31**	**3.24**	**3.17**
22	4.30	3.44	3.05	2.82	2.66	2.55	2.47	2.40	2.35	2.30	2.26	2.23
	7.94	**5.72**	**4.82**	**4.31**	**3.99**	**3.76**	**3.59**	**3.45**	**3.35**	**3.26**	**3.18**	**3.12**
23	4.28	3.42	3.03	2.80	2.64	2.53	2.45	2.38	2.32	2.28	2.24	2.20
	7.88	**5.66**	**4.76**	**4.26**	**3.94**	**3.71**	**3.54**	**3.41**	**3.30**	**3.21**	**3.14**	**3.07**
24	4.26	3.40	3.01	2.78	2.62	2.51	2.43	2.36	2.30	2.26	2.22	2.18
	7.82	**5.61**	**4.72**	**4.22**	**3.90**	**3.67**	**3.50**	**3.36**	**3.25**	**3.17**	**3.09**	**3.03**
25	4.24	3.38	2.99	2.76	2.60	2.49	2.41	2.34	2.28	2.24	2.20	2.16
	7.77	**5.57**	**4.68**	**4.18**	**3.86**	**3.63**	**3.46**	**3.32**	**3.21**	**3.13**	**3.05**	**2.99**
26	4.22	3.37	2.89	2.74	2.59	2.47	2.39	2.32	2.27	2.22	2.18	2.15
	7.72	**5.53**	**4.64**	**4.14**	**3.82**	**3.59**	**3.42**	**3.29**	**3.17**	**3.09**	**3.02**	**2.96**
27	4.21	3.35	2.96	2.73	2.57	2.46	2.37	2.30	2.25	2.20	2.16	2.13
	7.68	**5.49**	**4.60**	**4.11**	**3.79**	**3.56**	**3.39**	**3.26**	**3.14**	**3.06**	**2.98**	**2.93**
28	4.20	3.34	2.95	2.71	2.56	2.44	2.36	2.29	3.24	2.19	2.15	2.12
	7.64	**5.45**	**4.57**	**4.07**	**3.76**	**3.53**	**3.36**	**3.23**	**3.11**	**3.03**	**2.95**	**2.90**
29	4.18	3.33	2.93	2.70	2.54	2.43	2.35	2.28	2.22	2.18	2.14	2.10
	7.60	**5.52**	**4.54**	**4.04**	**3.73**	**3.50**	**3.33**	**3.20**	**3.08**	**3.00**	**2.92**	**2.87**
30	4.17	3.32	2.92	2.69	2.53	2.42	2.34	2.27	2.21	2.16	2.12	2.09
	7.56	**5.39**	**4.51**	**4.02**	**3.70**	**3.47**	**3.30**	**3.17**	**3.06**	**2.98**	**2.90**	**2.84**
32	4.15	3.30	2.90	2.67	2.51	2.40	2.32	2.25	2.19	2.14	2.10	2.07
	7.50	**5.34**	**4.46**	**3.97**	**3.66**	**3.42**	**3.25**	**3.12**	**3.01**	**2.94**	**2.86**	**2.80**
34	4.13	3.28	2.88	2.65	2.49	2.38	2.30	2.23	2.17	2.12	2.08	2.05
	7.44	**5.29**	**4.42**	**3.93**	**3.61**	**3.38**	**3.21**	**3.08**	**2.97**	**2.89**	**2.82**	**2.76**
36	4.11	3.26	2.86	2.63	2.48	2.36	2.28	2.21	2.15	2.10	2.06	2.03
	7.39	**5.25**	**4.38**	**3.89**	**3.58**	**3.35**	**3.18**	**3.04**	**2.94**	**2.86**	**2.78**	**2.72**
38	4.10	3.25	2.85	2.62	2.46	2.35	2.26	2.19	2.14	2.09	2.05	2.02
	7.35	**5.21**	**4.34**	**3.86**	**3.54**	**3.32**	**3.15**	**3.02**	**2.91**	**2.82**	**2.75**	**2.69**
40	4.08	3.23	2.84	2.61	2.45	2.34	2.25	2.18	2.12	2.07	2.04	2.00
	7.31	**5.18**	**4.31**	**3.83**	**3.51**	**3.29**	**3.12**	**2.99**	**2.88**	**2.80**	**2.73**	**2.66**
42	4.07	3.22	2.83	2.59	2.44	2.32	2.24	2.17	2.11	2.06	2.02	1.99
	7.27	**5.15**	**4.29**	**3.80**	**3.49**	**3.26**	**3.10**	**2.96**	**2.86**	**2.77**	**2.70**	**2.64**
44	4.06	3.21	2.82	2.58	2.43	2.31	2.23	2.16	2.10	2.05	2.01	1.98
	7.24	**5.12**	**4.26**	**3.78**	**3.46**	**3.24**	**3.07**	**2.94**	**2.84**	**2.75**	**2.68**	**2.62**
46	4.05	3.20	2.81	2.57	2.42	2.30	2.22	2.14	2.09	2.04	2.00	1.97
	7.21	**5.10**	**4.24**	**3.76**	**3.44**	**3.22**	**3.05**	**2.92**	**2.82**	**2.73**	**2.66**	**2.60**
48	4.04	3.19	2.80	2.56	2.41	2.30	2.21	2.14	2.08	2.03	1.99	1.96
	7.19	**5.08**	**4.22**	**3.74**	**3.42**	**3.20**	**3.04**	**2.90**	**2.80**	**2.71**	**2.64**	**2.58**
50	4.03	3.18	2.79	2.56	2.40	2.29	2.20	2.13	2.07	2.02	1.98	1.95
	7.17	**5.06**	**4.20**	**3.72**	**3.41**	**3.18**	**3.02**	**2.88**	**2.78**	**2.70**	**2.62**	**2.56**

Table X *F* Distribution (*continued*)
5% (roman type) and 1% (boldface type) points for the distribution of *F*

Degrees of freedom for denominator (ν_2)	Degrees of freedom for numerator (ν_1)											
	14	16	20	24	30	40	50	75	100	200	500	∞
21	2.20	2.15	2.09	2.05	2.00	1.96	1.93	1.89	1.87	1.84	1.82	1.81
	3.07	**2.99**	**2.88**	**2.80**	**2.72**	**2.63**	**2.58**	**2.51**	**2.47**	**2.42**	**2.38**	**2.36**
22	2.18	2.13	2.07	2.03	1.98	1.93	1.91	1.87	1.84	1.81	1.80	1.78
	3.02	**2.94**	**2.83**	**2.75**	**2.67**	**2.58**	**2.53**	**2.46**	**2.42**	**2.37**	**2.33**	**2.31**
23	2.14	2.10	2.04	2.00	1.96	1.91	1.88	1.84	1.82	1.79	1.77	1.76
	2.97	**2.89**	**2.78**	**2.70**	**2.62**	**2.53**	**2.48**	**2.41**	**2.37**	**2.32**	**2.28**	**2.26**
24	2.13	2.09	2.02	1.98	1.94	1.89	1.86	1.82	1.80	1.76	1.74	1.73
	2.93	**2.85**	**2.74**	**2.66**	**2.58**	**2.49**	**2.44**	**2.36**	**2.33**	**2.27**	**2.23**	**2.21**
25	2.11	2.06	2.00	1.96	1.92	1.87	1.84	1.80	1.77	1.74	1.72	1.71
	2.89	**2.81**	**2.70**	**2.62**	**2.54**	**2.45**	**2.40**	**2.32**	**2.29**	**2.23**	**2.19**	**2.17**
26	2.10	2.05	1.99	1.95	1.90	1.85	1.82	1.78	1.76	1.72	1.70	1.69
	2.86	**2.77**	**2.66**	**2.58**	**2.50**	**2.41**	**2.36**	**2.28**	**2.25**	**2.19**	**2.15**	**2.13**
27	2.08	2.03	1.97	1.93	1.88	1.84	1.80	1.76	1.74	1.71	1.68	1.67
	2.83	**2.74**	**2.63**	**2.55**	**2.47**	**2.38**	**2.33**	**2.25**	**2.21**	**2.16**	**2.12**	**2.10**
28	2.06	2.02	1.96	1.91	1.87	1.81	1.78	1.75	1.72	1.69	1.67	1.65
	2.80	**2.71**	**2.60**	**2.52**	**2.44**	**2.35**	**2.30**	**2.22**	**2.18**	**2.13**	**2.09**	**2.06**
29	2.05	2.00	1.94	1.90	1.85	1.80	1.77	1.73	1.71	1.68	1.65	1.64
	2.77	**2.68**	**2.57**	**2.49**	**2.41**	**2.32**	**2.27**	**2.19**	**2.15**	**2.10**	**2.06**	**2.03**
30	2.04	1.99	1.93	1.89	1.84	1.79	1.76	1.72	1.69	1.66	1.64	1.62
	2.74	**2.66**	**2.55**	**2.47**	**2.38**	**2.29**	**2.24**	**2.16**	**2.13**	**2.07**	**2.03**	**2.01**
32	2.02	1.97	1.91	1.86	1.82	1.76	1.74	1.69	1.67	1.64	1.61	1.59
	2.70	**2.62**	**2.51**	**2.42**	**2.34**	**2.25**	**2.20**	**2.12**	**2.08**	**2.02**	**1.98**	**1.96**
34	2.00	1.95	1.89	1.84	1.80	1.74	1.71	1.67	1.64	1.61	1.59	1.57
	2.66	**2.58**	**2.47**	**2.38**	**2.30**	**2.21**	**2.15**	**2.08**	**2.04**	**1.98**	**1.94**	**1.91**
36	1.89	1.93	1.87	1.82	1.78	1.72	1.69	1.65	1.62	1.59	1.56	1.55
	2.62	**2.54**	**2.43**	**2.35**	**2.26**	**2.17**	**2.12**	**2.04**	**2.00**	**1.94**	**1.90**	**1.87**
38	1.96	1.92	1.85	1.80	1.76	1.71	1.67	1.63	1.60	1.57	1.54	1.53
	2.59	**2.51**	**2.30**	**2.32**	**2.22**	**2.14**	**2.08**	**2.00**	**1.97**	**1.90**	**1.86**	**1.84**
40	1.95	1.90	1.84	1.79	1.74	1.69	1.66	1.61	1.59	1.55	1.53	1.51
	2.56	**2.49**	**2.37**	**2.29**	**2.20**	**2.11**	**2.05**	**1.97**	**1.94**	**1.88**	**1.84**	**1.81**
42	1.94	1.89	1.82	1.78	1.73	1.68	1.64	1.60	1.57	1.54	1.51	1.49
	2.54	**2.46**	**2.35**	**2.26**	**2.17**	**2.08**	**2.02**	**1.94**	**1.91**	**1.85**	**1.80**	**1.78**
44	1.92	1.88	1.81	1.76	1.72	1.66	1.63	1.58	1.56	1.52	1.50	1.48
	2.52	**2.44**	**2.32**	**2.24**	**2.15**	**2.06**	**2.00**	**1.92**	**1.88**	**1.82**	**1.78**	**1.75**
46	1.91	1.87	1.80	1.75	1.71	1.65	1.62	1.57	1.54	1.51	1.48	1.46
	2.50	**2.42**	**2.40**	**2.22**	**2.13**	**2.04**	**1.98**	**1.90**	**1.86**	**1.80**	**1.76**	**1.72**
48	1.90	1.86	1.79	1.74	1.70	1.64	1.61	1.56	1.53	1.50	1.47	1.45
	2.48	**2.40**	**2.28**	**2.20**	**2.11**	**2.02**	**1.96**	**1.88**	**1.84**	**1.78**	**1.73**	**1.70**
50	1.90	1.85	1.78	1.74	1.69	1.63	1.60	1.55	1.52	1.48	1.46	1.44
	2.46	**2.39**	**2.26**	**2.18**	**2.10**	**2.00**	**1.94**	**1.86**	**1.82**	**1.76**	**1.71**	**1.68**

Table X *F* Distribution (*continued*)
5% (roman type) and 1% (boldface type) points for the distribution of *F*

Degrees of freedom for denominator (ν_2)	Degrees of freedom for numerator (ν_1)											
	1	2	3	4	5	6	7	8	9	10	11	12
55	4.02	3.17	2.78	2.54	2.38	2.27	2.18	2.11	2.05	2.00	1.97	1.93
	7.12	**5.01**	**4.16**	**3.68**	**3.37**	**3.15**	**2.98**	**2.85**	**2.75**	**2.66**	**2.59**	**2.53**
60	4.00	3.15	2.76	2.52	2.37	2.25	2.17	2.10	2.04	1.99	1.95	1.92
	7.08	**4.98**	**4.13**	**3.65**	**3.34**	**3.12**	**2.95**	**2.82**	**2.72**	**2.63**	**2.56**	**2.50**
65	3.99	3.14	2.75	2.51	2.36	2.24	2.15	2.08	2.02	1.98	1.94	1.90
	7.04	**4.95**	**4.10**	**3.62**	**3.31**	**3.09**	**2.93**	**2.79**	**2.70**	**2.61**	**2.54**	**2.47**
70	3.98	3.13	2.74	2.50	2.35	2.32	2.14	2.07	2.01	1.97	1.93	1.89
	7.01	**4.92**	**4.08**	**3.60**	**3.29**	**3.07**	**2.91**	**2.77**	**2.67**	**2.59**	**2.51**	**2.45**
80	3.96	3.11	2.72	2.48	2.33	2.21	2.12	2.05	1.99	1.95	1.91	1.88
	6.96	**4.88**	**4.04**	**3.56**	**3.25**	**3.04**	**2.87**	**2.74**	**2.64**	**2.55**	**2.48**	**2.41**
100	3.94	3.09	2.70	2.46	2.30	2.19	2.10	2.03	1.97	1.92	1.88	1.85
	6.90	**4.82**	**3.98**	**3.51**	**3.20**	**2.99**	**2.82**	**2.69**	**2.59**	**2.51**	**2.43**	**2.36**
125	3.92	3.07	2.68	2.44	2.29	2.17	2.08	2.01	1.95	1.90	1.86	1.83
	6.84	**4.78**	**3.94**	**3.47**	**3.17**	**2.95**	**2.79**	**2.65**	**2.56**	**2.47**	**2.40**	**2.33**
150	3.91	3.06	2.67	2.43	2.27	2.16	2.07	2.00	1.94	1.89	1.85	1.82
	6.81	**4.75**	**3.91**	**3.44**	**3.13**	**2.92**	**2.76**	**2.62**	**2.53**	**2.44**	**2.37**	**2.30**
200	3.89	3.04	2.65	2.41	2.26	2.14	2.05	1.98	1.92	1.87	1.83	1.80
	6.76	**4.71**	**3.88**	**3.41**	**3.11**	**2.90**	**2.73**	**2.60**	**2.50**	**2.41**	**2.34**	**2.28**
400	3.86	3.02	2.62	2.39	2.23	2.12	2.03	1.96	1.90	1.85	1.81	1.78
	6.70	**4.66**	**3.83**	**3.36**	**3.06**	**2.85**	**2.69**	**2.55**	**2.46**	**2.37**	**2.29**	**2.23**
1000	3.85	3.00	2.61	2.38	2.22	2.10	2.02	1.95	1.89	1.84	1.80	1.76
	6.66	**4.62**	**3.80**	**3.34**	**3.04**	**2.82**	**2.66**	**2.53**	**2.43**	**2.34**	**2.26**	**2.20**
	3.84	2.99	2.60	2.37	2.21	2.09	2.01	1.94	1.88	1.83	1.79	1.75
	6.64	**4.60**	**3.78**	**3.32**	**3.02**	**2.80**	**2.64**	**2.51**	**2.41**	**2.32**	**2.24**	**2.18**

Table X *F* Distribution (*continued*)
5% (roman type) and 1% (boldface type) points for the distribution of *F*

Degrees of freedom for denominator (ν_2)	Degrees of freedom for numerator (ν_1)											
	14	16	20	24	30	40	50	75	100	200	500	∞
55	1.88	1.83	1.76	1.72	1.67	1.61	1.58	1.52	1.50	1.46	1.43	1.41
	2.43	**2.35**	**2.23**	**2.15**	**2.06**	**1.96**	**1.90**	**1.82**	**1.78**	**1.71**	**1.66**	**1.64**
60	1.86	1.81	1.75	1.70	1.65	1.59	1.56	1.50	1.48	1.44	1.41	1.39
	2.40	**2.32**	**2.20**	**2.12**	**2.03**	**1.93**	**1.87**	**1.79**	**1.74**	**1.68**	**1.63**	**1.60**
65	1.85	1.80	1.73	1.68	1.63	1.57	1.54	1.49	1.46	1.42	1.39	1.37
	2.30	**2.37**	**2.18**	**2.09**	**2.00**	**1.90**	**1.84**	**1.76**	**1.71**	**1.64**	**1.60**	**1.56**
70	1.84	1.79	1.72	1.67	1.62	1.56	1.53	1.47	1.45	1.40	1.37	1.35
	2.35	**2.28**	**2.15**	**2.07**	**1.98**	**1.88**	**1.82**	**1.74**	**1.69**	**1.63**	**1.56**	**1.53**
80	1.82	1.77	1.70	1.65	1.60	1.54	1.51	1.45	1.42	1.38	1.35	1.32
	2.32	**2.24**	**2.11**	**2.03**	**1.94**	**1.84**	**1.78**	**1.70**	**1.65**	**1.57**	**1.52**	**1.49**
100	1.79	1.75	1.68	1.63	1.57	1.51	1.48	1.42	1.39	1.34	1.30	1.28
	2.26	**2.19**	**2.06**	**1.98**	**1.89**	**1.79**	**1.73**	**1.64**	**1.59**	**1.51**	**1.46**	**1.43**
125	1.77	1.72	1.65	1.60	1.55	1.49	1.45	1.39	1.36	1.31	1.27	1.25
	2.23	**2.15**	**2.03**	**1.94**	**1.85**	**1.75**	**1.68**	**1.59**	**1.54**	**1.46**	**1.40**	**1.37**
150	1.76	1.71	1.64	1.59	1.54	1.47	1.44	1.37	1.34	1.29	1.25	1.22
	2.20	**2.12**	**2.00**	**1.91**	**1.83**	**1.72**	**1.66**	**1.56**	**1.51**	**1.43**	**1.37**	**1.33**
200	1.74	1.69	1.62	1.57	1.52	1.45	1.42	1.35	1.32	1.26	1.22	1.19
	1.17	**2.09**	**1.97**	**1.88**	**1.79**	**1.69**	**1.62**	**1.53**	**1.48**	**1.39**	**1.33**	**1.28**
400	1.72	1.67	1.60	1.54	1.49	1.42	1.38	1.32	1.28	1.22	1.16	1.13
	2.12	**2.04**	**1.92**	**1.84**	**1.74**	**1.64**	**1.57**	**1.47**	**1.42**	**1.32**	**1.24**	**1.19**
1000	1.70	1.65	1.58	1.53	1.47	1.41	1.36	1.30	1.26	1.19	1.13	1.08
	2.09	**2.01**	**1.89**	**1.81**	**1.71**	**1.61**	**1.54**	**1.44**	**1.38**	**1.28**	**1.19**	**1.11**
	1.69	1.64	1.57	1.52	1.46	1.40	1.35	1.28	1.24	1.17	1.11	1.00
	2.07	**1.99**	**1.87**	**1.79**	**1.69**	**1.59**	**1.52**	**1.41**	**1.36**	**1.25**	**1.15**	**1.00**

Table XI Exponential function

x	e^{-x}	*x*	e^{-x}	*x*	e^{-x}
.0	1.000	1.5	.223	3.0	.050
.1	.905	1.6	.202	3.1	.045
.2	.819	1.7	.183	3.2	.041
.3	.741	1.8	.165	3.3	.037
.4	.670	1.9	.150	3.4	.033
.5	.607	2.0	.135	3.5	.030
.6	.549	2.1	.122	3.6	.027
.7	.497	2.2	.111	3.7	.025
.8	.449	2.3	.100	3.8	.022
.9	.407	2.4	.091	3.9	.020
1.0	.368	2.5	.082	4.0	.018
1.1	.333	2.6	.074	4.5	.011
1.2	.301	2.7	.067	5.0	.007
1.3	.273	2.8	.061	6.0	.002
1.4	.247	2.9	.055	7.0	.001

Table XII Random numbers

31 75 15 72 60	68 98 00 53 39	15 47 04 83 55	88 65 12 25 96	03 15 21 91 21
88 49 29 93 82	14 45 40 45 04	20 09 49 89 77	74 84 39 34 13	22 10 97 85 08
30 93 44 77 44	07 48 18 38 28	73 78 80 65 33	28 59 72 04 05	94 20 52 03 80
22 88 84 88 93	27 49 99 87 48	60 53 04 51 28	74 02 28 46 17	82 03 71 02 68
78 21 21 69 93	35 90 29 13 86	44 37 21 54 86	65 74 11 40 14	87 48 13 72 20
41 84 98 45 47	46 85 05 23 26	34 67 75 83 00	74 91 06 43 45	19 32 58 15 49
46 35 23 30 49	69 24 89 34 60	45 30 50 75 21	61 31 83 18 55	14 41 37 09 51
11 08 79 62 94	14 01 33 17 92	59 74 76 72 77	76 50 33 45 13	39 66 37 75 44
52 70 10 83 37	56 30 38 73 15	16 52 06 96 76	11 65 49 98 93	02 18 16 81 61
57 27 53 68 98	81 30 44 85 85	68 65 22 73 76	92 85 25 58 66	88 44 80 35 84
20 85 77 31 56	70 28 42 43 26	79 37 59 52 20	01 15 96 32 67	10 62 24 83 91
15 63 38 49 24	90 41 59 36 14	33 52 12 66 65	55 82 34 76 41	86 22 53 17 04
92 69 44 82 97	39 90 40 21 15	59 58 94 90 67	66 82 14 15 75	49 76 70 40 37
77 61 31 90 19	88 15 20 00 80	20 55 49 14 09	96 27 74 82 57	50 81 60 76 16
38 68 83 24 86	45 13 46 35 45	59 40 47 20 59	43 94 75 16 80	43 85 25 96 93
25 16 30 18 89	70 01 41 50 21	41 29 06 73 12	71 85 71 59 57	68 97 11 14 03
65 25 10 76 29	37 23 93 32 95	05 87 00 11 19	92 78 42 63 40	18 47 76 56 22
36 81 54 36 25	18 63 73 75 09	82 44 49 90 05	04 92 17 37 01	14 70 79 39 97
64 39 71 16 92	05 32 78 21 62	20 24 78 17 59	45 19 72 53 32	83 74 52 25 67
04 51 52 56 24	95 09 66 79 46	48 46 08 55 58	15 19 11 87 82	16 93 03 33 61
15 88 09 22 61	17 29 28 81 90	61 78 14 88 98	92 52 52 12 83	88 58 16 00 98
71 92 60 08 19	59 14 40 02 24	30 57 09 01 94	18 32 90 69 99	26 85 71 92 38
64 42 52 81 08	16 55 41 60 16	00 04 28 32 29	10 33 33 61 68	65 61 79 48 34
79 78 22 39 24	49 44 03 04 32	81 07 73 15 43	95 21 66 48 65	13 65 85 10 81
35 33 77 45 38	44 55 36 46 72	90 96 04 18 49	93 86 54 46 08	93 17 63 48 51
05 24 92 93 29	19 71 59 40 82	14 73 88 66 67	43 70 86 63 54	93 69 22 55 27
56 46 39 93 80	38 79 38 57 74	19 05 61 39 39	46 06 22 76 47	66 14 66 32 10
96 29 63 31 21	54 19 63 41 08	75 81 48 59 86	71 17 11 51 02	28 99 26 31 65
98 38 03 62 69	60 01 40 72 01	62 44 84 63 85	42 17 58 83 50	46 18 24 91 26
52 56 76 43 50	16 31 55 39 69	80 39 58 11 14	54 35 86 45 78	47 26 91 57 47
78 49 89 08 30	25 95 59 92 36	43 28 69 10 64	99 96 99 51 44	64 42 47 73 77
49 55 32 42 41	08 15 08 95 35	08 70 39 10 41	77 32 38 10 79	45 12 79 36 86
32 15 10 70 75	83 15 51 02 52	73 10 08 86 18	23 89 18 74 18	45 41 72 02 68
11 31 45 03 63	26 86 02 77 99	49 41 68 35 34	19 18 70 80 59	76 67 70 21 10
12 36 47 12 10	87 05 25 02 41	90 78 59 78 89	81 39 95 81 30	64 43 90 56 14
09 18 82 00 97	32 82 53 95 27	04 22 08 63 04	83 38 98 73 74	64 27 85 80 44
90 04 58 54 97	51 98 15 06 54	94 93 88 19 97	91 87 07 61 50	68 47 66 46 59
73 18 95 02 07	47 67 72 62 69	62 29 06 44 64	27 12 46 70 18	41 36 18 27 60
75 76 87 64 90	20 97 18 17 49	90 42 91 22 72	95 37 50 58 71	93 82 34 31 78
54 01 64 40 56	66 28 13 10 03	00 68 22 73 98	20 71 45 32 95	07 70 61 78 13
08 35 86 99 10	78 54 24 27 85	13 66 15 88 73	04 61 89 75 53	31 22 30 84 20
28 30 60 32 64	81 33 31 05 91	40 51 00 78 93	32 60 46 04 75	94 11 90 18 40
53 84 08 62 33	81 59 41 36 28	51 21 59 02 90	28 46 66 87 95	77 76 22 07 91
91 75 75 37 41	61 61 36 22 69	50 26 39 02 12	55 78 17 65 14	83 48 34 70 55
89 41 59 26 94	00 39 75 83 91	12 60 71 76 46	48 94 97 23 06	94 54 13 74 08

Table XII Random numbers (*continued*)

77 51 30 38 20	86 83 42 99 01	68 41 48 27 74	51 90 81 39 80	72 89 35 55 07
19 50 23 71 74	69 97 92 02 88	55 21 02 97 73	74 28 77 52 51	65 34 46 74 15
21 81 85 93 13	93 27 88 17 57	05 68 67 31 56	07 08 28 50 46	31 85 33 84 52
51 47 46 64 99	68 10 72 36 21	94 04 99 13 45	42 83 60 91 91	08 00 74 54 49
99 55 96 83 31	62 53 52 41 70	69 77 71 28 30	74 81 97 81 42	43 86 07 28 34
60 31 14 28 24	37 30 14 26 78	45 99 04 32 42	17 37 45 20 03	70 70 77 02 14
49 73 97 14 84	92 00 39 80 86	76 66 87 32 09	59 20 21 19 73	02 90 23 32 50
78 62 65 15 94	16 45 39 46 14	39 01 49 70 66	83 01 20 98 32	25 57 17 76 28
66 69 21 39 86	99 83 70 05 82	81 23 24 49 87	09 50 49 64 12	90 19 37 95 68
44 07 12 80 91	07 36 29 77 03	76 44 74 25 37	98 52 49 78 31	65 70 40 95 14
41 46 88 51 49	49 55 41 79 94	14 92 43 96 50	95 29 40 05 56	70 48 10 69 05
94 55 93 75 59	49 67 85 31 19	70 31 20 56 82	66 98 63 40 99	74 47 42 07 40
41 61 57 03 60	64 11 45 86 60	90 85 06 46 18	80 62 05 17 90	11 43 63 80 72
50 27 39 31 13	41 79 48 68 61	24 78 18 96 83	55 41 18 56 67	77 53 59 98 92
41 39 68 05 04	90 67 00 82 89	40 90 20 50 69	95 08 30 67 83	28 10 25 78 16
25 80 72 42 60	71 52 97 89 20	72 68 20 73 85	90 72 65 71 66	98 88 40 85 83
06 17 09 79 65	88 30 29 80 41	21 44 34 18 08	68 98 48 36 20	89 74 79 88 82
60 80 85 44 44	74 41 28 11 05	01 17 62 88 38	36 42 11 64 89	18 05 95 10 61
80 94 04 48 93	10 40 83 62 22	80 58 27 19 44	92 63 84 03 33	67 05 41 60 67
19 51 69 01 20	46 75 97 16 43	13 17 75 52 92	21 03 68 28 08	77 50 19 74 27
49 38 65 44 80	23 60 42 35 54	21 78 54 11 01	91 17 81 01 74	29 42 09 04 38
06 31 28 89 40	15 99 56 93 21	47 45 86 48 09	98 18 98 18 51	29 65 18 42 15
60 94 20 03 07	11 89 79 26 74	40 40 56 80 32	96 71 75 42 44	10 70 14 13 93
92 32 99 89 32	78 28 44 63 47	71 20 99 20 61	39 44 89 31 36	25 72 20 85 64
77 93 66 35 74	31 38 45 19 24	85 56 12 96 71	58 13 71 78 20	22 75 13 65 18
91 30 70 69 91	19 07 22 42 10	36 69 95 37 28	28 82 53 57 93	28 97 66 62 52
68 43 49 46 88	84 47 31 36 22	62 12 69 84 08	12 84 38 25 90	09 81 59 31 46
48 90 81 58 77	54 74 52 45 91	35 70 00 47 54	83 82 45 26 92	54 13 05 51 60
06 91 34 51 97	42 67 27 86 01	11 88 30 95 28	63 01 19 89 01	14 97 44 03 44
10 45 51 60 19	14 21 03 37 12	91 34 23 78 21	88 32 58 08 51	43 66 77 08 83
12 88 39 73 43	65 02 76 11 84	04 28 50 13 92	17 97 41 50 77	90 71 22 67 69
21 77 83 09 76	38 80 73 69 61	31 64 94 20 96	63 28 10 20 23	08 81 64 74 49
19 52 35 95 15	65 12 25 96 59	86 28 36 82 58	69 57 21 37 98	16 43 59 15 29
67 24 55 26 70	35 58 31 65 63	79 24 68 66 86	76 46 33 42 22	26 65 59 08 02
60 58 44 73 77	07 50 03 79 92	45 13 42 65 29	26 76 08 36 37	41 32 64 43 44
53 85 34 13 77	36 06 69 48 50	58 83 87 38 59	49 36 47 33 31	96 24 04 36 42
24 63 73 87 36	74 38 48 93 42	52 62 30 79 92	12 36 91 86 01	03 74 28 38 73
83 08 01 24 51	38 99 22 28 15	07 75 95 17 77	97 37 72 75 85	51 97 23 78 67
16 44 42 43 34	36 15 19 90 73	27 49 37 09 39	85 13 03 25 52	54 84 65 47 59
60 79 01 81 57	57 17 86 57 62	11 16 17 85 76	45 81 95 29 79	65 13 00 48 60
94 01 54 68 74	32 44 44 82 77	59 82 09 61 63	64 65 42 58 43	41 14 54 28 20
74 10 88 82 22	88 57 07 40 15	25 70 49 10 35	01 75 51 47 50	48 96 83 86 03
62 88 08 78 73	95 16 05 92 21	22 30 49 03 14	72 87 71 73 34	39 28 30 41 49
11 74 81 21 02	80 58 04 18 67	17 71 05 96 21	06 55 40 78 50	73 95 07 95 52
17 94 40 56 00	60 47 80 33 43	25 85 25 89 05	57 21 63 96 18	49 85 69 93 26

Table XII Random numbers (*continued*)

66 06 74 27 92	95 04 35 26 80	46 78 05 64 87	09 97 15 94 81	37 00 62 21 86
54 24 49 10 30	45 54 77 08 18	59 84 99 61 69	61 45 92 16 47	87 41 71 71 98
30 94 55 75 89	31 73 25 72 60	47 67 00 76 54	46 37 62 53 66	94 74 64 95 80
69 17 03 74 03	86 99 59 03 07	94 30 47 18 03	26 82 50 55 11	12 45 99 13 14
08 34 58 89 75	35 84 18 57 71	08 10 55 99 87	87 11 22 14 76	14 71 37 11 81
27 76 74 35 84	85 30 18 89 77	29 49 06 97 14	73 03 54 12 07	74 69 90 93 10
13 02 51 43 38	54 06 61 52 43	47 72 46 67 33	47 43 14 39 05	31 04 85 66 99
80 21 73 62 92	98 52 52 43 35	24 43 22 48 96	43 27 75 88 74	11 46 61 60 82
10 87 56 20 04	90 39 16 11 05	57 41 10 63 68	53 85 63 07 43	08 67 08 47 41
54 12 75 73 26	26 62 91 90 87	24 47 28 87 79	30 54 02 78 86	61 73 27 54 54
33 71 34 80 07	93 58 47 28 69	51 92 66 47 21	58 30 32 98 22	93 17 49 39 72
85 27 48 68 93	11 30 32 92 70	28 83 43 41 37	73 51 59 04 00	71 14 84 36 43
84 13 38 96 40	44 03 55 21 66	73 85 27 00 91	61 22 26 05 61	62 32 71 84 23
56 73 21 62 34	17 39 59 61 31	10 12 39 16 22	85 49 65 75 60	81 60 41 88 80
65 13 85 68 06	87 64 88 52 61	34 31 36 58 61	45 87 52 10 69	85 64 44 72 77
38 00 10 21 76	81 71 91 17 11	71 60 29 29 37	74 21 96 40 49	65 58 44 96 98
37 40 29 63 97	01 30 47 75 86	56 27 11 00 86	47 32 46 26 05	40 03 03 74 38
97 12 54 03 48	87 08 33 14 17	21 81 53 92 50	75 23 76 20 47	15 50 12 95 78
21 82 64 11 34	47 14 33 40 72	64 63 88 59 02	49 13 90 64 41	03 85 65 45 52
73 13 54 27 42	95 71 90 90 35	85 79 47 42 96	08 78 98 81 56	64 69 11 92 02
07 63 87 79 29	03 06 11 80 72	96 20 74 41 56	23 82 19 95 38	04 71 36 69 94
60 52 88 34 41	07 95 41 98 14	59 17 52 06 95	05 53 35 21 39	61 21 20 64 55
83 59 63 56 55	06 95 89 29 83	05 12 80 97 19	77 43 35 37 83	92 30 15 04 98
10 85 06 27 46	99 59 91 05 07	13 49 90 63 19	53 07 57 18 39	06 41 01 93 62
39 82 09 89 52	43 62 26 31 47	64 42 18 08 14	43 80 00 93 51	31 02 47 31 67
59 58 00 64 78	75 56 97 88 00	88 83 55 44 86	23 76 80 61 56	04 11 10 84 08
38 50 80 73 41	23 79 34 87 63	90 82 29 70 22	17 71 90 42 07	95 95 44 99 53
30 69 27 06 68	94 68 81 61 27	56 19 68 00 91	82 06 76 34 00	05 46 26 92 00
65 44 39 56 59	18 28 82 74 37	49 63 22 40 41	08 33 76 56 76	96 29 99 08 36
27 26 75 02 64	13 19 27 22 94	07 47 74 46 06	17 98 54 89 11	97 34 13 03 58
38 10 17 77 56	11 65 71 38 97	95 88 95 70 67	47 64 81 38 85	70 66 99 34 06
39 64 16 94 57	91 33 92 25 02	92 61 38 97 19	11 94 75 62 03	19 32 42 05 04
84 05 44 04 55	99 39 66 36 80	67 66 76 06 31	69 18 19 68 45	38 52 51 16 00
47 46 80 35 77	57 64 96 32 66	24 70 07 15 94	14 00 42 31 53	69 24 90 57 47
43 32 13 13 70	28 97 72 38 96	76 47 96 85 62	62 34 20 75 89	08 89 90 59 85
64 28 16 18 26	18 55 56 49 37	13 17 33 33 65	78 85 11 64 99	87 06 41 30 75
66 84 77 04 95	32 35 00 29 85	86 71 63 87 46	26 31 37 74 63	55 38 77 26 81
72 46 13 32 30	21 52 95 34 24	92 58 10 22 62	78 43 86 62 76	18 39 67 35 38
21 03 29 10 50	13 05 81 62 18	12 47 05 65 00	15 29 27 61 39	59 52 65 21 13
95 36 26 70 11	06 65 11 61 36	01 01 60 08 57	55 01 85 63 74	35 82 47 17 08
40 71 29 73 80	10 40 45 54 52	34 03 06 07 26	75 21 11 02 71	36 63 36 84 24
58 27 56 17 64	97 58 65 47 16	50 25 94 63 45	87 19 54 60 92	26 78 76 09 39
89 51 41 17 88	68 22 42 34 17	73 95 97 61 45	30 34 24 02 77	11 04 97 20 49
15 47 25 06 69	48 13 93 67 32	46 87 43 70 88	73 46 50 98 19	58 86 93 52 20
12 12 08 61 24	51 24 74 43 02	60 88 35 21 09	21 43 73 67 86	49 22 67 78 37

Table XII Random numbers (*continued*)

03 99 11 04 61	93 71 61 68 94	66 08 32 46 53	84 60 95 82 32	88 61 81 91 61
38 55 59 55 54	32 88 65 97 80	08 35 56 08 60	29 73 54 77 62	71 29 92 38 53
17 54 67 37 04	92 05 24 62 15	55 12 12 92 81	59 07 60 79 36	27 95 45 89 09
32 64 35 28 61	95 81 90 68 31	00 91 19 89 36	76 35 59 37 79	80 86 30 05 14
69 57 26 87 77	39 51 03 59 05	14 06 04 06 19	29 54 96 96 16	33 56 46 07 80
24 12 26 65 91	27 69 90 64 94	14 84 54 66 72	61 95 87 71 00	90 89 97 57 54
61 19 63 02 31	92 96 26 17 73	41 83 95 53 82	17 26 77 09 43	78 03 87 02 67
30 53 22 17 04	10 27 41 22 02	39 68 52 33 09	10 06 16 88 29	55 98 66 64 85
03 78 89 75 99	75 86 72 07 17	74 41 65 31 66	35 20 83 33 74	87 53 90 88 23
48 22 86 33 79	85 78 34 76 19	53 15 26 74 33	35 66 35 29 72	16 81 86 03 11
60 36 59 46 53	35 07 53 39 49	42 61 42 92 97	01 91 82 83 16	98 95 37 32 31
83 79 94 24 02	56 62 33 44 42	34 99 44 13 74	70 07 11 47 36	09 95 81 80 65
32 96 00 74 05	36 40 98 32 32	99 38 54 16 00	11 13 30 75 86	15 91 70 62 53
19 32 25 38 45	57 62 05 26 06	66 49 76 86 46	78 13 86 65 59	19 64 09 94 13
11 22 09 47 47	07 39 93 74 08	48 50 92 39 29	27 48 24 54 76	85 24 43 51 59
21 44 58 27 93	24 83 19 32 41	14 19 97 62 68	70 88 36 80 02	03 82 91 74 43
72 51 37 64 00	52 22 59 23 48	62 30 89 84 81	29 74 43 31 65	33 14 16 10 20
71 47 94 50 27	76 16 05 74 11	13 78 01 36 32	52 30 87 77 62	88 87 43 36 97
83 21 05 14 66	09 08 85 03 95	26 74 30 53 06	21 70 67 00 01	99 43 98 07 67
68 74 99 51 48	94 89 77 86 36	96 75 00 90 24	94 53 89 11 43	96 69 36 18 86
05 18 47 57 63	47 07 58 81 58	05 31 35 34 39	14 90 80 88 30	60 09 62 15 51
13 65 16 25 46	96 89 22 52 40	47 51 15 84 83	87 34 27 88 18	07 85 53 92 69
00 56 62 12 20	00 29 22 40 69	25 07 22 95 19	52 54 85 40 91	21 28 22 12 96
50 95 81 76 95	58 07 26 89 90	60 32 99 59 55	71 58 66 34 17	35 94 76 78 07
57 62 16 45 47	46 85 03 79 81	38 52 70 90 37	64 75 60 33 24	04 98 68 36 66
09 28 22 58 44	79 13 97 84 35	35 42 84 35 61	69 79 96 33 14	12 99 19 35 16
23 39 49 42 06	93 43 23 78 36	94 91 92 68 46	02 55 57 44 10	94 91 54 81 99
05 28 03 74 70	93 62 20 43 45	15 09 21 95 10	18 09 41 66 13	78 23 45 00 01
95 49 19 79 76	38 30 63 21 92	82 63 95 46 24	72 43 49 26 06	23 19 17 46 93
78 52 10 01 04	18 24 87 55 83	90 32 65 07 85	54 03 46 62 51	35 77 41 46 92
96 34 54 45 79	85 93 24 40 53	75 70 42 08 40	86 58 38 39 44	52 45 67 37 66
77 96 33 11 51	32 36 49 16 91	47 35 74 03 38	23 43 52 40 65	08 45 89 53 66
07 52 01 12 94	23 23 80 17 48	41 69 06 73 28	54 81 43 77 77	10 05 74 23 32
38 42 30 23 09	70 70 38 57 36	46 14 81 42 58	29 23 61 21 52	05 08 86 58 25
02 46 36 55 33	21 19 96 05 55	33 92 80 18 17	07 39 68 92 15	30 72 22 21 02
83 76 16 08 73	43 25 38 41 45	60 83 32 59 83	01 29 14 13 49	20 36 80 71 26
14 38 70 63 45	80 85 40 92 79	43 52 90 63 18	38 38 47 47 61	41 19 63 74 80
51 32 19 22 46	80 08 87 70 74	88 72 25 67 36	66 16 44 94 31	66 91 93 16 78
72 47 20 00 08	80 89 01 80 02	94 81 33 19 00	54 15 58 34 36	35 35 25 41 31
05 46 65 53 06	93 12 81 84 64	74 45 79 05 61	72 84 81 18 34	79 98 26 84 16
39 52 87 24 84	82 47 42 55 93	48 54 53 52 47	18 61 91 36 74	18 61 11 92 41
81 61 61 87 11	53 34 24 42 76	75 12 21 17 24	74 62 77 37 07	58 31 91 59 97
07 58 61 61 20	82 64 12 28 20	92 90 41 31 41	32 39 21 97 63	61 19 96 79 40
90 76 70 42 35	13 57 41 72 00	69 90 26 37 42	78 46 42 25 01	18 62 79 08 72
40 18 82 81 93	29 59 38 86 27	94 97 21 15 98	62 09 53 67 87	00 44 15 89 97

Table XII Random numbers (*continued*)

34 41 48 21 57	86 88 75 50 87	19 15 20 00 23	12 30 28 07 83	32 62 46 86 91
63 43 97 53 63	44 98 91 68 22	36 02 40 08 67	76 37 84 16 05	65 96 17 34 88
67 04 90 90 70	93 39 94 55 47	94 45 87 42 84	05 04 14 98 07	20 28 83 40 60
79 49 50 41 46	52 16 29 02 86	54 15 83 42 43	46 97 83 54 82	59 36 29 59 38
91 70 43 05 52	04 73 72 10 31	75 05 19 30 29	47 66 56 43 82	99 78 29 34 78
19 61 27 84 30	11 66 19 47 70	77 60 36 56 69	86 86 81 26 65	30 01 27 59 89
39 14 17 74 00	28 00 06 42 38	73 25 87 17 94	31 34 02 62 56	66 45 33 70 16
64 75 68 04 57	08 74 71 28 36	03 46 95 06 78	03 27 44 34 23	66 67 78 25 56
92 90 15 18 78	56 44 12 29 98	29 71 83 84 47	06 45 32 53 11	07 56 55 37 71
03 55 19 00 70	09 48 39 40 50	45 93 81 81 35	36 90 84 33 21	11 07 35 18 03
98 88 46 62 09	06 83 05 36 56	14 66 35 63 46	71 43 00 49 09	19 81 80 57 07
27 36 98 68 82	53 47 30 75 41	53 63 37 08 63	03 74 81 28 22	19 36 04 90 88
59 06 67 59 74	63 33 52 04 83	43 51 43 74 81	58 27 82 69 67	49 32 54 39 51
91 64 79 37 83	64 16 94 90 22	98 58 80 94 95	49 82 95 90 68	38 83 10 48 38
83 60 59 24 19	39 54 20 77 72	71 56 87 56 73	35 18 58 97 59	44 90 17 42 91
24 89 58 85 30	70 77 43 54 39	46 75 87 04 72	70 20 79 26 75	91 62 36 12 75
35 72 02 65 56	95 59 62 00 94	73 75 08 57 88	34 26 40 17 03	46 83 36 52 48
14 14 15 34 10	38 64 90 63 43	57 25 66 13 42	72 70 97 53 18	90 37 93 75 62
27 41 67 56 70	92 17 67 25 35	93 11 95 60 77	06 88 61 82 44	92 34 43 13 74
82 07 10 74 29	81 00 74 77 49	40 74 45 69 74	23 33 68 88 21	53 84 11 05 36

Table XIII Data on census tracts

The following data are from a sample of 110 tracts from Los Angeles County for 1970. The variable x_1 denotes the median family income for 1969, x_2 the median value of owner homes, x_3 the median value of rent per month paid by renters, and x_4 the percentage of families owning homes. Data are from the 1970 censuses of housing and population.

Serial number	x_1	x_2	x_3	x_4
1	11,301	23,900	110	28
2	15,126	37,200	172	82
3	8,490	18,300	91	27
4	14,722	30,200	221	95
5	8,128	17,400	90	53
6	9,318	21,400	124	18
7	11,904	22,600	128	69
8	11,022	26,300	123	43
9	8,561	22,600	90	17
10	10,855	24,500	121	27
11	10,522	22,300	110	56
12	17,171	33,400	188	93
13	9,717	18,800	127	37
14	22,679	47,100	300	98
15	5,155	17,600	74	12
16	12,484	24,900	163	87
17	13,910	26,900	137	84
18	10,534	29,600	134	43
19	12,866	26,700	130	78
20	18,375	40,300	128	67
21	12,197	29,100	150	39
22	14,954	32,000	214	93
23	16,356	45,200	142	60
24	14,550	27,600	98	73
25	22,832	50,000	127	81
26	9,174	28,400	100	34
27	8,917	20,100	83	55
28	16,320	45,300	156	64
29	7,707	22,200	98	15
30	6,911	17,500	82	22
31	6,879	18,100	81	15
32	7,407	17,200	96	40
33	15,831	35,400	277	33
34	6,925	18,100	89	28
35	10,006	30,000	138	75
36	6,602	15,400	76	48
37	13,798	28,600	164	72
38	15,476	21,400	105	71
39	12,469	25,400	135	92
40	11,193	20,600	114	88

Table XIII (*continued*)

Serial number	x_1	x_2	x_3	x_4
41	14,487	29,100	215	94
42	10,502	20,400	129	64
43	13,668	29,200	210	84
44	11,387	20,800	146	91
45	11,229	19,600	118	73
46	12,564	23,800	155	84
47	12,344	21,300	135	91
48	12,774	26,700	98	86
49	19,195	42,700	181	90
50	10,398	21,900	115	58
51	10,693	19,500	122	70
52	13,943	25,300	139	86
53	11,969	28,600	128	44
54	6,580	13,200	60	80
55	11,000	16,400	95	86
56	9,097	17,500	97	69
57	9,902	17,800	101	64
58	11,037	30,200	117	36
59	8,726	18,000	89	46
60	10,267	22,900	107	40
61	7,574	17,100	80	43
62	8,423	17,500	95	40
63	12,681	36,300	106	37
64	12,000	19,300	145	85
65	8,791	19,300	83	39
66	11,955	21,900	152	85
67	9,946	17,800	132	82
68	10,484	18,700	105	75
69	9,375	18,900	110	42
70	16,467	36,700	271	95
71	11,012	22,800	124	78
72	14,256	32,400	148	76
73	12,537	22,800	178	92
74	11,915	31,300	131	44
75	9,900	28,900	123	29
76	10,947	22,500	132	63
77	15,776	34,600	159	71
78	21,429	44,700	300	94
79	9,730	19,900	78	58
80	5,741	26,100	95	8

Table XIII (*continued*)

Serial number	x_1	x_2	x_3	x_4
81	11,037	26,800	90	40
82	9,354	19,000	125	46
83	24,302	50,000	259	3
84	10,524	45,000	128	4
85	10,893	34,000	141	47
86	8,711	23,100	100	26
87	5,570	13,300	67	26
88	5,691	15,400	74	33
89	10,120	22,100	97	79
90	9,192	17,000	87	74
91	10,917	28,300	134	51
92	16,843	23,800	174	77
93	8,512	18,400	87	38
94	13,505	29,200	137	74
95	8,965	20,500	85	53
96	12,416	22,700	141	79
97	13,215	26,500	118	80
98	8,596	16,100	107	40
99	9,904	18,400	99	56
100	11,137	30,400	119	40
101	11,637	22,800	114	66
102	8,776	18,200	109	61
103	8,819	21,800	90	61
104	11,298	33,900	112	81
105	18,620	32,700	207	89
106	22,433	47,100	251	86
107	11,899	20,300	140	90
108	9,427	23,400	96	57
109	8,633	18,700	88	31
110	13,881	31,100	105	91

Answers

Numerical answers depend upon the extent to which the calculations are rounded off and upon the order of operations; consequently, the student should not expect to agree precisely with all of the following answers. Some of the answers were obtained with the aid of a hand electronic calculator, others were obtained without such equipment; therefore, some of these answers may be slightly in error. Answers of a nonnumerical nature are included to give the student a rough idea of the kind of answer expected for such exercises.

Chapter 2

3. (*a*) (*i*) discrete (*ii*) discrete (*iii*) continuous (*b*) (*i*) All the customers who pass through the checkout stations during the week. (*ii*) Choosing instants of time when observations may be made, all the customers in the shortest queue at those times. (*iii*) The same as in (*i*).
5. If .01 is chosen as the class interval, the boundaries will be .4205–.4305 and .5605–.5705, and the class marks will be .4255 and .5655.
7. Continuous-variable treatment would be more accurate.
11. Class boundaries are 0, 1, 2, ...
13. Class boundaries are 15, 16, . . . , 22.
15. Skewed to the right because students with less than a C average will drop out or be dismissed from school, and the mean is probably not higher than halfway between a C and a B average.
17. 4.43
19. 175.3
21. 8.85 percent
23. A large percentage of the black families in the U.S. are living in the South Atlantic states, where their fertility rate is considerably higher than elsewhere and considerably higher than the white fertility rate. Hence, in calculating the weighted mean, the contribution of the South Atlantic states dominates the black fertility rate.
25. (*a*) $\bar{x} = 44$ (*b*) $s = 5.2$
27. $s = 2.03$
29. $(\bar{x} \pm s) = (2.4, 6.5)$. Interpolation gives 71 percent and 96 percent.
31. The distribution may be heavily skewed, with a large percentage of the shots close to the center but with a small percentage lying some distance away.
33. (*a*) $\bar{x}_A = 1106.7$, $\bar{x}_B = 1070.0$; hence A averages 36.7 more hours than B. (*b*) $s_A = 200$, $s_B = 137$; hence B has more uniform quality than A. Choice would depend on whether uniformity of quality is preferable to increased length of life.

35. (*a*) The mean will be increased by 10 points but the standard deviation will not be changed. (*b*) The mean and standard deviation will both be increased by 10 percent.
37. $\overline{x} = 0$, $s = 1.005$; hence $\overline{x} \pm s$ includes all the data. The usual interpretation of s does not apply to a pathological distribution such as this one.
39. (*a*) mean $= 4.5$ (*b*) median $= 4$, mode $= 6$, range $= 8$
43. mean $= 2.29$; median $= 2.5$
45. mean $= 44.2$; median $\doteq 43.5$
47. There is no unique mode because 38, 42, and 46 all occur twice. The median is 44 and the range is 16. The mean deviation is 4.2.
49. 2.4
51. (*a*) 10 (*b*) 20 (*c*) 18 (*d*) 30 (*e*) 60 (*f*) 86
55. (*a*)

x	8.55	10.55	. . .	32.55
f	6	4	. . .	1

(*c*) 16 (*d*) $\overline{x} = 15.9$ (*e*) $\overline{x} = 16.0$ (*f*) 6 or 7 (*g*) 6.7 (*h*) $s = 6.6$
(*i*) guess 11, 14, 19

Chapter 3

1. HHHH, HHHT, HHTH, HHTT, HTHH, HTHT, THHH, THHT, HTTH, HTTT, THTH, THTT, TTHH, TTHT, TTTH, TTTT
3. (*a*) (*b*)

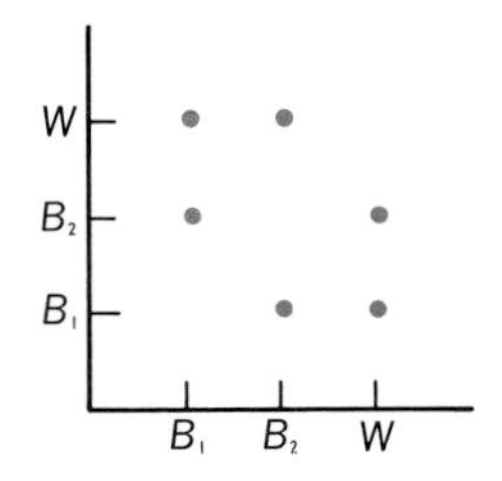

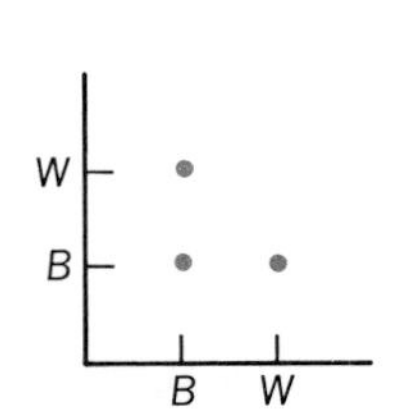

5. $\frac{1}{6}$, $\frac{1}{9}$
7. $\frac{4}{10}$ $\frac{3}{10}$ $\frac{3}{10}$ — e_1 e_2 e_3
9. $\frac{2}{6}$
11. (*a*) $\frac{1}{12}$ (*b*) $\frac{1}{12}$ (*c*) $\frac{1}{2}$
13. (*a*) $\frac{17}{18}$ (*b*) $\frac{4}{9}$ (*c*) $\frac{4}{9}$ (*d*) $\frac{5}{6}$ (*e*) $\frac{4}{9}$
15. $\frac{6}{36}$, $\frac{8}{36}$, $\frac{3}{36}$
17. (*a*) .81 (*b*) .05 (*c*) 15 years or more
19. $\frac{8}{9}$
21. $\frac{7}{16}$

23. (*a*) No, because the events are not independent. Homes with an automatic washer are more likely to have a television set than those without such a washer. (*b*) No, because the events are not mutually exclusive.
25. Solution of $p^5 = .90$, which is approximately $p = .98$.
27. (*a*) .59 (*b*) .33
29. (*a*) $\frac{1}{32}$ (*b*) $\frac{1}{64}$ (*c*) $\frac{3}{32}$
31. .864
33. (*a*) $\frac{1}{10}$ (*b*) $\frac{1}{8}$ (*c*) $\frac{4}{5}$
35. $\frac{14}{17}$
37. $P\{RR \mid R_1\} = \frac{2}{3}$; hence you would be foolish to take an even bet.
39. (*a*) .78 (*b*) .81
41. (*a*) 6,720 (*b*) 32,768
43. 120
45. $\frac{3243}{10,829}$
47. .58
49. (*a*) .44 (*b*) .70 (*c*) .0128 (*d*) $\overbrace{.00 \cdots 0157}^{11}$
51. (*a*) .002 (*b*) .086 (*c*) .00022
53. (*a*) $\frac{1}{10}$ (*b*) $\frac{2}{5}$ (*c*) $\frac{7}{10}$ (*d*) $\frac{9}{10}$ (*e*) 1 (*f*) $\frac{1}{2}$ (*g*) $\frac{5}{6}$ (*h*) $\frac{1}{2}$

Chapter 4

1. $P\{0\} = \frac{1}{4}, P\{1\} = \frac{1}{2}, P\{2\} = \frac{1}{4}$
3. $P\{5\} = \frac{1}{3}, P\{6\} = \frac{1}{3}, P\{7\} = \frac{1}{3}$
5. $\mu = 6, \sigma = \sqrt{4/3} = 1.15$
7. $\mu = 4.5, \sigma = \sqrt{8.25} = 2.87$
9. (*a*) $P\{0\} = \frac{9}{24}, P\{1\} = \frac{8}{24}, P\{2\} = \frac{6}{24}, P\{3\} = 0, P\{4\} = \frac{1}{24}$ (*c*) $\mu = 1$, $\sigma = 1$
11. $E[x] = .10, E[100x] = 100E[x] = 10$
13. $E[A] = 690, E[N] = 700$; hence do not advertise.
15. $E[S] = 41\frac{2}{3}$
19. (*a*) $\frac{91}{216}$ (*b*) $E[x] = -\frac{17}{216}$; hence he will average a loss of about 8 cents per game.

Chapter 5

1. $P\{4\} = \frac{1}{9}, P\{5\} = \frac{2}{9}, P\{6\} = \frac{3}{9}, P\{7\} = \frac{2}{9}, P\{8] = \frac{1}{9}$
3. $P\{0\} = \frac{1}{16}, P\{1\} = \frac{4}{16}, P\{2\} = \frac{6}{16}, P\{3\} = \frac{4}{16}, P\{4\} = \frac{1}{16}$
5. $P\{0\} = \frac{1}{32}, P\{1\} = \frac{5}{32}, P\{2\} = \frac{10}{32}, P\{3\} = \frac{10}{32}, P\{4\} = \frac{5}{32}, P\{5\} = \frac{1}{32}$
7. $\frac{9}{256}$
9. $\mu = 2, \sigma = \sqrt{\frac{4}{3}}$
13. No. The trials are not independent and the probability is not constant. Stock price behavior in the past is no assurance of similar behavior in the future.

15. (*a*) .1587 (*b*) .3085 (*c*) .3085 (*d*) .4013 (*e*) .5328
17. 21 percent
19. $P\{x < 16\} = .0082$
21. (*a*) $P\{x > 41\} = .034$ (*b*) $P\{20 < x < 40\} = .90$
23. $P\{8 < x < 9\} = .1359$
25. (*a*) exact value $= \frac{7}{32}$, approximate value $= .22$ (*b*) exact value $= \frac{37}{256}$, approximate value $= .14$
27. .04
29. .015
31. $P\{0\} = .04$. Since negative values are impossible, this fairly large probability for the smallest possible value of x suggests that the left tail of a normal curve would fit poorly.
33. $n = 200$, $p = .99$
35. $\bar{x} = .58$; hence use $E[x] = 227\dfrac{e^{-.58}(.58)^x}{x!}$. This gives

x	0	1	2	3	4
$E[x]$	127	74	21	4	1

These frequencies are sufficiently close to the observed frequencies to conclude that a Poisson model is satisfactory.

37. (*a*) .368 (*b*) .004
39. $\left[\binom{10}{9}\binom{15}{11} + \binom{15}{10}\right] \Big/ \binom{25}{20} = .313$
41. .55
43. A guarantee of 11.4 years will satisfy; hence 12 years are necessary.
45. (*a*) .023 (*b*) $\mu = 12.1$; the present setting is therefore satisfactory.
47. (*a*) $P\{x\} = \dfrac{8!}{x!(8-x)!}(.4)^x(.6)^{8-x}$ (*b*) .63 (*c*) .64 (*d*) .07
49. (*a*) $P\{x\} = \binom{5}{x}\binom{95}{10-x} \Big/ \binom{100}{10}$

    (*b*) $P\{x\} = \dfrac{10!}{x!(10-x)!}(.05)^x(.95)^{10-x}$ (*c*) $P\{x\} = \dfrac{e^{-.5}(.5)^x}{x!}$

    (*d*) $\mu = .5$, $\sigma = .66$, (*f*) The normal approximation is poor here.

Chapter 6

1. Registrar's card files and random numbers or, say, select twenty 10 o'clock classes at random and select 5 students at random from each class.
3. (*a*) The number of words on each of the pages. (*b*) The complete set of the number of words on each page.
5. Any alphabetical listing in which all letters are included in the sample. If a smaller or larger sample is desired, it is necessary to shift from every twenty-fifth to a more convenient proportion.

7. Most people do not write to congressman, even if they have strong views. Those who do write are often highly in favor or highly opposed. Pressure groups often request their members to write. The ordinary citizen is seldom heard.
9. Select 2 three-digit random numbers and use them as the two coordinates of a point in the square that is 1000 feet on a side and has two of its sides on the positive x- and y-axes. This locates a random point at which to center the hoop.
11. Assign each man a number from 1 to 6. Choose any column from Table XII and read off successive integers in it, discarding the integers 0, 7, 8, and 9.
15. (*a*) .84 (*b*) .001 (*c*) .84 (*d*) .01 (*e*) .04
17. The mean curve should be three times as tall at 6 and about one-third as wide.
21. .05
23. The number 70 represents the accuracy of the carbon dating technique. It probably corresponds to one standard deviation of the error of the technique. On this basis it appears quite likely that the Vikings left by the year 1150.
25. $P\{\Sigma_{i=1}^{50} x_i > 105\} = .04$
27. .58
29. (*a*) 1.35 (*b*) .013
31. .022
33. $P\{|\bar{x} - \mu| > 1\} = .026$
35. (*a*) .86 (*b*) .00 (*c*) .86 (*d*) .003 (*e*) .037

Chapter 7

1. (*a*) $P\{|\text{error}| < 4.8\} = .95$ (*b*) 4.8 would be replaced by 2.4; hence twice the accuracy.
3. (*a*) $n = 138$ using $z_0 = 1.96$ and $n = 144$ using $z_0 = 2$. (*b*) $n = 553$ using $z_0 = 1.96$ and $n = 576$ using $z_0 = 2$.
5. $20.1 < \mu < 29.9$
7. (*a*) $135.3 < \mu < 144.7$ (*b*) $136.1 < \mu < 143.9$
9. (*a*) $P\{\text{error} < 1.4\} = .95$ (*b*) $n \doteq 96$ (*c*) $40.6 < \mu < 43.4$
11. (*a*) $1.18 < \mu < 1.30$ (*b*) Yes, because it is highly probable that $\mu > 1.18$, and therefore it is quite certain that $\mu > 1.00$.
13. $P\{\text{error} < .09\} = .95$
15. $.55 < p < .65$
17. $n \doteq 3458$
19. (*a*) $.18 < p < .22$ (*b*) $.084 < p < .116$
21. (*a*) $n \doteq 384$ (*b*) Yes, because the size sample needed will never exceed the value calculated for $p = .5$.
23. Confidence interval for p is given by $.234 < p < .310$; hence limits for Np are given by $23{,}400 < Np < 31{,}000$.
25. $.76 < p < .84$

27. (*a*) $17.1 < \mu < 22.9$ (*b*) $15.9 < \mu < 24.1$
29. $20.0 < \mu < 24.0$
31. (*a*) $\bar{x} = 4.5$, $s^2 = 7$ (*b*) $.3 < \mu < 8.7$ (*c*) $1.9 < \mu < 7.1$
33. (*a*) .03 (*b*) .017 (*c*) $.01 < p < .08$ (*d*) $-.003 < p < .063$; the normal-curve approximation is poor here.
35. Limits for p are given by $.00383 < p < .0162$; hence limits for Np are given by $7{,}660 < Np < 32{,}400$.
39. $-.11 < p_1 - p_2 < .05$
41. $.005 < p_1 - p_2 < .031$. No, because 0 is not inside this interval.
43. Limits for $p_1 - p_2$ are given by $-.055 < p_1 - p_2 < -.045$. Since 0 is not inside this interval, at least one of the samples is not a random sample.
45. (*a*) $774 < \mu < 834$ (*b*) $.135 < p < .171$
47. (*a*) $n \doteq 576$ (*b*) Since $P\{\hat{p} > .30\} = .000$, it is almost certain that production will never be stopped when $p = .20$.
49. $.004 < p_1 - p_2 < .044$. Since 0 is not inside this interval, there appears to be a real sex difference with respect to theft.

Chapter 8

1. Assuming an individual is innocent until proved guilty, the hypothesis to be tested is that an individual is innocent; hence the type I error is convicting an innocent individual and the type II error is letting a thief go free. Society considers a type I error here more serious than a type II error.
3. $\alpha = \frac{1}{4}$, $\beta = \frac{3}{4}$
5. $\alpha = .25$, $\beta = .51$
7.

x	0	1	2	3	4	5	6	7	8
$P_0\{x\}$	.004	.031	.109	.219	.273	.219	.109	.031	.004
$P_1\{x\}$	.000	.001	.010	.047	.136	.254	.296	.198	.058

$\alpha = .144$, $\beta = .448$

9. $z = -1.33$; hence accept H_0.
11. $z = 3.67$; hence reject H_0. These students are undoubtedly superior in English.
13. $z = .40$; hence accept the hypothesis that $\mu = 4.5$. Retraining is not justified.
15. $z = 1.74$; hence no.
17. $z = 2.86$; hence the mixture is too strong.
19. There is no assurance that the tosses were random. It is very difficult to eliminate human error in sequences of trials.
21. $z = -3.0$; hence reject $H_0{:}p = .96$ in favor of $H_1{:}p < .96$. The quality is undoubtedly less than that claimed.
23. (*a*) $z = 1.43$; hence an investigation is not warranted. (*b*) $x_0 = 61.5$; hence 62.
25. Limits for percentages are 1.05 and 4.05. Days numbered 18, 22, and 38 appear to be out of control.

27. $z = 1.71$; hence accept H_0.
29. $z = 3.3$; hence reject H_0.
31. $z = -1.20$; hence accept $H_0: \mu_1 = \mu_2$, that is, accept the hypothesis that heavy doses of vitamin C had no affect on the average number of colds caught.
33. $z = 1.22$; hence accept $H_0: p_1 = p_2$. No.
35. $z = -2.68$; hence reject the hypothesis of no difference.
39. $z = 1.67$. Since $z_0 = 1.64$ for a one-tailed test, reject H_0 in favor of H_1, which means that the difference is at least 10 percentage points.
41. $t = -1.79$. For $\nu = 19$, $t_0 = 2.09$ for a two-tailed test; hence accept H_0.
43. $t = -3.09$; hence reject H_0. The auditors are justified in claiming a negative mean error of some amount. A confidence interval would be helpful here.
45. $t = .92$, $t_0 = 2.10$; hence accept the hypothesis of no difference.
47. $t = -1.80$, $t_0 = -2.12$; hence accept the hypothesis of no difference. No.
49. $t = 7.83$; hence reject $H_0: \mu = 0$. The vitamin is undoubtedly beneficial.
51.

p	0	.1	.3	.5	.7	.9	1.0
β	1	.99	.91	.75	.51	.19	.0

53. (*a*) .66 (*b*) .41

(*c*)

p	.3	.4	.5	.6	.7	.8	.9
$P\{x = 0\}$	.24	.13	.06	.03	.01	.00	.00

55. $z = 2.59$; hence reject $H_0: p = .25$ in favor of $H_1: p > .25$. This implies that there are too many admirals coming out of the upper 25 percent of classes to be attributed to chance.
57. $n \doteq 137$
59. $z = 5.0$; hence reject the hypothesis of no difference. Older men undoubtedly have a stronger belief in inheritance as a cause.

Chapter 9

1. guess $r = .8, .1, -.95$
3. (*a*) .4 (*b*) $-.6$ (*c*) $-.5$
5. $r = .48$
7. $r = .74$
9. The number of accidents per year for drivers who have accidents decreases with age.
11. Gross sales is the sum of wholesale and retail sales. If wholesale and retail sales are independent, as it appears in this problem, one would expect wholesale and gross sales to be positively correlated due to wholesale sales being common to both variables.
13. $r = -.16$ as compared with $r = .48$. Throwing out small and large values of x reduced the data to a set of points that appear to be a random set.
15. $r_0 = \pm.396$ for a two-tailed test. Since $r = .35$, accept $H_0: \rho = 0$.
17. $r = .61$. Since $r_0 = .378$ for a one-tailed test, reject $H_0: \rho = 0$.
19. $r = .74$. Since $r_0 = .306$ for a one-tailed test, reject $H_0: \rho = 0$.

21. Calculate the value of r between wholesale (x) and retail (y) prices with retail prices chosen i units further along in time than wholesale prices. Do this for $i = 0, 1, 2, \ldots$ and choose the value of i that yields the largest value of r.
23. (*a*) $r_0 = .553$ for a two-tailed test; hence accept $H_0{:}\rho = 0$.

Chapter 10

3. (*a*) \$600, \$1200 (*b*) The formula was undoubtedly derived for average families with regular incomes, and therefore is not applicable to values outside their range.
5. (*c*) 24.8 (*d*) 26.7 (*e*) A straight line is not satisfactory if large incomes are included. The point (11.776, 40.2) will be far off the line in (*d*). A parabola would probably give a better fit.
7. (*a*) $\hat{y} = .157 + 1.383\,x$ (*b*) $\hat{y}(38.9) = 53.97$ (*c*) A 39 percent increase.
9. $\hat{y} = 330 + .122x$
11. (*a*) $\hat{y} = .573 + .0162x$ (*b*) $\hat{y}(400) = 7.05$; hence about \$7. (*c*) Replace x by $x + 100$ and subtract the two values. This gives 1.62; hence the price should increase about \$1.62.
13. Errors are: $-.41$, .24, .01, .27, $-.58$, $-.66$, $-.46$, .23, .03, $-.19$, $-.20$, .21, $-.16$, .03, .28, $-.08$, .27, $-.05$, $-.07$, $-.11$, .10, .83, .47. The errors have a random-appearing graph; hence they are likely to be independent of time.
15. $r^2 = .47$; hence the sum of the squares of the errors has been cut about in half. The relationship, however, is not sufficiently strong to enable one to predict tensile strength from hardness with much accuracy.
17. (*a*) $\hat{y} = 20.33 + .730x$ (*b*) $s_e = 8.24$ (*c*) $r^2 = .55$ shows that the square of the error of prediction is cut about in half. From (*b*) it is to be expected that one should be able to guess a student's mathematical aptitude score to within about 10–15 points if his verbal test score is known.
19. $s_e = 27.9$; twenty-five percent exceed s_e, and 0 percent exceed $2s_e$.
21. (*a*) $\hat{y} = 1.30 + .7925x$ (*b*) $\hat{y}(50) = 40.9$, or 41 (*c*) $s_e = 5.4$ (*d*) $\hat{y}(60) \doteq 49$. No, because it is highly likely that more than 50 will show when 49 is the mean number predicted and the standard deviation is as large as 5.4.
23. (*a*) $\hat{y} = 1460 - .40x$ (*b*) Yes, because the regression coefficient is negative and the standard deviation is fairly small, indicating that the estimates are likely to be good. (*c*) If x is replaced by $x - 10$, the values of $\hat{y}$ will differ by 40; hence reducing the price by \$100 will increase sales, on the average, by about 40 cars.
25. (*b*) $\hat{y} = -3.824 + 1.951x$ (*c*) $1.66 < \beta < 2.24$
27. $.55 < \beta < 3.95$
29. $t = -.34$; hence accept $H_0{:}\beta = 2.5$.
31. (*a*) $t = 1$; hence accept $H_0{:}\beta = 1.01$. (*b*) $t = 5.05$; hence reject

$H_0{:}\alpha = 0$. The device is off by an additive amount, which implies that the y-values tend to be too large.

37. (*a*) $\hat{y} = -199.6 + 3.15x$ (*b*) $s_e = 18.3$ (*c*) $2.57 < \beta < 3.73$
41. (*b*) $\hat{y} = 5.78 + .570x$ (*d*) $-1.0, .3, -.5, .0, .2, .6, 1.3, .1, -.2, .0, .3, -.3, .8$ (*e*) $\hat{y}(16.5) = 15.2$ (*f*) $s_e = .63$ (*h*) $t = 1.50$; hence accept $H_0{:}\beta = .5$ (*i*) $.49 < \beta < .65$

Chapter 11

1. (*a*) $y(x_1 + 1) - y(x_1) = 3.3$; hence the increase would be 330 pounds per acre. (*b*) $y(x_2 + k) - y(x_2) = 1$; hence k must satisfy this equation. The solution is $k = 250$. The accumulated temperature would need to increase 250 units. (*c*) From (*a*) and (*b*) it is clear that rainfall is far more important than temperature. From (*b*) a modest 100-pound increase would require an extremely large temperature change, in view of the fact that the mean is 594.
3. $y = 5.18 + 1.50x_1 + .877x_2$
5. (*c*) A parabola would probably fit well. If values of x larger than 9,000 were available, the curve would probably flatten out, with the result that one branch of a hyperbola with a horizontal asymptote might be a better model.
7. (*a*) Errors for first are: $-.1, 1.4, -1.2, -.2, -.9, .1, 1.5, -.6, -.3, .1, .1$. Errors for second are: $.7, 1.2, .2, -.8, -2.0, -.1, 2.6, .2, -.8, -.2, 1.0$. (*b*) $s_e(1) = .93$, $s_e(2) = 1.30$ (*c*) Since the standard error is increased about 40 percent, C is certainly useful in assisting to predict P.
9. (*a*) .32 (*b*) If two students with the same intelligence test series are compared, the student with the higher reading rate will have a slightly lower grade-point average. This is not a paradox because doing well on an intelligence test depends somewhat on one's reading speed; therefore, a slightly less intelligent student who is a fast reader will tend to have as high an intelligence test score as one who is more intelligent but reads slower. (*c*) Since the coefficient of x_2 is so small numerically, and therefore the influence of x_2 in prediction is minor, it is not surprising that the coefficient of x_1 is about the same in the two equations. If the raw data were available, it would be possible to calculate s_e for both regressions and observe whether the contribution of x_2 is negligible.

Chapter 12

1. $\chi^2 = 26.6$, $\chi_0^2 = 16.9$; hence reject the hypothesis of uniform conditions.
3. $\chi^2 = 35$, $\chi_0^2 = 7.8$; hence not the same distribution of blood types.
5. $\chi^2 = 13.6$, $\chi_0^2 = 6.0$; hence reject $H_0{:}p_1 = p_2 = p_3 = \frac{1}{3}$.
7. (*a*) $\chi^2 = 5.8$, $\chi_0^2 = 12.6$; hence accept the hypothesis of a uniform rate. (*b*) $\chi^2 = 4.7$, $\chi_0^2 = 3.8$; hence reject H_0 in favor of a higher weekend rate.
9. $\chi^2 = 5.33$, $\chi_0^2 = 3.84$; hence reject the postulated proportions.
11. $\chi^2 = 4.7$, $\chi_0^2 = 3.84$; hence reject the hypothesis that the probabilities are correct in favor of the payoffs for the difficult holes being too low.

13. Combining the last two cells gives $\chi^2 = .7$. Here $\nu = 1$ and $\chi_0^2 = 3.84$. The observed frequencies are compatible with those for a Poisson model.
15. Expected frequencies are given by $32\frac{4!}{x!(4-x)!}\left(\frac{1}{2}\right)^4$. This gives

o_i	4	9	8	8	3
e_i	2	8	12	8	2

Combining the first two cells and the last two cells gives $\chi^2 = 2.3$. Here $\nu = 2$ and $\chi_0^2 = 5.99$; hence the binomial model is satisfactory.
17. $\chi^2 = 1.6$, $\chi_0^2 = 5.99$; hence accept the hypothesis of no difference.
19. $\chi^2 = 4.8$, $\chi_0^2 = 3.84$; hence reject the hypothesis of independence. A comparison of observed and expected frequencies shows that the treatment is actually harmful.
21. $\chi^2 = 5.3$, $\chi_0^2 = 3.8$; hence reject the hypothesis of independence. Drunkeness appears to be more common in single-car fatalities than in multiple-car fatalities.
23. $\chi^2 = 24.2$, $\chi_0^2 = 9.49$; hence reject the hypothesis that the drawings were random. Capsule mixing and drawing is a poor way of obtaining random samples.
25. $\chi^2 = 5.2$, $\chi_0^2 = 3.8$; hence reject independence. There appear to be too many alcoholic children from alcoholic parents.
27. $\chi^2 = 27.4$, $\chi_0^2 = 26.1$; hence reject $H_0:\sigma = 5$. A two-tailed test should be used here.
29. (*a*) $5.1 < \sigma < 11.0$ (*b*) $4.7 < \sigma < 13.0$
31. $\chi^2 = 6$, $\chi_0^2 = 12.4$; hence reject H_0. Here χ^2 falls in the left-tail part of the critical region.
33. (*a*) $\chi^2 = 29.3$. The critical values are $\chi_1^2 = 10.3$ and $\chi_2^2 = 35.5$; hence accept H_0. (*b*) $.45 < \sigma < .84$
35. $\sigma_{\hat{p}_1-\hat{p}_2} = .030$, $z = 2.2$; hence reject $H_0:p_1 = p_2$.
39. $\chi^2 = 6.4$, $\chi_0^2 = 3.84$; hence reject independence. There is a slight tendency for those having insurance to be good loan risks.
41. The observed frequencies are given by

3313	1176	466
3628	1478	659
789	504	337
158	202	145

Calculation of expected frequencies will show that the value of χ^2 will be so large that there is no point in calculating it. One does not need a test for such large sets of data. Estimating comparative percentages would be more useful here.

Chapter 13

1. $F = 1.60$, $\nu_1 = 3$, $\nu_2 = 36$, $F_0 = 2.86$; hence accept the hypothesis of no real differences due to the modifications.

3.

Source	Sum of squares	Degrees of freedom	Mean square	F
Means	18.05	3	6.02	.56
Error	128.35	12	10.70	

Since $F < 1$, the hypothesis of equal means is accepted, which implies that the various dusting methods are equally effective.

5.

Source	Sum of squares	Degrees of freedom	Mean square	F
Means	26,208	3	8736	12.0
Error	11,642	16	728	

Since $\nu_1 = 3$, $\nu_2 = 16$, $F_0 = 3.24$, the hypothesis of no differences in the schemes is rejected.

7.

30	2	15	1.5
90	9	10	
120	11		

9. (*a*) $F_c = 18.4$, $\nu_1 = 3$, $\nu_2 = 12$, $F_0 = 3.49$; hence conclude that types differ. (*b*) $F_r = 6.6$, $\nu_1 = 4$, $\nu_2 = 12$, $F_0 = 3.26$; hence conclude that workmen differ.

11. $F = 10.8$, $\nu_1 = 3$, $\nu_2 = 12$, $F_0 = 3.49$; hence reject the hypothesis of no plot differences.

13. (*a*)

Source	Sum of squares	Degrees of freedom	Mean square	F
Columns	128	2	64	13.7
Rows	18	3	6	1.3
Error	28	6	$\frac{14}{3}$	
	174	11		

(*b*) Since $F_c = 13.7$ and $F_0 = 5.14$, reject the hypothesis of no differences due to raw materials. (*c*) Since $F_r = 1.3$ and $F_0 = 4.76$, accept the hypothesis of no differences in the brands.

15. $F = 11.6$, $\nu_1 = 2$, $\nu_2 = 6$, $F_0 = 5.14$; hence reject the hypothesis of no differences.

17. (*a*) $F_c = 11$, $\nu_1 = 3$, $\nu_2 = 6$, $F_0 = 4.8$; hence reject the hypothesis of no differences due to the catalyst. (*b*) $F_r = 12$, $\nu_1 = 2$, $\nu_2 = 6$, $F_0 = 5.1$; hence reject the hypothesis of no differences due to the temperature.

19. (*a*) $n_j = r$; therefore r can be factored out and placed in front of the sum. (*b*) This is a weighted sum of squares of the deviations of the column means from the grand mean, the weight being proportional to the sample size determining each mean.

Chapter 14

1. No. Homes with more than one phone would have a larger probability of being selected. A home here may represent a business.
3. (*a*) A simple scheme would be to select random addresses and then survey all housing units at those addresses. (*b*) When an address is selected, all housing units served by that address, whether on separate meters or not, should be surveyed.
5. Households differ in size with respect to adults; therefore households with a large number of adults will not be adequately represented.
11. (*a*) $V_R(\hat{p}) = .0006223$ (*b*) $\Sigma(p_h - p)^2(\pi_h/400) = .0000222$; hence $V_P(\hat{p}) = .000600$ and $V_P(\hat{p})/V_R(\hat{p}) = .96$. There is practically no improvement.
13. (*a*) A two-stage sampling scheme. (*b*) No, there is variability from case to case. This is a typical problem of simple cluster sampling.
15. (*a*) $\hat{p} = .375$ (*b*) $V_c(\hat{p}) = \sigma_c^2/20$, $\sigma_c^2 \doteq s_c^2 = .1020$; hence $\sigma_c(\hat{p}) \doteq .0714$. (*c*) $\sigma_{\hat{p}} = .0541$; the cost of sampling 80 cabs compared to sampling only 20 cabs undoubtedly would more than compensate for the lower precision of the cluster sample.
17. (*a*) Approximately 15%, 64%, and 20% if a proportional sample is to be taken with respect to the number of elevators in the three strata. (*b*) $\bar{x} = .15\bar{x}_1 + .64\bar{x}_2 + .20\bar{x}_3$. Its variance is given by $V_P(\bar{x}) = \frac{1}{100}[.15(20.1)^2 + .64(8.4)^2 + .20(2.5)^2] = 1.07$; hence $\sigma_{\bar{x}} = 1.03$.
19. (*a*) $\hat{p} = .083$ (*b*) $\sigma_{\hat{p}} = .0072$

Chapter 15

1. (*a*) Florida 275.1, California 100.0, Texas 3.6, Arizona 6.6 (*b*) No. Linked relatives are useful for comparing relative change over time but are meaningless here.
3. Index numbers such as these give rates of increase only; hence one cannot use them to compare rental costs in the two cities.
5.

Year	1963	1964	1965	1966	1967	1968	1969	1970
Price	100	102	104	115	120	125	131	147

7. (*a*) $I_p = .83$; hence a 17 percent decrease. (*b*) $I_p = .78$; hence a 22 percent decrease. (*c*) $I_q = .93$; hence a 7 percent decrease. (*d*) $I_v = .72$; hence a 28 percent decrease.
11. Index values are 100, 63, 130, 38.
15. $I_p(L) = k$, $I_p(P) = k$; hence $I_p(F) = \sqrt{k \cdot k} = k$.

17. (*a*)

87.8	94.3	115.0
88.9	96.9	120.2
90.0	100.0	124.2
91.2	104.0	131.2
92.5	109.0	

21. A 15-point gain.
25. (*a*) and (*b*); (*e*) is also often considered to be an index.
27. (*a*) 1.2; hence a 20 percent increase in real wages. (*b*) .80; hence it is worth 80 cents.

Chapter 16

1. $r = -.001$. The hypothesis $H_0{:}\rho = 0$ will obviously be accepted here without the necessity of a test.
3. $r = .629$ for lag 1. The acceptance region for $H_0{:}\rho = 0$ is the interval $(-.331, .279)$; hence reject H_0.
5. Rounding off to one decimal place gives: −.2, −.2, .6, .9, .6, .3, 0, −.1, −.4, .4, .6, .1, .3, .6, 1.2, 1.8, 2.5, 3.4, 3.4, 2.9, 2.5, 1.8, 2.1, 1.5, 1.2, .9, 1.7, 2.5, 2.2, 2.0, 2.0.
7. Rounding off to one decimal place gives: .5, .2, .3, .8, .3, −.1, 0, .2, .1, .1, .5, .6, .6, 1.2, 2.0, 2.5, 2.9, 3.1, 2.9, 2.5, 2.3, 1.7, 1.5, 1.3, 1.7, 1.7, 1.8, 2.1, 2.4.
9. Rounding off to the nearest integer gives: 14, 16, 11, 15, 17, 19, 21, 16, 16, 11, 12, 13, 15, 14, 10, 8, 12, 16, 16, 12, 10, 13, 14, 16, 15, 16, 16, 17, 21, 19, 20, 21, 19, 20, 16, 18, 15, 11, 9, 8, 9, 11, 15, 14, 14, 10, 13, 13, 15, 12, 11, 7, 10, 11, 12, 16, 16, 21, 15, 19, 17, 17, 11, 11.
11. Letting the x- values run from 1 to 33, $y = -.093 + .0814x$. There is a slight upward trend; the value of y increases about 2.6 units when x goes from 1 to 33, according to the least-squares line.
13. (*a*) $\hat{y} = 13.05 + .461\ (x - 1960)$
15. 299, 297, 297, 302, 298, 294, 304, 300, 305, 310, 309, 305, 327, 329, 333, 334, 337, 341, 336, 338, 343, 333, 339, 348
17. (*a*) 107,200 (*b*) 177,600
19. (*a*) An upward trend, but otherwise it looks like a random series. (*b*) $\hat{y} = 5.08 + .174x$, where $x = 0$ for 1950. (*c*) $\hat{y} = 8.04$, as compared with 7.4.
21. Linear projection does not look good. A parabola or an exponential function would probably be better.
23. $r = .77$
29. (*b*) The relationship will be linear from 1945 to 1965; however, after that there will be a decided curvature to the relationship.

31. (*b*) 5.4, 5.4, 5.4, 6.0, 6.0, 5.4, 5.4, 5.2, 5.0, 5.2, . . . , 6.2, 6.2, 5.8, 5.4, 5.2 5.2, 5.2, 5.4, 5.8, 5.8, 5.6, 5.8, 5.6, 5.0, 5.2, 5.0, 5.0, 5.2, 5.6, 5.6, 6.0, 6.4, 6.6, 6.4, 5.8, 5.6, 5.6, 5.6, 5.4, 6.0, 6.2, 6.2, 6.4, 6.8, 6.8, 6.4, 6.4, 5.8, 6.0, 6.0, 5.8, 5.8, 6.4, 6.0, 5.6, 6.2, 6.2, 6.2, 6.2 (*c*) The first half appears to be at a lower level than the second half.
33. $r = .13$, $r_0 = .259$, and the acceptance interval is $(-.277,.241)$; hence accept the hypothesis that $\rho = 0$ for a lag 11.
35. $\hat{y} = 26.10 + .413(x - 1940)$; this gives $r_1 = .17$. Since $r_0 \doteq .30$, accept the hypothesis of no correlation. The deviations appear to form a random sequence.

Chapter 17

1. $E[B] = 12{,}500$, $E[S] = 12{,}950$; hence buy stocks.
3. $p = .7$
5. $E[A] = 445$, $E(B] = 441$; hence choose A.
7.

		Parts ordered			
		0	1	2	3
Parts needed	0	0	−100	−200	−300
	1	−300	−100	−200	−300
	2	−600	−400	−200	−300
	3	−900	−700	−500	−300

9. $E[0] = 0$, $E[1] = 8$, $E[2] = 12$, $E[3] = 6$, $E[4] = -2$; hence prepare 2 bouquets.
11. (*a*)

Event	Action A	Action B	p
$S_A S_B$	0	30,000	(.35)(.5)
$S_A F_B$	0	65,000	(.35)(.5)
$F_A S_B$	32,000	0	(.65)(.5)
$F_A F_B$	0	3,000	(.65)(.5)

$E[A] = 10{,}400$, $E]B] = 17{,}600$; hence choose A. (*b*) Cost of uncertainty is $10,400.
13. Cost of uncertainty is $7.
15. (*a*) $E[E] = 13{,}750$, $E[N] = 11{,}000$; hence make the exchange. (*b*) $E[E] = 4{,}750$, $E[N] = 7{,}500$; hence make the exchange. The difference in each case is 2,750.
17. $E[L] = 10{,}000$. Perfect information gives $E = 100{,}000$; hence the value of perfect information is $90,000.
19. (*a*) $E[A] = 3.5$, $E[B] = -.9$, $E[C] = 4.75$, $E[D] = 2.00$; hence choose C. (*b*) $E = 5.85$; hence the value of perfect information is $1.10.

21. $p = .55$
23. $0, \frac{3}{20}, \frac{7}{20}, \frac{10}{20}, \frac{12}{20}, \frac{16}{20}$, and $\frac{20}{20}$
25. $E[L] = -30{,}000$, $E[N] = 0$, $E[FP] = 10{,}500$. (*a*) Follow the geologist's advice. (*b*) $10,500
27. $P\{0 \mid S\} = .41$
29. $P\{R \mid S\} = \frac{1}{3}$; hence decide coin is two-headed. Then $E = \frac{2}{3}$. Using prior probabilities, $E = 0$; hence the demonstration is worth the price.
31. (*b*) $\frac{33}{34}, \frac{51}{70}$, and $\frac{1}{2}$, respectively.
33. $\mu_1 = 4.09$; therefore he should buy.
35. (*a*) $E_0[P] = 1{,}000{,}000$ (*b*) $\delta = .004$ and $\mu_1 = 4502$; hence $E_1[P] = -494{,}000$. The firm should not proceed with the venture.
37.

Weather	Fixed	Variable
Sunny	600	500
P-cloudy	100	0
Cloudy	−200	−100

39. Calculating expectations for partly cloudy operations only, $E[F] = 150$ and $E[V] = 100$. Since both are positive, it pays to fly under a partly cloudy forecast; however now $E[F] = 635$ and $E[V] = 495$.

Index